Advances in Visual Data Compression and Communication

Meeting the Requirements of New Applications

Multimedia Computing, Communication and Intelligence

Series Editors: Chang Wen Chen and Shiguo Lian

Advances in Visual Data Compression and Communication: Meeting the Requirements of New Applications
Feng Wu
ISBN: 978-1-4822-3413-8

Effective Surveillance for Homeland Security: Balancing Technology and Social Issues
Edited by Francesco Flammini, Roberto Setola, and Giorgio Franceschetti
ISBN: 978-1-4398-8324-2

Music Emotion Recognition
Yi-Hsuan Yang and Homer H. Chen
ISBN: 978-1-4398-5046-6

Optimal Resource Allocation for Distributed Video and Multimedia Communications
Yifeng He, Ling Guan, and Wenwu Zhu
ISBN: 978-1-4398-7514-8

TV Content Analysis: Techniques and Applications
Edited by Yiannis Kompatsiaris, Bernard Merialdo, and Shiguo Lian
ISBN: 978-1-4398-5560-7

Advances in Visual Data Compression and Communication

Meeting the Requirements of New Applications

Feng Wu

CRC Press
Taylor & Francis Group
Boca Raton London New York

CRC Press is an imprint of the
Taylor & Francis Group, an **informa** business
AN AUERBACH BOOK

CRC Press
Taylor & Francis Group
6000 Broken Sound Parkway NW, Suite 300
Boca Raton, FL 33487-2742

Printed on acid-free paper
Version Date: 20140501

International Standard Book Number-13: 978-1-4822-3413-8 (Hardback)

Library of Congress Cataloging-in-Publication Data

Wu, Feng (Computer scientist)
Advances in visual data compression and communication : meeting the requirements of new applications / author, Feng Wu.
pages cm -- (Multimedia computing, communication and intelligence ; 5)
Includes bibliographical references and index.
ISBN 978-1-4822-3413-8 (hardback)
1. Image compression. I. Title.

TA1638.W83 2014
006.6--dc23 2014016291

Visit the Taylor & Francis Web site at
http://www.taylorandfrancis.com

and the CRC Press Web site at
http://www.crcpress.com

To the 15 years I spent with Microsoft Research Asia.

Contents

Part VII Future Work

Preface

I still clearly remember that it was September 13, 1999, when I started my job at Microsoft Research Asia (MSRA), formerly named Microsoft Research China. It was the first year the lab was established. Twenty-eight Chinese Ph.D.s passed the strict interviews and were hired as associate researchers by MSRA from among more than 1000 applicants. Even so, at that time we were not formal employees at Microsoft and only had a two-year contract probably because we received our Ph.D. degrees in China. Thus, we had to prove that we had the ability to do excellent research. All of us worked very hard, usually from 9 A.M. to 9 P.M. every day, seven days a week. When I recall the early years I spent at MSRA, I feel like it may have been the most wonderful time of my life.

At that time, Dr. Ya-Qin Zhang was the chief scientist at MSRA and the manager of the Internet Media Group (IMG). There were three researchers in this group who had come back from the United States, Dr. Jin Li, Dr. Shipeng Li, and Dr. Wenwu Zhu. And among the 28 associate researchers, four were in IMG, Dr. Huirong Shao, Dr. Hongwu Wang, Dr. Qian Zhang, and myself. Although I had several discussions with Ya-Qin Zhang regarding my research project, I mainly worked with Shipeng Li. Almost 15 years have passed, Li and I are still with MSRA. Zhang is now a Microsoft corporate vice president, and Shipeng is the president of Microsoft R&D in China. Jin Li works at Microsoft Research Redmond. Wenwu Zhu is a full professor at Tsinghua University and Qian is a full professor at Hong Kong University of Science and Technology. Huirong Shao and Hongwu Wang left MSRA early and work in high-tech companies in Silicon Valley, California.

In my Ph.D. thesis, I studied object-based video coding. Since it was difficult to automatically and accurately extract visual objects from image and video, the research topic looked less promising in real applications after I graduated. As I was trying to find a new research area, Dr. Weiping Li, who was Shipeng's Ph.D. supervisor, visited MSRA in October 1999 after the Moving Picture Experts Group (MPEG) meeting in Melbourne, Australia. At that time, he was actively developing a fine granularity scalable (FGS) video coding standard in MPEG. He introduced us to his bit-plane coding technology. He also told us that the coding efficiency of FGS was considerably low because it used low-quality reference for motion compensation in order to avoid drift error. Shipeng and I felt that it was a good research topic and immediately started to study the problem. This was my first research project at MSRA.

After two months, Shipeng and I brought our first technical proposal, the earliest version of progressive fine granularity scalable (PFGS) video coding, to the MPEG meeting in Maui, Hawaii, in December 1999. It was the first time that I went abroad. By using multiple references, our PFGS scheme significantly improved the performance of FGS coding. Although MPEG experts were very interested in our technology, they were concerned about the complexity of using multiple references. Thus, we continued to improve our PFGS scheme after the MPEG meeting. The scheme was simplified by using only one additional reference and its performance was even better than when multiple references were used. In March 2000, we proposed the improved PFGS scheme to MPEG. At the meeting, the FGS standard would be promoted as a committee draft and afterward, no new technical changes would be allowed. It was our last chance to get the PFGS scheme into the standard. Unfortunately, because of the complexity our effort failed again.

In the subsequent two to three years, I continued to improve the PFGS scheme and proposed the improved technologies to MPEG as an advanced FGS standard. At the same time, I also visited the Windows Media Player group several months every year to write our PFGS scheme into Windows Media Video 8 (WMV8) and Windows Media Video 9 (WMV9) for shipping in Microsoft products. Both efforts failed to make substantial progress. In order to prove the advantages of our PFGS scheme, we developed a real-time video streaming system using our scalable codec. We showed the system in MPEG over the Internet by streaming video content from Beijing to Washington DC. The performance was much better than that of Windows Media Player, which used a nonscalable codec. My research into PFGS continued until the emergence of the motion compensated temporal filtering (MCTF) technology. Although these efforts were not successful, our lab recognized my research capability through this project. I became a formal employee in May 2000 and was then promoted to a researcher in February 2001.

In addition to the PFGS scheme, Shipeng Li and Jin Li began to study how to integrate motion alignment into temporal wavelet transforms very early at MSRA. Their work used the lifting-based wavelet for temporal transform and motion vectors among frames. It was the same as the famous MCTF technology. I was not involved in this project in the beginning. When I started to mentor our interns who worked on this project in 2002, from my experiences in developing the PFGS scheme, I immediately realized that it was a better solution than the PFGS scheme for scalable video coding. We continued to improve the solution in various detailed technologies. All of these technologies made up our Barbell-based lifting coding scheme. When MPEG launched a call for proposals in 2003 to collect scalable video coding schemes and evaluate their performance, our Barbell-based lifting coding scheme ranked first and third among 21 schemes in the two testing scenarios, respectively. It was adopted as common software by MPEG for further exploration of wavelet-based video coding. Our research on scalable video coding is presented in Part II of this book.

Our research in directional transforms was actually inspired by our Barbell-based lifting coding scheme. I remembered one time when we brainstormed about the lifting-based temporal wavelet transform. I suddenly thought that we could apply the idea to the 2D spatial wavelet transform. In MCTF, each temporal lifting step can use motion vectors that indicate which pixels in different frames are transformed together. In the 2D spatial transform, an angle can also be used to indicate which neighboring pixels in the given direction are transformed. With this idea, I rapidly implemented our lifting-based directional wavelet transform in the JPEG 2000 reference software and observed that the energy of the transform coefficients was significantly reduced with directional adaptation. Later, my student continued to complete the coding process and showed a more than 2 dB gain over JPEG 2000 on images with rich orientation features.

Although there were many research papers on directional transforms already published, one obvious advantage of our lifting-based directional transform was that transformed coefficients have a similar structure and distribution to that of the non-directional counterpart. Therefore, existing quantization and entropy coding technologies can be applied efficiently without any changes. More importantly, the idea of constructing directional transforms through the incorporation of directional operators into the lifting structure offers a unified and efficient way to enable directional adaptation in various existing nondirectional transforms. Subsequently, we further developed the directional discrete cosine transform (DCT) by incorporating directional operators into its lifting implementation. Applying the proposed directional DCT in JPEG gets a similar performance gain to the directional wavelet in JPEG 2000. Furthermore, we theoretically analyzed the order of two 1D transforms in a 2D directional transform and proposed the order-adaptive directional lifting transform, which is more suitable for integration in the intra-frame coding of H.264. Our research into directional transforms is presented in Part III of this book.

After the first part of the H.264 standard was completed in 2003, I analyzed the developments in image and video compression. In most existing schemes and standards, the statistical correlation among pixels is considered a dominant factor in designing prediction, transforms, and entropy coding. During the past three decades, these coding technologies have evolved a great deal. A few coding standards were made and led to products. However, it is viewed to be a difficult challenge to further improve the coding technologies. In order to improve the coding performance, the computational complexity of encoding has been and will continue to increase significantly. This has been further verified by the new High Efficient Video Coding (HEVC) standard. So, at that time we started to look for other technologies that could potentially lead to a breakthrough. A possible option was to explore the visual correlation in image and video compression. Although using visual correlation has been recognized for a while, it faces great challenges in compression.

Thanks to the strong research group in computer vision at MSRA, we rapidly became familiar with the advances in computer vision. Our first attempt was image inpainting, which is an advanced vision-based synthesis technology. We proposed incorporating image inpainting into conventional coding schemes to synthesize structure regions by using coded edges. It not only introduced inpainting in compression to reduce the total amount of coded information, but it also enhanced the robustness and commonness of inpainting by explicitly coding important information. Toward the convergence of signal-processing-based compression and vision-based analysis/synthesis for image coding, this interaction offered new opportunities to break the waveform compression bottleneck. We thus received the best paper award at the IEEE Transactions on Circuits and Systems for Video Technology (CSVT) in 2009.

With the rapid development of digital cameras and other mobile devices equipped with cameras, there is a large number of images and videos available on the Internet. When you randomly take a picture with your phone on the street, you can often find some highly correlated images on the Internet that were taken at the same location from different viewpoints and angles, focal lengths, and illuminations. I personally think that large-scale data is an important factor that can bring significant changes in image and video compression. However, conventional compression technologies such as prediction and transform make it difficult to take advantage of highly correlated external data. Another attempt in vision-based coding was a cloud-based image coding scheme, which is different from current image coding even on the ground. The proposed scheme no longer compressed images pixel by pixel and instead tried to describe images and reconstruct them from a large-scale image database via the descriptions. It can achieve a thousand-to-one compression ratio with good visual quality.

Previous compression schemes and standards mainly target the transmission of images and video. However, the storage of large-scale images and video is becoming a valuable research topic with the emergence of applications in photo sharing and video surveillance. Our third attempt at vision-based coding was the compression of large-scale images for storage. We first analyzed their correlations and organized them as multiple trees. All images described by one tree have strong correlations. The correlations are represented by the relationship among parent and children nodes in the tree. According to the relationship, we can process correlated images as a sequence and jointly optimize the coding of the sequence. Our experimental results have shown that the size coding them as a sequence is only one-tenth that of coding them one by one. Large-scale storage is an important application that supports the use of external data for compression. Furthermore, the storage-oriented compression may not need to be standardized because images and video can be transcoded to a common format before they are output. It provides more flexibility for developing innovative compression technologies. Our research in vision-based coding is presented in Part IV of this book.

Before Chong Luo, who was an associate researcher at MSRA, asked to transfer to my group in 2008, I had never worked on communications and had no knowledge of even the basic concepts. Although my group is named Internet Media, all the researchers in this group only had expertise in media. Luo's participation provided me a good opportunity to learn communications. I have found that the best way of learning about a field is working with an expert in that field. Soon I could discuss detailed communication technologies with Luo and others. Luo previously worked on application-layer multicasts. I felt the work was too engineering focused and the technology was less likely to impact practical applications. We should look for more prospective research topics.

At that time, I had already spent a year studying the methodology of low-density parity-check (LDPC) codes for the performance analysis of compressive sensing with binary input. There are many similarities between LDPC codes and compressive sensing. Why didn't we use compressive sensing in communications such as LDPC codes? One obvious advantage is that compressive sensing provides an inherent compression when the source is sparse. Our first work applied compressive sensing for data gathering in a sensor network. The proposed compressive data gathering was able to reduce the global communications cost by using the correlation among sensor readings without introducing intensive computations or complicated transmission controls. This work was submitted to MobiCom 2009. At that time, I did not know how difficult it was for a paper to be accepted by the conference. Fortunately, our paper received high review scores and was accepted.

We next applied binary-input compressive sensing in the physical layer of a wireless network. In contrast to LDPC codes that use logical XOR, compressive sensing uses an arithmetic weighted sum to generate coded symbols from input bits. Thus it can be viewed as arithmetic codes. Even for binary input, the values of coded symbols are in a finite alphabet. This mapping from binary to a finite alphabet matches the requirements of modulation. In our proposed compressive modulation, the constellation usually contains hundreds of points, which is much denser than that used in current wireless networks. Received symbols are still decodable even in poor channel conditions by jointly decoding the received symbols together. The greatest advantage of our proposed compressive modulation is that it can adapt to a wide range of channel conditions without changing the coding and constellation, which is desirable for rate adaptation in wireless communications.

Although Shannon's information theory tells us that source and channel coding can be separate, they often need to be considered jointly in practical applications due to the limited length of data. However, source coding is achieved by entropy coding, and channel coding by linear codes, cyclic codes, or convolutional codes. They are hard to jointly optimize because of the different properties. Another important advantage of our proposed compressive modulation is that it provides a nice solution for joint source and channel coding in the physical layer. The source sparsity can be utilized to reduce the number of coded symbols or help the decoding of source bits in worse channel conditions. The performance of source coding provided by our proposed compressive modulation is close to that of ideal source coding. Our research into applying compressive sensing to communications is generally called compressive communications, which is presented in Part V of this book.

In digital systems, the redundancy in image and video is fully removed by compression, whereas in analog systems the redundancy in image and video is not removed at all. They present two extreme cases. Digital systems have good performance for image and video communication. But in order to ensure a nearly error-free transmission, the sending rate has to adapt to actual channel conditions by adjusting the channel coding rate and modulation. On the one hand, if the sending rate is too low, channel capacity cannot be fully utilized. On the other hand, if the sending rate is too high, the decoded bit error rate (BER) will increase dramatically. This phenomenon is commonly called the cliff effect in digital communications. However, it is widely known that wireless channels vary quickly and drastically. It is a challenge to instantly make an accurate estimation of channel conditions. In addition, digital systems cannot support image and video multicasts well. Although analog systems have been discarded by academia and the industry, the processing still has its own advantages especially in channel adaptation and multicast.

Recently, I have become interested in looking for a communications system between digital and analog systems, where images and video are not compressed completely and still have some redundancy. Such redundancy can be used for resisting channel noise, finding external correlated images and video, and introducing signal processing in the transmission. It provides an opportunity to make a breakthrough in visual data communications. Our first effort was motivated by SoftCast, which was proposed by Professor D. Katabi et al. at MIT. In SoftCast, video coefficients are directly transmitted without entropy coding by simulating the analog processing, although the scheme is implemented in a digital manner. However, the temporal transform in SoftCast is inefficient at exploiting the correlation among frames. We proposed introducing distributed video coding instead of the 3D transform and thus our scheme was called DCast. It can better utilize the temporal correlation through motion alignment at the receiver.

When data transmitted over a channel is no longer compressed into random bits, it actually enables us to introduce more processing in communications far beyond channel coding. Our second effort was to introduce the denoising technology in the physical layer of wireless transmissions. Denoising is a classical signal processing technology and seems irrelevant to communications. However, if transmitted data is correlated and added channel noise is random, denoising can be applied to attenuate and even remove channel noise by utilizing the data correlation. In our proposed scheme, we apply MCTF to remove the temporal correlation among frames but keep the spatial correlation in every frame. The received frames are first denoised by using the spatial correlation before reconstruction. Denoising is only an example and we strongly believe that more intellectual processing can be applied to communications in the future.

Multiple-input multiple-output (MIMO) technologies have become building blocks for high capacity links in contemporary and future wireless networks. Any new communications system cannot be successful if it cannot fully take advantage of MIMO. With mobile devices equipped with different numbers of antennae, broadcasting has to handle both channel and receiver antenna heterogeneity. It is a big challenge for existing transmission technologies to achieve adequate performance when matching the antenna number and channel conditions of every receiver. We have observed that when the source is not compressed completely, it is more flexible and suitable to be transmitted through MIMO. Furthermore, we have also proposed using compressive sensing to combine multiple data elements as one transmitting symbol. Our results have shown that all receivers with different numbers of antennae can achieve good performance. In this category of our research, since the transmission simulates the processing of analog systems but is still implemented in a digital manner, the proposed schemes are called pseudo-analog transmission, which is presented in Part VI of this book.

How time flies! It is my fifteenth year with MSRA. When I started my first research project on scalable video coding, I did not know what I would do next. Doing research is just like playing Go. After I did some in-depth research into scalable video coding, which is similar to an occupied corner in Go, I started to look for new opportunities and extended my research. Every extension seems unintended, but finally, when I started to prepare my material for IEEE Fellow in 2012, I realized that the research is very relevant and systematic. I would summarize my research as "Visual data compression and communication." Although I completed my education in China and did not go abroad before I joined MSRA, I was elected an IEEE Fellow as quickly as others. I appreciate that MSRA has provided me a totally open research atmosphere. I am granted the freedom to work on my preferred research topics and projects, and to pursue long-term goals that extend far beyond current product cycles.

When this book was almost completed, I made the biggest decision in my career–to leave Microsoft Research Asia and join the University of Science and Technology of China (USTC) as a professor. It is definitely a new page to write in my life.

Feng Wu, Beijing

Acknowledgments

I would like to first thank Prof. Jinpei Xu at the Harbin Institute of Technology (HIT) and Prof. Wen Gao at Peking University, although they did not directly contribute to the research presented in this book. Prof. Xu was my supervisor when I was a master's student at HIT. My first research project, continuous speech recognition using neural networks, was done with Prof. Xu. Every week we spent several hours discussing all the details, including the initial idea, the implementation, the experimental results, and planning for the next step. I really appreciate what I learned from Prof. Xu about the basic skills for doing research.

Prof. Gao was my Ph.D. supervisor. When I completed my courses and began my research, Prof. Gao had already moved from HIT to the Institute of Computing Technology (ICT) at the Chinese Academy of Science (CAS). Prof. Gao originally arranged for me to continue to work on speech recognition. I told him that I would like to change my research to video coding because my Chinese pronunciation has a severe southern accent. The training models could not recognize my speech using the data from others, and vice versa. Prof. Gao respected my choice and I started research on video coding. At that time, Prof. Gao was the head of the China delegation at the International Organization for Standardization (ISO)/International Electrotechnical Commission (IEC) MPEG and attended MPEG meetings every three months. Thus, I was well aware of the latest progress in video coding and received information as fast as my foreign peers. It was very helpful for my research. The problems I studied were almost at the same line as foreign experts. I think that this may be a reason why I was able to pass the MSRA interview, even though I did not have any international journal or conference papers published at that time.

Dr. Shipeng Li is my supervisor at MSRA and has further trained me. When I worked on the PFGS coding scheme, we often had some creative discussions in his office. The key progress in our PFGS scheme could not happen without his contributions. Dr. Li also trained me in how to write high-quality papers in English. My first journal paper, published in the IEEE Transaction on Circuits and Systems for Video Technology, was revised by Dr. Li. When I got the revised manuscript from him almost every paragraph had been revised by him. I carefully studied the revisions one by one and learned how to effectively and accurately express what I wanted to say in English. These writing skills were very helpful for my career.

I would like to thank Dr. Lin Luo, Dr. Jizheng Xu, and Dr. Ruiqin Xiong for their contributions to our research on scalable video coding. At that time, Luo was a Ph.D. student at the University of Science and Technology of China (USTC) and had worked as an intern at MSRA for several years. After graduation, she joined the IBM research lab in Beijing. Xu was an undergraduate student at USTC when he started to work as an intern at MSRA. Later, he got his master's degree at ICT with Dr. Ya-Qin Zhang and then joined MSRA as an assistant researcher. His research in this period was mainly related to scalable video coding. Xiong was a Ph.D. student at ICT with Ya-Qin Zhang as his supervisor. After graduation, he worked on a postdoctoral in Australia for two years and then came back as a faculty at Peking University.

I would like to thank Dr. Wenpeng Ding, Prof. Xiaolin Wu, Dr. Hao Xu, Dr. Xiulian Peng, and Dr. Jizheng Xu for their contributions to our research in directional transforms. At that time, Ding was a Ph.D. student supervised by me at USTC and he joined Beijing University of Technology as a faculty after graduation. Prof. Xiaolin Wu contributed the algorithm for estimating directions in directional wavelet transforms when he worked as a visiting professor at MSRA. Xu was a Ph.D. student at USTC and was a postdoctoral researcher in Hong Kong after graduation. Peng was a Ph.D. student supervised by me at USTC and joined MSRA as an associate researcher after graduation. When Xu worked on directional transforms, he was an associate researcher at MSRA and was also pursuing a Ph.D. degree co-supervised by me at Shanghai Jiaotong University.

I would like to thank Dr. Dong Liu, Dr. Lican Dai, Huanjin Yue, Zhongbo Shi, and Dr. Xiaoyan Sun for their contributions to our research on vision-based coding. At that time, Liu was a Ph.D. student at USTC supervised by me and he then joined Nokia Research Beijing after graduation. Recently, he joined USTC as an associate professor. Dai was a Ph.D. student at USTC and joined a high-tech company in Chengdu after graduation. Shi is my Ph.D. student at USTC and Yue is my Ph.D. student at Tianjing University. Sun is a lead researcher at MSRA. She leads our research efforts in this field.

I would like to thank Prof. Xiaopeng Fan, Hao Cui, Dr. Xiaolin Liu, Prof. Chang Wen Chen, and Dr. Chong Luo for their contributions to our research on compressive communications and pseudo-analog communications. Fan is an associate professor at HIT working on distributed video coding. When he visited MSRA for six months, I mentored him and extended his research from coding to communications. DCast was done while he was at MSRA. Cui is a Ph.D. student at USTC and Prof. Chang Wen Chen is his supervisor. He works at MSRA as an intern. Liu was a Ph.D. student at USTC supervised by me. Prof. Chen closely collaborates with me at MSRA. He visited us two to three weeks every year under a consulting contract. Luo joined my group as an associate researcher and then got her Ph.D. co-supervised by me at Shanghai Jiaotong University.

Finally, I would like to thank all the members of the Internet Media Group at MSRA. Most of them have worked with me for more than five years. Dr. Jizheng Xu started his internship at MSRA in 1999. He joined MSRA as an assistant researcher in September 2004 and now is a lead researcher. He is conducting our research projects on signal-processing-based compression and video coding standards. Dr. Xiaoyan Sun started her internship at MSRA in 2000. She joined MSRA as an associate researcher in November 2003 and is now a lead researcher. She is conducting our research projects on vision-based compression. Luo joined MSRA as an assistant researcher in September 2004 and is now a lead researcher. She is conducting our research projects on media communications. In addition, Dr. Zhiwei Xiong, Dr. Xiulian Peng, Dr. Bin Li, and Dr. Cuilin Lan joined MSRA as associate researchers in the past three years after they completed their Ph.D. research at MSRA.

Acronyms

ADC	Analog-to-Digital Conversion
ADL	Adaptive Directional Lifting
AEP	Asymptotic Equipartition Property
AGC	Automatic Gain Control
AM	Amplitude Modulation
AMC	Adaptive Modulation and Coding
AMVP	Advanced Motion Vector Prediction
APP	A Posteriori Probability
ASK	Amplitude-Shift Keying
AVC	Advanced Video Coding
AWGN	Additive White Gaussian Noise
BDF	Bidirectional Filtering
BER	Bit Error Rate
BM3D	Block Matching with 3D Filtering
BRP	Block Removal Pattern
CABAC	Context Adaptive Binary Arithmetic Coding
CAVLC	Context Adaptive Variable Length Coding
CB	Coding Block
CD	Compact Disc
CDD	Curvature Driven Diffusion
CDG	Compressive Data Gathering
CG	Computer Graphics
CHoG	Compressed Histogram of Gradient
CIF	Common Intermediate Format
CM	Compressive Modulation
CME	Correlated Motion Estimation
CTB	Coding Tree Block
CTU	Coding Tree Unit
CU	Coding Unit
DAC	Digital-to-Analog Conversion
DBF	Deblocking Filter

DCT Discrete Cosine Transform
DFD Displaced Frame Difference
dFT directional Filtering Transform
DSCQS Double Stimulus Continuous Quality Scale
DST Discrete Sine Transform
DVC Distributed Video Coding
DVD Digital Video Disk
DWT Discrete Wavelet Transform
EBCOT Embedded Block Coding with Optimized Truncation
ESCOT Embedded Subband Coding with Optimized Truncation
EXIT Extrinsic Information Transfer
EZBC Embedded Zero Block Coding
FDM Frequency Division Multiplexing
FGS Fine Granularity Scalable
FIR Finite Impulse Response
FLC Fixed Length Coding
FM Frequency Modulation
FSK Frequency-Shift Keying
GOP Group of Pictures
GPU Graphic Processing Unit
HD High Definition
HEC Header Extension Code
HEVC High Efficient Video Coding
HM Hierarchical Modulation
HVS Human Visual System
HVSBM Hierarchical Variable Size Block Matching
i.i.d. Independent and identically distributed
ISDN Integrated Services Digital Network
JPEG Joint Photographic Experts Group
JSCC Joint Source-Channel Coding
JVT Joint Video Team
LDPC Low-Density Parity-Check
LLR Logarithm of Likelihood Ratio
LLSE Linear Least Square Estimator
LMMSE Linear Minimum Mean Square Error
LTE Long-Term Evolution
MAC Medium Access Control
MAP Maximum A Posteriori
MB Macroblock
MC Motion Compensation
MCE Motion Compensated Extrapolation
MCI Motion Compensated Interpolation
MCTF Motion Compensated Temporal Filtering
ME Motion Estimation
MIMO Multiple-Input Multiple-Output

MISO Multiple-Input Single-Output
ML Maximum Likelihood
MPE Minimum Probability of Error
MPEG Moving Picture Experts Group
MSB Most Significant Bit
MSE Mean Square Error
MSRA Microsoft Research Asia
MST Minimum Spanning Tree
MT Motion Threading
MV Motion Vector
MVD Motion Vector Difference
OBMC Overlapped Block Motion Compensation
P2P Peer-to-Peer
PAM Pulse Amplitude Modulation
PB Prediction Block
PCA Principal Components Analysis
PDE Partial Differential Equation
PDF Probability Density Function
PDM Pulse Duration Modulation
PFGS Progressive Fine Granularity Scalable
PM Phase Modulation
PMF Probability Mass Function
PPM Pulse Position Modulation
PSD Power Spectral Density
PSK Phase-Shift Keying
PSNR Peak Signal-to-Noise Ratio
PU Prediction Unit
QAM Quadrature Amplitude Modulation
QCIF Quarter Common Intermediate Format
RDO Rate Distortion Optimization
RP Random Projcction
RSC Recursive Systematic Convolutional
SAD Sum of Absolution Difference
SAO Sample Adaptive Offset
SFMST SIFT Feature-Based Minimum Spanning Tree
SIFT Scale Invariant Feature Transform
SNR Signal-to-Noise Ratio
SSB Single Sideband
SSD Sum of Squared Difference
STBC Space Time Block Code
SURF Speed Up Robust Feature
SVC Scalable Video Coding
TB Transform Block
TDM Time Division Multiplexing
TDMA Time Division Multiple Access

TU Transform Unit
TV Television
UDF Unidirectional Filtering
UHD Ultra High Definition
URQ Uniform Reconstruction Quantization
VCEG Video Coding Experts Group
VHS Video Home System
VLC Variable Length Coding
VLD Variable Length Decoding
VoD Video on Demand

Part I
Basis for Compression and Communication

This part provides a basis for our advanced research on visual data compression and communication both theoretically and technically.

Chapter 1 discusses Shannon's information theory, which is the theoretical basis for visual data compression and communication. First, we introduce the concept of entropy, which is the minimum length for lossless compression of a source, according to Shannon's source coding theorem. Huffman coding and arithmetic coding are taken as examples to explain what source coding is and how a source is compressed toward its entropy. The rate distortion theorem is also discussed because of its importance to guiding the lossy compression of a source. Second, we introduce the concept of channel capacity, which is the maximum rate that a channel can be used with negligible transmission errors, according to Shannon's channel coding theorem. Hamming code is taken as an example to explain how channel coding works. Finally, the joint source and channel coding is introduced. The conditions for source channel separation are discussed. It is the theoretical justification why data compression and transmission can be studied independently.

As a concrete application of source coding, Chapter 2 discusses the basic framework of video compression and the major developments in recent decades. This serves as the technical basis for our research presented in Parts II, III, and IV of this book. First, we introduce the hybrid prediction and transform framework of video compression and the basic technologies used in the framework. They were established in the 1980s without fundamental changes until now. Second, the developments of compression technologies are discussed according to the evolution of video coding standards, from the earliest H.261, MPEG-1, MPEG-2, MPEG-4, H.264/MPEG-4 Advanced Video Coding (AVC) until the latest High Efficient Video Coding (HEVC) standard. These technical developments have improved the performance of video compression about one order of magnitude. Finally, to better understand modern compression technologies, the H.264/MPEG-4 AVC and HEVC standards are discussed in detail.

As a concrete application of channel coding, Chapter 3 discusses analog and digital transmission technologies. It is the technical basis for our research presented in Parts V and VI of this book. The analog transmission is introduced, although it is almost thrown away by academics and industry experts because of the dominance of digital systems. But some recent research shows that using the idea of analogy methods applied to modern digital communication have some unique advantages, especially for visual data communication. The analog modulation and multiplexing technologies are briefly discussed here. Then the digital transmission is introduced. We mainly focus on the technologies used in the physical layer, which consists of channel coding and modulation. The capacity-approached low-density parity-check (LDPC) codes and Turbo codes with digital modulation are discussed in detail.

Chapter 1
Information Theory

1.1 Introduction

Over the past few decades, with the rapid development of digital cameras and various mobile devices equipped with cameras, which are convenient to use anywhere and anytime, we have witnessed the magnitude of visual data (images, photos, and videos), produced by humans in their daily lives, increasing explosively beyond expectations. In 2012, Facebook announced that the number of photos uploaded by its users had exceeded 220 billion [1]. Furthermore, the amount is still increasing at an average rate of several hundred million per day. Additionally, the magnitude of visual data has increased due to increasing resolution. The resolution of images and photos has been increased from several hundred thousand pixels to several million pixels. Video resolution has been increasing from early quarter common intermediate format (QCIF) to current high definition (HD).

The increasing magnitude of visual data is also verified by wired and wireless network traffic. According to the Cisco visual networking index: forecast and methodology (2011–2016) [2], Internet video traffic will comprise 55% of all Internet traffic by 2016, rising from 51% in 2011. This does not include videos exchanged through peer-to-peer (P2P) file sharing. The sum of all forms of video, including television (TV), video on demand (VoD), Internet, and P2P, will be approximately 86% of all Internet traffic by 2016. The study further indicates that mobile video traffic exceeded 50% for the first time in 2012 [3]. Mobile video traffic will be 66.5% of all mobile traffic in 2016.

Therefore, no matter how fast wired and wireless network bandwidth is being increased, highly efficient compression of visual data and its highly efficient transmission will need to continuously improve in quality. The theoretical basis of compression and transmission of visual data is Shannon's information theory [4]. The typical scenario studied in information theory is shown in Figure 1.1. A transmitter sends a source S over a noisy channel to a receiver. The channel adds a certain noise to the transmitted source. The receiver has to recover the source from the received

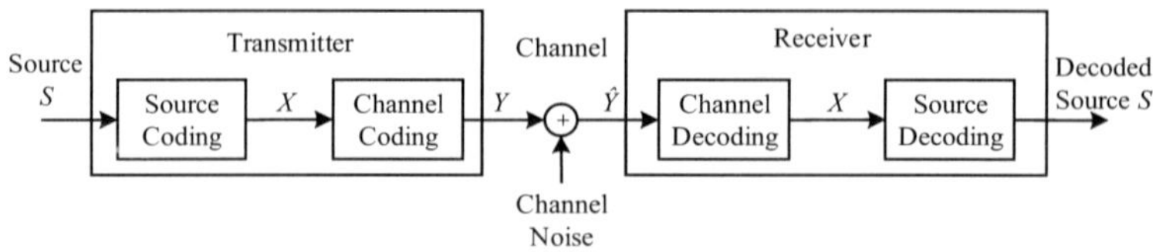

Figure 1.1 The basic framework for source and channel coding.

data with noise. To efficiently achieve it, there are two core technologies needed: source coding and channel coding.

Source usually contains a certain redundancy and thus directly transmitting it would cost more channel bandwidth than what is really needed. In the transmitter, source coding tries to remove the redundancy from the source when possible, by using various compression technologies. The output after source coding should be randomly distributed bits. Since the channel is noisy, the compressed bits will be corrupted by adding channel noise. Channel coding tries to introduce a certain redundancy back to the source, which protects the compressed bits. At the receiver, channel decoding corrects the received errors by using redundancy. Source decoding recovers the source from the compressed bits.

This chapter is organized according to T. M. Cover and J. A. Thomas's book *Elements of Information Theory* [5], which is a comprehensive book about Shannon's information theory. We do not pay much attention to either the completeness of the theory or the details of how to prove the theorems step by step. Instead, we try to highlight the core ideas behind the theory and make them more intuitively understood for further practical research on visual data compression and communication. The readers, who are interested in the entire information theory, can find information on this subject by Shannon [4] and Cover and Thomas [5].

Let us assume a finite symbol alphabet $\mathscr{A} = \{a_0, a_1, \cdots, a_{q-1}\}$, whose probabilities of occurrence are $\mathbf{p} = \{p_0, p_1, \cdots, p_{q-1}\}$, satisfying $\sum_i p_i = 1$. These probabilities are known. Let S be a discrete random variable, taking the symbol from the alphabet $\mathscr{A}$ according to the probability $p(s = a_i) = p_i$, where s is an instance of S. For different probability distributions, the information carried by S is different. Shannon's information theory defines entropy to measure the information.

Definition 1 *The entropy of a discrete random variable S is defined as*

$$H(S) = -\sum_i p_i \log p_i. \tag{1.1}$$

$H()$ plays a central role in the information theory as measures of information and uncertainty. If and only if all the p_i but one are zero, the entropy $H(S) = 0$. In this case, S always takes a_i whose probability is 1. The uncertainty is zero and thus the entropy is zero. If $p_0 = p_1 = \cdots = p_{q-1} = 1/q$, S has the same probability taking any a_i. The entropy $H(S)$ is the maximum, which is equal to $\log_2 q$.

The definition of entropy can be extended to a pair of random variables S_i and S_j. In information theory, the relationship between two random variables is described by joint entropy and conditional entropy.

Definition 2 *The joint entropy $H(S_i,S_j)$ of a pair of discrete random variables (S_i,S_j) with a joint distribution $p(s_i,s_j)$ is defined as*

$$H(S_i,S_j) = -\sum_{s_i\in\mathscr{A}}\sum_{s_j\in\mathscr{A}} p(s_i,s_j)\log p(s_i,s_j) = -E\log p(S_i,S_j). \tag{1.2}$$

The conditional entropy of S_i given S_j is defined as the expected value of the entropies of the conditional distributions.

Definition 3 *If $(S_i,S_j)\sim p(s_i,s_j)$, the conditional entropy $H(S_i \mid S_j)$ is defined as*

$$\begin{aligned} H(S_i \mid S_j) &= \sum_{s_j\in\mathscr{A}} p(s_j)H(S_i \mid S_j = s_j), && (1.3)\\ &= -\sum_{s_j\in\mathscr{A}} p(s_j)\sum_{s_i\in\mathscr{A}} p(s_i \mid s_j)\log p(s_i \mid s_j), && (1.4)\\ &= -\sum_{s_j\in\mathscr{A}}\sum_{s_i\in\mathscr{A}} p(s_i,s_j)\log p(s_i \mid s_j), && (1.5)\\ &= -E_{p(s_i,s_j)}\log p(S_i \mid S_j). && (1.6)\end{aligned}$$

As we have discussed, entropy is a measure of the uncertainty of a random variable. It is also a measure of the amount of information required on average to describe the random variable. However, mutual information is a measure of the amount of information that one random variable contains about another random variable. It is the reduction in the uncertainty of one random variable due to the knowledge of the other.

Definition 4 *Assume two random variables S_i and S_j with a joint probability mass function (PMF) $p(s_i,s_j)$. Their marginal probability mass functions are $p(s_i)$ and $p(s_j)$, respectively. The mutual information $I(S_i;S_j)$ is defined as*

$$\begin{aligned} I(S_i;S_j) &= \sum_{s_i\in\mathscr{A}}\sum_{s_j\in\mathscr{A}} p(s_i,s_j)\log\frac{p(s_i,s_j)}{p(s_i)p(s_j)}, && (1.7)\\ &- \sum_{s_i\in\mathscr{A}}\sum_{s_j\in\mathscr{A}} p(s_i,s_j)\log\frac{p(s_i \mid s_j)}{p(s_i)}, && (1.8)\\ &= -\sum_{s_i\in\mathscr{A}}\sum_{s_j\in\mathscr{A}} p(s_i,s_j)\log p(s_i) + \sum_{s_i\in\mathscr{A}}\sum_{s_j\in\mathscr{A}} p(s_i,s_j)\log p(s_i \mid s_j), && (1.9)\\ &= -\sum_{s_i\in\mathscr{A}} p(s_i)\log p(s_i) - \left(-\sum_{s_i\in\mathscr{A}}\sum_{s_j\in\mathscr{A}} p(s_i,s_j)\log p(s_i \mid s_j)\right), && (1.10)\\ &= H(S_i) - H(S_i \mid S_j). && (1.11)\end{aligned}$$

Next, we consider a set of random variables $S^n = \{S_1, S_2, \cdots, S_n\}$. All random variables have the same distribution $p(s)$. Furthermore, for any $i \neq j$, S_i is independent of S_j. In other words, S^n is an independent and identically distributed (i.i.d.) source. In information theory, the analog of the law of large numbers is the asymptotic equipartition property (AEP).

Definition 5 *If $S_1, S_2, \cdots, S_n$ are i.i.d. $\sim p(s)$, the AEP indicates*

$$-\frac{1}{n}\log p(S_1, S_2, \cdots, S_n) \to H(S) \tag{1.12}$$

in probability.

It is a direct consequence of the weak law of large numbers. The law of large numbers states that, for i.i.d. variables, $1/n\sum_i S_i$ is close to its expected value $E(S)$ for large n. The AEP states that $-1/n\log p(S_1, S_2, \cdots, S_n)$ is close to the entropy $H(S)$, where $p(S_1, S_2, \cdots, S_n)$ is the probability of observing the sequence $S_1, S_2, \cdots, S_n$. Thus, the probability $p(S_1, S_2, \cdots, S_n)$ will be close to $2^{-nH(S)}$.

Finally, we discuss the continuous random variable. Let S be a random variable with a cumulative distribution $F(s) = Pr(S \leq s)$. If $F(s)$ is continuous, the random variable is said to be continuous. Let $f(s) = F'(s)$ when the derivative exists. If $\int_{-\infty}^{\infty} f(s) = 1$, $f(s)$ is called the probability density function (PDF) for S. The set where $f(s) > 0$ is called the support set of S.

Definition 6 *The differential entropy $h(S)$ of a continuous random variable S with a PDF $f(s)$ is defined as*

$$h(S) = -\int_{\Omega} f(s)\log f(s)ds, \tag{1.13}$$

where Ω is the support set of the random variable.

If S is a zero-mean Gaussian variable, that is, $S \sim f(s) = (1/\sqrt{2\pi\sigma^2}) \times e^{-s^2/2\sigma^2}$,

$$\begin{aligned}
h(S) &= -\int f(s)\ln f(s)ds, & (1.14)\\
&= -\int f(s)\left[-\frac{s^2}{2\sigma^2} - \ln\sqrt{2\pi\sigma^2}\right]ds, & (1.15)\\
&= \frac{ES^2}{2\sigma^2} + \frac{1}{2}\ln 2\pi\sigma^2, & (1.16)\\
&= \frac{1}{2} + \frac{1}{2}\ln 2\pi\sigma^2, & (1.17)\\
&= \frac{1}{2}\ln e + \frac{1}{2}\ln 2\pi\sigma^2, & (1.18)\\
&= \frac{1}{2}\ln 2\pi e\sigma^2. & (1.19)
\end{aligned}$$

Changing the base of the logarithm, the differential entropy for a zero-mean Gaussian variable is $h(S) = 1/2\log 2\pi e\sigma^2$.

1.2 Source Coding

With the above basic information theory definitions, we are ready to discuss source coding. The objective of the so-called source coding is to find a source code C mapping all possible values of a random variable S in the alphabet $\mathscr{A}$ to a set of finite length of binary strings (also called codewords). The mapping must be one-to-one so that we can only recover a unique value from one given codeword. Let $C(a_i)$ denote the codeword corresponding to $s = a_i$ and $l(a_i)$ denote the length of $C(a_i)$. Therefore, the expected length $L(S)$ is given by

$$L(S) = \sum_i p_i l(a_i). \tag{1.20}$$

Shannon's first theorem states the relationship between the expected length and the entropy. It is described here.

Theorem 1.1. *Let l_0^*, l_1^*, $\cdots$, l_{q-1}^* be the optimal codeword lengths for a source distribution* $\mathbf{p}$, *and let L^* be the associated expected length of the optimal code ($L^* = \sum_i p_i l_i^*$), then*

$$H(S) \leq L^* < H(S) + 1. \tag{1.21}$$

This theorem is the binary version of Theorem 5.4.1 proved by Cover and Thomas [5]. The optimal expected length L^* is impossibly less than the entropy $H(S)$. In other words, the lossless compression of a random variable cannot be less than its entropy. Note that there is an overhead which is at most 1 bit because of the fact that $\log 1/p_i$ is not always an integer.

Let us take a simple example with four symbols $\mathscr{A} = \{a_0, a_1, a_2, a_3\}$. Their probabilities are $\{0.5, 0.25, 0.125, 0.125\}$. According to Eq. (1.1), the entropy $H(S)$ is equal to

$$H(S) = -(\frac{1}{2}\log\frac{1}{2} + \frac{1}{4}\log\frac{1}{4} + \frac{2}{8}\log\frac{1}{8}) = 1.75 \; bits.$$

The most straightforward way to conduct source coding is to assign 2 different bits to every symbol. It is so-called fixed length coding (FLC). The expected length $L(S)$ is 2 bits too. It is larger than the entropy by 0.25 bits and thus FLC is not optimal. This simple example shows the importance of source coding in the information theory. According to the theory, shorter codewords should be assigned to symbols with larger probabilities, while longer codewords to symbols with smaller probabilities. This is the basis of variable length coding (VLC). How to design the set of codewords and how to assign them to the symbols are just the problems to be solved in source coding. The optimization of source coding aims to make the excepted coding length $L(S)$ equal to the entropy $H(S)$. There are several source coding methods that are optimal and can reach information entropy. We will discuss Huffman coding and arithmetic coding in the following sections.

1.2.1 Huffman Coding

An optimal prefix code for a given distribution can be constructed by a simple algorithm proposed by Huffman [6], widely known as Huffman coding. In Huffman coding, a binary tree of nodes is first created. The size is equal to the number of symbols to code, q. Initially, all nodes are leaf nodes containing the probabilities of the symbols they represent. The coding process essentially begins with the leaf nodes. A new node whose children are the two nodes with the smallest probability is created. The new node's probability is equal to the sum of the children's probabilities. With the previous two nodes merged into one node, they are no longer considered. Instead the new node is considered. This procedure is repeated until only one node remains. As a common convention, bit "0" represents the left child and bit "1" represents the right child. A finished tree has up to q leaf nodes and $q-1$ internal nodes. This Huffman tree produces the optimal code length.

For the simple example with four symbols, a binary tree is established according to the algorithm in Figure 1.2a. The bits "0" and "1" are assigned in Figure 1.2b. Therefore, the codewords for the symbols a_0, a_1, a_2, and a_3 are "1," "01," "001," and "000," respectively. The shortest codeword "1" is assigned to the symbol a_0. Using Eq. (1.20) to calculate the expected length of the codewords, it is 1.75 bits, which is exactly equal to the entropy.

1.2.2 Arithmetic Coding

If the alphabet $\mathscr{A}$ is small (e.g., only two symbols $a_0 = 0$ and $a_1 = 1$), although their probabilities are different, Huffman codes still use one bit to represent them because

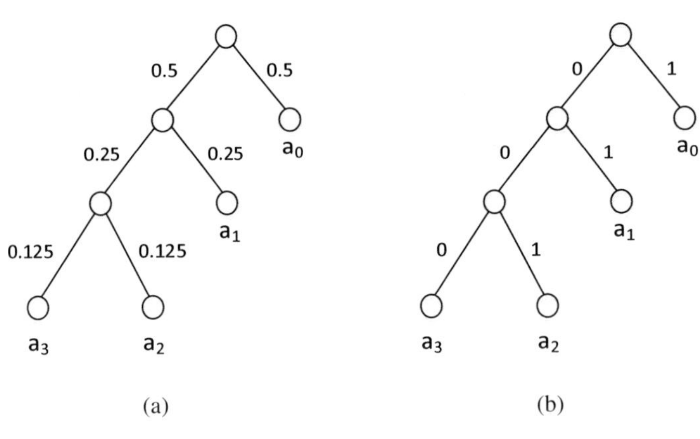

Figure 1.2 Huffman coding. (a) The process of building a binary tree; (b) the designed Huffman codes.

they cannot adjust codeword lengths at fractional bit precision. In this case, Huffman codes are not optimal. One alternative is to jointly code a source sequence instead of the individual source. According to the AEP, we can achieve an expected length per every source close to the entropy of $H(S)$. Therefore, it is desirable to have an efficient coding procedure that works for a long block of source letters. Huffman coding is not ideal for this situation, since it is a bottom-up procedure that requires the calculation of the probabilities of all source sequences of a particular block length and the construction of the corresponding complete code tree. Thus the block length is limited. A better scheme is one which can be easily extended to longer block lengths without having to redo all the calculations. Arithmetic coding achieves this goal.

Assume $s^n = s_1 s_2 \cdots s_n$ is a source sequence. The essential idea of arithmetic coding is to efficiently calculate the PMF $p(s^n)$ and the cumulative distribution function $F(s^n) = \sum_{u^n \leq s^n} p(u^n)$ for the source sequence s^n. We can use a number in the interval $[(F(s^n) - p(s^n)], [F(s^n)]$ as the codeword for s^n. The codewords for various sequences are different because they are within different interval $[(F(s^n) - p(s^n)], [F(s^n)]$. A simplified version of the arithmetic coding is described here to illustrate the primary.

$\mathscr{A}$ is assumed to be binary. A simple procedure is assumed to calculate $p(s_1, s_2, \cdots, s_n)$ for any string $s^n = s_1 s_2 \cdots s_n$. The natural lexicographic order is used on strings, so that a string s^n is greater than a string u^n if $s_i = 1$, $u_i = 0$ for the first i such that $s_i \neq u_i$. Equivalently, $s^n > u^n$ if $\sum_i s_i 2^{-i} > \sum_i u_i 2^{-i}$. We can arrange the strings as the leaves of a tree of depth n, where each level of the tree corresponds to one bit. Such a tree is illustrated in Figure 1.3. In Figure 1.3, the ordering $s^n > u^n$ corresponds to the fact that s^n is to the right of u^n on the same level of the tree.

We need to find $p(u^n)$ for all $u^n \leq s^n$ and use that to calculate $F(s^n)$. Looking at the tree in Figure 1.3, we might suspect that we need to calculate the probabilities of

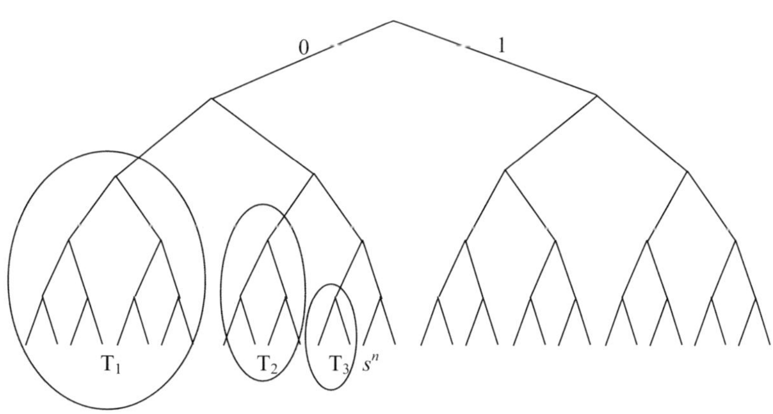

Figure 1.3 Tree of strings for arithmetic coding.

all the leaves to the left of s^n to find $F(s^n)$. The sum of these probabilities is the sum of the probabilities of all the subtrees to the left of s^n. Let $T_{s_1 s_2 \cdots s_{i-1}0}$ be a subtree starting with $s_1 s_2 \cdots s_{i-1}0$. The probability of this subtree is

$$p(T_{s_1 s_2 \cdots s_{i-1}0}) = \sum_{u_{i+1} \cdots u_n} p(s_1 s_2 \cdots s_{i-1} 0 u_{i+1} \cdots u_n) = p(s_1 s_2 \cdots s_{i-1} 0). \tag{1.22}$$

It can be calculated easily. Therefore, we can write $F(s^n)$ as

$$F(s^n) = \sum_{u^n \le s^n} p(u^n) = \sum_{T:T \text{ is to the left of } S} p(T) = \sum_{i:s_i=1} p(s_1 s_2 \cdots s_{i-1} 0). \tag{1.23}$$

Thus, $F(s^n)$ can be calculated quickly from $p(s^n)$. To encode the next bit of the source sequence, we need only calculate $p(s^n s_{n+1})$ and update $F(s^n s_{n+1})$ using the above method. Encoding can be done sequentially, by looking at the bits as they come in.

To decode the sequence, we use the same procedure to calculate the cumulative distribution function and check whether it exceeds the value corresponding to the codeword. We then use the tree in Figure 1.3 as a decision tree. At the top node, we check if the received codeword $F(s^n)$ is greater than $p(0)$. If it is, then the subtree starting with 0 is to the left of s^n and hence $s_1 = 1$. Continuing this process down the tree, we can decode the bits in sequence. Thus, we can compress and decompress a source sequence in a sequential manner.

The above procedure depends on a model for which we can easily compute $p(s^n)$. Two examples of such models are i.i.d. sources, where

$$p(s^n) = \prod_{i=1}^{n} p(s_i), \tag{1.24}$$

and the Markov source, where

$$p(s^n) = p(s_1) \prod_{i=2}^{n} p(s_i \mid s_{i-1}). \tag{1.25}$$

In both cases, we can easily calculate $p(s^n s_{n+1})$ from $p(s^n)$.

Note that it is not essential that the probabilities used in the encoding be equal to the true distribution of the source. In some cases, such as in image and video compression, it is difficult to describe a true distribution for the source. Even then, it is possible to apply the above arithmetic coding procedure. The procedure will be efficient only if the model distribution is close to the empirical distribution of the source. A more sophisticated use of arithmetic coding is to change the model dynamically to adapt to the source, which has been adopted in the recent video coding standards. Adaptive models work well for large classes of sources.

1.2.3 Rate Distortion Theory

In Huffman coding and arithmetic coding, the coding process does not lose any source information and the source can be exactly reproduced from coded data, known as lossless coding. In many cases, such as with image and video compression, some source information has to be discarded during the coding process so that the coding rate is not larger than a given rate. The decoded source always contains a certain difference from the original source. Therefore, this coding is called lossy coding. Rate distortion theory provides the theoretical principle for optimizing lossy coding.

A classical example of lossy coding is the alphabet $\mathscr{A}$ as an arbitrary real number. It will require an infinite number of bits to represent it. So a finite representation of a continuous random variable can never be perfect. How well can we do in getting as close to perfect as possible? To answer this question, it is necessary to define a distortion measure, which is a measure of distance between the random variable and its representation. The basic problem with the rate distortion theory can be stated as follows: given a source distribution and a distortion measure, what is the minimum expected distortion achievable at a particular rate? Or, what is the minimum rate description required to achieve a particular distortion?

We first consider the problem of representing a single continuous random variable S. Let the representation of S be denoted as $\hat{S} = Q(S)$, where $Q()$ is a quantization and de-quantization function. If we are given R bits to represent S, $\hat{S}$ can take on 2^R values. Here, we assume that all values have the same probability to occur and thus the coding process is not considered. The problem is finding the optimum set of values for $\hat{S}$ (called the reproduction points) and the regions that are associated with each value $\hat{S}$.

For example, let $S \sim N(0, \sigma_s^2)$ and assume a squared distortion measure. If 1 bit is given to represent S, it is clear that the bit should distinguish whether $S > 0$ or not. To minimize the squared error, each reproduced symbol should be at the conditional mean of its region, thus

$$Q(s) = \begin{cases} \sqrt{\frac{2}{\pi}}\sigma_s, & if\ s \geq 0 \\ -\sqrt{\frac{2}{\pi}}\sigma_s, & if\ s < 0 \end{cases}. \tag{1.26}$$

If we are given 2 bits to represent the source, the situation is not simple. Clearly, we want to divide the real line into four regions and use a point within each region to represent the sample. But it is no longer immediately obvious what the representation regions and the reconstruction points should be.

For general cases, let us assume a set of n i.i.d. random variables drawn according to a given distribution. These random variables are to be represented using nR bits. Since the source is i.i.d., the samples are independent. It may appear that the representation of each sample is an independent problem to be treated separately. But this is not true as we will show at the end of this section. The entire sequence is represented by a single index taking 2^{nR} values. This treatment of entire sequences immediately achieves a lower distortion for the same rate than independent quantization of the individual samples.

A $(2^{nR}, n)$ rate distortion code consists of an encoding function

$$f_n: \quad S^n \to 1, 2, \cdots, 2^{nR}, \tag{1.27}$$

and a decoding function,

$$g_n: \quad 1, 2, \cdots, 2^{nR} \to \hat{S}^n. \tag{1.28}$$

The distortion associated with the $(2^{nR}, n)$ code is defined as

$$D = Ed(S^n, g_n(f_n(S^n))), \tag{1.29}$$

where the expectation is with respect to the probability distribution on S^n, that is,

$$D = \sum_{s^n} p(s^n) d(s^n, g_n(f_n(s^n))). \tag{1.30}$$

The distortion function $d(s^n, \hat{s}^n)$ is defined as the squared error $d(s^n, \hat{s}^n) = 1/n \sum_i (s_i - \hat{s}_i)^2$.

A rate distortion pair (R, D) is said to be achievable if there exists a sequence of $(2^{nR}, n)$ rate distortion codes (f_n, g_n) with $\lim_{n\to\infty} Ed(S^n, g_n(f_n(S^n))) \leq D$. A rate distortion region for a source is the closure of the set of achievable rate distortion pairs (R, D). The rate distortion function $R(D)$ is the infimum of rates R such that (R, D) is in the rate distortion region of the source for a given distortion D. The distortion rate function $D(R)$ is the infimum of all distortions D such that (R, D) is in the rate distortion region of the source for a given rate R.

The main theorem in the rate distortion theory can now be stated as follows:

Theorem 1.2. *The rate distortion function for an i.i.d. source S^n with distribution $p(s^n)$ and bounded distortion function $d(s^n, \hat{s}^n)$ is equal to the associated information rate distortion function. Thus*

$$R(D) = \min_{p(\hat{s}^n|s^n): \sum_{(s^n, \hat{s}^n)} p(s^n) p(\hat{s}^n|s^n) d(s^n, \hat{s}^n) \leq D} I(S^n; \hat{S}^n) \tag{1.31}$$

is the minimum achievable rate at distortion D.

This is Theorem 13.2.1 proven by Cover and Thomas [5]. From the theorem, for a given distortion D, the minimum rate is the minimum mutual information between S^n and $\hat{S}^n$ for all pairs of $(s^n, \hat{s}^n)$, which the distortions are less than or equal to D.

With the rate distortion theorem, we can derive the close-form rate distortion function of a single Gaussian source. Let S be $\sim N(0, \sigma_s^2)$. According to the rate distortion theorem, we have

$$R(D) = \min_{f(\hat{s}|s): E(\hat{S}-S)^2 \leq D} I(S; \hat{S}). \tag{1.32}$$

We first find a lower bound for the above rate distortion function. Since $E(S - \hat{S})^2 \leq D$, we observe

$$
\begin{aligned}
I(S;\hat{S}) &= h(S) - h(S \mid \hat{S}), && (1.33)\\
&= \frac{1}{2}\log(2\pi e)\sigma_s^2 - h(S-\hat{S} \mid \hat{S}), && (1.34)\\
&\geq \frac{1}{2}\log(2\pi e)\sigma_s^2 - h(S-\hat{S}), && (1.35)\\
&\geq \frac{1}{2}\log(2\pi e)\sigma_s^2 - h(N(0,E(S-\hat{S})^2)), && (1.36)\\
&= \frac{1}{2}\log(2\pi e)\sigma_s^2 - \frac{1}{2}\log(2\pi e)E(S-\hat{S})^2, && (1.37)\\
&\geq \frac{1}{2}\log(2\pi e)\sigma_s^2 - \frac{1}{2}\log(2\pi e)D, && (1.38)\\
&= \frac{1}{2}\log\frac{\sigma_s^2}{D}, && (1.39)
\end{aligned}
$$

where Eq. (1.35) follows from the fact that conditioning reduces entropy and Eq. (1.36) follows from the fact that normal distribution maximizes entropy for a given second moment. Hence

$$
R(D) \geq \frac{1}{2}\log\frac{\sigma_s^2}{D}. \tag{1.40}
$$

To find the conditional density $f(\hat{s} \mid s)$ that achieves this lower bound Eq. (1.40), it is usually more convenient to look at the conditional density $f(s \mid \hat{s})$, which is sometimes called the test channel. If $D \leq \sigma_s^2$, we choose

$$
S = \hat{S} + Z, \quad \hat{S} \sim N(0, \sigma_s^2 - D), \quad Z \sim N(0, D), \tag{1.41}
$$

where $\hat{S}$ and Z are independent. For this joint distribution, we can calculate

$$
I(S;\hat{S}) = \frac{1}{2}\log\frac{\sigma_s^2}{D}, \tag{1.42}
$$

and $E(S-\hat{S})^2 = D$, thus achieving the bound in Eq. (1.40). If $D > \sigma_s^2$, we choose $\hat{S} = 0$ with probability 1, achieving $R(D) = 0$. Hence, the rate distortion for a Gaussian source with the squared error distortion is

$$
R(D) = \begin{cases} \frac{1}{2}\log\frac{\sigma_s^2}{D}, & 0 \leq D \leq \sigma_s^2 \\ 0, & D > \sigma_s^2 \end{cases}. \tag{1.43}
$$

We can rewrite Eq. (1.43) to express the distortion in terms of the rate,

$$
D(R) = \sigma_s^2 2^{-2R}. \tag{1.44}
$$

Each bit of description reduces the expected distortion by a factor of 4. With a 1 bit description, the best expected squared error is $\sigma_s^2/4 = 0.25\sigma^2$. We can compare this distortion with the quantization and de-quantization function given in Eq. (1.26).

According to the reproduction points, the expected distortion can be calculated as $(\pi - 2)\sigma_s^2/\pi \approx 0.3633\sigma_s^2$. Obviously, it is larger than the distortion given by Eq. (1.44). This is because the result in Eq. (1.44) is derived from the long block length. However, the result in Eq. (1.26) only considers one random variable. Therefore, the rate distortion limit $R(D)$ should be achieved by considering long block lengths.

1.3 Channel Coding

As shown in Figure 1.1, after source coding, coded source X no longer contains any redundancy. If X is a k-dimensional bit vector, the space 2^k will be fully occupied. Every value represents a valid message. It is very sensitive to channel noise because any corrupted bit in x will change x to another valid message. In this case, it is difficult to tell which x is really sent out. Therefore, in channel coding X is mapped into some sequence Y of channel codewords. Y is usually redundant because there are no other valid codewords in a neighboring space of every valid codeword. $\hat{Y}$ is the output of the channel. The output sequence $\hat{Y}$ is random but has a distribution that depends on the input sequence Y. From the output sequence, we attempt to recover the transmitted source.

Each of the possible input sequences induces a probability distribution on the output sequences. Since two different input sequences may give rise to the same output sequence, the inputs are confusable. In channel coding, we try to choose a non-confusable subset of input sequences so that with high probability, there is only one highly likely input that could cause a particular output. We can then reconstruct the input sequence at the output with a negligible probability of error.

1.3.1 Capacity

We define a discreet channel to be a system consisting of an input alphabet $\mathscr{Y}$ and output alphabet $\hat{\mathscr{Y}}$ and a probability transition matrix $p(y \mid \hat{y})$ that expresses the probability of observing the output symbol $\hat{y}$ given that we send the symbol y. The channel is said to be without memory if the probability distribution of the output depends only on the input at that time and is conditionally independent of previous channel inputs or outputs.

Definition 7 *The channel capacity of a discrete channel without memory is defined as*

$$C = \max_{p(y)} I(Y; \hat{Y}), \tag{1.45}$$

where the maximum is taken over by all possible input distributions $p(y)$.

The channel capacity is actually evaluated by the entropy $H(Y)$. The entropy is reduced from $H(Y)$ to $I(Y;\hat{Y}) = H(Y) - H(Y \mid \hat{Y})$ because we have $\hat{Y}$. In other words, the entropy reduction is equal to what we get from the channel.

An intuitive idea is described here about why we can transmit C bits of information over a channel. The basic idea is that, for large block lengths, every channel looks like the noisy typewriter channel and the channel has a subset of inputs that produce essentially disjointed sequences at the output. For each input n-sequence, we wish to ensure that no two Y sequences produce the same $\hat{Y}$ output sequence. Otherwise, we will not be able to decide which Y sequences was sent. The total number of possible $\hat{Y}$ sequences is $\approx 2^{nH(\hat{Y})}$. This set has to be divided into sets of size $2^{nH(\hat{Y}|Y)}$ corresponding to the different input Y sequences. The total number of disjointed sets is less than or equal to $2^{n(H(\hat{Y})-H(\hat{Y}|Y))} = 2^{nI(Y;\hat{Y})}$. Hence, we can send at most $\approx 2^{nI(Y;\hat{Y})}$ distinguishable sequences of length n.

The most important continuous alphabet channel is the Gaussian channel. This is a time discrete channel with output $\hat{Y}_i$ at time i, where $\hat{Y}_i$ is the sum of the input Y_i and the noise Z_i. The noise Z_i is drawn i.i.d. from a Gaussian distribution with variable σ^2. Thus

$$\hat{Y}_i = Y_i + Z_i, \quad Z_i \sim N(0, \sigma^2). \tag{1.46}$$

The noise Z_i is assumed to be independent of the signal Y_i. Without further conditions, the capacity of this channel may be infinite. If the noise variance is zero, then the receiver receives the transmitted symbol perfectly. If the noise variance is nonzero and there is no constraint on the input, we can choose an infinite subset of inputs arbitrarily far apart, so that they are distinguishable at the output with an arbitrarily small probability of error. Such a scheme has an infinite capacity as well. Thus if the noise variance is zero or the input is unconstrained, the capacity of the channel is infinite.

The most common limitation on the input is an energy or power constraint. We assume an average power constraint. For any codeword $(y_1, y_2, \ldots, y_n)$ transmitted over the channel, we require

$$\frac{1}{n} \sum_{i=1}^{n} y_i^2 \leq P. \tag{1.47}$$

This communication channel models many practical channels, including radio and satellite links. The additive noise in such channels may be due to a variety of reasons. However, by the central limit theorem, the cumulative effect of a large number of small random effects will be approximately normal, so the Gaussian assumption is valid in a large number of situations.

The information capacity of the Gaussian channel with power constraint P is

$$C = \max_{p(y): EY^2 \leq P} I(Y;\hat{Y}). \tag{1.48}$$

We can calculate the information capacity as follows. Expanding $I(Y;\hat{Y})$, we have

$$\begin{aligned} I(Y;\hat{Y}) &= h(\hat{Y}) - h(\hat{Y} \mid Y), \\ &= h(\hat{Y}) - h(Y+Z \mid Y), \\ &= h(\hat{Y}) - h(Z \mid Y), \\ &= h(\hat{Y}) - h(Z), \end{aligned} \tag{1.49}$$

since Z is independent of Y. Now, $h(Z) = 1/2\log 2\pi e\sigma^2$. Also, we have

$$E\hat{Y}^2 = E(Y+Z)^2 = EY^2 + 2EYEZ + EZ^2 = P + \sigma^2, \tag{1.50}$$

since Y and Z are independent and $EZ = 0$. Given $E\hat{Y}^2 = P + \sigma^2$, the entropy of $\hat{Y}$ is bounded by $1/2\log 2\pi e(P+\sigma^2)$ because the normal maximizes the entropy for a given variance. Applying this result to bound the mutual information, we obtain

$$I(Y;\hat{Y}) = h(\hat{Y}) - h(Z), \tag{1.51}$$

$$\leq \frac{1}{2}\log 2\pi e(P+\sigma^2) - \frac{1}{2}\log 2\pi e\sigma^2, \tag{1.52}$$

$$= \frac{1}{2}\log(1+\frac{P}{\sigma^2}). \tag{1.53}$$

Hence the information capacity of the Gaussian channel is

$$C = \max_{p(y):EY^2\leq P} I(Y;\hat{Y}) = \frac{1}{2}\log(1+\frac{P}{\sigma^2}). \tag{1.54}$$

1.3.2 Coding Theorem

An (M,n) code for the channel $(\mathscr{Y}, p(\hat{y} \mid y), \hat{\mathscr{Y}})$ consists of the following:

1. An index set $\{1,2,\cdots,M\}$.
2. An encoding function $Y^n : \{1,2,\cdots,M\} \rightarrow \mathscr{Y}^n$, yielding codewords $Y^n(1)$, $Y^n(2)$, $\cdots$, $Y^n(M)$. The set of codewords is called the codebook.
3. A decoding function $g : \hat{\mathscr{Y}}^n \rightarrow \{1,2,\cdots,M\}$, which is a deterministic rule to assign a guess to each possible received vector.

Let

$$\lambda_i = Pr(g(\hat{Y}^n) \neq i \mid Y^n = Y^n(i)) = \sum_{\hat{y}^n} p(\hat{y}^n \mid y^n(i)) I(g(\hat{y}^n) \neq i) \tag{1.55}$$

be the conditional probability of error given that index i was sent, where $I()$ is the indicator function. The maximal probability of error $\lambda^{(n)}$ for an (M,n) code is defined as

$$\lambda^{(n)} = \max_{i\in\{1,2,\cdots,M\}} \lambda_i. \tag{1.56}$$

The rate R of an (M,n) code is

$$R = \frac{\log M}{n}. \tag{1.57}$$

Shannon's second theorem states the relation between channel capacity and the maximum transmitted rate.

Theorem 1.3. *All rates below capacity C are achievable. Specifically, for every rate $R < C$, there exists a sequence of $(2^{nR}, n)$ codes with maximum probability of error $\lambda^{(n)} \to 0$. Conversely, any sequence of $(2^{nR}, n)$ codes with $\lambda^{(n)} \to 0$ must have $R \leq C$.*

This is the Theorem 8.7.1 proved by Cover and Thomas [5]. The channel coding theorem promises the existence of block codes that will allow us to transmit information at rates below capacity with an arbitrarily small probability of error if the block length is large enough. In this chapter, we will not discuss the capacity-achieving channel codes. Instead, we will use simple Hamming codes to further explain channel coding.

1.3.3 Hamming Codes

The object of channel coding is to introduce redundancy so that even if some of the information is lost or corrupted, it will still be possible to recover the message at the receiver. The most obvious channel coding scheme is to repeat information. For example, to send a 1, we send 11111, and to send a 0, we send 00000. This scheme uses 5 symbols to send 1 bit, and therefore has a rate of 1/5 bits per symbol. If 3 or more bits are 1, we decode the block as a 1; otherwise we decode it as 0. An error occurs if and only if more than 3 of the bits are changed. By using longer repetition codes, we can achieve an arbitrarily low probability of error. But the rate of the code also goes to zero with block length. It is not a very useful code.

Hamming codes are a family of linear error-correcting block codes that generalized the Hamming (7,4) code invented by Richard Hamming in 1950 [7]. A (7.4) Hamming code produces 7 bits of output for every 4 bits of input. Hamming codes are linear block codes, which means that the encoding operation can be described in terms of a 4×7 generator matrix, such as

$$G = \begin{bmatrix} 1\,1\,0\,1\,0\,0\,0 \\ 0\,1\,1\,0\,1\,0\,0 \\ 0\,0\,1\,1\,0\,1\,0 \\ 0\,0\,0\,1\,1\,0\,1 \end{bmatrix}. \tag{1.58}$$

The codewords are obtained as a linear combination of the rows of G, where all the operations are computed as modulo 2 in each vector element. That is, the code is the row space of G. For a source vector $\mathbf{x} = [x_1, x_2, x_3, x_4]$, the codeword is

$$\mathbf{y} = \mathbf{x}G. \tag{1.59}$$

For example, if $\mathbf{x} = [0,0,1,1]$, the codeword is

$$\mathbf{y} = [0,0,1,1,0,1,0] + [0,0,0,1,1,0,1] = [0,0,1,0,1,1,1].$$

All codewords for this code are

$$\begin{matrix} [0,0,0,0,0,0,0], [1,1,0,1,0,0,0], [0,1,1,0,1,0,0], [1,0,1,1,1,0,0] \\ [0,0,1,1,0,1,0], [1,1,1,0,0,1,0], [0,1,0,1,1,1,0], [1,0,0,0,1,1,0] \\ [0,0,0,1,1,0,1], [1,1,0,0,1,0,1], [0,1,1,1,0,0,1], [1,0,1,0,0,0,1] \\ [0,0,1,0,1,1,1], [1,1,1,1,1,1,1], [0,1,0,0,0,1,1], [1,0,0,1,0,1,1] \end{matrix}. \tag{1.60}$$

It can be verified that the number of different bits between any two codewords is 3. This is known as Hamming distance. Therefore, Hamming $(7,4)$ codes can detect up to two bits of error and correct up to one bit of error.

Every (n,k) linear block code has associated with a $(n-k) \times n$ matrix H called the parity check matrix, which has the property

$$\mathbf{y}H^T = 0, \tag{1.61}$$

if and only if the vector $\mathbf{y}$ is a codeword. The parity check matrix is not unique. For the generator G of Eq. (1.58), the parity check matrix can be written as

$$H = \begin{bmatrix} 1\,0\,1\,1\,1\,0\,0 \\ 0\,1\,0\,1\,1\,1\,0 \\ 0\,0\,1\,0\,1\,1\,1 \end{bmatrix}. \tag{1.62}$$

It can be verified that $GH^T = 0$. The matrix H is expressed in terms of its column as

$$H = [\mathbf{h}_1\ \mathbf{h}_2\ \mathbf{h}_3\ \mathbf{h}_4\ \mathbf{h}_5\ \mathbf{h}_6\ \mathbf{h}_7]. \tag{1.63}$$

We can observe that the columns of H consist of the binary representations of the numbers 1 through 7, though not in numerical order. On the basis of this observation, we can generalize other Hamming codes. Hamming codes of length $n = 2m - 1$ and dimension $k = 2m - m - 1$ exist for every $m \geq 2$, having parity check matrices whose columns are binary representations of the numbers from 1 through n.

Supposing that a codeword $\mathbf{y}$ is sent and the received vector is

$$\hat{\mathbf{y}} = \mathbf{y} + \mathbf{z}\ (addition\ modulo\ 2). \tag{1.64}$$

The decoding computes the syndrome

$$\mathbf{d} = \hat{\mathbf{y}}H^T = (\mathbf{y}+\mathbf{z})H^T = \mathbf{z}H^T. \tag{1.65}$$

Because of the property in Eq. (1.61), the syndrome depends only on the error Z and not on the transmitted codeword. If $\mathbf{d} = 0$, then the decoded codeword is $\mathbf{y}$. If $\mathbf{d} \neq 0$, then let i denote the column of H, which is equal to $\mathbf{d}^T$. There is an error in position

i of $\hat{\mathbf{y}}$. The decoded codeword is $\mathbf{y} = \hat{\mathbf{y}} + \mathbf{z}_i$, where $\mathbf{z}_i$ is a vector which is all zeros except for a 1 in the ith position.

Hamming codes demonstrate how channel coding works in communication. Although Hamming codes are very simple in both encoding and decoding, they are not capacity-approached channel codes. In other words, using Hamming codes cannot approach the channel capacity. The capacity-approached channel codes will be discussed in Chapter 3.

1.4 Joint Source and Channel Coding

It is now time to combine the two main results that we have so far: source coding ($R > H$: Theorem 1.1) and channel coding ($R < C$: Theorem 1.3). Is the condition $H < C$ necessary and sufficient for sending a source over a channel?

For example, consider sending digital image or video over a discrete channel that is without memory. We could design a code to map the sequence of visual samples directly into the input of the channel, or we could compress the visual samples into its most efficient representation, then using the appropriate channel code to send it over the channel. It is not immediately clear that we are not losing something by using the two-stage method, since the data compression does not depend on the channel and the channel coding does not depend on the source distribution.

To answer this question, let us take the scenario shown in Figure 1.1. We have a source S that generates symbols from the alphabet $\mathscr{A}$. It is from a finite alphabet and satisfies the AEP. The sequence of symbols $S^n = \{S_1, S_2, \cdots, S_n\}$ is sent over the channel so that the receiver can reconstruct the sequence. Assume the one stage coding is considered. We map the sequence onto a codeword $Y^n(S^n)$ and send the codeword over the channel. The receiver looks at his received sequence $\hat{Y}^n$ and makes an estimation $\hat{S}^n$ of the sequence S^n that was sent. The receiver makes an error if $\hat{S}^n \neq S^n$. We define the probability of error $P_e^{(n)}$ as

$$P_e^{(n)} = Pr(S^n \neq \hat{S}^n) = \sum_{\hat{y}^n} \sum_{s^n} p(s^n) p(\hat{y}^n \mid y^n(s^n)) I(g(\hat{y}^n) \neq s^n), \tag{1.66}$$

where I is the indicator function and $g(\hat{y}^n)$ is the decoding function.

The joint source channel coding theorem is stated here:

Theorem 1.4. *If $S_1, S_2, \cdots, S_n$ is a finite alphabet stochastic process that satisfies the AEP, then there exists a source channel code with $P_e^{(n)} \to 0$ if $H(S) < C$. Conversely, for any stationary stochastic process, if $H(S) > C$, the probability of error is bounded away from zero, and it is not possible to send the process over the channel with an arbitrarily low probability of error.*

This theorem is Theorem 8.13.1 proved by Cover and Thomas [5]. It was proven using the two-stage method. In other words, the source sequence S^n is first compressed into its most efficient representation and then the coded source sequence is protected by the appropriate channel code. If the source sequence is AEP and the block length n is large enough, the two-stage method is as good as any other method of transmitting information over a noisy channel. It is known as Shannon's separate source channel.

This separate result has some important practical implications. It implies that we can consider the design of a communications system as a combination of two parts, source coding and channel coding. We can design source codes for the most efficient representation of data. We separately and independently design channel codes appropriate for the channel. The combination will be as efficient as anything we could design by considering both problems together. This offers an enormous reduction in complexity. Therefore, in practical applications, source coding and channel coding are often studied separately.

However, it may be appropriate to point out that this is not always true. There are examples of multiuser channels where decomposition breaks down. Another simple example is that of sending English text over an erasure channel. We can look for the most efficient binary representation of the text and send it over the channel. But the errors will be very difficult to decode. If however we send the English text directly over the channel, we can lose up to roughly half the letters and yet be able to make sense of the message. A similar phenomenon happens on the analog and digital systems for TV transmission.

Chapter 2
Hybrid Video Coding

Image and video compression is a concrete application of Shannon's source coding. For example, the entropy coding used in image and video compression is variants of Huffman coding and arithmetic coding that were discussed in Sections 1.2.1 and 1.2.2, respectively. The quantization is also developed from Shannon's rate distortion theory that was discussed in Section 1.2.3. However, the redundancy of image and video is more complicated than the sources studied in information theory. It is hard to characterize it by simple statistical models. Therefore, more compression technologies have to be developed to handle the complicated redundancy.

2.1 Hybrid Coding Framework

Let us take video as an example to discuss redundancy in visual data. Video consists of a series of continuously captured pictures. Each individual picture can be viewed as an image or photo. Visual data typically contains four different types of redundancies that are summarized as follows.

1. *Temporal redundancy* — Adjacent pictures look very similar. The differences among them are often caused by a camera moving and/or moving objects in the scene. Most background regions are the same after compensating for camera motion. Moving objects can be accommodated by some motion models.
2. *Spatial redundancy* — Neighboring pixels in a picture have similar values because the captured objects and background in a scene usually have some texture consistent regions. The strong correlation among pixels indicate that they can be efficiently represented in frequency domain.
3. *Visual redundancy* — Human visual system (HVS) is insensitive to a certain loss in visual data. It indicates that we can change visual data to some extent so that it can be compressed more efficiently. The distortion incurred by such information loss is often invisible to HVS.

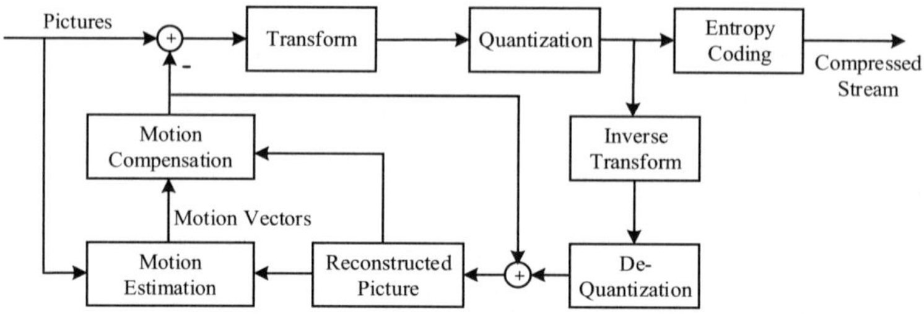

Figure 2.1 The framework of hybrid video coding.

4. *Statistic redundancy* — The symbols that describe visual information are not random. They present a strong nonuniform distribution. According to what we have discussed in Section 1.2, they can be efficiently represented by source coding.

All image and video compression schemes are designed in an effort to fully exploit the four types of redundancies. The target is to minimize distortion for a given compression rate or minimize compression rate for a given distortion. From the earliest video compression standard H.261 to the latest standard High Efficient Video Coding (HEVC), although compression technologies have been developed over several decades and coding efficiency has improved more than ten times, they share the same framework, as shown in Figure 2.1. This framework is featured in hybrid motion compensation and transform. So it is called hybrid video coding.

The temporal redundancies among video frames are exploited by motion estimation (ME) and motion compensation (MC). In the hybrid coding framework, every coded picture has to be reconstructed at both the encoder and the decoder. Furthermore, the reconstructed picture is required to be exactly the same at both the encoder and the decoder. It is an important assumption in hybrid video coding. For every input block, the motion estimation module in Figure 2.1 will find the most matched block in the reconstructed picture and record the motion vector (MV). The motion compensation module takes the MVs and the reconstructed picture to generate a predictive picture. After subtracting the predictive picture, the input picture becomes a residual picture that is much easier to compress than the original picture.

The process of motion estimation and motion compensation is shown in Figure 2.2. Figure 2.2a is one input picture in the Foreman sequence. The reconstructed picture is shown in Figure 2.2b. We can observe that the reconstructed picture has some slight block artifacts because of compression. Although the input picture and the reconstructed picture look similar, the head in the input picture has a rotation with respect to the reconstructed picture. Figure 2.2c shows the moving displacement of every input 8×8 block. It is a 2D MV, whose direction indicates the moving direction and magnitude is the moving distance. Figure 2.2d is the prediction by mapping

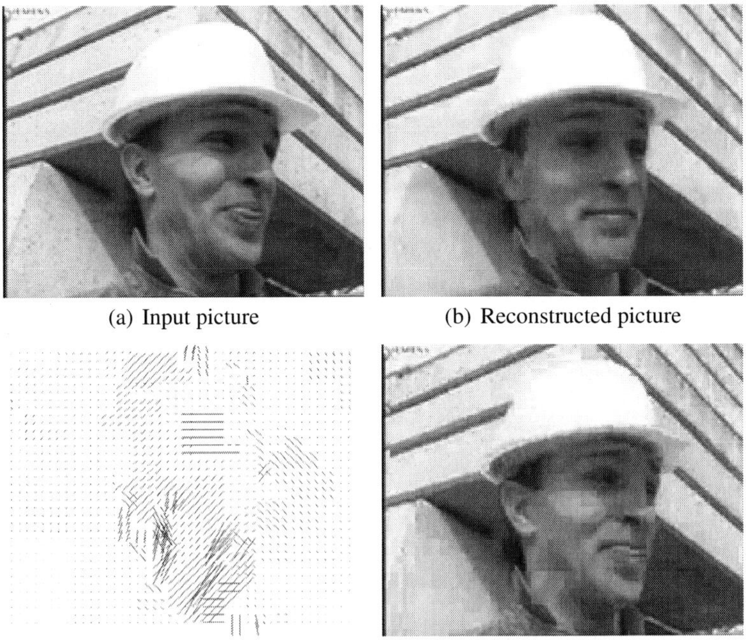

(a) Input picture (b) Reconstructed picture (c) Motion vectors (d) Prediction

Figure 2.2 An example of motion estimation and compensation.

the corresponding blocks in the reconstructed picture toward the input picture. We can observe that the prediction is much more similar to the input picture than the reconstructed picture.

For the first picture, there is no reconstructed picture available and thus it is compressed directly without motion compensation. This picture is called an intra picture. Correspondingly, the pictures coded by using motion compensation are called inter pictures. As we have discussed, neighboring pixels have similar values. For inter pictures, although they have been predicted by the reconstructed pictures, neighboring residual pixels still have similar values, as shown in Figure 2.3a. The spatial redundancies are exploited by transform. Discrete cosine transform (DCT) is widely used in most existing image and video coding standards. Its transform matrix T is obtained as a function of cosines

$$[t_{i,j}] = \begin{cases} \sqrt{\frac{1}{N}} \cos \frac{(2j+1)i\pi}{2N} & i=0, j=0,1,\cdots,N-1 \\ \sqrt{\frac{2}{N}} \cos \frac{(2j+1)i\pi}{2N} & i=1,2,\cdots,N-1, j=0,1,\cdots,N-1 \end{cases} . \tag{2.1}$$

The transform tries to describe visual data in frequency domain instead of pixel domain. The result of 8×8 DCT transform on the motion compensated picture

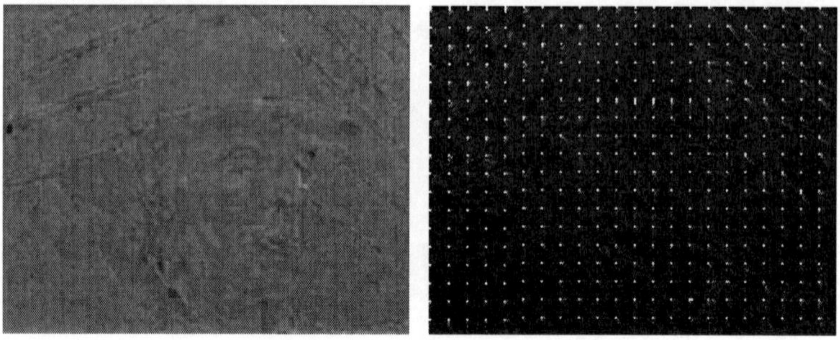

(a) Residual picture after motion compensation (b) The results of 8 × 8 DCT transform

Figure 2.3 An example of DCT transform on a motion compensated picture.

is shown in Figure 2.3b. We observe that every block in the residual picture has a distributed energy. After DCT transform, the energy of every block has been concentrated to the upper left region. The DCT coefficients in the upper left region are of low frequency, whereas the coefficients in the bottom right region are of high frequency. In other words, the energy of every block becomes more compact after transform. This is convenient for subsequent quantization and entropy coding.

Quantization in the hybrid video coding exploits visual redundancy. After DCT transform, each coefficient needs more than 10 bits to represent. The quantization objective is to map DCT coefficients from a large alphabet to a small alphabet. A simple uniform scalar quantization is shown in Figure 2.4. For all input coefficients,

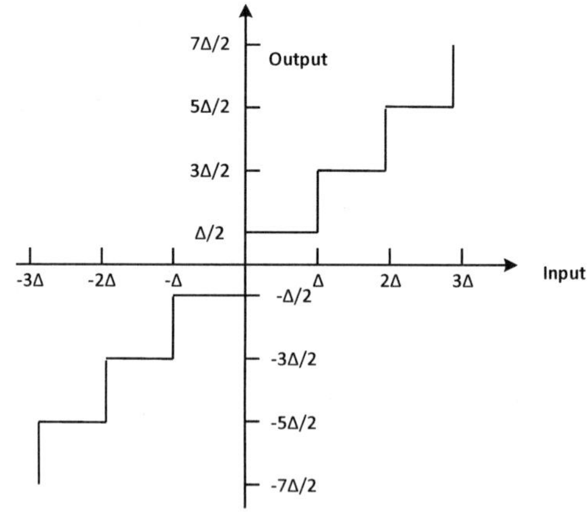

Figure 2.4 A scalar quantization.

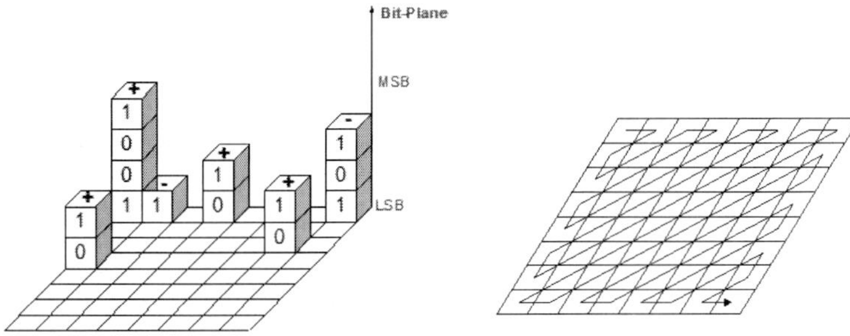

(a) An 8×8 DCT block (b) Scan order

Figure 2.5 The ZigZag scan.

the minimum unit is quantized to Δ. The parameter Δ is selected according to the compression rate. If the compression rate is low, the parameter Δ is large and the reconstructed pictures have large distortion. If the compression rate is high, the parameter Δ is small and we cannot observe the compression distortion. In addition, the quantization in Figure 2.4 assumes that the coefficients have uniform distribution. If the distribution is different (e.g., Gaussian), the different quantization needs to be designed.

In hybrid video coding, every 2D block has to be converted into a 1D array by the ZigZag scan. To clearly explain this procedure, Figure 2.5a is an 8×8 DCT block after quantization. In statistics, DCT coefficients have a larger amount of energy in low frequency than in high frequency. After quantization, low-frequency coefficients have a large probability as nonzero. Many high-frequency coefficients may be quantized to zero. Therefore, the scan order is a ZigZag from low frequency to high frequency as shown in Figure 2.5b. For the DCT block in Figure 2.5a, after the ZigZag scan the array of DCT coefficients is

$$9,\ -1,\ 0,\ 0,\ 0,\ 0,\ 2,\ 0,\ 0,\ 2,\ 0,\ \cdots,\ 0,\ -5,\ 0,\ \cdots,\ 0,\ 2,\ 0,\ \cdots,\ 0.$$

Finally, the 1D array is organized into (*run*, *level*) symbols. *run* indicates the number of continuous zeros and *level* indicates the value of nonzero coefficients. For our example, the generated symbols are

$$(0,9),(0,1),(4,2),(2,2),(18,5),(12,2)$$

The distribution of symbols are not uniform. Therefore, a VLC (similar to Huffman coding) is designed for the symbols or an arithmetic code is selected. The sign bit of every nonzero coefficient is coded with the symbol. In addition to the compressed coefficients, motion vectors and some metadata (like sequence headers, frame headers, and block headers) are also compressed by VLC. All these parts form the final compressed stream.

2.2 Technical Evolution

Although the framework of hybrid video coding has changed little in the past few decades, compression technologies of motion compensation, transform, quantization, and entropy coding have been continuously improving. Furthermore, new compression technologies such as intra prediction, deblocking filtering, and sample adaptive offset (SAO) filtering are introduced into the framework. Next, we will compare the compression technologies used in H.261 [8], MPEG-1 [9], MPEG-2 [10], MPEG-4 [11], H.264 [12, 13], and HEVC [14]. By the comparison, we can clearly see the trend of the developments on video compression.

2.2.1 H.261

H.261 is the first standard based on the hybrid video coding framework, ratified in November 1988. It was originally designed for transmission over Integrated Services Digital Network (ISDN) lines on which data rates are multiples of 64 kbit/s. The input video resolutions are common intermediate format (CIF) and QCIF using a 4:2:0 format. The compression of H.261 is very simple and is similar to the framework shown in Figure 2.1.

In H.261, only one reconstructed picture is used as reference for motion compensation. It is known as predictive (P) picture coding. The motion estimation and compensation are carried out at 8×8 blocks and each block has one motion vector at the integer pixel precision. Transform is the 8×8 DCT. H.261 designed 32 different quantizers. One quantizer with a step size of 8 is reserved for the intra DCT coefficients. The other 31 quantizers with a step size of an even value between 2 and 62 are used for other coefficients. In entropy coding, the 20 most commonly occurring (run, level) symbols are coded with a single VLC. All other combinations are coded with a 20-bit codeword, made up of a 6-bit escape code, a 6-bit run, and an 8-bit level. One difference from the framework shown in Figure 2.1 is that H.261 uses a loop filter. Before the temporal prediction is subtracted from the input picture, it can be filtered by a 2D filter with the 1D coefficients as $[1,2,1]/4$.

2.2.2 MPEG-1

MPEG-1 is a standard for lossy compression of video and audio. It is designed to compress video home system (VHS) quality raw digital video and compact disc (CD) audio down to 1.5 Mbit/s without excessive quality loss. Its video compression ratio is about 26:1. MPEG-1 was the first video standard to be widely adopted by the industry. It provides video and audio compression for CD.

Compared with H.261, MPEG-1 first introduces bidirectionally predictive coding. It is known as B picture coding, where two reconstructed pictures are used for motion compensation. One is the most recently reconstructed picture and another is the closest future reconstructed picture. By using both past and future pictures for prediction, MPEG-1 can get better coding performance. MPEG-1 also organizes four luma 8×8 blocks and two 8×8 chroma blocks as a 16×16 macroblock (MB). It was used as a basic coding unit until H.264/MPEG-4 Advanced Video Coding (AVC). P pictures have 1 motion vector per macroblock. B pictures have two motion vectors: one for the previous picture and one for the future picture. Motion vector is allowed at the half-pixel precision. The transform is still the 8×8 DCT. A 8×8 quantization matrix is introduced in MPEG-1. For the same step size, it enables low-frequency coefficients with low distortion and high-frequency coefficients with relative large distortion. The entropy coding uses a 2D Huffman table to code (Run, Level) symbols.

2.2.3 MPEG-2

MPEG-2 is widely used as the format of digital television signals that are broadcast by terrestrial (over-the-air), cable, and satellite systems. It also specifies the format of movies and other programs that are distributed on digital video disk (DVD) and similar discs. TV stations, TV receivers, DVD players, and other equipment are often designed to this standard. MPEG-2 is forward compatible with MPEG-1. It means that MPEG-1 streams observing typical constraints (e.g., in frame sizes and data rates) can be decoded by MPEG-2 decoders. The VLC tables were extended in MPEG-2 for better compression performance in higher data rates and resolutions.

In the TV industry, in order to reduce the amount of data each picture is separated into two fields: the top field, which is comprised of the odd numbered horizontal lines, and the bottom field, which is the even numbered lines. The two fields are displayed alternately with the lines of one field interleaving between the lines of the previous field. This format is called interlaced video. Two successive fields can be combined to a picture. To support the properties of interlaced video, MPEG-2 specifies different methods for field/picture adaptive motion compensation (picture based, field based, and dual prime prediction modes), as well as switching between field-/picture-wise DCT. Further, it is possible to switch into a 16×8 prediction mode, where separate motion vectors can be defined for the top and bottom halves of a MB.

MPEG-2 is the first standard to support scalable video coding (SVC), which provides signal-to-noise ratio (SNR) scalability and spatial scalability over a limited number of layers. In this context, the bitstream is subdivided into two or three parts, which means that by retaining or receiving only core parts of the stream, it is possible to reconstruct pictures at lower quality and resolution. To encode DCT coefficients related to different resolutions, differently optimized variable length codes are provided in the spatial scalable mode. MPEG-2 also defines a method of temporal scalability, which allows prediction of additional inserted pictures either from the base layer sequence or from an enhancement layer sequence.

2.2.4 MPEG-4

MPEG-4 is an innovative video standard. It tries to make a breakthrough on the hybrid video coding framework by considering video contents. The most famous concept introduced by MPEG-4 is object-based coding, although it has never been adopted by the industry. MPEG-1 and MPEG-2 are only able to encode rectangular video pictures. However, MPEG-4 extends the encoding of video objects, which can have arbitrary shapes. Video scenes can be composed from several objects, which may change in position, appearance, size, and so on, independent of each other. Although the compression of visual objects is novel and the compression performance is good, until now it is still hard to extract visual objects from video accurately and automatically.

Compared with MPEG-1 and MPEG-2, MPEG-4 has made some progress on the hybrid video coding framework. First, a macroblock can have up to four motion vectors. In other words, each 8×8 block can have its own motion vector. The motion vector has the quarter-pixel precision. Both of them greatly improve the performance of motion compensation. Second, MPEG-4 introduces global motion compensation, which allows to express the effect of camera motion by using only a small number of parameters. Third, different VLC tables can be selected where the codes are designed for more efficiently encoding at ranges of lower or higher rates. The selection is controlled by the encoder and depends on the target rate. Finally, the direct mode can determine the motion vectors within B pictures by inferencing from the co-located motion vectors in P picture without rate overhead.

In addition to layered scalable video coding, MPEG-4 also specifies fine granularity scalable (FGS) coding. It enables the compressed stream to be truncated at any length, which better fits the bandwidth fluctuation in the Internet. In order to guarantee that the reconstructed pictures are the same at both the encoder and decoder, MPEG-4 FGS only uses the base layer for motion compensation in the enhancement layer. Therefore, the coding performance of MPEG-4 FGS is poor.

2.2.5 H.264/MPEG-4 AVC

H.264/MPEG-4 Advanced Video Coding (AVC) is a video coding standard developed by the ITU-T Video Coding Experts Group (VCEG) together with the ISO/IEC JTC1 Moving Picture Experts Group (MPEG). The project partnership effort is known as the Joint Video Team (JVT). The ITU-T H.264 standard and the ISO/IEC MPEG-4 AVC standard (formally, ISO/IEC 14496-10 MPEG-4 Part 10, Advanced Video Coding) are jointly maintained so that they have identical technical contents. Although H.264 still follows the hybrid video coding framework, many compression technologies have been significantly improved.

In motion compensation, H.264 uses variable block sizes of 16×16, 16×8, 8×16, 8×8, 8×4, 4×8, or 4×4. The number of motion vectors for a macroblock varies from 1 to 16. Motion vectors have the quarter-pixel precision with high-quality interpolation filters. Multiple reference picture prediction allows one to define references for prediction of any macroblock from one of up to five previously decoded pictures. The B-type slices are generally compared to previous standards, denoted as bipredictive instead of bidirectional. This allows one to define structures of prediction of individual regions from two previous or two subsequent pictures, provided that a causal processing order is available. Furthermore, the prediction of B pictures from other B pictures is possible, which allows implementation of a B-picture pyramid. Different weighting factors can be used for the reference pictures in the B pictures.

In transform, H.264 uses an integer transform of block size 4×4 and 8×8. The transform design is not exactly a DCT, but could be interpreted as an integer approximation. As an entire processing of transform and quantization, implementation by 16-bit integer arithmetic precision is possible both for encoding and decoding. In contrast to previous standards based on the DCT, there is no dependency on a floating point implementation, such that no drift between the encoder and decoder picture representations can occur in the error-free transmission.

H.264 defines two different entropy coding methods, one of which is context adaptive variable length coding (CAVLC), and the other is context adaptive binary arithmetic coding (CABAC). Both are universally applicable to all elements of the code syntax, which are based on a systematic construction of VLC tables. By proper definition of the contexts, it is possible to exploit nonlinear dependencies between the different elements to be encoded. CABAC is a coding method for binary signals, and a binarization of multilevel values such as transform coefficients and motion vectors must be performed before CABAC can be applied. Unary codes, truncated unary codes, Exp-Golomb codes, or fixed-length codes are used for this purpose. Four different basic context models are defined, where the usage depends on the specific values to be encoded.

H.264 is the first to introduce prediction in intra coding, which is performed by first predicting the entire block from boundary pixels of adjacent blocks. Prediction is possible for 4×4, 8×8, or 16×16 blocks, where for the 16×16 case only horizontal, vertical, direct current (DC), and planar prediction are allowed. In the 8×8 and 4×4 cases, nine prediction types are supported (DC and eight directional spatial prediction modes).

Unlike the loop filter in H.261, an adaptive deblocking filter (DBF) is applied in the prediction loop. The adaptation process of the filter is nonlinear, with the low-pass strength of the filter steered by the quantization parameter and by syntax under the control of the encoder. Further parameters considered in the filter selection are the difference between motion vectors at the respective block edges, the coding mode used (e.g., stronger filtering is made for the intra mode), the presence of coded coefficients, and the differences between reconstruction values across the block boundaries.

2.2.6 HEVC

High Efficient Video Coding (HEVC) is a video compression standard, a successor to H.264/MPEG-4 AVC, under the joint development by the ISO/IEC JTC1 MPEG and ITU-T VCEG as ISO/IEC 23008-2 MPEG-H Part 2 and ITU-T H.265. MPEG and VCEG have established a Joint Collaborative Team on Video Coding (JCT-VC) to develop the HEVC standard. HEVC can double the data compression ratio compared to H.264/MPEG-4 AVC, and can support the compression of 8K ultra high definition (UHD) video and resolutions up to 8192×4320.

HEVC replaces macroblocks, which were used by previous standards, with a new coding scheme that uses larger block structures of up to 64×64 pixels and can better subpartition the picture into variable sized structures. HEVC initially divides the picture into Coding tree units (CTUs), which are then divided for each luma/chroma component into coding tree blocks (CTBs). A CTB can be 64×64, 32×32, or 16×16 with a larger block size usually increasing the coding efficiency. CTBs are then divided into Coding units (CUs). The arrangement of CUs within a CTB is known as a quadtree since a subdivision results in four smaller regions. CUs are then divided into Prediction units (PUs) of either intra-picture or inter-picture prediction, which can vary in size from 64×64 to 4×4.

HEVC defines a signed 16-bit range for both horizontal and vertical MVs. HEVC horizontal/vertical MVs have a range of -32768 to 32767, which given the quarter-pixel precision used by HEVC allows for a MV range of -8192 to 8191.75 luma samples. This compares to H.264/MPEG-4 AVC, which allows for a horizontal MV range of -2048 to 2047.75 luma samples and a vertical MV range of -512 to 511.75 luma samples. HEVC allows for two MV modes, which are advanced motion vector prediction (AMVP) mode and merge mode. AMVP mode uses data from the reference picture and can also use data from adjacent prediction blocks. The merge mode allows for the MVs to be inherited from neighboring prediction blocks. Merge mode in HEVC is similar to the skipped and direct motion inference modes in H.264/MPEG-4 AVC but with two improvements. The first improvement is that HEVC uses index information to select one of several available candidates. The second improvement is that HEVC uses information from the reference picture list and reference picture index.

HEVC specifies four Transform unit (TU) sizes of 4×4, 8×8, 16×16, and 32×32 to code the predicted residual. A CTB may be recursively partitioned into four or more TUs. TUs use integer transform that are similar to DCT. In addition, 4×4 luma transform blocks that belong to an intra-coded region are transformed using an integer transform that is derived from discrete sine transform (DST). This provides a 1% bit-rate reduction but was restricted to 4×4 luma transform blocks due to marginal benefits for the other transform cases. Chroma uses the same TU sizes as luma so there is no 2×2 transform for chroma.

HEVC uses a CABAC algorithm that is fundamentally similar to CABAC in H.264/MPEG-4 AVC. CABAC is the only entropy coding method that is allowed in HEVC, while there are two entropy coding methods allowed by H.264/MPEG-4

AVC. CABAC in HEVC is designed for higher throughput. For instance, the number of context coded bins has been reduced by a factor of 8 and the CABAC bypass-mode has been improved in terms of its design to increase throughput. Another improvement with HEVC is that the dependencies between the coded data have been changed to further increase throughput. Context modelling in HEVC has also been improved so that CABAC can better select a context that increases efficiency when compared to H.264/MPEG-4 AVC.

HEVC specifies 33 directional modes for intra prediction compared to the 8 directional modes for intra prediction specified by H.264/MPEG-4 AVC. HEVC also specifies planar and DC intra-prediction modes. The intra-prediction modes use data from neighboring prediction blocks that have been previously decoded.

HEVC specifies two loop filters that are applied in order, with the DBF applied first and the SAO filter applied afterward. Both loop filters operate during the inter-picture prediction loop. The DBF is similar to the one used by H.264/MPEG-4 AVC but with a simpler design and better support for parallel processing. The SAO filter is applied after the DBF and is made to allow for better reconstruction of the original signal amplitudes by using offsets from a transmitted look-up table. Per CTB the SAO filter can be disabled or applied in one of two modes: edge offset mode or band offset mode. The edge offset mode operates by comparing the value of a pixel to two of its eight neighbors using one of four directional gradient patterns. Based on a comparison with these two neighbors, the pixel is classified into one of five categories: minimum, two types of edges, maximum, or neither. For each of the first four categories an offset is applied. The band offset mode applies an offset based on the amplitude of a single pixel. The pixel is categorized by its amplitude into one of 32 bands. Offsets are specified for four consecutive of the 32 bands, because in flat areas which are prone to banding artifacts, pixel amplitudes tend to be clustered in a small range. The SAO filter was designed to increase picture quality, reduce banding artifacts, and reduce ringing artifacts.

2.2.7 Performance versus Encoding Complexity

As we have discussed, the framework of image and video compression does not have any significant changes. All listed video standards share the common hybrid coding framework. However, compression technologies have developed significantly. The technical evolution in different video coding standards are summarized in Table 2.1. We clearly observe some trends on the development of compression technologies.

In all compression technologies, motion compensation plays a critical role in improving coding performance. To increase the accuracy of motion compensation, the precision of motion vectors increases from integer pixel to quarter pixel. In the current resolution of video, 1/8 pixel precision or more of motion vectors does not bring much benefit to coding performance relative to the increased computation and memory needed for finer interpolation. The block partition provides flexibility to use the number of motion vectors adaptive to video contents. We observe that they bring a

Table 2.1 Technical evolution with video coding standards.

	H.261	MPEG-1	MPEG-2	MPEG-4	H.264	HEVC
MVs per MB	4	1	1	4	16	16
MV precison	1	1/2	1/2	1/4	1/4	1/4
Sizes in MB	1	1	1	2	7	7
References	1	1	1	1	5	5
B-frame	No	Yes	Yes	Yes	Bipredictive	Bipredictive
Intra prediction	No	No	No	No	9	35
Large MB	No	No	No	No	No	Yes
Transforms	DCT	DCT	DCT	DCT	4×4, 8×8 Integer	4×4, 8×8, 16×16, 32×32 Integer and 4×4 DST
Entropy	VLC	VLC	VLC	VLC	CAVLC CABAC	CABAC
RDO	No	No	No	No	Yes	Yes

significant gain for H.264 and HEVC, respectively. Increasing the number of references and flexible prediction structures are also a way to improve motion compensation accuracy. For videos with complicated scene changing, they may still have potential for further optimization in the future. It is worth mentioning that the rate for coding partitions, motion vectors, and references is no longer a small percentage of the total rate in HEVC. It would be advantageous to be able to compress them more efficiently.

Intra prediction is an important technology for improving the performance of intra coding in H.264 and HEVC. H.264 uses only 8 directional spatial prediction modes, while HEVC already increases the directional modes to 33. The current intra prediction only exploits the correlation with neighboring pixels. It is well known that nonlocal pixels often have stronger correlation. However, nonlocal intra prediction introduces a relatively large overhead to describe displacement like motion vectors. In addition, a geometric transform may be needed to better match the current block with nonlocal blocks. It would be advantageous to develop a technology for a better trade-off between the accuracy of nonlocal intra prediction and overhead.

The transform technologies have had few changes in standards except for transform size. The entropy coding has converged to context-adaptive arithmetic coding. HEVC discards the context-adaptive VLC because even for current mobile devices the real-time arithmetic decoding of HD video contents is not a problem anymore. The sample adaptive offset filter provides a new way to represent visual data. Although it shows a certain performance gain in HEVC, it is still unclear if it has much potential in future video compression. Figure 2.6 shows the performance improvements with the development of these compression technologies. The vast improvements in the performance of video compression are quite remarkable.

What is the basis for supporting the development of compression technologies in the past few decades? Actually, it is the Moore's law on computing hardware, which states that the number of transistors on integrated circuits doubles approximately every two years. When H.261 was specified in 1988, the fastest personal computers

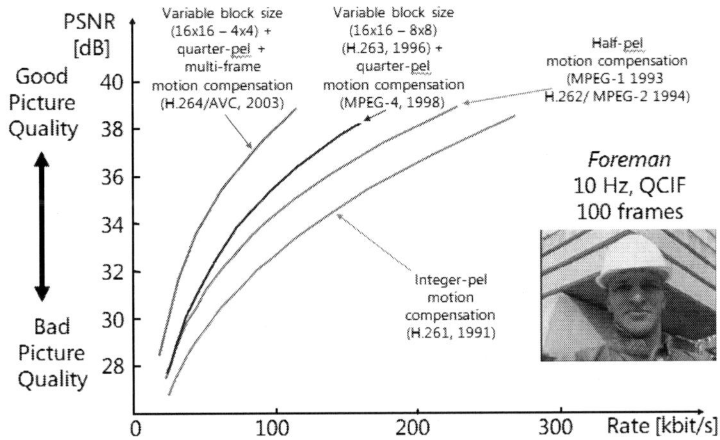

Figure 2.6 Performance improvements from H.261 to H.264.

at that time were even slower than current mobile phones. It is impossible to specify a video compression standard with similar computation as H.264 and HEVC. Even if some high-computation, high-efficiency compression technologies were available at that time, they could not be adopted by H.261. Therefore, every video coding standard actually tries to achieve a better trade-off between coding performance and coding complexity corresponding to the computation power available at that time.

The computation increasing for video encoding and decoding is not symmetric. The increased computation for decoding is very smaller to that for encoding. It is useful for practical applications. Some compression technologies will increase the decoding complexity. Fractional pixel motion compensation needs to perform interpolation, which increases computation and memory needed in the decoder. Multiple reference prediction mainly increases the decoding memory requirement. CABAC is the most intensive computation in decoding. Fortunately, with the development of computers and devices, it is not a problem to complete real-time decoding of HD video contents compressed by H.264 and HEVC.

The encoding complexity is mainly increased by using rate-distortion optimization to select coding parameters. We should note that H.264 already has many parameters for coding of a macroblock. The best way to select these parameters is to test all combinations. Therefore, for coding a macroblock, we have to evaluate hundreds of different coding selections. Even for H.264, many current software and hardware encoding implementations have to simplify the rate-distortion optimization. It will bring a certain performance loss. However, the parameter selection in HEVC is performed at 64×64 blocks. There are many thousands of parameter combinations. The fast and high-performance encoding is still a great challenge. Although HEVC's reference software is not yet optimized, it takes almost one day to compress a 10-second HD video sequence using even the fastest computer.

One solution to the above problem is to use many cores. From the research on H.264 encoding optimization, it is very difficult to use many cores for H.264 encoding because of the strong dependency among macroblocks. Furthermore, the macroblock size is a bit small for using many cores for coding of a macroblock in parallel. Fortunately, HEVC has changed the coding unit from a macroblock to a 64 × 64 block. It has the potential to use hundreds or more cores evaluating the coding parameter combinations within a 64 × 64 block. The latest graphic processing unit (GPU) NVIDIA GeForce GTX 690 has provided us with 3076 parallel cores for the task. The next urgent problem for video encoding optimization will be how to evaluate the coding parameter combinations in parallel. It is not easy because of the structure of the GPU, where 192 cores are organized as a group sharing the same program and only 64KB cache memory is available for a group of cores.

2.3 H.264 Standard

Here we will discuss the key compression technologies in H.264, which were briefly introduced in Section 2.2.5. This section is organized according to the contents in the H.264 overview paper [13]. The framework of H.264 is shown in Fig 2.7. Compared with the framework of hybrid video coding shown in Figure 2.1, it includes intra prediction and DBF. They are a complement to the framework of hybrid video coding. The module of coder control is the rate-distortion optimization, which selects the coding parameters macroblock by macroblock.

2.3.1 Motion Compensation

In H.264, each 16 × 16 macroblock can be partitioned into block sizes of 16 × 16, 16 × 8, 8 × 16, and 8 × 8 pixels, as shown in the upper part of Figure 2.8. In case

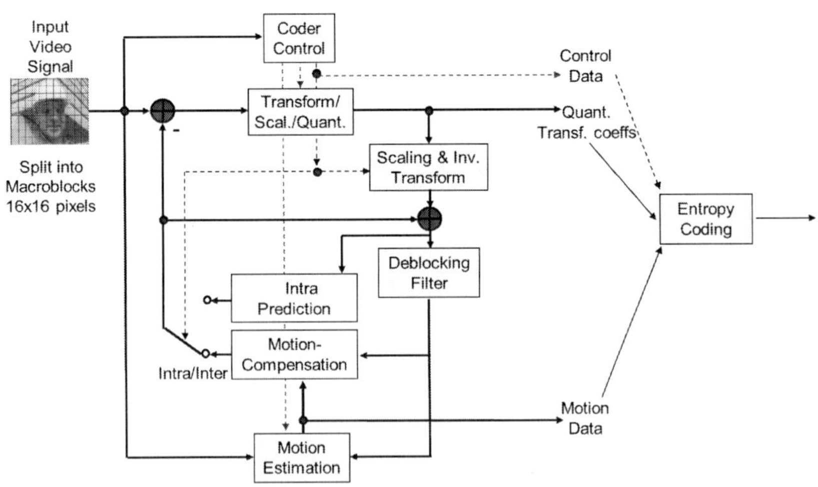

Figure 2.7 The framework of H.264.

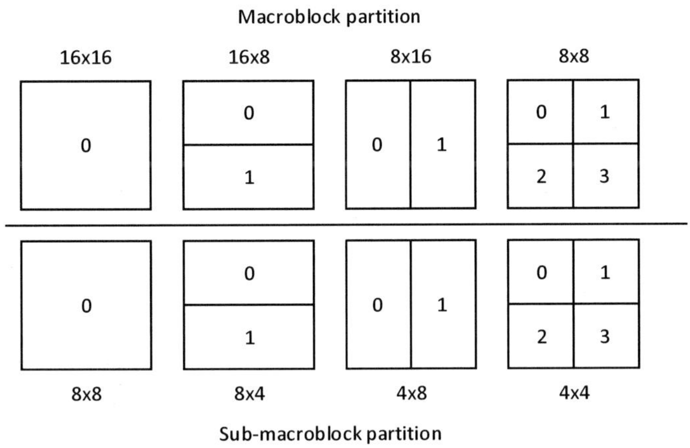

Figure 2.8 The macroblock and sub-macroblock partitions as defined in H.264.

partitions with 8×8 pixels are chosen, the block can be further partitioned into 8×4, 4×8, and 4×4, as shown in the bottom part of Figure 2.8. In other words, one macroblock in H.264 is allowed to have no more than 16 motion vectors. In H.264, no more than five references are allowed. The reference index parameter is transmitted for each motion-compensated 16×16, 16×8, 8×16, or 8×8 luma block. Motion compensation for smaller regions than 8×8 uses the same reference index for prediction of all blocks within the 8×8 region.

We note that the flexible macroblock partitions should be used with high precision of motion vectors together. The former provides a good adaptation to the complicated motion and the latter can fully take advantage of using multiple motion vectors. In H.264, the accuracy of motion compensation is in units of one quarter of the distance between luma pixels. Therefore, H.264 needs the quarter-pixel interpolation. It is completed in two steps. Half pixels are first generated by a 6-tap filter $\{1, -5, 20, 20, -5, 1\}$ and then quarter pixels are generated by averaging the two nearest pixels at integer and half positions. The prediction values for the chroma component are always obtained by bilinear interpolation.

The concept of B picture coding is generalized in H.264. The substantial difference between B and P pictures is that B pictures are coded in a manner in which some macroblocks or blocks may use a weighted average of two distinct motion-compensated prediction values for building the prediction signal. B pictures utilize two distinct lists of reference pictures, which are referred to as the first (list 0) and second (list 1) reference picture lists, respectively. Four different types of inter-picture predictions are supported: list 0, list 1, bipredictive, and direct prediction. The same partitions are also applied to B picture coding.

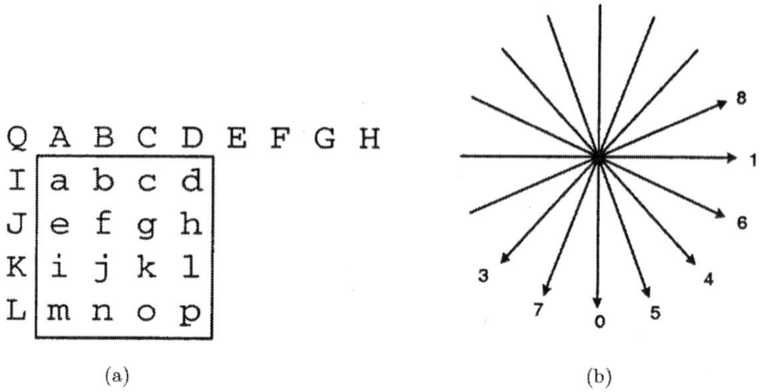

Figure 2.9 Intra 4 × 4 prediction. (a) the prediction of pixels a–p is conducted from pixels A∼Q; (b) Eight prediction directions for a 4 × 4 block.

2.3.2 Intra Prediction

Originally, H.264 had two different intra predictions: 4 × 4 prediction and 16 × 16 prediction. The 4 × 4 prediction is well suited for coding of parts of a picture with significant detail. The 16 × 16 prediction is more suited for coding very smooth areas of a picture. The 8 × 8 prediction is introduced later in the high profile.

When using the 4 × 4 prediction, each 4 × 4 block is predicted from spatially neighboring pixels as illustrated in Figure 2.9a. The 16 pixels of the 4 × 4 block which are labeled as a∼p are predicted using prior decoded pixels in adjacent blocks labeled as A–Q. For each 4 × 4 block, one of nine prediction modes can be utilized. In addition to the DC prediction (where one value is used to predict the entire 4 × 4 block), eight directional prediction modes are specified as illustrated in Figure 2.9b. Those modes are suitable to predict directional structures in a picture such as edges at various angles. The 8 × 8 intra prediction is similar to the 4 × 4 intra prediction expect for the larger block size.

When utilizing the 16 × 16 prediction, the whole luma component of a macroblock is predicted. Four prediction modes are supported. Prediction mode 0 (vertical prediction), mode 1 (horizontal prediction), and mode 2 (DC prediction) are specified similar to the modes in the 4 × 4 prediction except that instead of 4 neighbors on each side to predict a 4 × 4 block, 16 neighbors on each side to predict a 16 × 16 block are used. The prediction mode 4 is a plane prediction. It tries to generate the prediction by using a plane approximation.

2.3.3 Transform and Quantization

In H.264, the transformation is applied to 4 × 4 blocks, and instead of a 4 × 4 DCT, a separable integer transform with similar properties as a 4 × 4 DCT is used. The transform matrix is given as

$$H = \begin{bmatrix} 1 & 1 & 1 & 1 \\ 2 & 1 & -1 & -2 \\ 1 & -1 & -1 & 1 \\ 1 & -2 & 2 & -1 \end{bmatrix}. \tag{2.2}$$

Since the inverse transform is defined by exact integer operations, inverse-transform mismatches are avoided. Intra 16 × 16 prediction modes and chroma intra modes are intended for coding smooth areas. For that reason, the DC coefficients undergo a second transform with the resulting in transform coefficients covering the whole macroblock. An additional 2 × 2 Hadamard transform is also applied to the DC coefficients of the four 4 × 4 blocks of each chroma component.

A quantization parameter is used for determining the quantization of transform coefficients in H.264. The parameter can take 52 values. These values are arranged so that an increase of 1 in the quantization parameter means an increase of the quantization step size by approximately 12% (an increase of 6 means an increase of the quantization step size by exactly a factor of 2). A change of step size by approximately 12% also means roughly a reduction of bit rate by approximately 12%.

2.3.4 Entropy Coding

In H.264/AVC, two methods of entropy coding are supported. One is the CAVLC. In CAVLC, the number of nonzero quantized coefficients N, the actual values, and positions of the coefficients are coded separately. Without loss of generality, the coefficients for a luma 4 × 4 block are assumed as follows

$$8, 6, -2, 0, -1, 0, 0, 1, 0, 0, 0, 0, 0, 0, 0, 0.$$

The following data elements are used to convey information of the above quantized coefficients.

- Number of nonzero coefficients N and Trailing 1s: Trailing 1s (T1s) indicate the number of coefficients with absolute value equal to 1. In the example, $N = 5$ and $\text{T1s} = 2$. These two values are coded as a combined event. One out of four VLC tables is used based on the number of coefficients in neighboring blocks.
- Value of coefficients: The T1s need only sign specification since they all are equal to $+1$ or -1. Since the coefficient values have less spread for the last nonzero coefficients than for the first one, coefficient values are coded in reverse scan order. In the example, –2 is the first coefficient value to be coded. A starting VLC is used for that. When coding the next coefficient (having a value of 6 in the example), a new VLC may be used based on the just coded coefficient. In this way, adaptation is obtained by the use of VLC tables. Six exp-Golomb code tables are available for this adaptation.
- Signs: For T1s, this is sent as single bits. For the other coefficients, the sign bit is included in the exp-Golomb codes.

Positions of each nonzero coefficient are coded by specifying the positions of 0s before the last nonzero coefficient. It is split into two parts:

- *Total-zeros:* This codeword specifies the number of zeros between the last nonzero coefficient and its start. In the example, the Total-zeros is 3. Since it is already known that $N = 5$, the number must be in the range 0–11. Fifteen tables are available for N in the range 1–15. (If $N = 16$, there is no zero coefficient.)
- *RunBefore:* In the example it must be specified how the three zeros are distributed. First the number of 0s before the last coefficient is coded. In the example the number is 2. Since it must be in the range 0–3, a suitable VLC is used. Now there is only one 0 left. The number of 0s before the second to last coefficient must therefore be 0 or 1. In the example the number is 1. At this point, there are no 0s left and no more information is coded.

The efficiency of entropy coding can be improved further if the CABAC is used. On the one hand, the usage of arithmetic coding allows the assignment of a noninteger number of bits to each symbol of an alphabet, which is extremely beneficial for symbol probabilities that are greater than 0.5. On the other hand, the usage of adaptive codes permits adaptation to nonstationary symbol statistics. Another important property of CABAC is its context modelling. The statistics of already coded syntax elements are used to estimate conditional probabilities. These conditional probabilities are used for switching several estimated probability models. In H.264, the arithmetic coding core engine and its associated probability estimation are specified as multiplication-free low-complexity methods using only shifts and table look-ups. Compared to CAVLC, CABAC typically provides a reduction in bit rate between 5%–15%.

2.3.5 Deblocking Filtering

H.264 defines an adaptive in-loop deblocking filter. Figure 2.10 illustrates the principle of the deblocking filter using a visualization of a one-dimensional edge. Whether the pixels p_0 and q_0 as well as p_1 and q_1 are filtered is determined by using quantization parameter (QP) dependent on thresholds $\alpha(QP)$ and $\beta(QP)$. Thus, filtering of p_0 and q_0 only takes place if each of the following conditions is satisfied:

$$\begin{cases} |p_0 - q_0| < \alpha(QP) \\ |p_1 - p_0| < \beta(QP) \\ |q_1 - q_0| < \beta(QP) \end{cases}, \tag{2.3}$$

where $\beta(QP)$ is considerably smaller than $\alpha(QP)$. Accordingly, filtering of p_1 or q_1 takes place if the corresponding following condition is satisfied:

$$|p_2 - p_0| < \beta(QP) \quad or \quad |q_2 q_0| < \beta(QP). \tag{2.4}$$

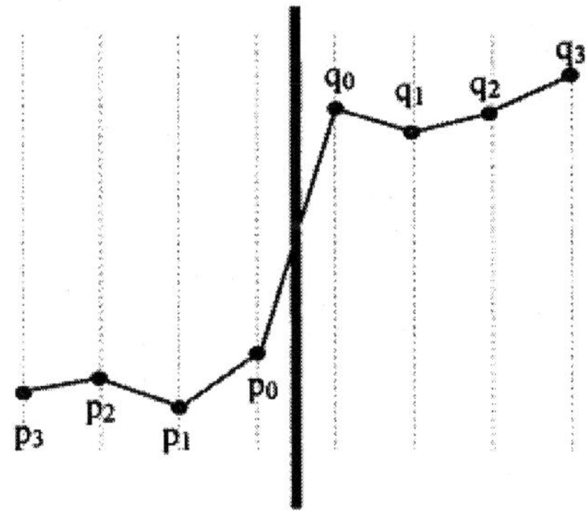

Figure 2.10 Principle of a deblocking filter.

The basic idea is that if a relatively large absolute difference between pixels near a block edge is measured, it is quite likely a blocking artifact and should therefore be reduced. However, if the magnitude of that difference is so large that it cannot be explained by the coarseness of the quantization used in the encoding, the edge is more likely to reflect the actual behavior of the source picture and should not be smoothed over. The blockiness is reduced, while the sharpness of the content is basically unchanged. Consequently, the subjective quality is significantly improved. The filter reduces the bit rate typically by 5%–10% while producing the same objective quality as the nonfiltered video.

2.3.6 Rate Distortion Optimization

As we have discussed, H.264 provides many coding parameters (intra modes, inter modes, motion vectors, and references) to select in coding a macroblock. According to Shannon's rate distortion theorem discussed in Section 1.2.3, the parameter selection should target the final rate distortion optimization. Assume that the video to code is S and the parameter set is P, for a given target rate R_0, the optimization can be formulated as

$$\min_{P} D(S,P), \quad subject\ to\ R(S,P) \leq R_0. \tag{2.5}$$

Here, $D(S,P)$ and $R(S,P)$ are the total distortion and rate, respectively, resulting from the quantization of S with a particular set of coding parameters P.

In practice, rather than solving the constrained problem in Eq. (2.5), an unconstrained formulation is employed, that is

$$P^* = \arg\min_P D(S,P) + \lambda R(S,P). \tag{2.6}$$

$\lambda \geq 0$ is the Lagrange parameter. In H.264, the Lagrange parameter for mode selection have been determined via experimental results

$$\lambda_{mode} = 0.85 \times 2^{(Q-12)/3}. \tag{2.7}$$

Q is the used quantizer value. The Lagrange parameter λ_{motion} is equal to $\sqrt{\lambda_{mode}}$ if the sum of absolution difference (SAD) is used as a measure in motion search, or is equal to λ_{mode} if the sum of squared difference (SSD) is used as a measure.

From the rate-distortion optimization, it is easy to understand why the H.264 encoding has high complexity. In theory, video should be encoded for any combinations of coding parameters to get the distortion and generated rate. And then, an optimal set of parameters is selected. However, it is impossible to check all combinations. H.264 usually use several steps to decide the coding parameters. It is a local optimization. Even so, the encoding complexity of H.264 is high. In the practical applications, only key parameters are selected by the rate-distortion optimization. The other parameters are selected by only comparing the distortion.

2.4 HEVC Standard

Here we will discuss the key compression technologies in HEVC, which have been briefly introduced in Section 2.2.6. This section is organized according to the contents in the HEVC overview paper [14]. The framework of the HEVC standard is shown in Figure 2.11. Compared with the H.264 framework in Figure 2.7, only the SAO filter module is introduced. However, since HEVC discards the macroblock unit that was widely used in H.264 and other previous standards, instead it uses larger block structures of up to 64 × 64 pixels and better subpartitions the picture into variable sized structures. It greatly changes the encoding and decoding process.

2.4.1 Motion Compensation

Coding tree units (CTUs) and coding tree blocks (CTBs) — Pictures in a video sequence are partitioned into the same size of CTUs. Each CTU contains a luma CTB and two chroma CTBs for the 4:2:0 format. In HEVC, the term "unit" means all color components and the term "block" means a color component. A luma CTB covers a square picture area of $L \times L$ pixels of the luma component and the corresponding chroma CTBs cover $L/2 \times L/2$ pixels of each of the two chroma components. The value of L may be equal to 16, 32, or 64. Compared with the traditional 16×16 macroblock, HEVC supports variable-size CTUs from 64 × 64 to 16 × 16. The support

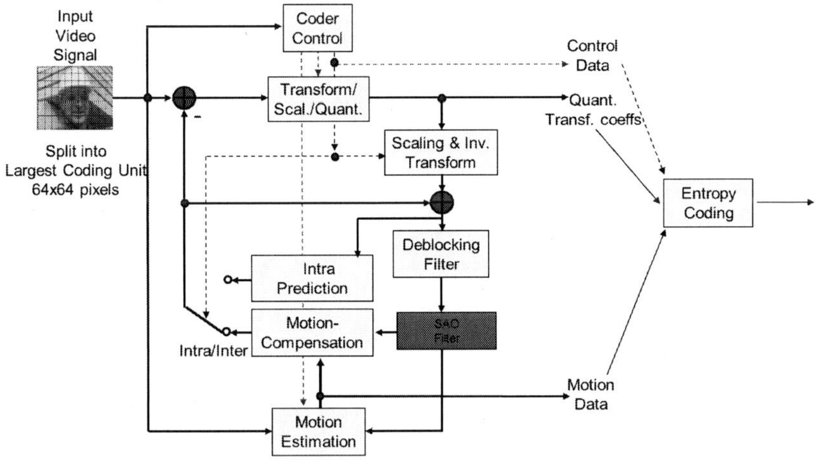

Figure 2.11 The framework of the HEVC standard.

of CTUs larger than in previous standards is particularly beneficial when encoding high-resolution video content.

Coding units (CUs) and coding blocks (CBs) — CTUs are further partitioned into CUs. As shown in Figure 2.12, the partitioning is achieved using quarter-tree structures and is iterated until the size for a luma CB reaches a minimum allowed CB size that is always 8×8 or larger. Thus CUs always cover a square picture area. Intra- or inter-prediction mode is determined at the CU level. For every CU, PU decides its prediction partition and TU decides its transform partition.

Prediction units (PUs) and prediction blocks (PBs) — As shown in Figure 2.13, HEVC defines eight different modes for splitting a CU into PUs. Note that the partition to a CU is not recursive and is only carried out once. In intra prediction, PB size is equal to CB size, or a CB is further partitioned into four PBs that each has their

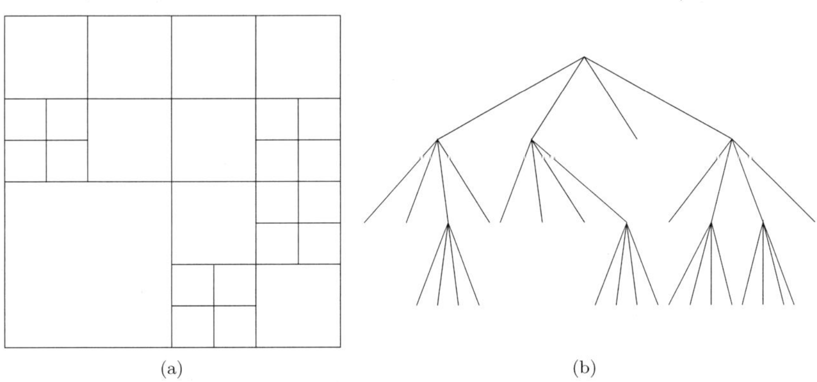

Figure 2.12 A CTB is partitioned into CBs and TBs. Solid lines indicate CB boundaries and dotted lines indicate TB boundaries. (a) CTB with its partitioning; (b) Corresponding quadtree.

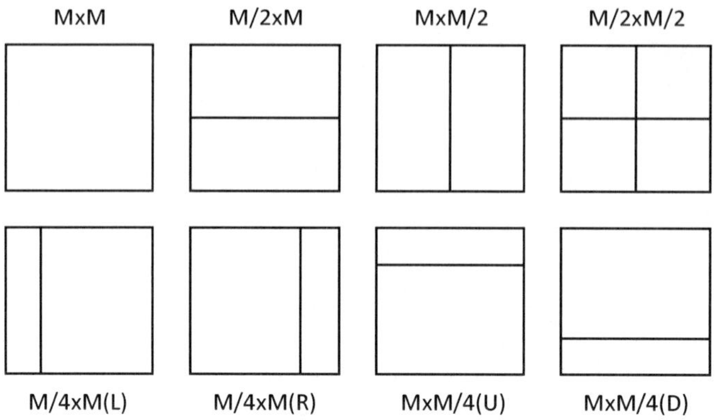

Figure 2.13 Modes for splitting a CU into PUs, subject to certain size constraints. For intra prediction, only M $\times$ M and M/2 $\times$ M/2 are supported.

own intra-prediction mode. The minimum PB size is 4×4. In inter prediction, it is specified whether the luma and chroma CBs are split into one, two, or four PBs. The splitting into four PBs is allowed only when the CB size is equal to the minimum allowed CB size. When a CB is split into four PBs, each PB covers a quadrant of the CB. When a CB is split into two PBs, as shown in Figure 2.13, six types of splitting are possible. Each PB is assigned one or two motion vectors and reference picture indices.

Transform units (TUs) and transform blocks (TBs) — Only square CB and TB partitioning are specified, where a block can be recursively split into quadrants, as illustrated in Figure 2.12. The partitioning is represented by a residual quadtree. The leaf node blocks resulting from the residual quadtree are the transform blocks that are further processed by transform coding. The encoder indicates the maximum and minimum luma TB sizes that it will use. Splitting is implicit when the CB size is larger than the maximum TB size. Not splitting is implicit when splitting would result in a luma TB size smaller than the indicated minimum. The chroma TB size is half of the luma TB size in each dimension, except when the luma TB size is 4×4, in which case a single 4×4 chroma TB is used for the region covered by four 4×4 luma TBs.

2.4.2 Intra Prediction

Intra prediction operates according to the TB size, and previously decoded boundary pixels from spatially neighboring TBs are used to form the prediction signal. Directional prediction with 33 different directional orientations is defined for square

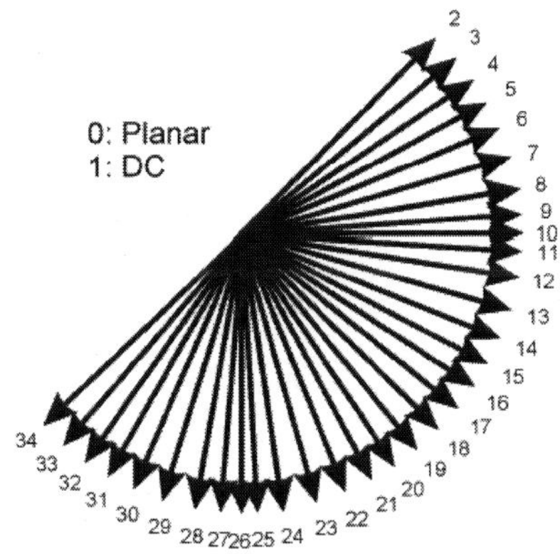

Figure 2.14 The directions of intra prediction in HEVC.

TB sizes from 4 × 4 up to 32 × 32. The possible prediction directions are shown in Figure 2.14. Alternatively, planar prediction (assuming an amplitude surface with a horizontal and vertical slope derived from the boundaries) and DC prediction (a flat surface with a value matching the mean value of the boundary pixels) can also be used. For chroma, the horizontal, vertical, planar, and DC prediction modes can be explicitly signalled, or the chroma prediction mode can be indicated as the same as the luma prediction mode.

In HEVC, the reference pixels used for the intra prediction are sometimes filtered by a three-tap $[1,2,1]/4$ smoothing filter in a manner similar to what was used for 8 × 8 intra prediction in H.264. HEVC applies smoothing operations more adaptively, according to the directionality, the amount of detected discontinuity, and block size. As in H.264, the smoothing filter is not applied for 4 × 4 blocks. For 8 × 8 blocks, only the diagonal directions with directions as 2, 18, or 34 use the reference pixel smoothing. For 16 × 16 blocks, the reference pixels are filtered for most directions except the near-horizontal and near-vertical directions in the range of 9–11 and 25–27. For 32 × 32 blocks, all directions except the exactly horizontal direction 10 and the exactly vertical direction 26 use the smoothing filter, and when the amount of detected discontinuity exceeds a threshold, bilinear interpolation from three neighboring region pixels is applied to form a smooth prediction. The intra-planar mode also uses the smoothing filter when the block size is greater than or equal to 8 × 8, and the smoothing is not used for the intra-DC mode.

2.4.3 Transform and Quantization

HEVC supports transform block sizes as 4 × 4, 8 × 8, 16 × 16, and 32 × 32. Two-dimensional transforms are computed by applying 1-D transforms in the horizontal and vertical directions. The elements of the core transform matrices were derived by approximating scaled DCT basis functions, under considerations such as limiting the necessary dynamic range for transform computation and maximizing the precision and closeness to orthogonality when the matrix entries are specified as integer values. For simplicity, only one integer matrix 32 points in length is specified, and subsampled versions are used for other sizes. For example, the matrix for the length-16 transform is shown here

$$H =$$

$$\begin{bmatrix}
64 & 64 & 64 & 64 & 64 & 64 & 64 & 64 & 64 & 64 & 64 & 64 & 64 & 64 & 64 & 64 \\
90 & 87 & 80 & 70 & 57 & 43 & 25 & 9 & -9 & -25 & -43 & -57 & -70 & -80 & -87 & 90 \\
89 & 75 & 50 & 18 & -18 & -50 & -75 & -89 & -89 & -75 & -50 & -18 & 18 & 50 & 75 & 89 \\
87 & 57 & 9 & -43 & -80 & -90 & -70 & -25 & 25 & 70 & 90 & 80 & 43 & -9 & -57 & -87 \\
83 & 36 & 36 & 83 & 83 & 36 & 36 & 83 & 83 & 36 & 36 & 83 & 83 & 36 & 36 & 83 \\
80 & 9 & 70 & 87 & 25 & 57 & 90 & 43 & 43 & 90 & 57 & 25 & 87 & 70 & 9 & 80 \\
75 & -18 & -89 & -50 & 50 & 89 & 18 & -75 & -75 & 18 & 89 & 50 & -50 & -89 & -18 & 75 \\
70 & -43 & -87 & 9 & 90 & 25 & -80 & -57 & 57 & 80 & -25 & -90 & -9 & 87 & 43 & -70 \\
64 & 64 & 64 & 64 & 64 & 64 & 64 & 64 & 64 & 64 & 64 & 64 & 64 & 64 & 64 & 64 \\
57 & 80 & 25 & 90 & 9 & 87 & 43 & 70 & 70 & 43 & 87 & 9 & 90 & 25 & 80 & 57 \\
50 & 89 & 18 & 75 & 75 & 18 & 89 & 50 & 50 & 89 & 18 & 75 & 75 & 18 & 89 & 50 \\
43 & 90 & 57 & 25 & 87 & 70 & 9 & 80 & 80 & 9 & 70 & 87 & 25 & 57 & 90 & 43 \\
36 & 83 & 83 & 36 & 36 & 83 & 83 & 36 & 36 & 83 & 83 & 36 & 36 & 83 & 83 & 36 \\
25 & -70 & 90 & -80 & 43 & 9 & -57 & 87 & -87 & 57 & -9 & -43 & 80 & -90 & 70 & -25 \\
18 & 50 & 75 & 89 & 89 & 75 & 50 & 18 & 18 & 50 & 75 & 89 & 89 & 75 & 50 & 18 \\
9 & 25 & 43 & 57 & 70 & 80 & 87 & 90 & 90 & 87 & 80 & 70 & 57 & 43 & 25 & 9
\end{bmatrix}$$

The matrices for the length-8 and length-4 transforms can be derived by using the first eight entries of rows 0, 2, 4, $\cdots$, and using the first four entries of rows 0, 4, 8, $\cdots$, respectively.

For the transform block size of 4×4, an alternative integer transform derived from a DST is applied to the luma residual blocks for intra-prediction modes, with the transform matrix

$$H = \begin{bmatrix} 29 & 55 & 74 & 84 \\ 74 & 74 & 0 & -74 \\ 84 & -29 & -74 & 55 \\ 55 & -84 & 74 & -29 \end{bmatrix}.$$

The basis functions of the DST better fit the statistical property that the residual amplitudes tend to increase as the distance from the boundary pixels. In terms of complexity, the 4×4 DST-like transform is not much more computationally demanding than the 4×4 DCT-like transform. The usage of the DST type of transform is restricted to only 4×4 luma transform blocks, since for other cases the additional coding efficiency improvement for including the additional transform type was found to be marginal.

For quantization, HEVC uses essentially the same uniform reconstruction quantization (URQ) scheme controlled by a quantization parameter QP as in H.264. The range of the QP values is defined from 0 to 51, and an increase by 6 doubles the quantization step size such that the mapping of QP values to step sizes is approximately logarithmic. Quantization scaling matrices are also supported. To reduce the memory needed to store frequency-specific scaling values, only quantization matrices of sizes 4×4 and 8×8 are used. For the larger transforms of 16×16 and 32×32, an 8×8 scaling matrix is sent and is applied by sharing values within 2×2 and 4×4 coefficient groups in frequency subspaces except for the value at the DC position, for which a distinct value is sent and applied.

2.4.4 Sample Adaptive Offset Filter

SAO is a process that modifies the decoded pixels by conditionally adding an offset value to each pixel after the application of DBF, based on values in look-up tables transmitted by the encoder. SAO filtering is performed on a region basis, based on a filtering type selected per CTB. A value of 0 for SAO type indicates that the SAO filter is not applied to the CTB, and the values 1 and 2 signal the use of the band offset and edge offset modes, respectively.

In the band offset mode, the selected offset value directly depends on the pixel amplitude. In this mode, the full pixel amplitude range is uniformly split into 32 segments called bands, and the pixel values belonging to four of these bands (which are consecutive within the 32 bands) are modified by adding transmitted values denoted as band offsets, which can be positive or negative. The main reason for using four consecutive bands is that in the smooth areas where banding artifacts can appear, the pixel amplitudes in a CTB tend to be concentrated in only a few of the bands. In addition, the design choice of using four offsets is unified with the edge offset mode, which also uses four offset values.

In the edge offset mode, SAO edge class with values from 0 to 3 signals whether a horizontal, vertical, or one of two diagonal gradient directions is used for the edge offset classification in a CTB. Figure 2.15 shows the four gradient patterns used for the respective SAO edge class. Each pixel in the CTB is classified into one of five categories by comparing the pixel value p located at some position with the values n_0 and n_1 of two pixels located at neighboring positions as shown in Table 2.2. This classification is done for each pixel based on decoded pixel values, so no additional signalling is required for the edge classification. Depending on the edge category at the pixel position, for edge categories from 1 to 4, an offset value from a transmitted look-up table is added to the pixel value. The offset values are always positive for categories 1 and 2 and negative for categories 3 and 4 — thus the filter generally has a smoothing effect in the edge offset mode.

For SAO modes 1 and 2, a total of four amplitude offset values are transmitted to the decoder for each CTB. For mode 1, the sign is also encoded. The offset values, SAO mode, and SAO class are determined by the encoder — typically using criteria that optimize rate-distortion performance. The SAO parameters can be indicated to be inherited from the left or above CTB using a merge flag to make the signalling

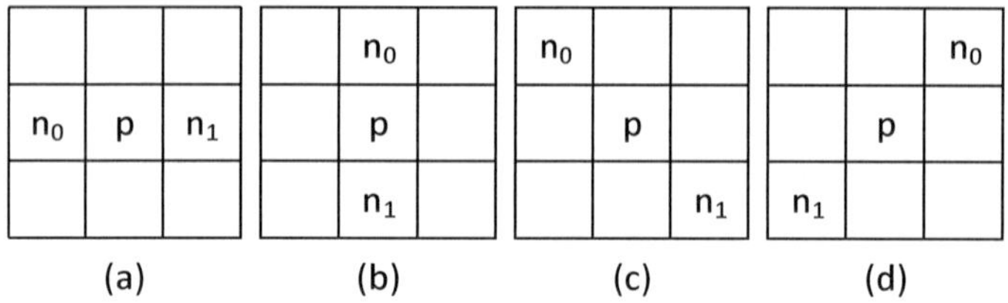

Figure 2.15 Four gradient patterns used in SAO. Sample labeled p indicates a center pixel to be considered. Two pixels labeled n_0 and n_1 specify two neighboring pixels along the (a) horizontal, (b) vertical, (c) 135 diagonal, and (d) 45 gradient patterns.

Table 2.2 Pixel edge categories in the SAO edge offset mode.

Edge Category	Condition	Meaning
0	Cases not listed below	Monotonic area
1	$p < n_0$ and $p < n_1$	Local min
2	$p < n_0$ and $p = n_1$ or $p < n_1$ and $p = n_0$	Edge
3	$p > n_0$ and $p = n_1$ or $p > n_1$ and $p = n_0$	Edge
4	$p > n_0$ and $p > n_1$	Local max

efficient. In summary, SAO is a nonlinear filtering operation which allows additional refinement of the reconstructed signal, and it can enhance the signal representation in both smooth areas and around the edges.

Chapter 3
Communication

In this chapter, we first introduce analog communication, which has been almost forgotten by both academics and industry experts for many years because of the dominance of digital communication. But our recent research discovers that simulating the behavior of analog communication in digital video communication can provide many interesting features that do not exist in digital communication. This is the reason why we discuss analog transmission technologies here. We also discuss digital communication in the physical layer (PHY), which consists of channel coding and modulation.

3.1 Analog Communication

The content in this section is organized according to the book *Analog Communication* [15]. Television (TV) was invented in 1923 and the color TV began in 1954. The core of these early TV systems is an analog transmission system. The analog system is simple and is designed for converting real-value video and audio samples into continuous high-frequency electromagnetic waves that are sent through the air or other medium. The analog system mainly consists of amplifier circuits. They amplify input signals before transmission, which help in accurate reception of the transmitted information at the receiver.

In general, video and audio signals are not suitable for direct transmission over a medium. For example, voice signals cannot travel longer distance in air and the signal gets attenuated rapidly. Hence, video and audio signals have to be converted to high-frequency electronic signals for transmission. Modulation is exactly the technology best suited for the conversion. The advantages of using modulation technologies includes reducing the size of the antenna, avoiding mixing signals, increasing communication distance, allowing multiplexing of signals, adjusting bandwidth, and so on.

3.1.1 Analog Modulation

Modulation technologies can be categorized as continuous-wave modulation or pulse modulation. In the continuous-wave modulation, a sinusoidal wave is used as the carrier. When the amplitude of the carrier is varied in accordance with the information signal, it is an amplitude modulation (AM), and when the angle of the carrier is varied, it is angle modulation. The angle modulation is further subdivided into frequency modulation (FM) and phase modulation (PM), in which the instantaneous frequency and phase of the carrier are varied according to the information signal, respectively. In a pulse module, the carrier consists of a periodic sequence of rectangular pulses. The pulse modulation is further subdivided into analog and digital type. In analog pulse modulation, the amplitude, duration, or position of a pulse is varied in accordance with sample values of the signal to have pulse amplitude modulation (PAM), pulse duration modulation (PDM), or pulse position modulation (PPM), respectively.

Let us define information signal and carrier signal as $s(t) = A_s \cos(2\pi f_s t)$ and $c(t) = A_c \cos(2\pi f_c t)$, respectively. A_s and A_c are the amplitude of the information and carrier signals. f_s and f_c are the frequency of the information and carrier signals. The standard form of an amplitude modulated wave is defined as

$$m(t) = A_c\,[1 + K_a s(t)]\,cos(2\pi f_c t), \tag{3.1}$$

where K_a is a constant called the amplitude sensitivity of the modulator. In order to faithfully recover the information signal, the absolute value of $K_a s(t)$ must be less than 1. The frequency f_c should be far larger than the signal frequency f_s. Substituting $s(t)$ into Eq. (3.1), we get

$$\begin{aligned} m(t) &= A_c\,[1 + \mu \cos(2\pi f_s t)] \cos(2\pi f_c t), \\ &= A_c \cos(2\pi f_c t) + \mu A_c \cos(2\pi f_s t) cos(2\pi f_c t), \\ &= A_c \cos(2\pi f_c t) + \frac{\mu A_c}{2} \cos\left(2\pi(f_c - f_s)t\right) - \frac{\mu A_c}{2} \cos\left(2\pi(f_c + f_s)t\right), \end{aligned} \tag{3.2}$$

where $\mu = K_a A_s$.

As shown in Eq. (3.2), the modulated carrier has new signals at different frequencies that are called as side frequencies or sidebands. They occur in the frequency spectrum directly above and below the carrier frequency

$$\begin{cases} f_{USB} = f_c + f_s \\ f_{LSB} = f_c - f_s \end{cases}. \tag{3.3}$$

The upper sideband is called f_{USB} and the lower sideband is called f_{LSB}. We know that bandwidth can be measured by subtracting the lowest frequency of the signal from the highest frequency of the signal. For the amplitude modulated wave it is

$$BW = f_{USB} - f_{LSB} = 2f_s = 2W, \tag{3.4}$$

where W is the signal bandwidth. The bandwidth required for the amplitude modulation is twice the bandwidth of the information signal. Thus AM is inefficient in bandwidth and power usage. The single sideband (SSB) modulation is an improvement over AM by suppressing one sideband and keeping another sideband. It is more efficient in using transmitter bandwidth and power because it avoids bandwidth doubling and the power wasted on the carrier.

FM conveys information over a carrier wave by varying its instantaneous frequency. This contrasts with AM, in which the amplitude of the carrier is varied while its frequency remains constant. The difference between the spontaneity and the frequency of the carrier is directly proportional to the instantaneous value of the amplitude of information signal. The frequency modulation is formulated as

$$\begin{aligned} m(t) &= A_c \cos\left(2\pi \int_0^t f(\tau)d\tau\right) \\ &= A_c \cos\left(2\pi \int_0^t [f_c + f_\Delta s(\tau)]\, d\tau\right) \\ &= A_c \cos\left(2\pi f_c t + 2\pi f_\triangle \int_0^t s(\tau)d\tau\right). \end{aligned} \tag{3.5}$$

In Eq. (3.5), $f(\tau)$ is the instantaneous frequency of the carrier and f_Δ is the frequency deviation, which represents the maximum shift away from f_c. Here we assume $A_s = 1$. Substituting $s(t)$ into Eq. (3.5), we can get

$$m(t) = A_c \cos\left(2\pi f_c t + \frac{f_\Delta}{f_s} sin(2\pi f_s t)\right). \tag{3.6}$$

PM is a form of modulation that represents information as variations in the instantaneous phase of a carrier wave. Unlike the more popular FM, PM is not widely used for radio transmissions. This is because it tends to require more complex receiving hardware and there can be ambiguity problems in determining whether, for example, the signal has changed phase by $+180°$ or $-180°$. The modulation can be formulated

$$m(t) = A_c \cos(2\pi f_c t + s(t) + \phi_c). \tag{3.7}$$

This shows how to modulate the phase — the greater $s(t)$ is at a point in time, the greater the phase shift of the modulated signal at that point. It can also be viewed as a change of the frequency of the carrier signal. Thus PM can be considered a special case of FM in which the carrier frequency modulation is given by the time derivative of the phase modulation.

3.1.2 Multiplexing

More efficient communications systems can be obtained if more than one signal is transmitted by a transmitter on the same carrier and on the same channel, or

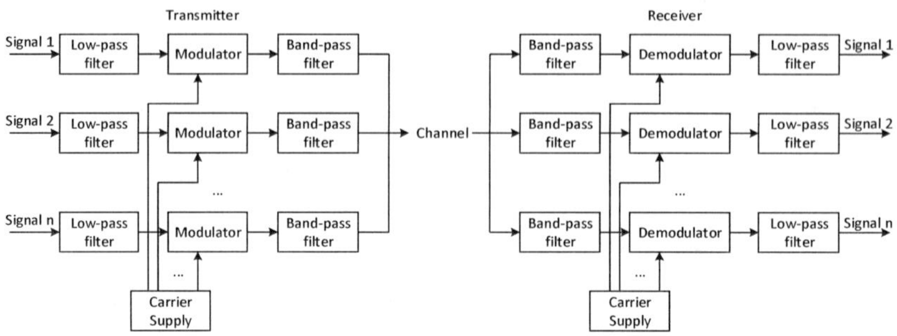

Figure 3.1 Block diagram of an FDM system.

multiple transmitters are transmitting simultaneously on the same channel. This process is known as “multiplexing.” Multiplexing requires that the signals be kept apart so that they do not interfere with each other, and thus they can be separated at the receiving end. This is accomplished by separating the signals by either frequency or time. There are two types of multiplexing in analog transmissions: frequency division multiplexing (FDM) and time division multiplexing (TDM).

Figure 3.1 shows the block diagram of an FDM system. As shown in the figure, signals, assumed to be low-pass, are passed through low-pass filters. This filtering removes high-frequency components that do not contribute significantly to signal representation but may disturb other signals that share a common channel. The filtered signals are then modulated with necessary carrier frequencies with the help of modulators. The most common method of modulation in FDM is SSB modulation, which requires a bandwidth that is approximately equal to that of the original signal. Then band-pass filters following the modulators are used to restrict the band of each modulated wave to its prescribed range. The outputs of the band-pass filters are combined in parallel, which form the input to the common channel. At the receiving end, the band-pass filters connected to the common channel separate the message signals on the frequency occupancy basis. Finally, the original signals are recovered by individual demodulators.

In TDM, each signal transmitted is sampled sequentially and the resulting pulse code is used to modulate the carrier. The same carrier frequency is used to transmit different pulses sequentially, one after another, to be transmitted, each of which is allotted a given time slot. Since only one signal modulates the carrier at any time, no added equipment and no increase in bandwidth are needed when multiplexing. The number of sequential channels that can be handled is limited by the time span required by any one channel pulse and interval between samples. Thus, in TDM, each signal occupies the entire bandwidth of the channel. However, each signal is transmitted for only a short period of time.

Figure 3.2 shows the block diagram of a TDM system, which is used to multiplex three signals. Each signal is allowed to use the channel for a fixed interval of time, called a time slot. The three signals use the channel sequentially one after the

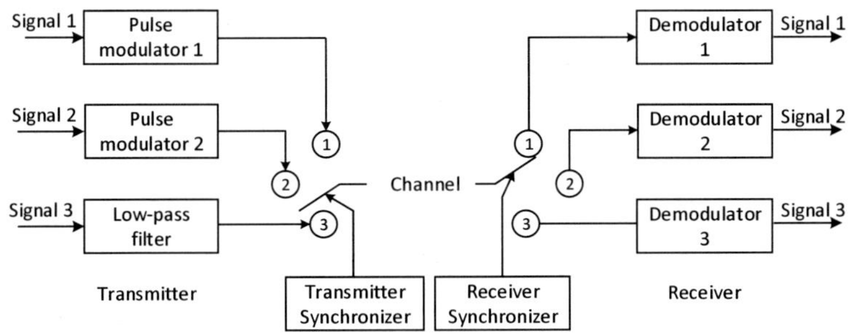

Figure 3.2 Block diagram of a TDM system.

other. One transmission of each channel completes one cycle of operation, called a frame. Once all the signals have been transmitted, the cycle repeats again and again, at a high rate of speed. A rotating switch called a commutator connects the output of each channel modulator to the input of the communication channel in turn. The commutator is realized with electronic switches since it has to rotate at high speeds. The commutator remains at each contact for a fixed interval of time, which is the time slot allotted for each channel. At the receiver, another switch is used, rotating in synchronization with the commutator in the sending end. This switch connects the pulses received to the appropriate demodulator. For the proper operation of the system, absolution synchronization is essential between the transmitter and the receiver.

3.2 Digital Communication

The digital communication in the physical layer consists of channel coding and digital modulation. The former introduces redundancy to transmitted data, which can be used to correct transmission errors at the receiver. The channel coding has been discussed by using Hamming codes as an example in Section 1.3.3. But Hamming codes are not capacity-achieving channel codes. Here, we will discuss low-density parity-check (LDPC) codes [16, 17] and Turbo codes [18, 19]. They can approach the Shannon limit quite closely. The latter multiplexes multiple channel coded bits into a symbol for transmission, which increases the transmission rate.

3.2.1 Low-Density Parity-Check (LDPC) Codes

LDPC codes were invented by Gallager in his doctoral thesis in 1961 [16]. These codes were almost forgotten until the mid-1990s. After the introduction of Turbo codes, they were independently rediscovered by MacKay et al. [17]. Because of the simple structure, they have been the focus of many theoretical analyses. They have been proven to be capable of approaching the Shannon limit more closely than any other class of codes.

Similar to Hamming codes, LDPC codes are linear block codes. They can be represented as (n,k) codes, where n is the code length and k is the information dimension. Here we consider binary LDPC codes only. For a LDPC code (n,k), the source is a $k \times 1$ vector and a codeword is a $n \times 1$ vector. The generator matrix G is $n \times k$ and the parity check matrix H is $m \times n$, where $m = n - k$. The LDPC code can be specified by the parity check matrix H. Before we discuss how to generate an H, we introduce the concept of weight of a binary vector as the number of nonzero elements in it. The column weight w_c of a column of a matrix is the weight of the column; similar for row weight w_r. An LDPC code is regular if the column weights are all the same and the row weights are all the same; otherwise it is irregular. For a regular LDPC code, $w_c n = w_r m$. This structure says that every bit participates in w_c checks and each check involves w_r bits. The rate of a regular LDPC code is $R = 1 - w_c/w_r$.

One way to construct a (w_c, w_r) parity matrix is as follows. First, the matrix H_0 is constructed as

$$H_0 = \begin{bmatrix} 1\,1\cdots 1 & & & \\ & 1\,1\cdots 1 & & \\ & & \ddots & \\ & & & 1\,1\cdots 1 \end{bmatrix}, \tag{3.8}$$

with m/w_c rows and n columns. There are w_r's 1 in every row in Eq. (3.8). This defines a $(1, w_r)$ regular parity check code. Then we form H by stacking permutations of H_0,

$$H = \begin{bmatrix} \pi_1(H_0) \\ \pi_2(H_0) \\ \vdots \\ \pi_{w_c}(H_0) \end{bmatrix}, \tag{3.9}$$

where each $\pi_i(H_0)$ denotes a matrix obtained by permuting the columns of H_0. Obviously, the choice of the permutations determines the distance structure of the code. However, a random choice of permutation will produce a good code on average. Gallager showed that if each permutation is chosen at random out of the $n!$ possible permutations, then the average minimum distance increases linearly with n. In many LDPC codes, n is quite large (such as $n > 10000$).

Now we have a parity check matrix H. It is expedient for encoding to determine the corresponding generator matrix G. A systematic generator matrix may be found as follows. Using Gaussian elimination (with binary arithmetic) determines an $m \times m$ matrix H_p^{-1} so that

$$H = H_p^{-1} H = [I \; H_2]. \tag{3.10}$$

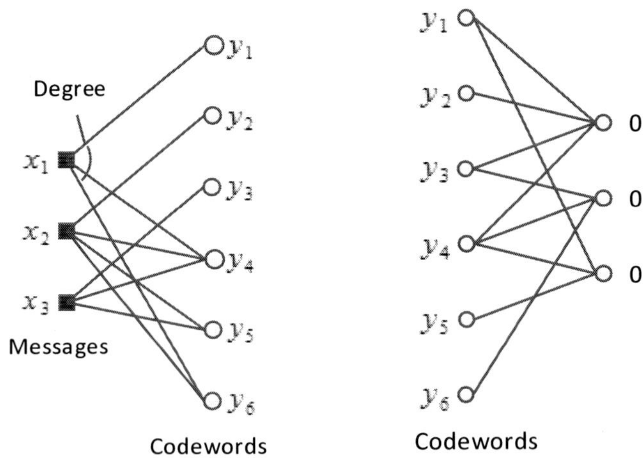

(a) Encoding (b) Parity check

Figure 3.3 An example of graph code.

If such a matrix H_p does not exist, it indicates that H is rank deficient $rank(H) < M$. The matrix H with the form of Eq. (3.10) is called systematic. We can directly write a systematic G from the systematic parity matrix H as

$$G = \begin{bmatrix} H_2 \\ I \end{bmatrix}. \tag{3.11}$$

The $HG = 0$, so G is a generator matrix for H.

LDPC codes can be represented by bipartite graphs. Figure 3.3 shows a simple LDPC code. Figure 3.3a is the encoding process and Figure 3.3b is the parity-check process. The generator matrix and the parity-check matrix are represented by edges, where the nonzero elements indicate edges. If nonzero elements are dominant in the matrix, the codes are referred to as dense; otherwise they are referred to as sparse (also referred to as low density). One important parameter to characterize the LDPC codes in bipartite graphs is degree, which corresponds to row and column weights. The node degree is defined as the number of edges incident to one node. As shown in Figure 3.3a, the degree of left node x_1 is 3. The edge of degree on the right and left is defined as the degree of its incident node at the right and left sides, respectively.

For a k-dimension binary source $\mathbf{x} = [x_1, x_2, \cdots, x_k]^T$, the LDPC codeword is generated by

$$\mathbf{y} = G\mathbf{x}. \tag{3.12}$$

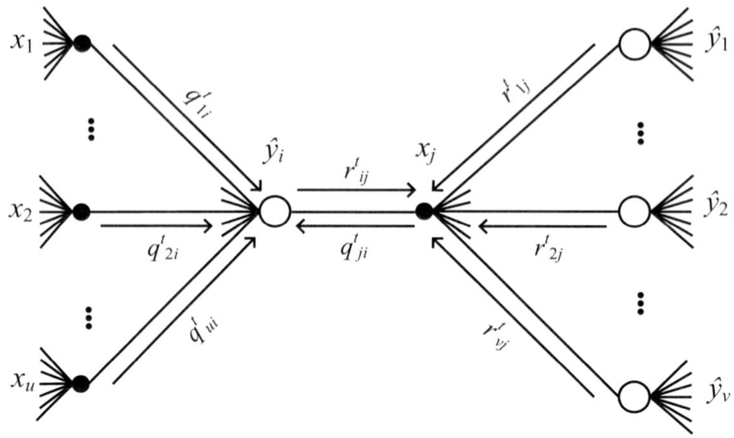

Figure 3.4 The graph fragment of the message passing algorithm.

The calculation is carried out in binary and so the result is a binary vector $\mathbf{y} = [y_1, y_2, \cdots, y_n]^T$. The codeword $\mathbf{y}$ will be transmitted over the noisy channel. If the channel is additive white Gaussian noise (AWGN), the received codeword is described as

$$\hat{\mathbf{y}} = \mathbf{y} + \mathbf{z}, or \ \ \hat{y}_i = y_i + z_i. \tag{3.13}$$

z_i is a zero-mean Gaussian noise with variation σ^2. The decoding can be formulated as

$$\begin{aligned} \hat{\mathbf{x}} = & \ \arg\max_{\mathbf{x} \in GF(2)^k} P(\mathbf{x}|\hat{\mathbf{y}}) \\ \text{s.t.} & \ \ \hat{\mathbf{y}} = \mathbf{G}\mathbf{x}^T + \mathbf{z} \end{aligned} \tag{3.14}$$

The decoding algorithm for Eq. (3.14) is an example of a message passing algorithm. Messages are passed among the nodes in the bipartite graph shown in Figure 3.3a, where the left nodes are source bits and the right nodes are coded bits. In the horizontal step, messages in the form of probability are passed to the source nodes. In the vertical step, messages are passed to the check nodes. Figure 3.4 shows one iteration in the message passing decoding algorithm. Let us take a connecting pair of source bit x_j and check bit $\hat{y}_i$ as an example. $q^t_{ji}()$ is the probability message sent from x_j to $\hat{y}_i$ in the tth iteration and $r^t_{ij}()$ is the message from $\hat{y}_i$ to x_j. Denote R_i as the set of source bits connected to a check bit $\hat{y}_i$, and $R_i \setminus j$ as set R_i excluding source bit x_j. Denote C_j as the set of check bits connected to x_j, and $C_j \setminus i$ as set C_j excluding symbol $\hat{y}_i$.

The standard message passing algorithm for LDPC decoding is described as follows.

(1) Initialization

The initial message sent from x_j to $\hat{y}_i$ is $q_{ji}^0(x_j = 0) = p_j$ and $q_{ji}^0(x_j = 1) = 1 - p_j$, where p_j is an estimation from the channel for a systematic LDPC code

$$p_j = Pr(x_j = 0 \mid \hat{y}_j) = \frac{1}{1 + e^{-2\hat{y}_j/\sigma^2}}, \tag{3.15}$$

or the prior probability.

(2) Horizontal processing

The message sent from $\hat{y}_i$ to x_j in the tth iteration is

$$\begin{cases} r_{ij}^t(x_j = 0) = \frac{1}{2} + \frac{1}{2}\prod_{m \in R_i \setminus j}\left(1 - 2q_{mi}^{t-1}(x_j = 1)\right) \\ r_{ij}^t(x_j = 1) = 1 - r_{ij}^t(x_j = 0) \end{cases}. \tag{3.16}$$

(3) Vertical processing

The message sent from x_j to $\hat{y}_i$ in the tth iteration is

$$\begin{cases} q_{ji}^t(x_j = 0) = K_{ji} \cdot p_j \cdot \prod_{m \in C_j \setminus i} r_{mj}^t(x_j = 0) \\ q_{ji}^t(x_j = 1) = K_{ji} \cdot (1 - p_j) \cdot \prod_{m \in C_j \setminus i} r_{mj}^t(x_j = 1) \end{cases}, \tag{3.17}$$

where K_{ji} is the normalization constant to make $q_{ji}^t(x_j = 0) + q_{ji}^t(x_j = 1) = 1$.

After T iterations, the message passing algorithm stops and makes a hard decision on values of each source bit x_j by calculating

$$\begin{cases} q_{ji}^T(x_j = 0) = K_{ji} \cdot p_j \cdot \prod_{m \in C_j} r_{mj}^T(x_j = 0) \\ q_{ji}^T(x_j = 1) = K_{ji} \cdot (1 - p_j) \cdot \prod_{m \in C_j} r_{mj}^T(x_j = 1) \end{cases}. \tag{3.18}$$

Then the values of input binary digits are estimated by:

$$x_j = \begin{cases} 0 & q_{ji}^T(x_j = 0) \geq q_{ji}^T(x_j = 1) \\ 1 & \text{otherwise} \end{cases} \tag{3.19}$$

3.2.2 Turbo Codes

Shannon's channel coding theorem implies good coding performance for random codes as the code block length increases, but increasing block length typically implies an exponentially increasing decoding complexity. In 1993, an approach for error correction coding was introduced, which provided very long codewords with only (relatively) modest decoding complexity. These codes were called Turbo codes [27, 28].

The Turbo code encoder consists of two (or more) systematic block codes which share message data via interleaving. In its most conventional realization, the codes are obtained from recursive systematic convolutional (RSC) codes — but other codes

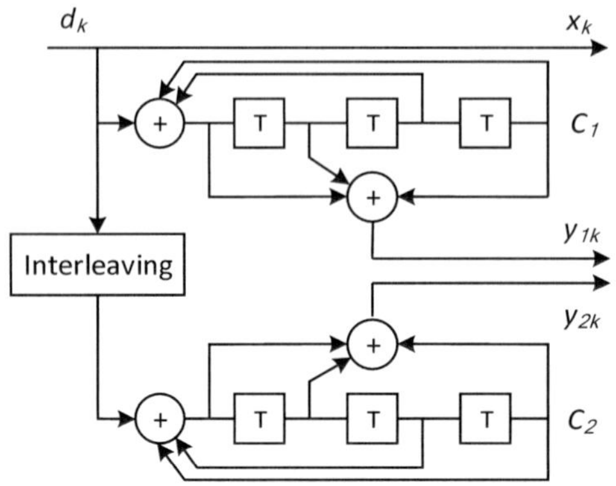

Figure 3.5 An example of a Turbo encoder.

can be used as well. Figure 3.5 shows a Turbo code adopted in the long-term evolution (LTE) wireless communication standard. The data flow (d_k at time k) goes directly to a first elementary RSC encoder C_1. After interleaving, it feeds (d_n at time k) a second elementary RSC encoder C_2. Although these two encoders are identical in Figure 3.5, it is not necessary for Turbo codes. Both C_1 and C_2 encoders have the constraint length $K = 4$ and memory $v = K - 1$. The transfer function of each RSC encoder is the rational function

$$G(x) = \frac{1+D^2+D^3}{1+D+D^3}, \tag{3.20}$$

where D indicates a delay.

Data d_k is systematically transmitted as symbol x_k and redundancies y_{1k} and y_{2k} produced by C_1 and C_2 may be completely transmitted for an $R = 1/3$ encoding or punctured for high rates. The two elementary coding rates R_1 and R_2 associated with C_1 and C_2 after puncturing may be different, but for the best decoding performance, they will satisfy $R_1 \leq R_2$. The global rate R of the composite code, R_1 and R_2, are linked by

$$\frac{1}{R} = \frac{1}{R_1} + \frac{1}{R_2} - 1. \tag{3.21}$$

A key development in Turbo codes is the iterative decoding algorithm. In the iterative decoding algorithm, the decoders take turns operating on the received data. Each decoder produces an estimate of the probabilities of the transmitted symbols. Probabilities of the symbols from one decoder known as extrinsic probabilities are passed to the other decoder in the symbol order appropriate for the encoder, where they are used as prior probabilities for the other decoder. The decoder thus passes probabilities back and forth between the decoders, with each decoder combining the

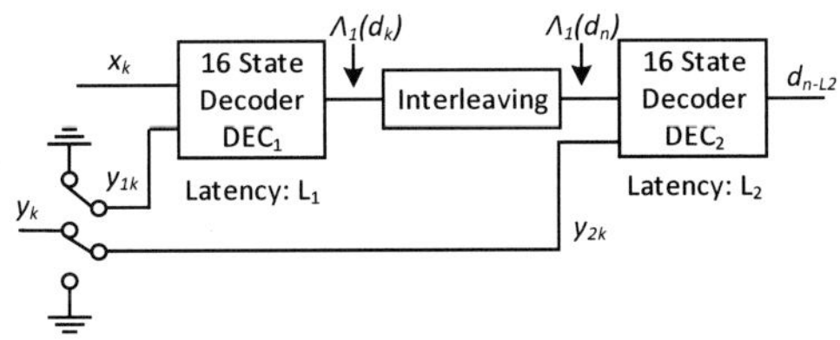

Figure 3.6 The decoding of Turbo codes.

evidence it receives from the incoming prior probabilities with the parity information provided by the code. After a number of iterations, the decoder converges to an estimate of the transmitted codeword.

The decoder of Turbo codes is shown in Figure 3.6, which is made up of two elementary decoders (DEC_1 and DEC_2) in a serial concatenation scheme. The first elementary decoder DEC_1 is associated with the lower rate R_1 encoder C_1 and yields a weighted decision. For a discrete Gaussian channel without memory, the decoder input is made up of a couple $\mathbf{r}_k$ of two random variables $\hat{x}_k$ and $\hat{y}_k$, at time k

$$\begin{cases} \hat{x}_k = x_k + i_k \\ \hat{y}_k = y_k + q_k \end{cases}, \tag{3.22}$$

where i_k and q_k are two independent noises with the same variance σ^2. The redundant information $\hat{y}_k$ is demultiplexed and sent to decoder DEC_1 when $y_k = y_{1k}$ and to decoder DEC_2 when $y_k = y_{2k}$. When the redundant information of a given encoder (C_1 or C_2) is not emitted, the corresponding decoder input is set to zero.

Consider an RSC code with the constraint length K. At time k, the encoder state $\mathbf{S}_k$ is represented by

$$\mathbf{S}_k = (a_k, a_{k-1}, \cdots, a_{k-K+2}).$$

Let us also suppose that the information bit sequence $\{d_k\}$ is made up of N independent bits d_k, taking values zero and one with equal probability. The encoder initial state $\mathbf{S}_0$ and final state $\mathbf{S}_N$ are both equal to zero

$$\mathbf{S}_0 = \mathbf{S}_N = (0, 0, \cdots, 0) = \mathbf{0}.$$

The encoder outputs codeword sequence, noted $\mathbf{C}_1^N = \{\mathbf{c}_1 \cdots \mathbf{c}_k \cdots \mathbf{c}_N\}$, where $\mathbf{c}_k = (x_k, y_k)$ is the input to a discrete Gaussian memoryless channel. The channel output is the sequence $\mathbf{R}_1^N = \{\mathbf{r}_1 \cdots \mathbf{r}_k \cdots \mathbf{r}_N\}$, where $\mathbf{r}_k = (\hat{x}_k, \hat{y}_k)$ is defined by Eq. (3.22).

The a posteriori probability (APP) of a decoded bit d_k can be derived from the joint probability $\lambda_k^i(\mathbf{m})$ defined by

$$\lambda_k^i(\mathbf{m}) = Pr\{d_k = i, \mathbf{S}_k = \mathbf{m} \mid \mathbf{R}_1^N\}. \tag{3.23}$$

Thus, the APP of a decoded bit d_k is equal to

$$Pr\{d_k = 1 \mid \mathbf{R}_1^N\} = \sum_{\mathbf{m}} \lambda_k^i(\mathbf{m}) \quad i = 0, i. \tag{3.24}$$

The logarithm of likelihood ratio (LLR) $\Lambda(d_k)$ associated with a decoded bit d_k can be written as

$$\Lambda(d_k) = \log \frac{\sum_{\mathbf{m}} \lambda_k^1(\mathbf{m})}{\sum_{\mathbf{m}} \lambda_k^0(\mathbf{m})}. \tag{3.25}$$

Finally, the decoder can make a decision by comparing $\Lambda(d_k)$ to a threshold equal to zero

$$\begin{cases} \hat{d}_k = 1, \; if \Lambda(d_k) \geq 0 \\ \hat{d}_k = 0, \; if \Lambda(d_k) < 0 \end{cases}. \tag{3.26}$$

According to the algorithm proposed by Bahl et al. [20], the LLR $\Lambda(d_k)$ in Eq. (3.25) can be calculated as

$$\Lambda(d_k) = \log \frac{\sum_{\mathbf{m}} \sum_{\mathbf{m}'} Pr\{d_k = 1, \mathbf{S}_k = \mathbf{m}, \mathbf{S}_{k-1} = \mathbf{m}', \mathbf{R}_1^{k-1}, \mathbf{r}_k, R_{k+1}^N\}}{\sum_{\mathbf{m}} \sum_{\mathbf{m}'} Pr\{d_k = 0, \mathbf{S}_k = \mathbf{m}, \mathbf{S}_{k-1} = \mathbf{m}', \mathbf{R}_1^{k-1}, \mathbf{r}_k, R_{k+1}^N\}}. \tag{3.27}$$

Using the Bayes rule and taking into account that events after time k are not influenced by observation $\mathbf{R}_1^k$ and bit d_k, if state $\mathbf{S}_k$ is known, the LLR $\Lambda(d_k)$ is equal

$$\begin{aligned} \Lambda(d_k) = & \log \\ & \frac{\sum_{\mathbf{m}} \sum_{\mathbf{m}'} Pr\{\mathbf{R}_{k+1}^N \mid \mathbf{S}_k = \mathbf{m}\} Pr\{\mathbf{S}_{k-1} = \mathbf{m}' \mid \mathbf{R}_1^{k-1}\}}{\sum_{\mathbf{m}} \sum_{\mathbf{m}'} Pr\{\mathbf{R}_{k+1}^N \mid \mathbf{S}_k = \mathbf{m}\} Pr\{\mathbf{S}_{k-1} = \mathbf{m}' \mid \mathbf{R}_1^{k-1}\}} \\ & \times \frac{Pr\{d_k = 1, \mathbf{S}_k = \mathbf{m}, \mathbf{r}_k \mid \mathbf{S}_{k-1} = \mathbf{m}'\}}{Pr\{d_k = 0, \mathbf{S}_k = \mathbf{m}, \mathbf{r}_k \mid \mathbf{S}_{k-1} = \mathbf{m}'\}} \end{aligned} \tag{3.28}$$

In order to compute the LLR $\Lambda(d_k)$, let us introduce the probability functions $\alpha_k(\mathbf{m})$, $\beta_k(\mathbf{m})$, and $\gamma_i(\mathbf{r}_k, \mathbf{m}', \mathbf{m})$ defined by

$$\alpha_k(\mathbf{m}) = Pr\{\mathbf{S}_k = \mathbf{m} \mid \mathbf{R}_1^k\}, \tag{3.29a}$$

$$\beta_k(\mathbf{m}) = \frac{Pr\{\mathbf{R}_{k+1}^N \mid \mathbf{S}_k = \mathbf{m}\}}{Pr\{\mathbf{R}_{k+1}^N \mid \mathbf{R}_1^k\}}, \tag{3.29b}$$

$$\gamma_i(\mathbf{r}_k, \mathbf{m}', \mathbf{m}) = Pr\{d_k = i, \mathbf{S}_k = \mathbf{m}, \mathbf{r}_k \mid \mathbf{S_{k-1}} = \mathbf{m}'\}. \tag{3.29c}$$

Substituting Eq. (3.29) into Eq. (3.27), $\Lambda(d_k)$ is equal to

$$\Lambda(d_k) = \log \frac{\sum_{\mathbf{m}} \sum_{\mathbf{m}'} \gamma_1(\mathbf{r}_k, \mathbf{m}', \mathbf{m}) \alpha_{k-1}(\mathbf{m}') \beta_k(\mathbf{m})}{\sum_{\mathbf{m}} \sum_{\mathbf{m}'} \gamma_0(\mathbf{r}_k, \mathbf{m}', \mathbf{m}) \alpha_{k-1}(\mathbf{m}') \beta_k(\mathbf{m})}, \tag{3.30}$$

where probabilities $\alpha_k(\mathbf{m})$ and $\beta_k(\mathbf{m})$ can be recursively calculated from probability $\gamma_i(\mathbf{r}_k, \mathbf{m}', \mathbf{m})$ as

$$\alpha_k(\mathbf{m}) = \frac{\sum_{\mathbf{m}'}\sum_{i=0}^{1}\gamma_i(\mathbf{r}_k,\mathbf{m}',\mathbf{m}\alpha_{k-1}(\mathbf{m}')}{\sum_{\mathbf{m}}\sum_{\mathbf{m}'}\sum_{i=0}^{1}\gamma_i(\mathbf{r}_k,\mathbf{m}',\mathbf{m})\alpha_{k-1}(\mathbf{m}')}, \tag{3.31}$$

$$\beta_k(\mathbf{m}) = \frac{\sum_{\mathbf{m}'}\sum_{i=0}^{1}\gamma_i(\mathbf{r}_{k+1},\mathbf{m}',\mathbf{m})\beta_{k+1}(\mathbf{m}')}{\sum_{\mathbf{m}}\sum_{\mathbf{m}'}\sum_{i=0}^{1}\gamma_i(\mathbf{r}_{k+1},\mathbf{m}',\mathbf{m})\alpha_k(\mathbf{m}')}. \tag{3.32}$$

The probability $\gamma_i(\mathbf{r}_k,\mathbf{m}',\mathbf{m})$ can be determined from transition probabilities of the discrete Gaussian memoryless channel and transition probabilities of the encoder trellis. $\gamma_i(\mathbf{r}_k,\mathbf{m}',\mathbf{m})$ is given by

$$\begin{aligned}\gamma_i(\mathbf{r}_k,\mathbf{m}',\mathbf{m}) =& p(\mathbf{r}_k \mid d_k = i, \mathbf{S}_k = \mathbf{m}, \mathbf{S}_{k-1} = \mathbf{m}') \\ &\times q(d_k = i \mid \mathbf{S}_k = \mathbf{m}, \mathbf{S}_{k-1} = \mathbf{m}') \times \pi(\mathbf{S}_k = \mathbf{m} \mid \mathbf{S}_{k-1} = \mathbf{m}').\end{aligned} \tag{3.33}$$

$p(\cdot \mid \cdot)$ is the transition probability of the discrete Gaussian memoryless channel. Conditionally to $(d_k = i, \mathbf{S}_k = \mathbf{m}, \mathbf{S}_{k-1} = \mathbf{m}')$, $\hat{x}_k$ and $\hat{y}_k(\mathbf{r}_k = (\hat{x}_k, \hat{y}_k))$ are two uncorrelated Gaussian variables and thus we obtain

$$\begin{aligned}&p(\mathbf{r}_k \mid d_k = i, \mathbf{S}_k = \mathbf{m}, \mathbf{S}_{k-1} = \mathbf{m}') = \\ &p(\hat{x}_k \mid d_k = i, \mathbf{S}_k = \mathbf{m}, \mathbf{S}_{k-1} = \mathbf{m}')p(\hat{y}_k \mid d_k = i, \mathbf{S}_k = \mathbf{m}, \mathbf{S}_{k-1} = \mathbf{m}').\end{aligned} \tag{3.34}$$

Since the convolutional encoder is a deterministic machine $q(d_k = i \mid \mathbf{S}_k = \mathbf{m}, \mathbf{S}_{k-1} = \mathbf{m}')$ is equal to 0 or 1. The transition state probabilities $\pi(\mathbf{S}_k = \mathbf{m} \mid \mathbf{S}_{k-1} = \mathbf{m}')$ of the trellis are defined by the encoder input statistic. Generally, $Pr\{d_k = 1\} = Pr\{d_k = 0\} = 1/2$ and since there are two possible transitions from each state, $\pi(\mathbf{S}_k = \mathbf{m} \mid \mathbf{S}_{k-1} = \mathbf{m}') = 1/2$ for each of these transitions.

Since the encoder is systematic ($x_k = d_k$), the transition probability $p(\hat{x}_k \mid d_k = i, \mathbf{S}_k = \mathbf{m}, \mathbf{S}_{k-1} = \mathbf{m}')$ in expression $\gamma_i(\mathbf{r}_k,\mathbf{m}',\mathbf{m})$ is independent of state values $\mathbf{S}_k$ and $\mathbf{S}_{k-1}$. Therefore, we can factorize this transition probability in the numerator and in the denominator of Eq. (3.30).

$$\Lambda(d_k) = \log\frac{p(\hat{x}_k \mid d_k = 1)}{p(\hat{x}_k \mid d_k = 0)} + \log\frac{\sum_{\mathbf{m}}\sum_{\mathbf{m}'}\gamma_1(\hat{y}_k,\mathbf{m}',\mathbf{m})\alpha_{k-1}(\mathbf{m}')\beta_k(\mathbf{m})}{\sum_{\mathbf{m}}\sum_{\mathbf{m}'}\gamma_0(\hat{y}_k,\mathbf{m}',\mathbf{m})\alpha_{k-1}(\mathbf{m}')\beta_k(\mathbf{m})}. \tag{3.35}$$

Conditional to $d_k = 1$ (resp. $d_k = 0$), variable x_k is Gaussian with mean 1 (resp. –1) and variance σ^2, thus the LLR $\Gamma(d_k)$ is still equal to

$$\Lambda(d_k) = \frac{2}{\sigma^2}\hat{x}_k + W_k, \tag{3.36}$$

where

$$W_k = \Lambda(d_k)\mid_{\hat{x}_k=0} = \log\frac{\sum_{\mathbf{m}}\sum_{\mathbf{m}'}\gamma_1(\hat{y}_k,\mathbf{m}',\mathbf{m})\alpha_{k-1}(\mathbf{m}')\beta_k(\mathbf{m})}{\sum_{\mathbf{m}}\sum_{\mathbf{m}'}\gamma_0(\hat{y}_k,\mathbf{m}',\mathbf{m})\alpha_{k-1}(\mathbf{m}')\beta_k(\mathbf{m})}. \tag{3.37}$$

Both decoders DEC_1 and DEC_2 now use the above algorithm. As shown by Eq. (3.36), the LLR at the decoder output can be expressed as a sum of two terms if the noises at the decoder inputs are independent at each time k. Hence, if the noise at the decoder DEC_2 inputs are independent, the LLR $\Lambda_2(d_k)$ at the decoder DEC_2 output can be written as

$$\Lambda_2(d_k) = f(\Lambda(d_k)) + W_{2k} \tag{3.38}$$

with

$$\Lambda_1(d_k) = \frac{2}{\sigma^2}\hat{x}_k + W_{1k}. \tag{3.39}$$

From Eq. (3.37), we can see that the decoder DEC_2 extrinsic information W_{2k} is a function of the sequence $\{\Lambda_1(d_n)\}_{n \neq k}$. Since $\Lambda_1(d_n)$ depends on observation $\mathbf{R}_1^N$, extrinsic information W_{2k} is correlated with observations $\hat{x}_k$ and $\hat{y}_{1k}$, regarding the noise. Nevertheless, from Eq. (3.37), the greater $\mid n-k \mid$ is, the less correlated are $\Lambda_1(d_n)$ and observations $\hat{x}_k, \hat{y}_k$. Thus, due to the presence of interleaving between decoders DEC_1 and DEC_2, extrinsic information W_{2k} and observations $\hat{x}_k, \hat{y}_{1k}$ are weakly correlated. Therefore, extrinsic information W_{2k} and observations $\hat{x}_k, \hat{y}_{1k}$ can be used jointly for carrying out a new decoding bit d_k.

3.2.3 Digital Modulation

In digital modulation, an analog carrier signal is modulated by a discrete signal. Digital modulation can be considered digital-to-analog conversion, and the corresponding demodulation or detection as analog-to-digital conversion. The changes in the carrier signal are chosen from a finite number of M alternative symbols (the modulation alphabet). The most fundamental digital modulation techniques are based on keying: phase-shift keying (PSK), frequency-shift keying (FSK), amplitude-shift keying (ASK), and quadrature amplitude modulation (QAM).

In all of the above methods, each of these phases, frequencies, or amplitudes are assigned a unique pattern of binary bits. Usually, each phase, frequency, or amplitude encodes an equal number of bits. This number of bits comprises the symbol that is represented by the particular phase, frequency, or amplitude. In QAM, an in-phase signal (the I signal, for example, a cosine waveform) and a quadrature phase signal (the Q signal, for example, a sine wave) are modulated with a finite number of amplitudes and added together. It can be seen as a two-channel system, each channel using ASK. The resulting signal is equivalent to a combination of PSK and ASK.

These are the general steps used by the modulator to transmit data: (1) Group the incoming data bits into codewords, one for each symbol that will be transmitted. (2) Map the codewords to attributes, for example, amplitudes of the I and Q signals, or frequency or phase values. (3) Adapt pulse shaping or some other filtering to limit the bandwidth and form the spectrum of the equivalent low-pass signal. (4) Perform digital-to-analog conversion (DAC) of the I and Q signals. (5) Generate a high frequency sine carrier waveform, and perhaps also a cosine quadrature component. Carrying out the modulation, for example, by multiplying the sine and cosine waveform with the I and Q signals, results in the equivalent low-pass signal being frequency shifting to the modulated passband signal or RF signal. (6) Amplification and analog band-pass filtering to avoid harmonic distortion and periodic spectrum.

At the receiver side, the demodulator typically performs the following steps: (1) Band-pass filtering. (2) Automatic gain control (AGC) to compensate for attenuation. (3) Frequency shifting of the RF signal to the equivalent baseband I and Q signals, or to an intermediate frequency (IF) signal, by multiplying the RF signal with a local oscillator sine wave and cosine wave frequency. (4) Sampling and analog-to-digital conversion (ADC). (5) Equalization filtering, for example, a matched filter, compensation for multipath propagation, time spreading, phase distortion, and frequency selective fading to avoid inter-symbol interference and symbol distortion. (6) Detection of the amplitudes of the I and Q signals, or the frequency or phase of the IF signal. (7) Quantization of the amplitudes, frequencies, or phases to the nearest allowed symbol values. (8) Mapping of the quantized amplitudes, frequencies, or phases to codewords (bit groups). (9) Parallel-to-serial conversion of the codewords into a bitstream. (10) Pass the resulting bitstream for further processing such as removal of any error-correcting codes.

The above modulation and demodulation processing can be described mathematically as follows. An N-dimensional signal constellation is denoted by

$$\mathscr{A} = \{\mathbf{y}_j, 1 \leq j \leq M\}. \tag{3.40}$$

Its M elements $\mathbf{y}_j$ will be called signal points, vectors, or N-tuples. The basic parameters of a signal constellation $\mathscr{A}$ are its dimension N; its size M (number of signal points); its average energy $E(\mathscr{A}) = 1/M\sum_j \parallel y_j \parallel^2$; and its minimum squared distance $d^2_{min}(\mathscr{A})$, which is an elementary measure of its noise resistance. A secondary parameter is the average number $K_{min}(\mathscr{A})$ of nearest neighbors (points at distance $d_{min}(\mathscr{A})$).

From these basic parameters we can derive such parameters as: the bit rate (nominal spectral efficiency) $\rho = (2/N)\log_2 M(b/2D)$; the average energy per two dimensions $E_s = (2/N)E(\mathscr{A})$, or the average energy per bit $E_b = E(\mathscr{A})/(\log_2 M) = E_s/\rho$; energy-normalized figures of merit such as $d^2_{min}(\mathscr{A})/E(\mathscr{A})$, $d^2_{min}(\mathscr{A})/E_s$, or $d^2_{min}(\mathscr{A})/E_b$, which are independent of scale. Figure 3.7 shows an example of

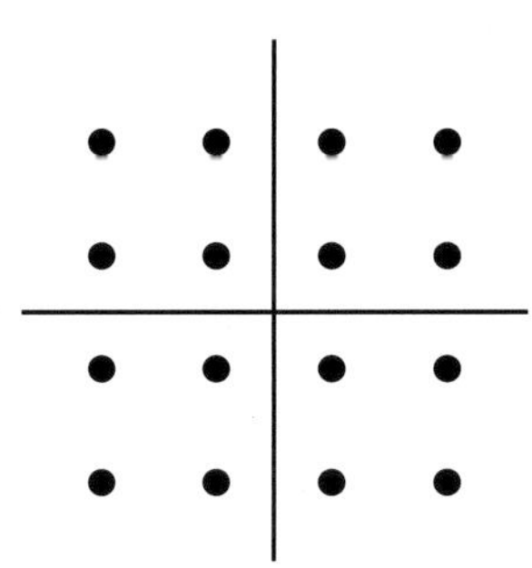

Figure 3.7 16-QAM.

16-QAM. It can carry $M = 16$ different symbols and a symbol consists of 4 bits. Thus the modulation rate is $\rho = 4$.

In digital communications, we are usually interested in the minimum probability of error (MPE) decision rule: given a received vector $\hat{\mathbf{y}}$, choose the signal point $\mathbf{y}^* \in \mathscr{A}$ to minimize the probability of decision error $Pr\{E\}$. Since the probability that a decision $\mathbf{y}^*$ is correct is simply the a posteriori probability $p(\mathbf{y}^* \mid \hat{\mathbf{y}})$, the MPE rule is equivalent to the maximum a posteriori (MAP) rule: choose the $\mathbf{y}^* \in \mathscr{A}$ such that $p(\mathbf{y}^* \mid \hat{\mathbf{y}})$ is maximum among all $p(\mathbf{y}_j \mid \hat{\mathbf{y}}), \mathbf{y}_j \in \mathscr{A}$. By Bayes law,

$$p(\mathbf{y}_j \mid \hat{\mathbf{y}}) = \frac{p(\hat{\mathbf{y}} \mid \mathbf{y}_j)p(\mathbf{y}_j)}{p(\hat{\mathbf{y}})}. \tag{3.41}$$

If the signals $\mathbf{y}_j$ are equiprobable, so $p(\mathbf{y}_j) = 1/M$ for all j, then the MAP rule is equivalent to the maximum likelihood (ML) rule: choose the $\mathbf{y}^* \in \mathscr{A}$ such that $p(\hat{\mathbf{y}} \mid \mathbf{y}^*)$ is maximum among all $p(\hat{\mathbf{y}} \mid \mathbf{y}_j), \mathbf{y}_j \in \mathscr{A}$. Using the noise PDF, we can write

$$p(\hat{\mathbf{y}} \mid \mathbf{y}_j) = \frac{1}{(2\pi\sigma^2)^{N/2}} e^{-\|\hat{\mathbf{y}}-\mathbf{y}_j\|^2/2\sigma^2}. \tag{3.42}$$

The symbol variance is $\sigma^2 = N0/2$.

Part II
Scalable Video Coding

This part studies scalable video coding (SVC). The fundamental problem with SVC is closely related to motion compensation. As we have discussed in Chapter 2, every input picture will subtract a predictive picture in motion compensation. The predictive picture must be exactly the same at the encoder and decoder. If there are some differences between the predictive pictures used in the encoder and the decoder, the differences in different pictures will be accumulated and thus the quality of decoded pictures will continuously drop, which is known as *drifting error*. In SVC, it is hard to keep the predictive picture the same in both the encoder and decoder because part of the compressed data may be dropped during transmission due to bandwidth variation. Thus, the encoder does not know how many bits are really decoded and exactly which predictive pictures are used in the decoder. This inevitably results in drift error in SVC as long as motion compensation exists. In the following three chapters, we will study how to reduce or eliminate the drift error in SVC, while keeping high coding efficiency.

Chapter 4 introduces SVC and MPEG-4 fine granularity scalable (FGS) coding. A basic framework for efficient scalable video coding, namely, progressive fine granularity scalable (PFGS) coding, is presented. The PFGS framework has all the features of MPEG-4 FGS, such as fine granularity scalability, channel adaptation, and error recovery. Unlike the FGS coding, the PFGS framework uses multiple layers of references with increasing quality to make motion compensation more accurate for improved coding efficiency. However, using multiple layers of references with different quality also creates several issues. First, extra picture buffers are needed for storing the multiple reconstructed reference layers. This would increase the memory cost and computational complexity of the PFGS scheme. Based on the basic framework, a simplified and efficient PFGS framework is further proposed. The simplified PFGS framework needs only one extra picture buffer with almost the same coding efficiency as the original framework. Second, there might be an undesirable increase and fluctuation of coefficients to be coded when switching from a low-quality reference to a high-quality one, which could partially offset the advantage of using a high-quality reference. A further improved PFGS scheme can eliminate the fluctuation of enhancement-layer coefficients when switching references by always using only one high-quality prediction reference for all enhancement layers.

Chapter 5 presents our research on using advanced motion threading technology for improved performance in 3D wavelet coding. First, we extend an original motion threading idea to a lifting-based implementation. Methods for enabling fractional-pixel alignment in motion threading and for processing many-to-one pixel mapping and non-referred pixels are proposed to reduce the wavelet boundary effects. Second, we devise an advanced motion threading technology, in which one set of motion vectors is generated for each temporal layer of wavelet coefficients for temporal scalability. In order to reduce the motion overhead information, especially at low bit rates, several correlated motion prediction modes at the macroblock level are developed to exploit the intra/inter-layer correlation in motion vector coding. Finally, rate-distortion optimization is utilized in motion estimation to select the best motion prediction mode for each macroblock.

Chapter 6 provides an overview of our Barbell-lifting coding scheme that was adopted as the common software by the MPEG ad hoc group on further exploration of wavelet-based video coding. The core technologies used in this scheme, such as Barbell lifting, layered motion coding, 3D entropy coding, and base-layer embedding, are discussed. The paper also analyzes and compares the proposed scheme with the H.264 SVC standard because the hierarchical temporal prediction technique used in H.264 SVC has a close relationship with motion compensated temporal filtering (MCTF) in wavelet coding. The commonalities and differences between these two schemes are exhibited for readers to better understand modern scalable video coding technologies. Several challenges that still exist in our Barbell-lifting coding scheme, for example, performance of spatial scalable coding and accurate motion compensated lifting, are also discussed. Two additional technologies are presented here, although they are not yet integrated into the common software. Finally, we compare it with H.264 SVC and another well-known 3D wavelet coding scheme, Motion Compensated Embedded Zero Block Coding (MC-EZBC).

Chapter 4
Progressive Fine Granularity Scalable (PFGS) Coding

4.1 Introduction

With recent developments in computing technology, such as high-capacity storage devices, high-speed wired and wireless networks, and compression and transmission technologies, more and more users expect to enjoy high-quality multimedia services over the Internet [21–23]. Typically, there are two approaches to providing multimedia services on-demand: offline downloading and online streaming. Since the streaming approach enables users to experience a multimedia presentation on the fly while it is being downloaded from the Internet, it has prevailed in both academia and industry circles. In virtue of the streaming technology, users no longer have to suffer from long and even unacceptable transmission time for full downloads.

Figure 4.1 exemplifies a typical scenario for streaming video contents to users. Raw video sequences are usually compressed in advance and then saved on the storage device. Upon the client's request, the streaming server retrieves compressed video data from the storage device and delivers it through the Internet that consists of many heterogeneous sub-networks. Receivers may use different devices for decoding and presenting received video data with different resolutions, frame rates, and qualities. Scalable video coding (SVC) provides a good compression solution for the above application. A video sequence is coded into a scalable stream, which consists of multiple layer streams. Only the base layer stream can be independently decoded and provides the lowest visual quality, whereas the enhancement layer streams are decoded together with the base layer stream and lower enhancement layer streams and enhance the base or lower layer video from frame rate, resolution, and quality, respectively. Therefore, SVC techniques are typically categorized into three types, that is, quality scalability, temporal scalability, and spatial scalability.

Quality scalability provides different quality representations of coded visual information within a single scalable stream [24–26]. The base layer is coarsely quantized, and the difference between the reconstructed base layer picture and original picture is coded into one or multiple enhancement layers with finer quantization. Temporal scalability is able to present coded visual information at varying frame rates from a

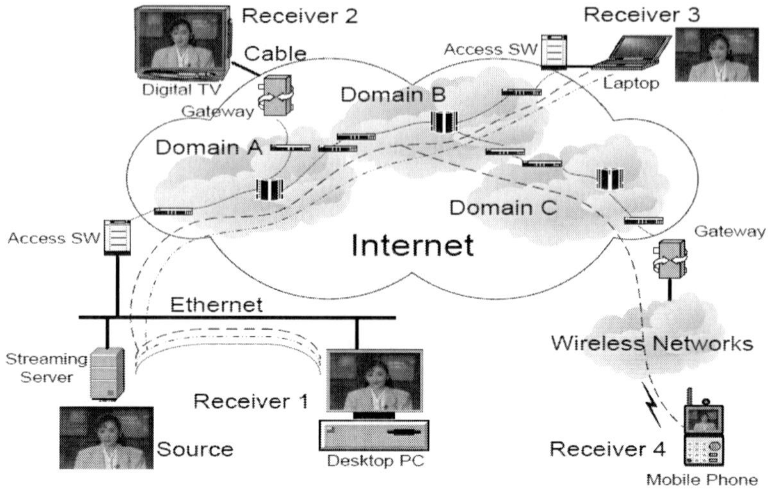

Figure 4.1 An example scenario of streaming video.

single scalable stream [27–30]. The base layer video is set at a low frame rate. By decoding all temporal pictures, users can get the smoothest presentation. Although the official emergence of temporal scalability in the video coding standard is from MPEG-2, it can be essentially fulfilled even in MPEG-1 by dropping B pictures. Spatial scalability provides a multi-resolution representation of coded visual information within a single scalable stream [31–34]. In general, the base layer represents a low-resolution video, and the enhancement layer provides additional data to reproduce video at high resolution. Such compression techniques are particularly useful for the streaming server to simultaneously serve PC devices and non-PC devices only with low-resolution screen and limited computational power.

All these scalable video coding techniques (quality, temporal, and spatial) provides only coarse granularity scalability, in which an enhancement layer is either available and decoded as a whole or not available at all. Accordingly, this is known as *layered coding*. These techniques only have limited adaptation on channel bandwidth variations. Furthermore, multiple-layer coding techniques suffer from coding efficiency loss due to inefficient motion compensation and overhead bits. This is the reason why the industry has chosen to use stream switching in commercial streaming systems instead of scalable video coding, although the latter has achieved great success in video coding standards, such as MPEG-2, MPEG-4, and H.264.

4.2 Fine Granularity Scalable Video Coding

Unlike the layered video coding techniques we have discussed, MPEG-4 FGS provides flexible and precise adaptation on channel bandwidth variations, where the enhancement layer streams can be arbitrarily truncated. The requirements on FGS video coding were brought forward in MPEG [35]. It is well known that motion compensation is extensively used to fully exploit temporal redundancy among pictures of a video for high coding efficiency. However, utilizing motion compensation usually creates a problem with providing fine granularity scalability due to temporal dependency of one picture on another. Any truncation on reference bits will bring drifting errors without exception.

When MPEG-4 launched FGS, it had already adopted a wavelet-based coding scheme for still texture. Matching pursuit [36–38] and bit-plane coding [39] techniques were being evaluated. For backwards compatibility, four candidate solutions were proposed by Chen et al. [40]: discrete cosine transform (DCT) base layer plus wavelet enhancement layer; DCT base layer plus matching pursuit enhancement layer; DCT base layer plus DCT enhancement layer; and matching pursuit base layer plus matching pursuit enhancement layer. The solution with wavelet-based coding techniques [41, 42] for both the base layer and the enhancement layer is out of the question, although it inherently provides fine granularity scalability. In the succeeding meetings, MPEG established a core experiment on fine granularity scalability [43]. Based on common conditions, experimental results of wavelet coding of image residues [44–46], matching pursuit coding of image residues [47, 48], and bit plane coding of DCT residues [49] were proposed to MPEG and evaluated. Finally, the bit plane coding of DCT residues was chosen as the baseline by MPEG-4 due to its comparable coding efficiency and implementation simplicity.

Figure 4.2 illustrates the block diagram of the MPEG-4 FGS encoder. The lower part outlined by a box is the base layer encoder. If an input picture is coded with temporal prediction, the displacement of each macroblock is first estimated based on the association with its reference stored in the Buffer. Each inter macroblock has either one or four MVs. The precision of a motion vector is up to a quarter of a pixel. Temporal prediction is generated by displacing the reference macroblocks or blocks, where integer pixels are directly copied from the reference and fractal pixels are obtained from neighboring integer pixels in terms of interpolation. The input picture for intra coding or the predictive residual picture obtained by subtracting the temporal prediction from the input picture is performed with a DCT at each 8×8 block. After quantization and ZigZag scanning, the obtained DCT data is coded with variable length coding (VLC) to form the output base layer stream. With de-quantization and an inverse DCT transform, the reconstructed picture or the reconstructed residual picture plus the temporal prediction is used to update the Buffer after clipping for the next picture coding.

The upper part in Figure 4.2 outlined by a box is the enhancement layer encoder. The quantized residues of the base layer video calculated in the image domain by subtracting the reconstructed image block from the input image block are coded with bit-plane coding. The obtained residues are then performed with a DCT transform on each 8×8 block. The DCT residues may be shifted up with frequency weighting and selective enhancement techniques for better visual quality and enhanced interested regions. Finally, the DCT residual picture is processed bit-plane by bit-plane to generate the output enhancement layer stream.

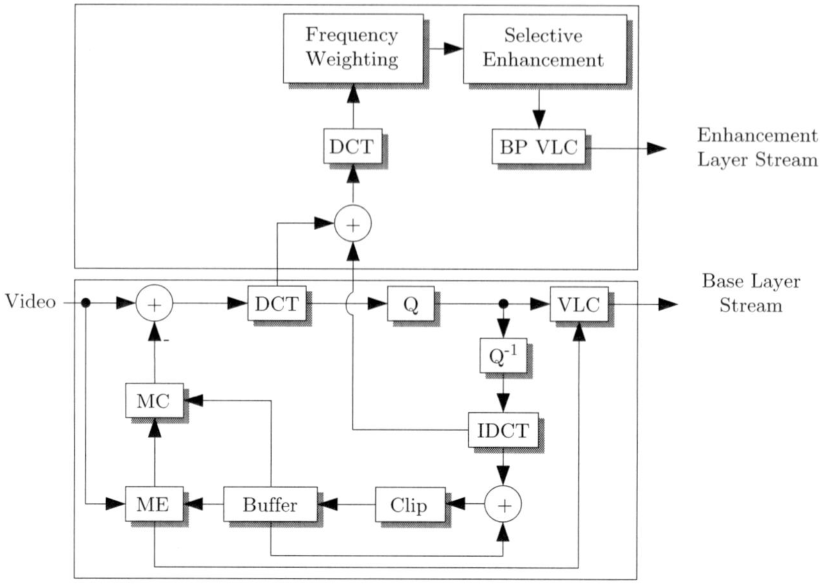

Figure 4.2 The block diagram of the MPEG-4 FGS encoder.

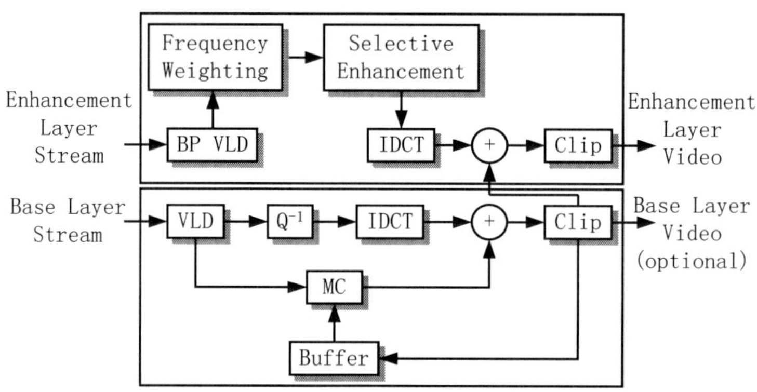

Figure 4.3 The block diagram of the MPEG-4 FGS decoder.

Figure 4.3 illustrates the block diagram of the MPEG-4 FGS decoder. The lower part outlined by a box is the base layer decoder, and the upper part is the enhancement layer decoder. The base layer stream is first decoded with variable length decoding (VLD) to reconstruct DCT data. After de-quantization and inverse DCT, the reconstructed picture or the reconstructed residual picture plus the temporal prediction forms the decoded base layer video. In general, it is stored in the Buffer for the next picture coding. It is also an optional display picture if the enhancement layer is not available.

To reconstruct the enhancement layer picture, the input stream is first decoded using bit-plane VLD. The decoded block bit-planes are used to compose DCT data according to their binary locations, which may be then shifted down based on the frequency weighting and selective enhancement shifting factors. After inverse discrete cosine transform (IDCT), the image domain residues are reconstructed. They are added to the reconstructed clipped base layer pixels to reconstruct the enhancement layer pixels. The reconstructed enhancement layer pixels are limited to a value range between 0 and 255 by the clipping process at the enhancement layer to generate the final enhancement layer picture.

4.3 Basic PFGS Framework

One major feature of the FGS coding scheme is that the base layer and all enhancement layers in a predicted picture are always predicted from the reconstructed version of the base layer in the reference picture. Therefore, the FGS coding scheme provides excellent error recovery from occasional data losses or errors in enhancement layers. By predicting all enhancement layers from the base layer, losses or corruptions of one or more enhancement layers during transmission have no effect on the pictures that follow. However, since the prediction is always based on the lowest quality base layer, the coding efficiency of the FGS scheme is not as good as, and sometimes much worse than, traditional SNR scalability schemes, such as in Macnicol et al. [50]. On the other hand, in traditional signal-to-noise ratio (SNR) scalability schemes, same layer references are used to provide better predictions, which in turn normally provide better coding efficiency. But once there is an error or packet loss in the enhancement layers, it would propagate to the end of a group of pictures (GOP) and would cause serious drifting errors in the higher layers of the following predicted pictures. Even though there may be sufficient bandwidth available later, the decoder could not recover to the highest quality until another GOP starts. Therefore, the traditional SNR scalability schemes are normally only suitable for simulcasting in stable channels.

In order to improve the coding efficiency of FGS, a basic framework for more efficient scalable video coding was first proposed to MPEG-4 by Li et al. [51], referred to as PFGS video coding. Similar to FGS, the PFGS coding scheme also encodes video pictures into multiple layers, including a base layer of relatively lower quality video and multiple enhancement layers of increasingly higher quality video. However, in the PFGS framework we try to use several high-quality references for the predictions in the enhancement-layer encoding rather than always using the base layer. Using high-quality references would make motion prediction more accurate and thus could improve coding efficiency. Our experimental results show that the PFGS scheme can achieve consistently better coding efficiency than the FGS scheme while keeping all the properties of FGS, such as fine granularity scalability, channel adaptation, and error recovery.

There are still several issues to be addressed in the basic PFGS framework proposed by Li et al. [51]. First, it needs multiple extra picture buffers to save multiple reconstructed layers as references, which increases the memory cost and computational complexity of the PFGS encoder and decoder. Fortunately, not every reference layer makes the same contribution to the improvement of coding efficiency. Only a few among the reference layers make significant contributions to improving coding efficiency, while others have little effect. How to choose a minimal number of reference layers to achieve high coding efficiency remains an open problem. Another problem is the fluctuation and increase of DCT coefficients when switching from a low-quality reference to a high-quality one. An efficient approach that can take full advantage of a high-quality reference without causing any fluctuation should be investigated to further improve the coding efficiency of the basic PFGS framework.

4.3.1 Basic Ideas to Build the PFGS Framework

As discussed in the previous section, the FGS video-coding scheme provides very good bandwidth adaptation and error recovery properties, but it sacrifices coding efficiency. On the other hand, the SNR scalability traditional schemes have good coding efficiency, but they lose bandwidth adaptation and error recovery properties. Is there a new framework that can balance between coding efficiency and scalability properties? In this section, we try to present such a general framework that can keep the properties of the FGS coding scheme, such as fine granularity scalability, channel adaptation, and error recovery, while using as many predictions from the same reference layer as possible as in the traditional SNR scalability schemes.

There are two key points in designing such a framework. The first point is to use as many predictions from the enhancement reference layers as possible (for coding efficiency), instead of always using the base layer as in the FGS scheme. Since the quality of an enhancement layer is better than that of the base layer, such a framework makes motion compensation as accurate as possible for any given video layer to maintain coding efficiency. The second point is to keep a prediction path from the base layer to the highest quality layers across several pictures (for error recovery and channel adaptation). This will make sure that the coding schemes can gracefully recover from losses or errors. Lost or erroneous higher quality enhancement layers may be automatically reconstructed from lower layers gradually over a few pictures with such a prediction path.

Figure 4.4 conceptually illustrates such an exemplary framework for efficient video coding with no drifting problem. In the illustrated framework, picture 2 is predicted from the base layer and even enhancement layers of picture 1 (i.e., the second and fourth enhancement layers). Picture 3 is predicted from the base layer and odd enhancement layers of picture 2 (i.e., the first and third enhancement layers). Picture 4 is again predicted from the base layer and even enhancement layers of picture 3,

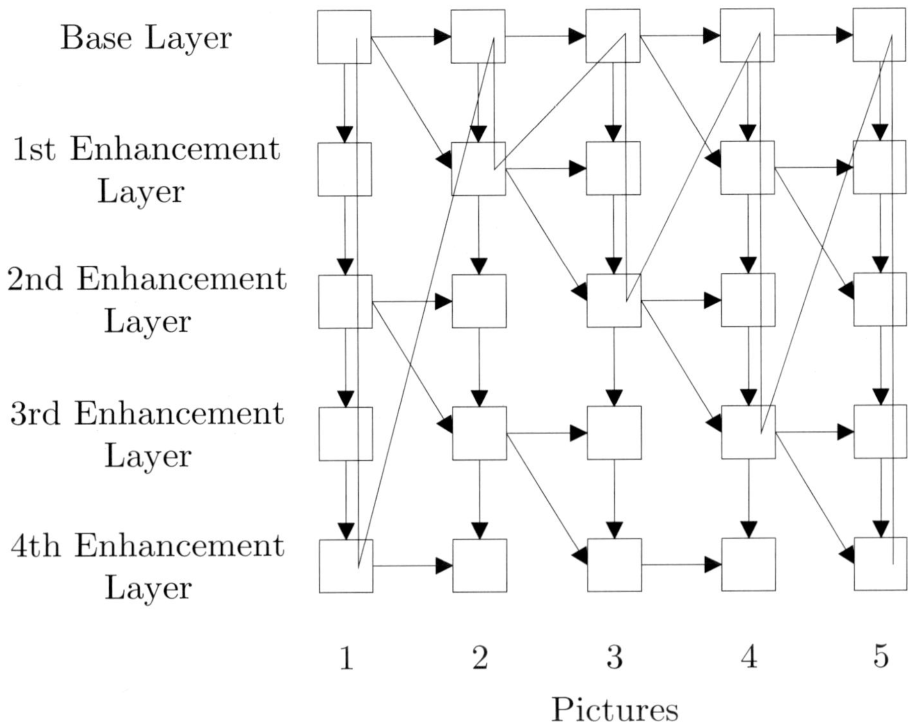

Figure 4.4 PFGS framework.

and so on. It is obvious that the three highest enhancement layers in picture 2 are predicted from high-quality reference layers in picture 1, rather than from the base layer in picture 1. Since the quality of an enhancement layer is higher than that of the base layer, the framework shown in Figure 4.4 provides more accurate motion prediction to improve coding efficiency. On the other hand, many prediction paths from the lowest quality layer to the highest quality layer are preserved. For instance, the base layer of picture 1, the first enhancement layer of picture 2, the second enhancement layer of picture 3, the third enhancement layer of picture 4, and the fourth enhancement layer of picture 5 constitute such a complete path.

The advantages of the PFGS framework are obvious when it is applied to video transmission over the Internet or wireless channels. The encoded bitstream can adapt to the available bandwidth of the channel without any drifting problem. Figure 4.4 shows an example of this bandwidth adaptation process. The dashed line traces the transmitted video layers. Note that at picture 2, there is a reduction in bandwidth. At this picture, the transmitter (server) simply drops the bits of higher layers (from the second to the fourth enhancement layers). However, after picture 2, as the bandwidth increases, the transmitter simply transmits more layers of video bits. After three pictures (at picture 5), the decoder side can obtain up to the highest quality video layer again. Note that in all these operations, no re-encoding or retransmission of the video bitstream is required. Similarly, when several enhancement layers in one or more pictures have packet losses or errors, the recovery process is the same as that of the bandwidth adaptation case. We can see that the processes of bandwidth adaptation and error recovery are graceful and gradual progressively recovering across several frames. Therefore, this framework is called PFGS video coding.

Figure 4.4 exemplifies a case where the group depth is 2. Group depth defines how many layers may refer back to a common reference layer. The group depth can be changed in each picture. If the group depth is 1, the PFGS framework essentially becomes the traditional SNR scalability schemes as in Macnicol et al. [50]. If the group depth is equal to the total number of layers, the PFGS framework essentially represents FGS in Ling et al. [52] and Li [49]. The above description is a special case of a more general case where in each picture the reference layers used for prediction can be randomly assigned as long as a prediction path from lowest layer to highest layer is maintained across several pictures.

4.3.2 The Simplified PFGS Framework

It is very clear that compared to FGS, the implementation of such a PFGS framework as shown in Figure 4.4 needs several extra picture buffers to save the reconstructed enhancement layers as references. In Figure 4.4, there are two extra buffers needed for encoding each picture. For example, the second and fourth enhancement layers in picture 1 are used as references for enhancement layer coding in picture 2, and the first and third enhancement layers in picture 2 are also used as references for the enhancement layer coding in picture 3, and so on. In fact, the number of extra picture buffers increases as the number of enhancement layers increases. If we could reduce the number of extra picture buffers to a minimum number while still maintaining almost the same coding efficiency, it will be a very significant complexity reduction for the PFGS implementation, especially for the hardware implementation.

Although the PFGS framework can use as many additional reference buffers as possible to achieve greater coding efficiency improvement, we still prefer a much more simplified PFGS framework that can provide a good trade-off between coding efficiency versus the memory cost and computational complexity. Not all enhancement layers are well suited to be used as references. Only a few enhancement layers as references can make a significant contribution to improving coding efficiency, whereas others have little effect. Generally, the lower enhancement layers are not good references. Since these layers may comprise errors with large magnitudes caused by motion across frames, the correlations of lower enhancement layers between adjacent pictures are weak. Here, the correlation between the two same layers in adjacent pictures is defined as the absolute sum of binary difference of two bit planes. The larger the absolute value is, the weaker the correlation between them. The higher enhancement layers are not good references either. First, the bit rate of the higher enhancement layers is too high for most applications. The gain by using a high-quality reference will appear only at a very high bit rate. Second, the small magnitude errors encoded in these layers may be just produced by noise. Therefore, the correlation of higher enhancement layers between adjacent pictures are weak too. Only by using the middle enhancement layers as references can we achieve maximum coding efficiency improvement, since the DCT coefficients in the middle enhancement layers show strong correlations between adjacent pictures.

A simplified PFGS framework with only two picture buffers could offer a good trade-off between coding efficiency versus the extra memory cost and computational complexity. In the simplified PFGS framework, the first picture buffer is used to save the reconstructed base layer in a previous picture as a reference for the base layer and the lower quality enhancement layers in a predicted picture. The second picture buffer is used to save a reconstructed enhancement layer in a previous picture as a reference for the higher quality enhancement layers. An exemplary framework is shown in Figure 4.5. The base layer and the first two enhancement layers of picture 2 are predicted from the base layer of picture 1, and other higher quality enhancement layers of picture 2 are predicted from the third enhancement layer of picture 1. Instead of using a fixed enhancement layer, the reference alternates between two different enhancement layers to form a complete prediction path from the base layer to the highest enhancement layer for error recovery and channel adaptation. Which enhancement layer is used as the high-quality reference depends on the contents of the video sequence and the bit rate of the base layer. How to choose such an enhancement layer is an optimization issue in the encoding process. The index of the enhancement layer selected as the reference can be encoded as part of the video bitstream.

Since the framework in Figure 4.5 uses two reference layers for the prediction, it produces two sets of predicted DCT coefficients: (1) a first set of predicted DCT coefficients are prediction errors formed by referencing the base layer, which is a low-quality reference layer and (2) a second set of predicted DCT coefficients are prediction errors formed by referencing a higher quality enhancement layer. The first set of predicted DCT coefficients are encoded in the base layer and lower enhancement layers, and the difference between the second set of predicted DCT coefficients and those reconstructed from the base layer and low enhancement layers is encoded to form higher enhancement layers.

4.4 Improvements to the PFGS Framework

4.4.1 Potential Coding Inefficiency Due to Two References

The goal of using the second high-quality reference is to reduce the bit rate of the higher enhancement layers. As shown in Figure 4.5, the second set of predicted DCT coefficients will have a statistically lower magnitude compared with the first set of predicted DCT coefficients because its reference is higher quality and hence closer to the original image. Theoretically, we also expect that the differences between the

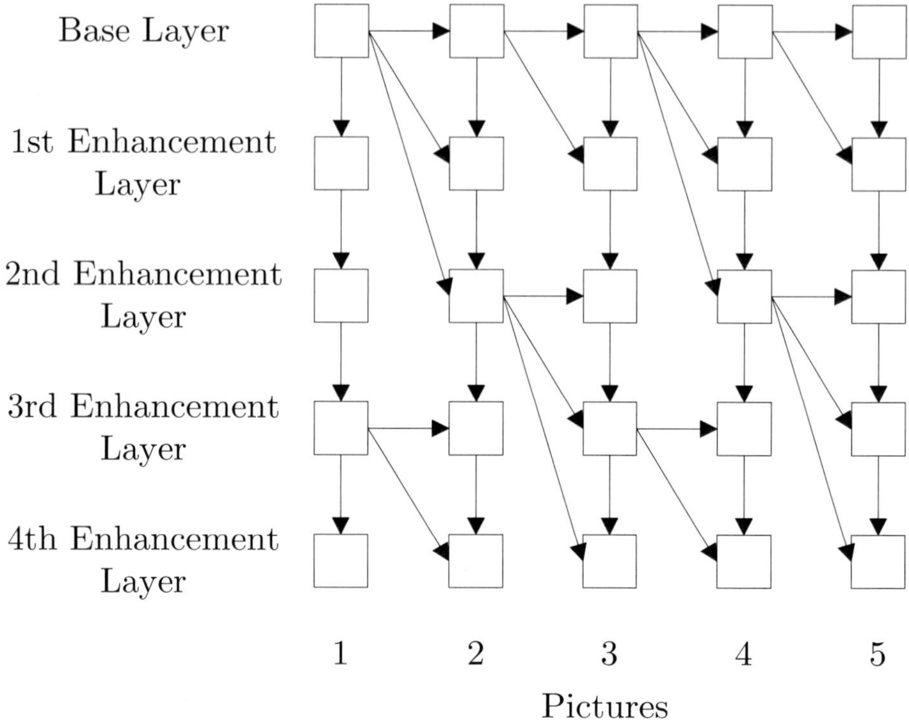

Figure 4.5 Simplified PFGS framework with only two buffers.

second set of predicted DCT coefficients and those reconstructed from the base layer and lower enhancement layers are smaller than the residues of the first set of predicted DCT coefficients after the base layer and lower enhancement layer encoding. Lower DCT differences translate into fewer coding layers, therefore resulting in better coding efficiency. However, the expectation is only valid statistically.

The displaced frame difference (DFD) image is defined as the difference between the original image and the motion predicted image. With the linear DCT transforms, the DCT coefficients of the DFD image are equal to the difference between the DCT coefficients of the original image and that of the predicted image. There are two kinds of differential operations in PFGS coding schemes for improving coding efficiency. The first kind of differential operations are essentially the normal motion compensation operations that are used to generate the DFD in order to reduce temporal redundancy. These subtractive processes are denoted with horizontal solid-line arrows between adjacent frames in Figure 4.5. The second kind of differential operations are performed on the predicted DCT coefficients within one picture after switching from a lower quality reference to a higher quality one to reduce DCT coefficient redundancy. Normally, the results of the second kind of differential operations are called residues. For convenience, we define the following terminology for referencing these differences:

- *LQPi*—Low-quality predicted image that is generated by motion compensation from the lower quality reference
- *LQPd*—Difference between LQPi and the original image
- *LQPD*—DCT coefficients of LQPd
- *LQBR*—DCT residues that are produced by subtracting the already encoded (quantized and then de-quantized) DCT coefficients in the base layer from LQPD coefficients
- *LQPR*—DCT residues that are produced by subtracting the already encoded DCT coefficients in the previous layers from the LQPD coefficients
- *HQPi*—A high-quality predicted image that is generated by motion compensation from the higher quality reference
- *HQPd*—Differences between HQPi and the original image
- *HQPD*—DCT coefficients of HQPd
- *HQPR*—DCT residues that are produced by subtracting the already encoded DCT coefficients in the previous layers from the HQPD coefficients
- *HQBR*—DCT residues that are produced by subtracting the already encoded DCT coefficients in the base layer from the HQPD coefficients

Specifically, Figure 4.6 illustrates the relationships between the above terminology. It is obvious that the HQPi will produce lower DFD DCT coefficients compared with the LQPi because the reference is of a higher quality. However, the dynamic range of the HQPD coefficients is not necessarily always less than that of the LQPD coefficients. For some instances, the magnitude of individual HQPD coefficients may actually increase compared with that of the LQPD coefficients due to non-ideal motion compensation. Moreover, in order to reduce the redundancy between LQPD coefficients and HQPD coefficients and to further improve coding efficiency, normally only the difference between the HQPD and the reconstructed low-layer LQPD; that is, the residue HQPR is coded in higher enhancement layers. Although doing so generally will reduce the energy of the coefficients to be coded, the dynamic range of the difference actually would increase (causing more fluctuation) and an additional sign map would be required to code the different HQPR.

Both the undesired fluctuation in magnitude and the additional sign bit are particularly inefficient for bit-plane coding. First, if the undesired fluctuation and increase in magnitude exceed the range represented by residual bit planes, it would seriously affect the coding efficiency of bit-plane coding. For example, assume that three bit planes are used to encode the LQPR in higher enhancement layers when referencing a low-quality layer. However, after switching to a higher quality reference, if the absolute value of an individual HQPR coefficient exceeds 7, one or multiple additional bit planes have to be inserted between lower and higher enhancement layers to represent the excess range. Second, the sign of HQPR may be different from that of LQBR. Since the sign bit of every coefficient is encoded after the MSB of that coefficient, for those coefficients whose most significant bits (MSBs) have been encoded in the base layer and lower enhancement layers, new signs have to be encoded again.

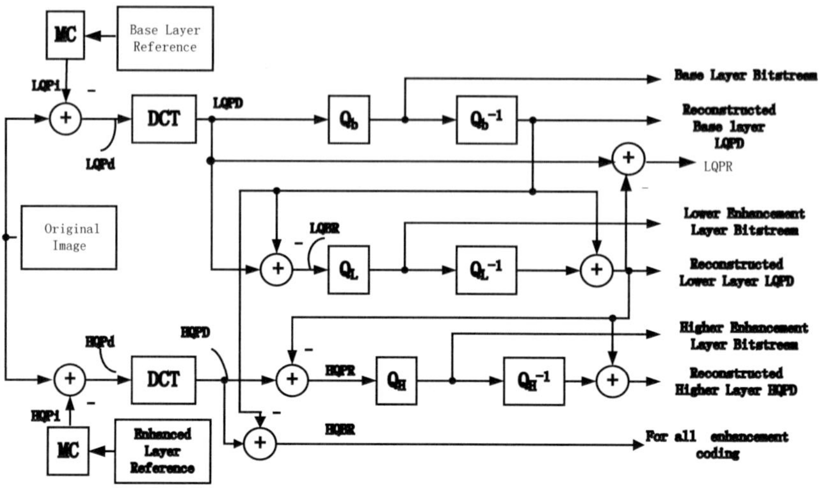

Figure 4.6 Illustration of the relationships of all the differences in the PFGS framework.

4.4.2 A More Efficient PFGS Framework

Ideally, in order to avoid excess fluctuation and increase in magnitude of the prediction residues mentioned above, the bit planes encoded in lower and higher enhancement layers should be from one set of prediction residues. In addition, the extra sign layer is also completely avoided by doing so. In the baseline FGS case, the bit planes of all enhancement layers are indeed always from the same set of residue LQBRs. However, since it always uses base layer video as references, the overall coding efficiency is limited, especially for higher enhancement layers. On the other hand, in the PFGS framework shown in Figure 4.5, only the lower enhancement layers include bit planes from the LQBR, while the higher enhancement layers encode bit planes from HQPR as shown in Figure 4.6, causing exactly the same problem.

A conditional replenishment method described by Tan et al. [53] can be used to eliminate this kind of fluctuation when switching from a lower quality reference to a higher quality one. In the conditional replenishment scheme, not all the LQPR coefficients are replaced by the HQPR coefficients. The lower layer prediction coefficients LQPR are conditionally replaced by the higher layer prediction coefficients HQPR depending on the values of the reconstructed lower layer LQPD. If the reconstructed lower layer coefficient LQPD is zero then the corresponding coefficient in LQPR is replaced with that in HQPR. If the reconstructed coefficient is not zero, then no replacement is done. Though the conditional replenishment can solve the fluctuation problem, it partially loses the advantages of using a high-quality reference. Nonzeros in the reconstructed lower layer coefficients LQPD are an essential condition to cause fluctuation, but not a sufficient one. In many cases, even though there are nonzeros in the reconstructed lower layer coefficients, the residues can still be reduced by replacing the prediction coefficients LQPR with HQPR. On the other hand, the conditional replenishment method essentially needs a new reference mixed with the lower quality reference and the higher quality reference in DCT domain. For the decoder, since only two predicted image LQPi and HQPi are available, two extra DCT transforms will be needed in order to get the corresponding LQPD and HQPD, thus the computational complexity of a decoder will be increased. Therefore, a more efficient and simple approach that can take full advantage of a high-quality reference without causing any fluctuation should be investigated to further improve the coding efficiency.

We propose solving this problem within the PFGS framework. The improved PFGS framework based on a new structure is given in Figure 4.7. In this improved framework, while the base layer encoding is still the same as that in the baseline FGS and PFGS framework in Figure 4.5, all enhancement layers encode differences between the HQPD and the de-quantized LQPD from the base layer, that is, HQBR. Note that now the lower enhancement layers contain the first few most significant bits in the HQBR coefficients instead of that of the LQBR coefficients as in the baseline FGS and the PFGS framework in Figure 4.5. Since all enhancement layers always encode the same set of coefficient HQBRs in the improved PFGS framework, there is neither sign change nor fluctuation among any enhancement layers.

The lower enhancement layers, which also encode the differences between HQPD and reconstructed base layer LQPD, would introduce several problems. It seems that this would cause drifting errors if the HQPi reference were not available at the decoder. Moreover, this seems to destroy the channel adaptation and error recovery properties with FGS/PFGS. Indeed, when the HQPi reference is not available at the decoder, we have to use the LQPi reference instead; this introduces some errors in lower enhancement layers due to the different references used in the encoder and the decoder. In fact, the overall quality loss in the lower enhancement layers is very small, because a better prediction compensates most of the loss caused by using different references.

How to minimize the quality loss in the lower enhancement layers is not important. It is much more important to find a way to prevent these errors in one picture from propagating to other pictures. Fortunately, the framework shown in Figure 4.7 suggests a scheme to solve the potential error-drifting problem completely. The key here is to make sure that the encoder and decoder have the same reconstructed references for any future picture prediction although the reconstructed reference may not have the best quality it could get if reconstructed using a high-quality reference. We will show this through an example. As in FGS, we still assume there is no error in the base layer. In the decoder end, if there are packet losses or errors in the third enhancement layer in picture 1, which is used in the encoder end to get the HQBR coefficients, all the enhancement layers in picture 2 will have to use the base layer in picture 1 as a reference. Of course, there would be some quality loss by doing so. However, as long as in both the encoder end and the decoder end the reconstruction of the second enhancement layer of picture 2 always uses the base layer of picture

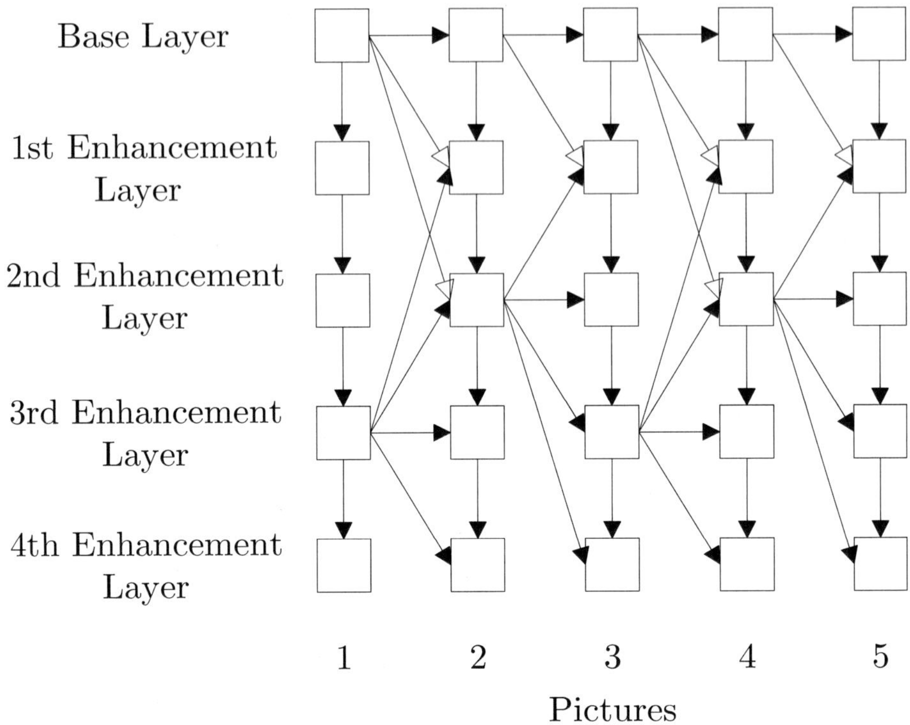

Figure 4.7 Improved PFGS framework. Solid arrows are for prediction references and hollow arrows with solid lines for reconstruction references, hollow arrows with dashed lines are for reconstruction of lower layers when the previous enhancement reference layer is not available.

1 as the reference, then the errors in the reconstruction of the second enhancement layer could not further propagate to any pictures followed. The unique feature of this improved framework is that in a prediction picture, the reference used for prediction could be different from the reference used for reconstruction. This feature prevents error drifting and preserves all bandwidth adaptation and error recovery features of PFGS. Moreover, it brings more coding efficiency gain in the higher enhancement layers. In Figure 4.7, for example, we notice that the second enhancement layer in picture 2 can always be reconstructed by referencing the base layer of picture 1 to prevent the error drifting into future pictures. However, if the third enhancement layer of picture 1 is available, a better quality second enhancement layer of picture 2 can still be reconstructed with it for display purpose only. The reconstruction of a display image can be different from that of reference image is yet another feature of the improved framework.

The architecture of encoding *B* pictures using the improved framework is shown in Figure 4.8. The bidirectional motion estimation determines the type of motion compensation and motion vectors by referencing the original picture of a previous intra *I* or prediction *P* picture and the original picture of the next *I* or *P* picture. The

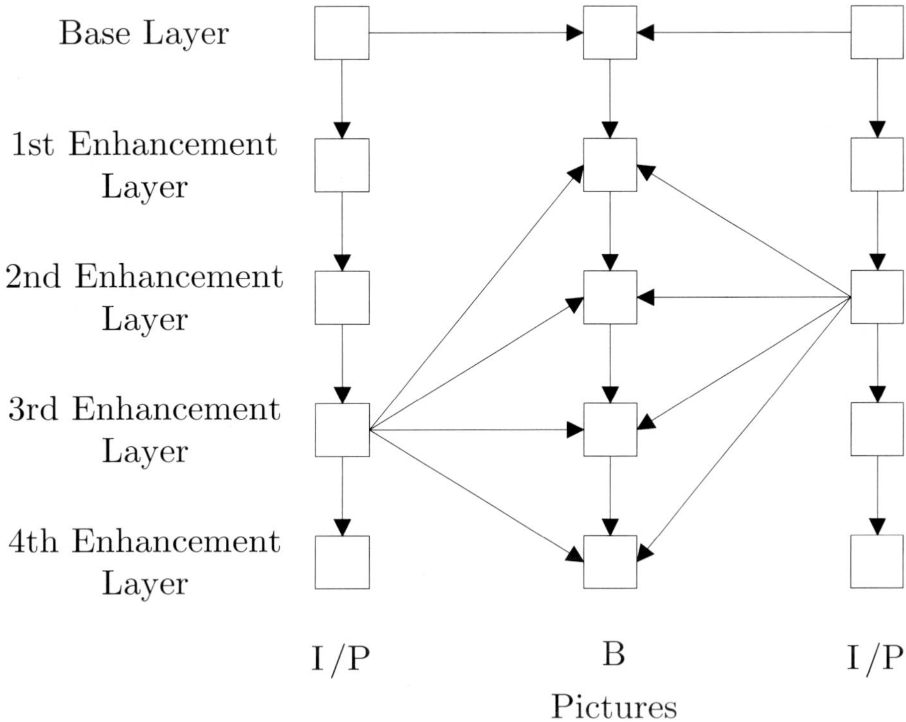

Figure 4.8 *B* pictures encoding in the improved PFGS framework.

LQPD coefficients of the *B* picture are formed by referencing the base layer in a forward reference picture and/or the base layer in a backward reference picture. At the same time, the HQPD coefficients of the *B* picture are formed by an enhancement layer in a forward reference picture and/or the enhancement reference in a backward reference picture. Similar to *P* pictures, all the enhancement layers in a *B* picture are encoded with the difference between HQPD and de-quantized LQPD from the base layer. Because none of the enhancement layers in a *B* picture is used as a reference for other pictures, the errors in lower enhancement layers of pictures have no effect to any other pictures. However, when there are enhancement reference layers available, they can be certainly used to produce better quality *B* pictures.

4.5 Implementation of the PFGS Encoder and Decoder

In the previous section, we discussed an improved PFGS framework. How to implement the encoder and decoder based on the improved PFGS framework will be the focus of this section. First, an encoder with two reference buffers for video prediction is given in Figure 4.9. Picture Buffer 0 is used to save the reconstructed base layer in a previous picture as a reference for the base layer coding. Picture Buffer 1 is used

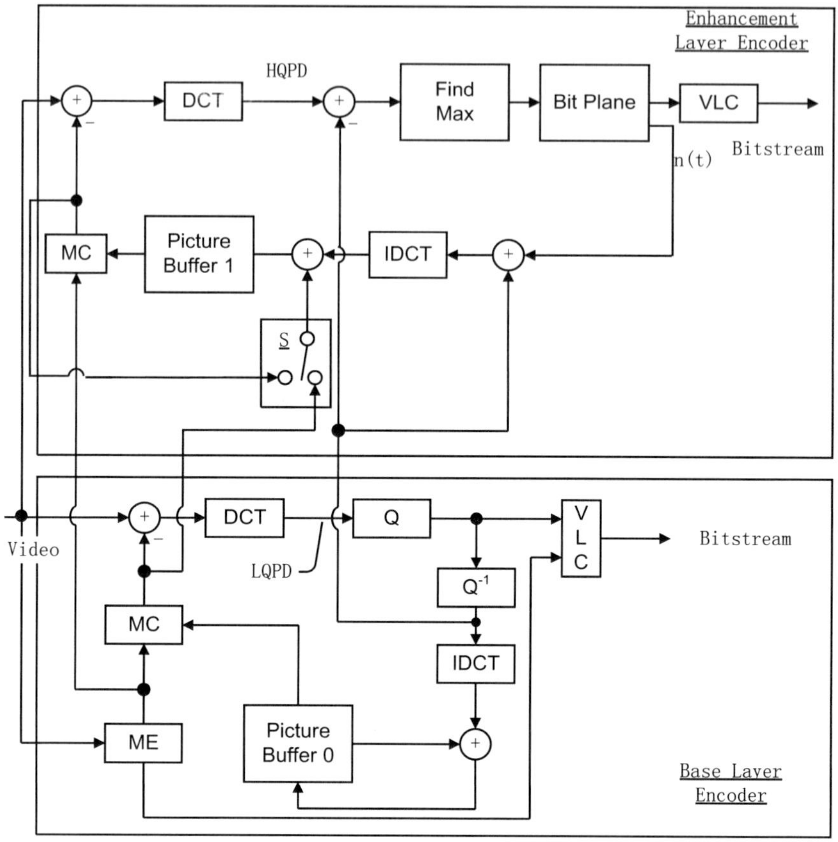

Figure 4.9 Encoding diagram of the improved PFGS framework.

to save a reconstructed enhancement layer in a previous picture as a reference for coding all enhancement layers.

The base layer encoding is the same as that of baseline FGS, which can be compatible with other standards, such as MPEG-2, MPEG-4, and H.263. The motion estimation (ME) module gets the motion vectors between two adjacent original pictures and outputs its results to two motion compensators (MCs). The first motion compensator predicts the picture by referencing the reconstructed base layer in Picture Buffer 0. The second motion compensator predicts the picture by referencing a reconstructed enhancement layer in Picture Buffer 1. After the first motion compensation and DCT transform of the DFD image, we obtain LQPD coefficients in the base layer encoder as shown in Figure 4.9. The LQPD coefficients are quantized by scalar quantization and compressed by VLC into the base layer bitstream. Generally, the step size of the scalar quantizer is large in order to generate a relatively short bitstream.

On the other hand, the second motion compensator and the second DCT module generate the HQPD, which are the DCT transforms of the DFD image with a high-quality reference. The differences (residues) between HQPD and the reconstructed LQPD from the base layer are encoded in all the enhancement layers. The bit plane coding technique is used in the enhancement layers to provide an embedded bitstream and fine granularity scalability. For the improved PFGS framework, the difference now is that all the enhancement layers always use an enhancement layer in the previous picture as a reference as opposed in the baseline FGS case that all enhancement layers always use the base layer in the previous picture as a reference. The maximum absolute value of the differences determines the maximum number of bit planes in a picture. The 64 absolute values in a 8×8 difference block are arranged in a ZigZag order into an array. A bit plane is defined as an array of 64 bits, which are taken from each of the 64 absolute values at the same significant bit position. This implies that the quantization steps of the enhancement layers are a series of factor 2^i, where i is from the maximum number of bit planes to 0. For the lowest enhancement layer, i equals the maximum number of bit planes and for the highest enhancement layer, i equals 0. For each bit plane of each block, (RUN, EOP) symbols are formed and encoded using variable length codes to produce the enhancement bitstream for that bit plane. EOP stands for the "End Of Plane." The sign bit of each difference coefficient is encoded with one bit following the MSB of that coefficient as the coefficient's MSB gets encoded. The binary "0" denotes a positive difference and binary "1" denotes a negative one.

The enhancement layer in the current picture used for the prediction of enhancement layers in the next picture is reconstructed using the first $n(t)$ bit planes in the bitstream and the reference of either the base layer or the enhancement reference layer in the previous picture, which is controlled by the switch $\underline{S}$. If the enhancement reference layer for the next picture is a higher layer than the enhancement reference layer for the current picture, that is, $n(t) > n(t-1)$, then the enhancement reference layer in the previous picture will be used to reconstruct the enhancement reference layer for the next picture. Otherwise, the base layer in the previous reference picture will be used to reconstruct the enhancement reference layer for the next picture.

Figure 4.10 gives a diagram of a PFGS video decoder with two reference buffers. The two-buffer configuration offers a good trade-off between coding efficiency and extra cost in memory and computational complexity. The first picture buffer is used to save the reconstructed base layer in a previous picture as the reference for the base layer. It can also be used as the reconstruction reference for some lower quality enhancement layers to generate a display image when the higher quality reference in a previous picture is not available due to errors or packet losses. A second picture buffer is used to save the reconstructed enhancement layer in a previous picture as the prediction reference for all enhancement layers. The decoder is very similar to a baseline FGS decoder but with additional modules to reconstruct and save a middle enhancement layer as a second reference. The switch $\underline{S}$ in the decoder is used to control which buffer to be used to form the next enhancement layer reference. Compared with FGS, an additional buffer, an additional MC module, and an additional IDCT are needed in the improved PFGS framework. In Figure 4.10, $n(t)$ denotes the

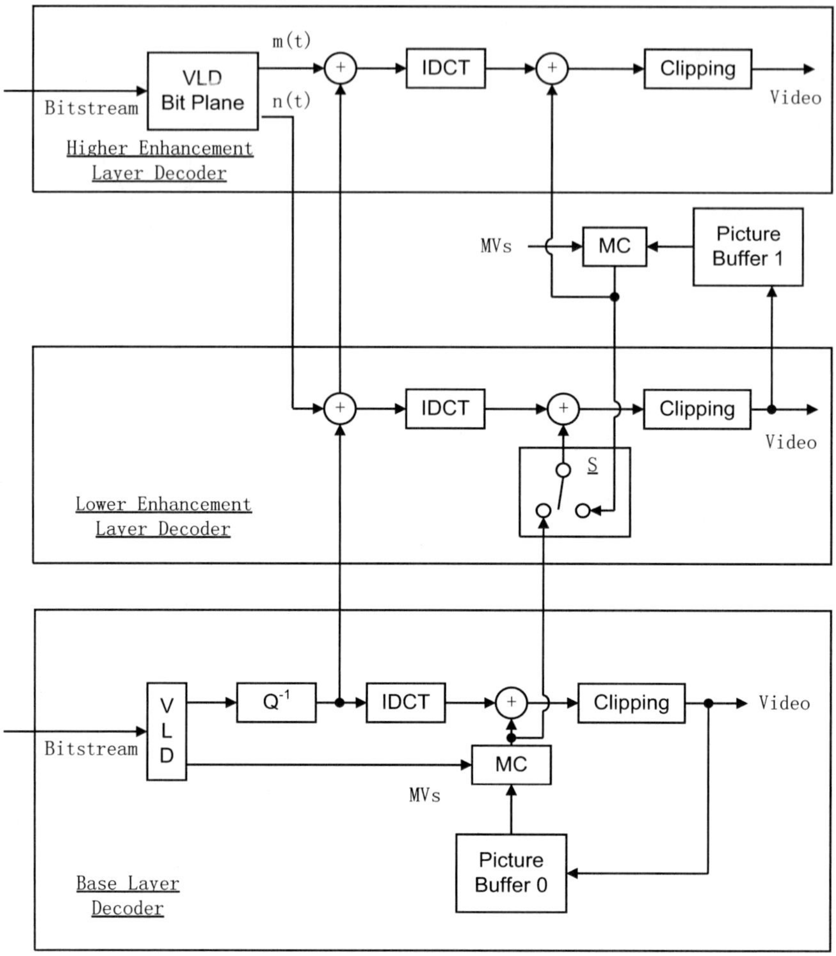

Figure 4.10 Decoding diagram of the improved PFGS framework.

number of bit-plane layers needed to reconstruct the next enhancement reference layer. $m(t)$ denotes the number of bit-plane layers used to reconstruct the current display image.

4.6 Experimental Results and Analyses

Extensive experiments and simulations have been performed to test the performance of the PFGS framework. First, a simple simulation experiment is designed to demonstrate that with only one extra picture buffer the simplified PFGS framework can still provide most of the coding efficiency gain obtained in the original PFGS framework with multiple buffers. The PFGS framework with multiple references is shown in Figure 4.4 and the simplified PFGS framework with only two references is shown in Figure 4.5. These two frameworks are almost the same except for the number of references. At the same time, this experiment also gives the FGS results at the same testing conditions.

All test conditions are the same as those specified in the MPEG-4 core experiments. The sequences Akiyo, Foreman, and Coastguard (CIF format) are used in this experiment. Only the first picture is encoded as an I picture and all the other pictures are encoded as *P* pictures. The base layer encoder is a predictive encoder that includes motion compensation and DCT transformation modules, and it could be compatible to other standards, such as H.263, MPEG-2, or MPEG-4. In this experiment, we use MPEG-4 baseline video coder for the base layer coding. A simple half-pixel motion estimation scheme using linear interpolation is implemented to extract the motion vectors between video pictures. The range of the motion vector is set to ± 31.5 pixels. The same motion vectors are applied to all MCs, which in turn produce the predicted images.

The bit rate of the base layer is 128 kbits/s with TM5 rate control, and the encoding frame rate is 10 Hz. The bit rate of the enhancement layers is not constrained. Since the enhancement layers produce an embedded bitstream, the streaming server can truncate it at any place to match the channel bandwidth. The truncating procedure can be independent of a decoder. In our simulation experiments, the enhancement layer bitstream is truncated at 64 kbits/s, 128 kbits/s, $\cdots$, until 384 kbits/s, with an interval of 64 kbits/s. In order to reduce fluctuation when switching between references with different quality, the conditional replenishment scheme is used in the two PFGS frameworks.

We define the least significant bit plane as index "0" and the next least significant bit plane as index "1," and so on. For the PFGS framework with multiple picture buffers, even pictures are predicted from the even layers of the previous picture, and the odd pictures are predicted from the odd layers of the previous picture. This alternating structure repeats throughout the encoding of the whole video sequence. For the simplified PFGS framework with only two picture buffers, we choose the most efficient enhancement layers as references through experiments. For the Akiyo sequence, the second reference alternates between the second bit plane and the third bit plane. For the Foreman sequence, the second reference alternates between the third bit plane and the fourth bit plane. For the Coastguard sequence, the second reference alternates between the fourth bit plane and the fifth bit plane.

Table 4.1 gives some experimental results of coding these video sequences using the methods described above. For the Akiyo and Coastguard sequences, the luminance PSNR in the PFGS framework with two picture buffers is almost the same as with multiple picture buffers. For the Foreman sequence, the luminance PSNR in PFGS framework with two picture buffers is about 0.15 dB less than with multiple picture buffers. It is clearly shown that by only using one additional middle enhancement layer as the reference, we can achieve almost the same coding efficiency as

Table 4.1 PSNR versus bit-rate comparison between FGS, PFGSM, and PFGST for the Akiyo, Coastguard, and Foreman sequences. PFGSM denotes the PFGS framework with multiple buffers and PFGST denotes the PFGS framework with two buffers.

Bit Rate (kbit/s)	Akiyo			Coastguard			Foreman		
	FGS	PFGSM	PFGST	FGS	PFGSM	PFGST	FGS	PFGSM	PFGST
192	42.10	42.12	42.10	27.75	27.86	27.84	31.18	31.32	31.30
256	42.75	42.77	42.76	28.52	28.72	28.70	31.89	32.14	32.12
320	43.13	43.21	43.20	29.27	29.54	29.52	32.51	32.86	32.83
384	43.47	43.60	43.59	29.79	30.16	30.14	33.15	33.60	33.54
448	43.84	44.03	44.02	30.19	30.62	30.60	33.70	34.27	34.15
512	44.27	44.50	44.48	30.62	31.10	31.08	34.17	34.76	34.61

using multiple references. On the other hand, the conditional replenishment approach used in the PFGS architecture with multiple references discards most parts of the higher quality references, since HQPR coefficients do not replace LQPR coefficients when the MSBs of the DCT coefficients have been encoded in previous enhancement layers. Meanwhile, from the results presented in Table 4.1, we can see that the two PFGS frameworks can only achieve up to 0.5 dB PSNR gain on average. The main reason is that the conditional replenishment scheme is not efficient although it can eliminate the fluctuation.

Extensive simulations have been conducted to test the performance of the improved PFGS framework. There are two picture buffers in the improved PFGS framework. One is used to save the reconstructed base layer, and another is used to save the reconstructed enhancement reference layer. The base layer is always predicted from the first buffer and all enhancement layers are predicted from the second buffer. The results of the improved PFGS framework are compared with that of FGS. In addition, the results of the nonscalable video-coding scheme are also presented to show the cost of implementing the fine granularity functionality. The nonscalable video scheme is exactly the same as the base layer of the FGS scheme, where TM5 controls the output bit rate.

In this experiment, each picture is followed by 59 predicted pictures including *P* pictures and *B* pictures. There are three *B* pictures between every two *P* pictures. The range of the motion vector is set to ± 31.5 pixels. The same motion vectors are applied to the two motion compensators, which in turn produce the predicted images. The bit rate of the base layer is 256 kbits/s with TM5 rate control, and the encoding frame rate is 30 Hz. The enhancement bitstreams are truncated at 128 kbits/s, 256 kbits/s, $\cdots$, until 2048 kbits/s, with an interval of 128 kbits/s. In the PFGS scheme, to get a constant video quality, the enhancement layer bits cannot be equally allocated to each picture. The reason is very simple. For instance, in Figure 4.7, picture 2 is predicted from the third enhancement layer of picture 1. Since the reference is of higher quality and is closer to the original image, a shorter bitstream will be generated due to the smaller prediction errors. However, picture 3 is predicted from the second enhancement layer with lower quality. Larger prediction errors would result in a longer bitstream in picture 3. Therefore, to get more or less the same quality across each picture, bits spent for encoding odd frames should be more than that for encoding even frames. On the other hand, if allocating a fixed bit rate for all pictures, for example, at a bit rate where the third enhancement layer in an even picture is partially decoded, the previous odd picture may be only decoded up to the second enhancement layer due to the fact that its bitstream is longer. In turn, the quality of even pictures may be limited since its higher layer reference (third enhancement layer) in odd pictures is not available. This will not only cause the fluctuation of PSNR in adjacent pictures, but also affect the overall coding efficiency of the video sequence. Fortunately, the PFGS scheme provides an embedded and fully scalable bitstream, so that the rate control does not have to occur at the encoding time. A simple rate allocation or truncation module can be used in the streaming server to obtain the optimized quality given a bit-rate constraint.

Some experimental results of the improved PFGS framework are given in Figure 4.11 for the Akiyo, Coastguard, and Foreman sequences, respectively. As we expected, generally, there is not too much performance gain at lower bit rates compared with FGS. Sometimes there is even quality loss since the reconstruction reference in the decoder may be different from that in the encoder for the lowest few enhancement layers. For the Akiyo sequence, there is only some small quality loss at lower bit rates. This is because in lower enhancement layers, we have to use LQPi instead of HQPi to reconstruct the display image since the HQPi is not available yet at lower bit rates. For the Coastguard and Foreman sequences, we cannot see the quality loss, because a better prediction compensates most of the loss caused by using different references. However, as the bit rate increases, we can see consistently significant coding efficiency gain. For the Coastguard sequence, the gain in Y component can be up to 1 dB. For the Foreman sequence, the gain in Y component can be up to 1.3 dB. The reason why the Akiyo sequence does not show significant gain is that the base layer bit rate is already very high for this sequence. The quality is already very good at that bit rate.

The experimental results also show that the coding efficiency gap between the FGS scheme and the nonscalable video coding exceeds 3.0 dB. Even though the PFGS scheme has significantly improved the coding efficiency of FGS, the gap between the PFGS scheme and the nonscalable video coding is still large. How to further improve the coding efficiency of the PFGS scheme is still an open question.

4.7 Simulation of Streaming PFGS Video over Wireless Channels

For most applications over the Internet, the fundamental technical problem is that the network bandwidth varies widely in range from one user to another (user variations) and from time to time (temporal variations). The major advantage of the PFGS scheme is its bandwidth adaptation capability that provides a good solution for bandwidth fluctuation. With the rapid development of wireless communications, the wireless channel becomes an increasingly popular and convenient means to access the Internet. As we know, wireless channels have different characteristics than wired Internet channels. They are typically noisy and suffer from a number of

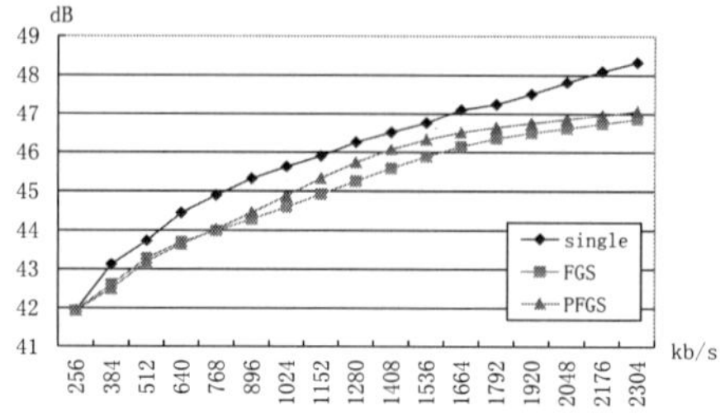

(a)

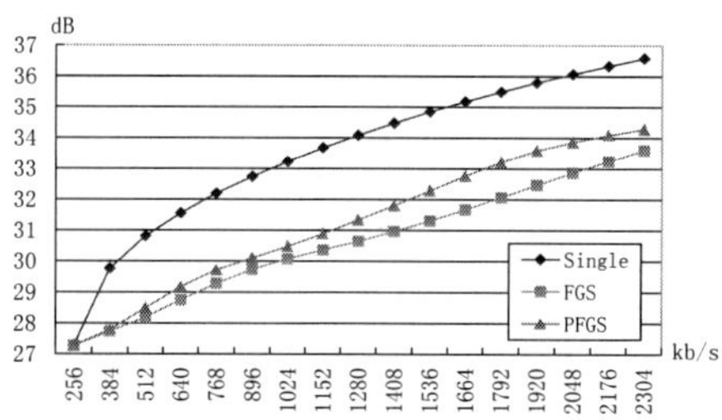

(b)

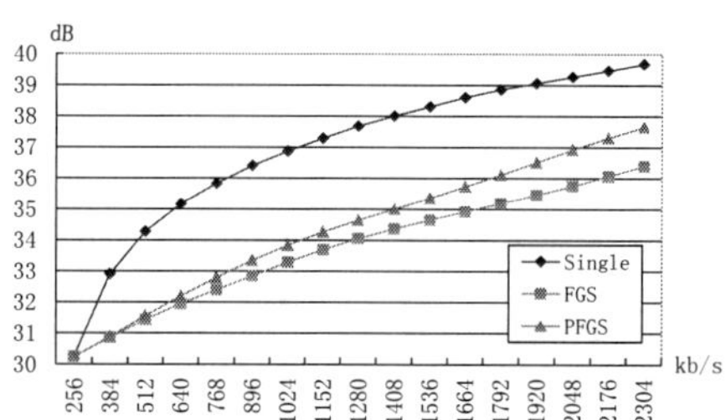

(c)

Figure 4.11 PSNR versus bit-rate comparison between FGS and PFGS for the *Y* component, (a) Akiyo, (b) Coastguard, (c) Foreman.

channel degradations, such as random bit errors and burst errors due to fading and multiple path reflections. When compressed video data is sent over these channels, the effect of channel errors on the compressed video bitstream can be severe. As a result, the video decoder that is decoding the corrupted video bitstream often loses synchronization. Moreover, predictive coding techniques such as motion compensation used in various video compression standards make the situation even worse. The decoders based on these techniques would quickly propagate the effects of channel errors across the video sequence and rapidly degrade video quality. To deal with this problem, some error resilience and concealment methods are introduced by Wang and Zhu [54] and Talluri [55].

For applications involving transmitting video over error-prone channels, such as wireless channels, the PFGS framework provides a more robust solution. PFGS provides an inherent error recovery feature that can gracefully recover from any enhancement layer errors. The PFGS framework also provides a layered bitstream structure with different importance at different layers. In this layered bitstream structure, the most important information can be sent separately and with increased error protection compared to the less important enhancement information. There are basically two bitstreams in the PFGS framework: the base layer bitstream and the enhancement layer bitstream. The base layer bitstream is very sensitive to channel errors. Any random errors or burst errors may cause the decoder to lose synchronization, and the decoded errors will propagate to the start of the next GOP. However, the enhancement layers can tolerate channel errors. When there are errors in the enhancement layer bitstream, a decoder can simply drop the rest of the enhancement bitstream of the given picture and search for the next synchronization marker. There should be neither obvious visual artifacts nor error propagation due to the error recovery feature of the PFGS scheme. Generally, since the bit rate of the base layer is very low, channel coding with the same number of bits can provide better protection to the base layer than to the overall bitstream. If the channel is a time-varying channel, as in many wireless communication situations, it is very feasible to adaptively adjust the rate allocation between the source and channel coding operations seamlessly with the PFGS scheme. Therefore, the quality of base layer can be maintained at a stable level.

In this section, we simulate the delivery of two kinds of video bitstreams over wireless channels. The first kind of bitstream is a MPEG-4 single-layer bitstream. Some error-resilience tools are applied from the source coding viewpoint, such as Resynchronization Marker, Data Partitioning, and header extension code (HEC). The second kind of bitstream is the PFGS bitstream. The same error-resilience tools are applied to the base layer, and the enhancement layers are encoded without any error protection and concealment. In the PFGS coding scheme, both the overall source bit rate and bit allocation between source and channel coding can be dynamically adjusted depending on the channel feedback. However, it is very difficult for the single-layer case to adapt to bandwidth fluctuation for bit re-allocation. Therefore, the channel bandwidth and bit allocation between source and channel coding are fixed in our simulation.

We use a two-state Markov model to simulate the wireless channel as in Figure 4.12 [56]. This model can characterize the error sequences generated by data transmission channels. In the good state, G errors occur with low probability α, while in the bad state B they occur with high probability β. The errors occur in clusters or bursts with relatively long error-free intervals (gaps) between them. The state

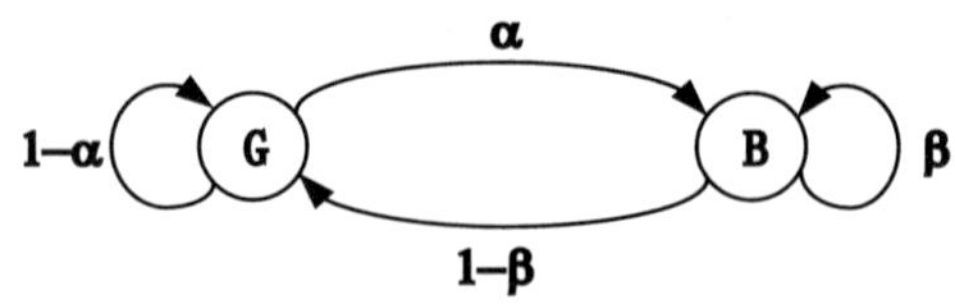

Figure 4.12 Gilbert model for simulating a wireless channel.

transitions are shown in Figure 4.12 and summarized by its transition probability matrix

$$M = \begin{bmatrix} 1-\alpha & \alpha \\ 1-\beta & \beta \end{bmatrix}$$

The model can be used to generate sequences of symbol error. In this case, it is common to set $\alpha \approx 0$ and $\beta \approx 0.5$. However, in situations where a ReedSolomon (RS) code over $GF(2^m)$ is to be used, it is more appropriate for the model to generate m-bit symbol errors. If $\beta = 0$, this model will simulate a random error, where α is the bit error rate.

The QCIF News sequence is used in our simulation, and the encoding frame rate is 10 Hz. Only the first picture is encoded as an I picture, and other pictures are encoded as P pictures. The total channel bandwidth is 64 kbits/s and 8 kbits/s of it is used for channel coding for each bitstream. For the MPEG-4 single-layer bitstream, the bit rate of source coding is 56 kbits/s. For the PFGS bitstream, 32 kbits/s is used for base layer source coding. The enhancement layer bitstream can be truncated to 24 kbits/s to fit in the channel bandwidth. It is obvious that the PFGS bitstream can be decoded from 32 kbits/s to 56 kbits/s. The channel parameter α randomly varies from 0.04 to 0.1, and parameter β also randomly varies from 0.3 to 0.6 during simulation. The channel coding is implemented with a general RS code. An RS (64, 56, 8) code is applied to the MPEG-4 single-layer bitstream (a block of 64 bytes with 56 bytes of data and 8 bytes of protection). And an RS (40, 32, 8) code is applied to the PFGS base layer bitstream (a block of 40 bytes with 32 bytes of data and 8 bytes of protection). Both decoders have included some basic error-concealment tools such as copying from motion-compensated areas in the previous picture.

The experimental results are shown in Figure 4.13. After going through the simulated wireless channel, it is clear that there are too many visual artifacts in the decoded images from the received MPEG-4 single-layer bitstream which contains errors now, while the results using the improved PFGS framework still provide very good image quality. The reason is that the PFGS framework provides a robust enhancement layer bitstream. Since only the sensitive base layer bitstream needs the protection of channel coding, the 8 kbits/s channel coding can provide stronger protection for the base layer bitstream. For the MPEG-4 single-layer bitstream, the whole bitstream is very sensitive to channel errors. The 8 kbits/s channel coding has to provide protection for every bit in that bitstream. Therefore, the channel coding

Figure 4.13 Simulation results of delivering different video bitstreams over wireless channels. The first four images are decoded from the received MPEG-4 single-layer bitstream. The latter four images are decoded from the received PFGS bitstream.

protection becomes relatively weak and more channel errors could not be corrected under the same channel condition.

4.8 Summary

In this chapter, a highly efficient scalable video-coding framework is first presented. Compared with the FGS scheme in MPEG-4, the PFGS framework tries to use some higher quality references to improve the coding efficiency, since higher quality references make motion predictions more accurate. In order to keep the scalability properties, such as fine granularity scalability, bandwidth adaptation, and error recovery, the basic PFGS framework preserves a complete prediction path from the lowest layer to the highest layer. However, multiple extra picture buffers are needed to save the reconstructed reference layers in the basic PFGS framework, which increases the memory cost and computational complexity. We simplify the basic PFGS framework from multiple extra picture buffers to one extra picture buffer, while still keeping almost the same coding efficiency. Moreover, an improved PFGS framework is presented to further improve coding efficiency by effectively eliminating the fluctuation at enhancement layers when switching references. While the original PFGS framework provides about 0.5 dB coding efficiency gain over the FGS in MPEG-4, the improved PFGS framework provides about 1 dB coding efficiency gain. The coding efficiency gap between scalable video coding and the nonscalable video coding is closing.

The advantages of the PFGS scheme, such as layered coding and error recovery, are further demonstrated through a wireless transmission simulation. The simulation results clearly show that under the same conditions in an error-prone channel, the PFGS scheme can produce much better decoded images than the single-layer nonscalable MPEG-4 scheme.

In the future, more studies need to be done on how to further improve the coding efficiency and robustness of the PFGS framework, and on how to optimally transport it through the Internet and wireless channels. The coding efficiency gap between the nonscalable video coding and the PFGS video coding is still big although it is closing. How can we design an optimized PFGS encoder with improved coding efficiency? How can we optimally allocate (truncate) bits among different frames under an overall bit-rate constraint? Sometimes, the base layer bitstream may still be too long for certain applications, so how shall we add spatial scalability in the PFGS framework? How can real-time video communication issues be addressed? What are the best network protocols for streaming the PFGS video over the Internet? These are all open questions that need to be answered.

Some simple error-detection and resynchronization tools should be added to the PFGS enhancement-layer bitstream to improve its robustness and efficiency. In the present enhancement-layer bitstream, once an error is detected, the rest of the bitstream in the current picture is simply dropped. If there are some resynchronization markers in the enhancement-layer bitstream, we can just discard the part of the bitstream between two resynchronization markers and continue decoding the rest of the bitstream to minimize the error effects. On the other hand, how to allocate bit rate between source and channel coding dynamically according to channel conditions is another topic that needs further consideration.

Chapter 5
Motion Threading for 3D Wavelet Coding

5.1 Introduction

Wavelet transform [57] provides a multi-scale representation of image and video signals in the space-frequency domain. Aside from energy compaction and decorrelation properties that facilitate efficient compression of natural images, a major advantage of wavelet representation is its inherent scalability. It endows compressed video and image streams with flexibility and scalability in adapting to heterogeneous and dynamic networks, diverse client devices, and the like. Furthermore, recent progresses in literature have shown that 3D wavelet-based video coding schemes are able to compete in performance with the conventional hybrid standard approaches (e.g., H.264/AVC [13]).

In 3D wavelet-based video coding, wavelet transforms are applied temporally across frames, and horizontally and vertically within each frame, respectively. The correlations among frames are exploited by a temporal wavelet transform operated on original frames instead of motion compensation from previously reconstructed frames. Here we will focus on the scenarios in which the temporal transform exists prior to 2D spatial transform, namely, the T+2D case. Due to the object and/or camera motion in a scene, the same point on a moving object can be located at different pixel positions in consecutive frames. To take full advantage of temporal correlation and the ability to achieve high coding efficiency, the temporal transform should be performed along motion trajectories. Due to complications in combining wavelet transform and motion alignment, the efficiency of a temporal transform can become a bottleneck in high performance 3D wavelet-based video coding schemes.

Some global and local motion models are proposed for motion alignment in temporal wavelet transform. Taubman and Zakhor [58] predistort the video sequence before the temporal transform by translating pictures relative to one another, while Wang et al. [59] use a mosaic technique to deform each video frame into a common coordinate system. Both schemes assume a global motion model, which may be inadequate for video sequences with local motion. To overcome this limitation, Ohm [60] proposes a block-matching technique that is similar to the one used in standard video coding schemes, while paying special attention to covered/uncovered and connected/unconnected regions. But it fails to achieve perfect reconstruction with motion alignment with respect to subpixel accuracy. Other related works also adopt similar motion models but focus on different aspects, such as tri-zerotree [61], rate allocation [62], and set partitioning in hierarchical trees (SPIHTs) [63].

Since 2001, several groups have looked into combining motion alignment with the lifting-structure of wavelet transform. These techniques are generally known as motion compensated temporal filtering (MCTF). One noteworthy work is by Pesquet-Popescu and Bottreau [64], in which the authors implement the first subpixel (1/2-pel) motion alignment with perfect reconstruction in the motion-compensated lifting framework. They are the first to incorporate overlapped-block motion alignment into the lifting-based temporal transform. However, only the Haar filters are used by Pesquet-Popescu and Bottreau [64]. Luo et al. first employ the biorthogonal 5/3 wavelet filters with 1/2-pel motion alignment [65]. Secker and Taubman use both the Haar and 5/3 filters again with 1/2-pel motion alignment [66]. Several follow-up works, for example, Chen and Woods [67] and Flierl and Girod [68], also demonstrate the advantage of the 5/3 filters for temporal transform. There are a number of other related publications with different focuses (e.g., using 1/4-pel motion alignment, 9/7 or longer filters, adaptive or low-complexity implementation, scalable motion coding, etc.) [69–73].

At that time, Moving Picture Experts Group (MPEG) was actively exploring and promoting inter-frame wavelet coding techniques. The Motion Compensated Embedded Zero Block Coding (MC-EZBC) coder [67, 69] proposed by Chen and Woods has become well known because of its good performance. In MC-EZBC, each pair of frames is motion estimated in a hierarchical fashion to achieve up to 1/8-pel accuracy before going through a motion-aligned lifting-based Haar transform. Additional wavelet decompositions along the temporal direction are performed on the low-pass frames by following the same procedure. Using sub-pel motion alignment, MC-EZBC performs about the same as the H.264 standard for some sequences. In the original MC-EZBC coder [69], the Haar filters do not fully exploit the long-term correlation across video frames. In addition, motion vectors at each temporal decomposition layer are estimated and coded independently without exploiting cross-layer correlations among them. In an improved version of MC-EZBC [67], the 5/3 filters are used in conjunction with improved motion vector coding.

5.2 Motion Threading

Xu et al. [74] propose a motion threading (MT) approach that employs longer wavelet filters to exploit the long-term correlation across frames along the motion trajectory. The baseline architecture of the original 3D MT wavelet coder is shown in Figure 5.1, where each column represents a frame. Backward motion estimation is performed on each pair of frames from $Frame_0$ to $Frame_1$ at the macroblock level. Pixels along the same motion trajectory are linked to form a nonoverlapping

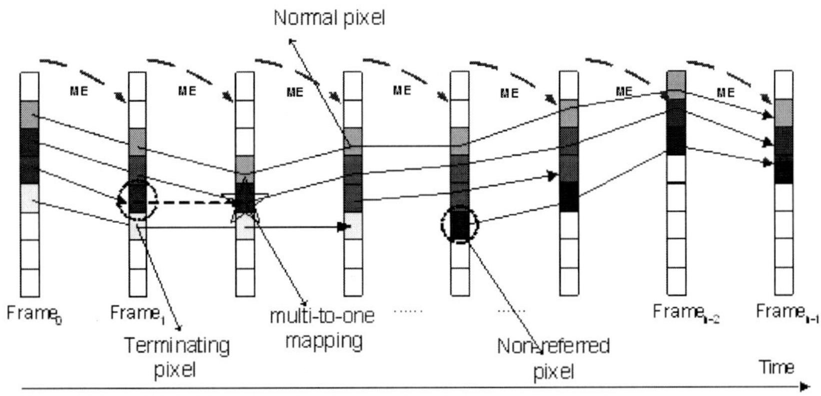

Figure 5.1 The original MT implementation by Xu et al. [74].

thread according to the motion vectors of the macroblocks they belong to. The bi-orthogonal 9-7 or 5-3 wavelet filters are used for transforming the threads before each resulting frame goes through a 2D spatial wavelet transform with the 9-7 filters. After the 3D wavelet transform, all wavelet coefficients are coded into one embedded bitstream.

Motion threading offers a flexible yet efficient way of incorporating motion alignment into the temporal-domain wavelet transform, with encouraging results reported by Xu et al. [74]. However, MT still has limitations. On the one hand, motion threads are not allowed to overlap so as to guarantee perfect reconstruction at the decoder side. If multiple pixels in the current frame are mapped to the same pixel in the next frame, that is, whenever a scenario of many-to-one mapping occurs, only one of them in the current frame can be linked to the one in the next frame. Each of the remaining pixels has to be marked as a terminating pixel to signify that its associated thread ends at the current frame. On the other hand, for pixels in the current frame that are not linked by those in the previous frame, that is, when there are non-referred pixels, we have to start a new motion thread for each one. When the input sequence has complex motion, the number of many-to-one mappings or non-referred pixels (or thread boundaries) will be large. Due to the boundary effects in wavelet reconstruction [75], these artificial thread boundaries will degrade the coding performance.

Motion threading is only performed at full-pel resolution by Xu et al. [74]. It is well known that motion estimation at 1/2- or 1/4-pel resolution can significantly improve coding efficiency in standard video coding. We thus aim to extend MT to sub-pel resolution here (as was done by Luo et al. [65]) while maintaining perfect reconstruction for improved 3D wavelet video coding. We first adopt a lifting-based wavelet transform for the motion threads. We carefully manage cases with many-to-one pixel mapping and non-referred pixels in MT in order to reduce the number of thread boundary pixels (hence the boundary effects). We propose techniques for 1/2- or 1/4-pel motion estimation and motion alignment under the perfect reconstruction constraint [76]. We also generate one set of motion vectors for each temporal layer of wavelet coefficients for temporal scalability. Since motion vectors in adjacent regions within a temporal layer and the same region in different layers have strong correlations, we devise a correlated motion estimation (CME) scheme to facilitate motion vector coding [77]. Eight modes are designed to capture the variation of local motion correlation at the macroblock level and a rate-distortion (R-D) optimized algorithm is employed to select the best mode for each macroblock.

Compared to Secker and Taubman [78], CME by Luo et al. [77] shares the same idea of exploiting the correlation among cross-layer motion vectors. However, the two approaches are different: CME reduces the number of motion vectors by defining eight modes, whereas the scheme by Luo et al. [77] aims to improve prediction accuracy in motion vector coding. Thus, although our CME technique is designed for MT, the idea equally applies to other 3D wavelet coding schemes, such as MC-EZBC [67] and those by Secker and Taubman [78, 79].

5.3 Advanced Motion Threading

The advanced MT technique assumes a lifting-based wavelet transform and it aims to reduce the number of artificially terminating/emerging threads in the original implementation of MT. The accuracy of motion alignment is increased to sub-pel level while maintaining perfect reconstruction.

5.3.1 Lifting-Based Motion Threading

The lifting structure is an efficient implementation of the wavelet transform with low memory and computational complexity [80, 81]. Every finite impulse response (FIR) wavelet filter can be factored into a few lifting stages [80]. As an example, the lifting steps of a one-level biorthogonal 5-3 wavelet transform are shown in Figure 5.2. The input signal $x_0, x_1, \cdots, x_6$ starts from Figure 5.2a, and the wavelet coefficients L_0, L_1, L_2 and H_0, H_1, H_2 come out at Figure 5.2b. From Figure 5.2, we have the lifting steps as:

$$\begin{cases} H_i = x_{2i+1} + a \times (x_{2i} + x_{2i+2}) \\ L_i = x_{2i} + b \times (H_{i-1} + H_i) \end{cases}, \tag{5.1}$$

with $a = -1/2, b = 1/4$. This equivalently implements the convolution kernel of 5-3 wavelet transform (up to scaling).

In the lifting-based implementation, each lifting step only updates half the nodes, and the original value of the updated nodes will not be needed in subsequent steps. The elementary lifting step is circled in Figure 5.2, which only involves three nodes. Thus the updated values can be saved in the same memory that holds the original

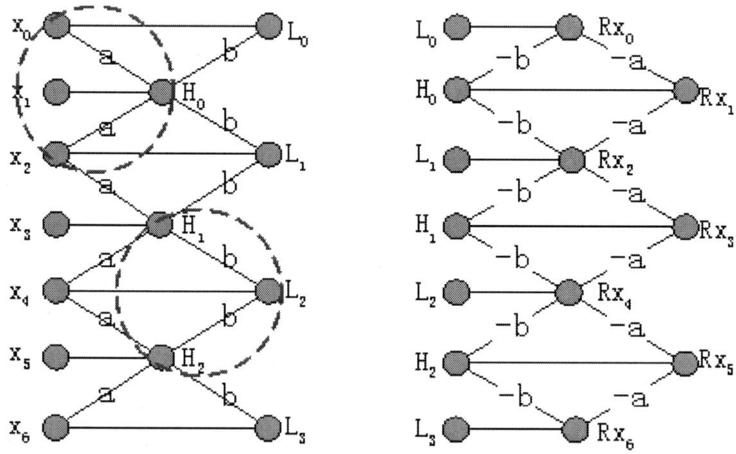

(a) (b)

Figure 5.2 Lifting-based wavelet implementation of the 5-3 biorthogonal wavelet transform. (a) forward transform with lifting steps shown in circles and (b) the inverse transform.

values with in-place computation. For the wavelet transform, low-pass and high-pass coefficients are interleaved. Each lifting step can be straightforwardly inversed with an inverse lifting unit. Thus the inverse transform structure is easy to obtain, as shown in Figure 5.2b.

We implement the wavelet transform of motion threads using lifting. Video frames go through the lifting process step by step. We first update the odd frames to become high-pass frames and then the even frames to become low-pass frames. Figure 5.3 shows the lifting-based 5-3 temporal wavelet structure, where each column is a frame, and each block represents a pixel. Macroblock-based motion estimation and alignment are always performed from an odd frame to the forward and backward adjacent even frames. In each lifting step, either the pixels in odd frames or those in even frames are lifted along the motion threads.

Therefore, different from Eq. (5.1), the lifting steps with taking motion threads into account can be formulated by

$$\begin{cases} H_i(x) = F_{2i+1}(x) + a \times \left(F_{2i}\left(MT_{2i+1,2i}(x)\right) + F_{2i+2}\left(MT_{2i+1,2i+2}(x)\right)\right) \\ L_i(x) = F_{2i}(x) + b \times \left(H_{i-1}\left(MT^{-1}_{2i-1,2i}(x)\right) + H_i\left(MT^{-1}_{2i+1,2i}(x)\right)\right) \end{cases}, \tag{5.2}$$

where $a = -1/2$, $b = 1/4$. F_{2i+1} is the pixel of the input frame. $H_i()$ and $L_i()$ are the generated high-pass and low-pass wavelet coefficients. x is the position of the pixel or wavelet coefficient.

The first equation in Eq. (5.2) describes the prediction step. $MT_{2i+1,2i}()$ and $MT_{2i+1,2i+2}()$ represent the motion alignment from the frame $2i+1$ to the frame $2i$ and from the frame $2i+1$ to the frame $2i+2$, respectively. In this step, motion estimation is first performed from an odd frame to the forward and backward adjacent even frames. The estimated motion vector is used to form the function $MT_{2i+1,2i}()$

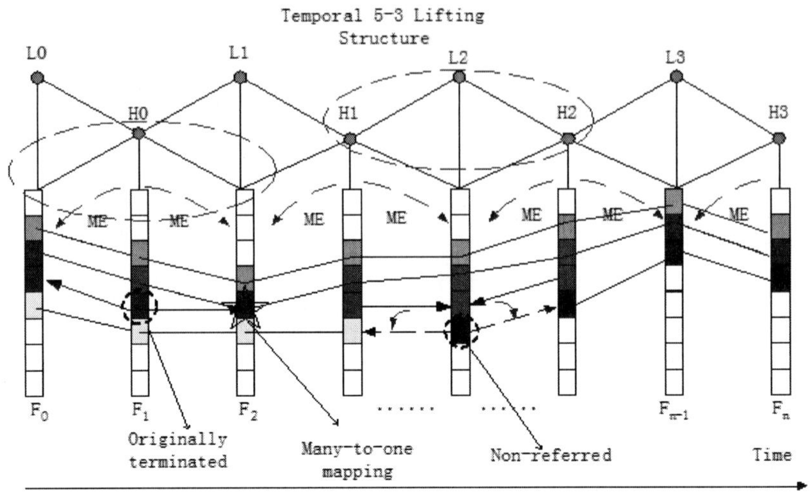

Figure 5.3 Lifting-based MT with bidirectional motion search.

and $MT_{2i+1,2i+2}()$ as follows:

$$\begin{cases} MT_{2i+1,2i}(x) = x + MV_{2i+1,2i}(x) \\ MT_{2i+1,2i+2}(x) = x + MV_{2i+1,2i+2}(x) \end{cases}, \tag{5.3}$$

where $MV_{2i+1,2i}(x)$ and $MV_{2i+1,2i+2}(x)$ are the motion vectors of the pixel x in the frame $2i+1$. In the proposed scheme, $MV_{2i+1,2i}(x)$ and $MV_{2i+1,2i+2}(x)$ are allowed to point out a noninteger pixel. $F_{2i}(MV_{2i+1,2i}(x))$ and $F_{2i+2}(MV_{2i+1,2i+2}(x))$ in the fractional-pixel position are calculated with Sinc interpolation [82].

The second equation in Eq. (5.2) describes the update step. Obviously, the update step needs the motion alignment $MT^{-1}_{2i-1,2i}()$ and $MT^{-1}_{2i+1,2i}()$ from an even frame to the forward and backward adjacent odd frames. In most 3D wavelet coding schemes, motion alignment of the update step is derived from the motion vectors of the prediction stage so as to save bits for coding motion vectors. We use $MT^{-1}_{2i+1,2i}()$ as a example to describe the motion alignment in the update step as follows:

$$MT^{-1}_{2i+1,2i}(x + \lfloor MV_{2i+1,2i}(x) \rfloor) = x + \lfloor MV_{2i+1,2i}(x) \rfloor - MV_{2i+1,2i}(x). \tag{5.4}$$

Since $MV_{2i+1,2i}(x)$ has the precision of fractional pixel, the operator $\lfloor\ \rfloor$ rounds every element of a vector to the nearest integer. In the fractional-pixel position $H_{i-1}(MT^{-1}_{2i-1,2i}(x))$ and $H_i(MT^{-1}_{2i+1,2i}(x))$ are calculated with Sinc interpolation.

Figure 5.4 assists in explaining Eq. (5.4). Black circles represent pixels at full-pel resolution, gray ones at 1/2-pel resolution, and white ones at 1/4-pel resolution. Solid curves with solid arrows represent motion vectors generated from motion

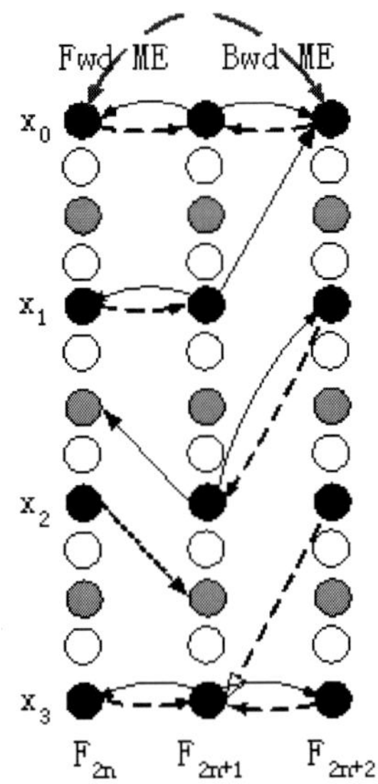

Figure 5.4 Motion alignment with 1/4-pel resolution.

estimation, and dashed curves with solid arrows represent motion vectors which are direct inverses of the solid curves. It is easy to understand Eq. (5.4) if $MV_{2i+1,i}(x)$ is an integer motion vector. For example, the motion vector of pixel x_2 in F_{2n+1} points to the pixel x_1 in F_{2n+2} in the prediction step. As a result, the pixel x_1 in F_{2n+2} is aligned to the high-pass coefficient of pixel x_2 in F_{2n+1} in the update step. In other words, $MT^{-1}_{2i+1,2i}()$ is the inverse function of $MT_{2i+1,2i}$ in this case. In another case, if the motion vector points to a fractional-pixel position, the corresponding pixel in the update step is first found with $x + \lfloor MV_{2i+1,2i}(x) \rfloor$ and its motion vector is the inverse version of $MV_{2i+1,2i}(x)$. The motion alignment of pixel x_2 from F_{2n+1} to F_{2n} in Figure 5.4 gives such an example.

Since the length of each wavelet filter is short, as long as needed frames have entered the input buffer, related lifting steps described in Eq. (5.2) can be executed without waiting for more frames to come in. Thus we implement a temporal extension of the line-based wavelet transform scheme used in JPEG 2000 [83] with minimum buffer size during encoding and decoding (see Hu et al. [75]).

5.3.2 Many-to-One Mapping and Non-Referred Pixels

Due to object movement, camera movement, and scene changes, there are usually different types of motion in real-world video. This would result in many-to-one mappings and non-referred pixels in MT. With lifting, we can carry on temporal filtering on pixels that are originally terminated by Xu et al. [74] due to many-to-one mappings. As shown in Figure 5.3, during each basic lifting step, the originally terminated pixel in $Frame_1$ can be updated using both its left (forward) and right (backward) connected pixels instead of being stopped on the right. When the corresponding pixel in $Frame_2$ is to be lifted, although several pixels in $Frame_1$ are mapped to it, only the first mapped one of them is used. For a non-referred pixel in an even frame, which originally indicates the boundary of a motion thread, it will be linked on both sides using the motion vectors from adjacent motion threads as shown by the dashed curve with a hollow arrow in Figure 5.4. This way, each pixel is guaranteed to be linked in both forward and backward directions. With these rules, Eq. (5.4) can still be used to derive the motion alignment of many-to-one mapping and non-referred pixels in the update step.

Although we can always link each pixel in the current frame with a pair of pixels in the forward and the backward frames, sometimes this arrangement is counterproductive. For example, when camera motion, occlusion, or scene change occurs, many pixels in the current frame might have good matches in the previous or next frame. Owing to the weak correlation among these pixels, linking them into motion threads will actually hurt the coding efficiency and result in ghost artifacts after the wavelet transform.

Since motion vectors are searched at the macroblock level in our implementation, mismatched pixels are also processed at the macroblock level. We mark a macroblock as a terminating one when it contains many terminating pixels. The terminating MB indicates the boundaries of the relative motion threads. The forward sum of absolution difference (SAD), backward SAD, and the average SAD values of the MB are used in mode decision. Accordingly, besides the normal Bid mode in which a pixel is bi-directionally linked, two additional MB modes, Fwd and Bwd, are used to represent the terminating status at the backward and forward directions, respectively.

For each MB, deciding the best mode among Bid, Fwd, and Bwd is an optimization problem. As we have discussed, the reconstruction error at the motion thread boundaries is larger than within the threads, hence the boundary effects. To avoid these boundary effects due to artificial thread terminations, we favor Bid in mode decision. Since terminating MBs only have unidirectional motion vectors, choosing the Fwd or Bwd mode saves the cost of motion vector coding. Details on mode selection are given later.

5.4 Multi-Layer Motion Threading

In our 3D wavelet video coder, temporal domain wavelet transform is applied before the spatial-domain wavelet transform. A 4-level dyadic wavelet transform structure along the temporal axis (or motion threads) is shown in Figure 5.5. In our original MT implementation [74], motion vectors are estimated and transmitted only at the highest layer. When performing the dyadic wavelet transform, we used different levels of

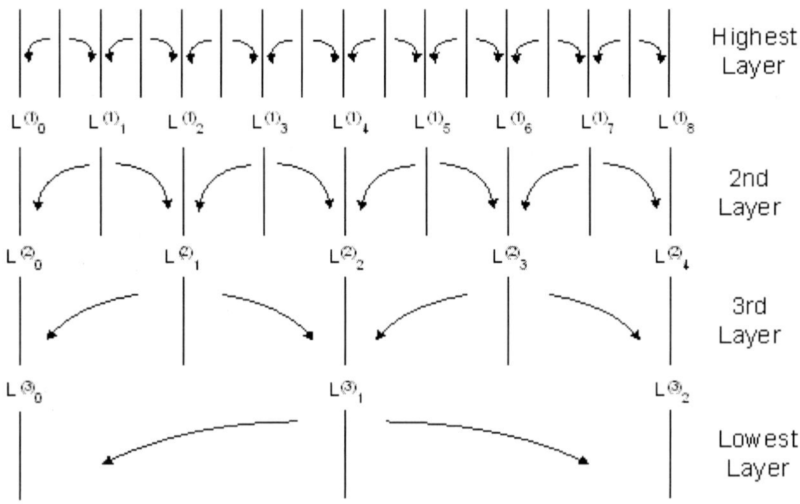

Figure 5.5 A 4-level dyadic wavelet transform structure along the temporal axis.

aggregation of the motion vectors generated at the highest layer. However, by simply aggregating motion vectors generated from the highest layer, motion alignment in lower frame rate layers may be inaccurate. In addition, in the temporal scalability mode when the frame rate is less than the full rate, motion vectors at the highest layer have to be fully transmitted to generate motion vectors at lower layers. This incurs too much overhead in motion vector coding.

To rectify this shortcoming by Xu et al. [74], we propose an advanced MT to estimate and transmit motion vectors at each temporal decomposition layer [76]. At each layer, motion estimation is performed on the original frames (after appropriate downsampling) instead of the wavelet transformed low-pass frames. With multiple sets of motion vectors, it is imperative to exploit correlation among them for efficient coding. We first examine the correlation of motion vectors between two adjacent frames. The bidirectional motion estimations are always from odd frames to two neighboring even frames as shown in Figure 5.6. When the video frames undergo a constant speed motion, an assumption can be made that if a block at a position (x,y) in $Frame_{2n+1}$ has a matching block at the position $(x+dx,y+dy)$ in $Frame_{2n}$; then using a uniform temporal motion extrapolation, the same block must move to position $(x-dx,y-dy)$ in $Frame_{2n+2}$. This assumption is also used in B-frames in traditional video coding. When encoding the motion vectors, the forward and backward motion vectors have the same absolute value but with opposite signs, thus only one directional motion vector needs to be coded. This type of correlation is represented by two direct inverse (FwdDir/BwdDir) MB modes in motion estimation.

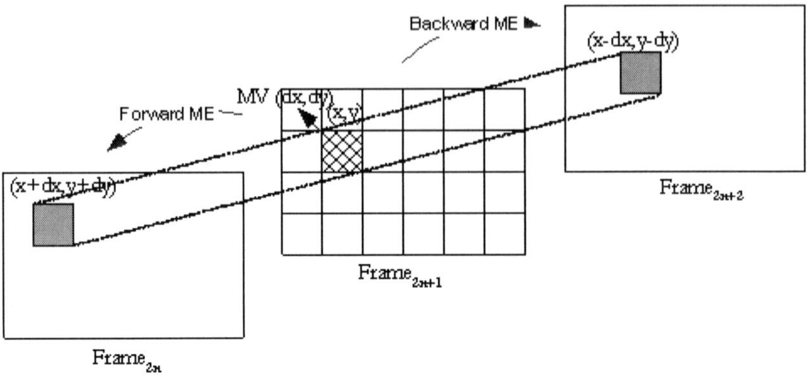

Figure 5.6 Motion correlation between two adjacent frames.

There are several different kinds of orders to perform motion vector estimation and coding of multiple layers. One is that both motion estimation and coding are performed from a high layer to a low layer, which first estimates the motion vectors at the highest layer using traditional block motion estimation. Then the lower layer motion vectors are predicted and encoded based on the higher layer. Similar to the temporal scalability problem we mentioned regarding the original MT, although this method ensures the motion accuracy, the higher layer motion vectors always needs to be transmitted to generate the lower layer one even at low frame-rate case. Hence the high-to-low method is not efficient for low bit-rate applications.

Another is that both motion estimation and coding are performed from a low layer to a high layer. Since lower layer motion vectors do not depend on higher layer ones, the latter do not have to be transmitted in down-sampled frame-rate applications. The disadvantage of this method lies in the precision of the motion vectors due to the long frame interval at lower level. For a 4-layer temporal decomposition, the interval is 8 frames. Furthermore, the prediction of the higher layer motion vectors may be affected as well. For achieving the temporal scalability, we adopt this order to utilize inter-layer correlation.

When the order of motion estimation is different from that of motion vector coding, another solution is that motion estimation is performed from a high layer to a low layer to obtain more accurate motion vectors and motion vectors are encoded from a low layer to a high layer to maintain temporal scalability. This solution should be further investigated in the future to achieve a better trade-off between the precision of motion vectors and the temporal scalability.

Figure 5.7 illustrates the inter-layer correlation of motion vectors between two adjacent temporal layers. The inter-layer correlative modes assume that the sum of

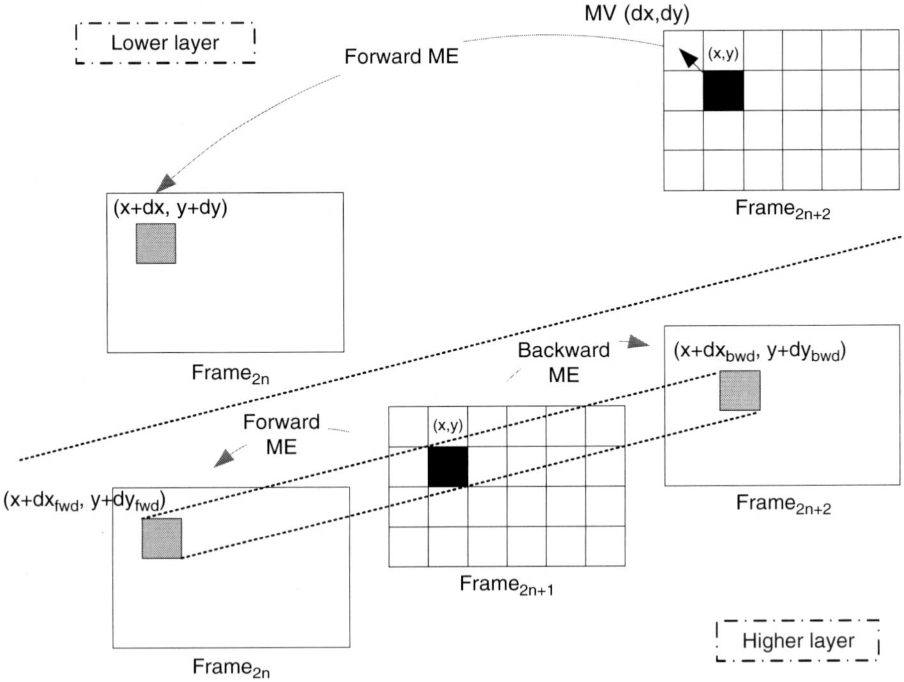

Figure 5.7 Correlation of motion vectors between two adjacent temporal layers.

the absolute value of the forward and the backward motion vectors are equal to the motion vector of the same MB location in the previous layer. Therefore, one motion vector can be saved within each forward-backward motion vector pair. An extreme case of this assumption is that the forward and the backward motion vectors are equal, while no motion vector in the higher layer needs to be transmitted. Accordingly, three modes are designed to represent these correlation types.

5.5 Correlated Motion Estimation with R-D Optimization

In the previous sections, we have discussed the advanced motion-threading scheme and motion vector correlations in multi-layer motion threading. In this section, eight block modes are designed to represent different motion and correlation types. An R-D optimized CME scheme is proposed to select from the designed modes based on a compound cost function [77]. The cost function considers not only the matching correctness but also the motion vector bits.

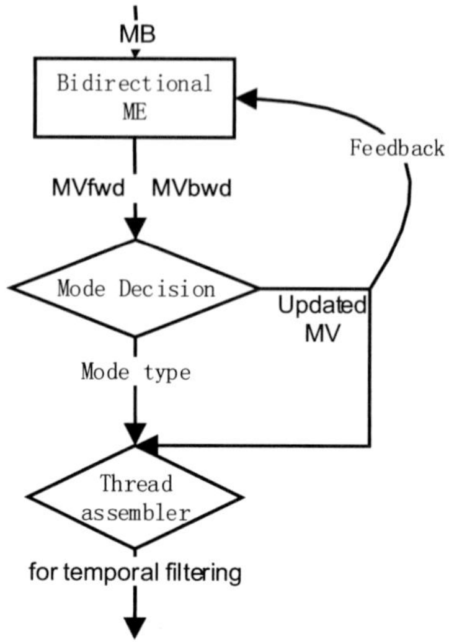

Figure 5.8 Flowchart of the correlated motion estimation.

5.5.1 Definition of the Mode Types

Figure 5.8 shows the flowchart of the proposed inter-layer motion estimation. Each MB in the odd frames is first bidirectionally motion searched using a motion search technique similar to H.264. Here, only the 16×16 MB mode is used. A pair of independent forward and backward MVs is then generated. After that, according to previous analysis, we define eight MB modes to represent the diverse local motion properties. The mode selection is based on an R-D optimized criterion. The forward or backward motion vectors are updated according to the chosen mode. The next MB within the same frame may use the updated motion vectors to initiate the center value in the motion search. Then motion threads are assembled according to the chosen MB mode.

As shown in Figure 5.9, each column denotes one frame, and each block can be considered a macroblock. The current frame F_{2n+1} is divided into MBs. A dark gray MB represents an anchor block referred by the current MB. With camera zooming, the mapping block of the dark gray MB may not exist, as shown by the light gray blocks in the Fwd and Bwd mode in Figure 5.9. Each mode is defined as follows:

DirL: An inter-layer correlated bidirectional mode. The forward and backward motion vectors use half the absolute value of the inter-layer predicting MV, with according inverse signs. No motion bits need to be transmitted in this mode. Some relatively smooth motion types, such as background motion, may belong to this mode.

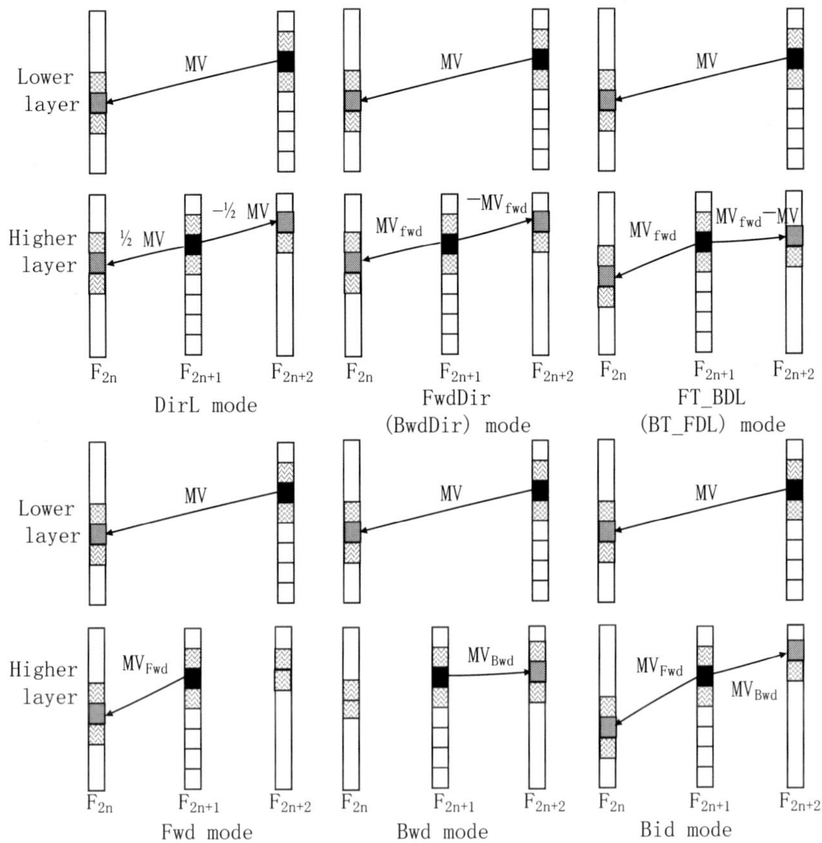

Figure 5.9 Eight motion modes in CME.

FT_BDL (BT_FDL): Two inter-layer correlated bidirectional modes. The forward (backward) MV is transmitted, while the relative backward (forward) MV is calculated as: MVbwd = MVfwd $\pm$ MVprevlayer (MVfwd = MVbwd $\mp$ MVprevlayer). The selection of + or - operator depends on the backward or forward direction of the MVprevlayer. These two modes imply that the absolute values of the bidirectional motion vectors are unequal. Only one motion vector needs to be sent. Between FT_BDL and BT_FDL modes, the one with less motion cost is chosen.

FwdDir (BwdDir): Two intra-layer correlated bidirectional mode, where the forward and backward MVs have the same absolute value and inverse signs: $MVbwd = -MVfwd$ or ($MVfwd = -MVbwd$). Only forward (backward) MV needs to be transmitted.

Fwd (Bwd): Two single-directional modes. Only the forward (backward) MV is transmitted, and the other direction of the thread is terminated. When some particular motion such as occlusion occurs, completely matched pixels may only exist either in the forward or in the backward frame. Therefore, the motion threads should be terminated in the mismatched direction.

Bid: The normal bidirectional macroblock mode. Both the forward and the backward MVs are transmitted.

The single direction prediction modes, such as Fwd and Bwd modes, will influence the prediction and update steps given in Eq. (5.2). When these two modes are selected, some pixels only have the prediction from one side. In the same way, the corresponding pixels derived from Eq. (5.4) may be updated from one side as well. They are treated as boundary pixels. The filtering parameters a and b in Eq. (5.2) are multiplied with two similar to symmetric boundary extensions.

5.5.2 R-D Optimized Mode Decision

After a pair of independent forward and backward MVs are generated, from the designed eight MB modes, the one with the smallest cost is selected based on an R-D optimized criterion. A mode symbol is attached to identify the selected mode. The mode selection criterion is defined as: $Cost = \eta SAD + \lambda Bits_{motion}$, which consists of two terms:

SAD is the sum of the absolute difference between the current macroblock and the motion compensated matching blocks. The SAD value evaluates the matching correctness of the forward and backward motion vectors. For the bidirectionally linked modes, such as Bid and Direct, the SAD term is the absolute difference between the original MB and the average of the forward and backward matching blocks. For the Fwd and Bwd modes, the SAD term takes only the one directional SAD value because the other direction is terminated. Pixels belonging to an MB with Fwd or Bwd modes are the boundaries of the relative motion threads. As we have mentioned, the wavelet synthesis error on the thread boundary is larger than the within-thread pixels. Therefore, we maintain a bias against the Fwd and Bwd modes as they introduce artificial boundaries in MT. We test several sequences and empirically set η to two for the Fwd and Bwd modes and one for the other six modes. peak signal-to-noise ratio (PSNR) for η as one shows about 0.2 dB loss.

$Bits_{motion}$ represents the bits for coding the motion vector difference (MVD). Since the mode distribution varies considerably for different sequence motion types and different temporal decomposition layers, the bits for identifying the mode symbol are Huffman coded accordingly after the motion estimated process. Thus the symbol bit cost is not counted in the $Bits_{motion}$ term during the motion estimation.

In Bid mode, both the forward and the backward motion vectors are coded, hence $Bits_{motion}$ consists of the bits for the bidirectional MVs; in the other modes, $Bits_{motion}$ only involves single directional motion bits. λ is the weight parameter to control the R-D contribution of the motion cost in the cost function. Generally, at the low bit rate, the slope of the rate-distortion curve is sharper than at the high bit rate, thus λ should be larger at a low bit rate to properly estimate the motion cost percentage. However, because of the non-recursive wavelet coding structure, motion estimation is performed prior to the transform and quantization. Therefore, λ can only be set to a fixed value for all the bit rates in our coder (we note that a novel scheme based on scalable motion vector coding was recently proposed by Secker and Taubman [72] that effectively accommodates different λ_s in one coder). In our experiments, we target at an average PSNR range of 30 dB to 40 dB with the mean of 34 dB. According to $PSNR = 10log_{10}(255^2/D)$, $D = Q^2/12$ under the high rate assumption for uniform scalar quantization, and the empirical formulae $\lambda = \sqrt{0.85} \times Q$ [84], we can relate λ to the target PSNR. For $PSNR$ = 34 dB, we have $\lambda \approx 16$ and we use $\lambda = 16$ in our experiments with six sequences.

For each MB, the mode which has the smallest total cost value is selected. Various mode distributions are calculated according to different sequence and temporal layers. During the MB mode decision, the occurrence of each mode is counted for each temporal decomposition layer. The probability distribution of each mode is used in training the Huffman tables. Then the selected mode of each MB is Huffman encoded according to the trained tables. The Huffman tables for each temporal layer are written into the header of the transmitted bitstream.

5.6 Experimental Results

5.6.1 Coding Performance Comparison

In this section, we compare the proposed advanced MT scheme with three benchmark coders: original MT [74], MC-EZBC [67], and H.264 [13] JM61e. The results from MC-EZBC are cited from the report in the Excel file shown by Chen and Woods [85]. Advanced MT uses all eight modes, and Bidmode MT means only the Bid mode selected. In advanced MT presented here and the original MT coder, the input sequence is temporally filtered as a whole with the extension of the line-based implementation in JPEG 2000, without being explicitly divided into a group of pictures (GOP). Since the interval of the four-layer decomposed low-pass frames is 16, the GOP size can be regarded as 16. The temporally filtered coefficient frames are further 3-layer spatially transformed with Spacl wavelet packet in JPEG 2000 [69], where the three first-layer spatial high-bands are further vertically and horizontally analyzed, and then entropy coded as the operation by Luo et al. [65]. In advanced MT, a quarter-pixel motion is used. Since the real motion range almost doubles when the temporal interval is doubled, the motion search size is set as 32 for the highest tem poral layer, and 64 for the second layer, and 128 for the rest of the layers. In H.264 coder, the GOP is set as a whole sequence with only one I-frame, and two B-frames are inserted into every two P-frames. Quarter-pixel prediction with search-size 32 and among 5 referenced frames is applied to H.264. CABAC and R-D optimization are also turned on. Experiments are performed on six MPEG standard CIF sequences. All the sequences are of 300 frames with the frame rate as 30 Hz.

We first compare the proposed advanced MT of eight-mode selection with that of only Bid mode. We intend to clarify the contribution of the mode selection in the

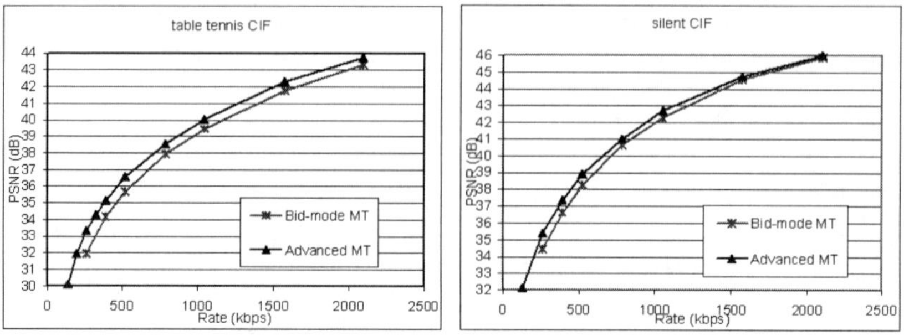

Figure 5.10 Comparison between Bid mode only and eight-mode advanced MT.

advanced MT. For table tennis in Figure 5.10, eight-mode selection improves about 1.3 dB at 250 kbps and 0.4 dB at 2 Mbps; for silence, the improvement is 1.0 dB at 250 kbps and 0.1 dB at 2Mbps. The improvement mainly comes from the reduction of motion cost, thus it is more obvious at lower bit rates and the two curves are convergent at the high end. The improvement for the table sequence is significant even at the high end because a scene change occurring at the 131th frame makes the bidirectional linkage improper.

Next, we compare the eight-mode advanced MT with three benchmark codecs. Figure 5.11 shows that the original MT scheme achieves a close performance to H.264 with Coastguard. However, with other sequences, the performance is not satisfactory. This is due to the boundary effect of a large number of truncated motion threads and the insufficient integer motion accuracy. The proposed advanced MT improves coding efficiency by up to 6.0 dB with Mobile and achieves a similar performance to H.264. The coding gain for Mobile mainly comes from enhanced motion accuracy because Mobile possesses many texture details. With Table, the advanced MT outperforms the original MT by more than 3 dB, and approaches the performance of H.264. With Silent, Foreman, and Stefan, where the irregular motion generates a lot of truncated motion threads in the original MT, advanced MT improves about 2.0–4.0 dB with the avoidance of truncation based on the temporal lifting structure. With these three sequences, advanced MT is still inferior to H.264 by about 1.5 dB due to the shortage in motion thread alignment for complex motion. However, we should note that the advanced MT can provide frame-rate and PSNR scalability, which is not achievable for the single-layer H.264 bitstream. With the Coastguard sequence, advanced MT improves by about 1.5 dB, and even outperforms H.264 by 0.6 dB. The improvement of the advanced MT is considerably significant.

5.6.2 Macroblock Mode Distribution

In this experiment, we investigate the performance of mode-selective motion estimation. Three standard sequences: Foreman, Coastguard, and Mobile are used to

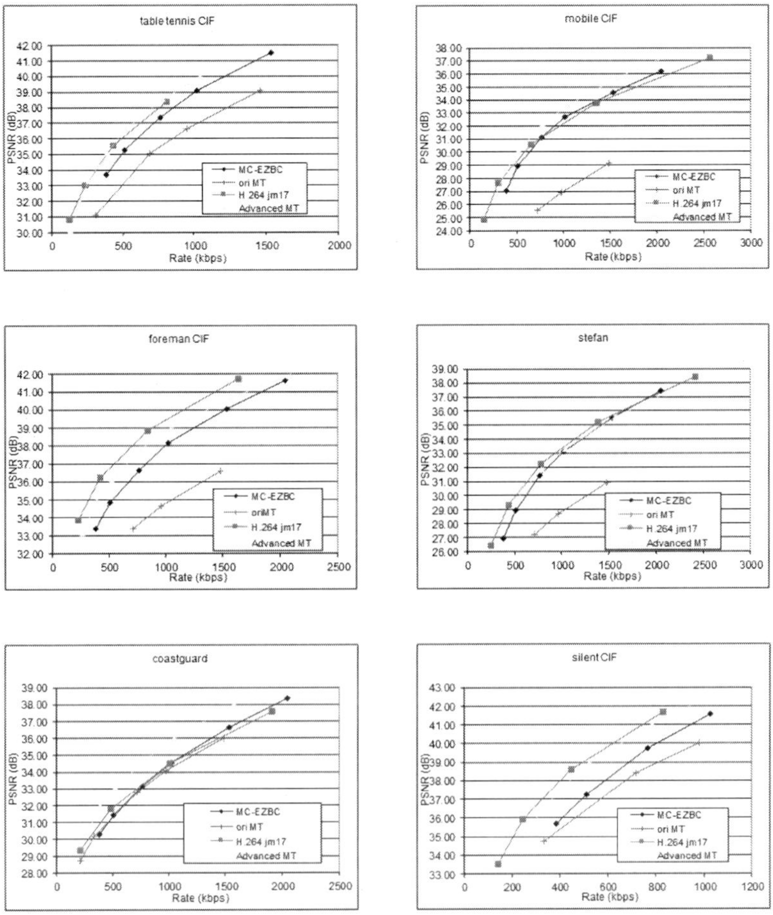

Figure 5.11 Comparisons among MC-EZBC, H.264, original MT, and advanced MT.

represent diverse motion types. All of them are in CIF format, 30 Hz, with 300 frames in all.

First, we examine the distribution of the modes. The test sequences are temporally decomposed into four layers. For the lowest layer, DirL, FT_BDL, and BT_FDL modes are disabled since the lower predictor is not available. Table 5.1 shows the distribution versus decomposition layer of the three test sequences.

In the lowest layer, Bid mode takes a large percentage. As the layer gets higher, the distribution becomes biased toward the inter-layer and intra-layer modes. At the highest layer, DirL is about 50% of the total modes. With the Foreman sequence, which contains more uneven and irregular face motions and camera motions, FT_BDL and BT_FDL modes are frequently used instead of DirL. In Mobile,

Table 5.1 Mode distribution percentage versus temporal layer.

Foreman				
Mode Type	Lowest Layer 4	Layer 3	Layer 2	Highest Layer 1
DirL	0	6.38	14.13	38.00
FT_BDL	0	34.72	36.14	23.85
BT_FDL	0	17.70	17.28	12.74
FwdDir	5.17	2.46	3.56	7.19
BwdDir	5.27	2.76	3.04	4.54
Fwd	3.44	2.06	1.21	0.41
Bwd	5.79	2.25	1.05	0.30
Bid	80.33	31.66	23.59	12.96
Mobile				
Mode Type	Lowest Layer 4	Layer 3	Layer 2	Highest Layer 1
DirL	0	33.95	60.37	67.94
FT_BDL	0	21.74	7.64	0.75
BT_FDL	0	8.20	3.36	0.55
FwdDir	12.70	8.87	14.85	24.57
BwdDir	5.54	4.44	6.30	5.42
Fwd	1.99	0.65	0.12	0.03
Bwd	0.58	0.28	0.10	0.02
Bid	79.19	21.87	7.25	0.72
Coastguard				
Mode Type	Lowest Layer 4	Layer 3	Layer 2	Highest Layer 1
DirL	0	9.59	24.78	57.63
FT_BDL	0	24.63	23.17	14.64
BT_FDL	0	11.19	9.93	6.08
FwdDir	14.49	13.13	13.14	9.73
BwdDir	15.08	11.55	8.50	4.44
Fwd	3.09	0.66	0.21	0.04
Bwd	1.39	0.97	0.25	0.03
Bid	65.95	28.28	20.00	7.40

the motion is more regular, thus the mode distribution concentrates around the DirL, FwdDir, and BwdDir modes where the forward and backward MVs are equal. The mode distribution of Coastguard is between Foreman and Mobile. The water wave, with which motion estimation fails to catch an accurate motion, leads to the relatively uniform mode distribution.

Next, we analyze the contribution of the layer correlated ME scheme. As shown in Table 5.2, intra layer denotes the scheme with only intra-layer modes such as FwdDir, BwdDir, Fwd, Bwd, and Bid modes; CME is the proposed scheme which considers both the intra-layer modes and the inter layer one such as DirL, FT BDL, and BT FDL. The total bits consist of the motion bits and the mode bits. With the same PSNR performance, the motion bits reduction is shown in Table 5.2. The bits reduction is 18.1% for Foreman, 19.2% for Coastguard, and 29.2% for Mobile. Since Mobile contains more regular motion, for most MBs, DirL mode is selected, thus the

Table 5.2 Motion bits reduction.

		Foreman	Mobile	Coastguard
Intra Layer (kbps)	MV bits	96.5	49.4	69.2
	Mode bits	15.0	12.5	14.7
	Total bits	111.5	61.9	83.9
CME (kbps)	MV bits	66.8	25.2	43.2
	Mode bits	24.5	18.6	24.6
	Total bits	91.3	43.8	67.8
Total Reduction	Bits (kbps)	20.2	18.1	16.1
	Percentage	18.1%	29.2%	16.1%

bits reduction is relatively larger. At the low bit rate (e.g., 384 kb/s), the bit reduction is equivalent to about 0.3 dB PSNR improvement.

5.7 Summary

An advanced motion threading technique is presented to improve the existing motion threading technique for wavelet video coding. With a lifting-based temporal wavelet structure, the boundary effect problem caused by artificial thread truncation in the original MT is solved well. Fractional-pixel motion can also be applied in the new structure. Multi-layer motion threading is used to achieve efficient frame-rate scalability. To efficiently reduce the motion cost as well as achieve good frame-rate scalability, we have analyzed both inter-layer and intra-layer correlations. A novel R-D optimized correlated motion estimation scheme is proposed for the motion estimation process. Experimental results show that the saving of the motion vector bits is up to 29.2% with the Mobile sequence. The lifting-based motion threading has many benefits. In future works, adaptive length wavelet filters can be applied into the lifting structure to adapt the various local properties along a video sequence.

Chapter 6
Barbell-Lifting Based 3D Wavelet Coding

6.1 Introduction

We started the study of 3D wavelet-based video coding from motion threading for exploiting long-term correlations across frames along motion trajectories [74, 86]. The work by Xu et al. [74] also proposes an efficient entropy coding, 3D embedded subband coding with optimized truncation (ESCOT) for 3D wavelet coefficients. However, the motion threading technique still has limitations on handling many-to-one mapping and non-referred pixels. To solve these problems, Luo et al. further develop the lifting-based motion threading technique [65, 76, 87].

Subsequently, additional effort has been invested in this area. Xiong et al. propose multiple modes with different block sizes (similar to that in H.264/AVC) for accurate motion alignment and overlapped block motion alignment to suppress the blocking boundaries in prediction frames [88]. Feng et al. propose an energy distributed update technique to eliminate mismatches between the prediction and update steps in the motion-aligned temporal lifting transform [89]. All these techniques are integrated into a general lifting framework called Barbell lifting [90]. In addition, to maintain high performance of 3D wavelet-based video coding in a broad range of bit rates, Xiong et al. also investigate layered motion vector estimation and coding [91]. Ji et al. propose an approach to incorporate a close-loop H.264/AVC into 3D wavelet video coding to improve performance at low bit rates [92].

Moving Picture Experts Group (MPEG) plays an important role in actively exploring and promoting scalable video coding technologies from Fine Granularity Scalable (FGS) coding to inter-frame wavelet coding. In 2003, a call for proposals (CfP) was released to collect scalable video coding technologies and evaluate their performances [93]. A total of 21 submissions were finally made, which met all the deadlines listed in the CfP, including our Barbell-lifting coding scheme [94]. There are two test scenarios in the CfP. For scenario 1, the working bandwidth range is large and three levels of temporal/spatial scalabilities are required. Scenario 2 contains a comparatively narrow working bandwidth range and only has two levels of temporal/spatial scalabilities. The Barbell-lifting coding scheme ranks first in scenario 1 and third in scenario 2 [95]. Finally, this scheme was adopted as common software by the MPEG ad hoc group on further exploration of wavelet video coding [96], and is, therefore, publicly available for all MPEG members.

6.2 Barbell-Lifting Coding Scheme

The overall block diagram of the Barbell-lifting coding scheme is depicted in Figure 6.1. First, to exploit the correlation among neighboring video frames, wavelet transform is temporally performed on the original frames to decompose them into low-pass frames and high-pass frames. To handle motion in video frames, motion compensation is incorporated with the Barbell lifting, which is a high-dimensional extension of the basic 1D lifting structure of wavelet transform [80]. Second, to further exploit the spatial correlation within the resulting temporal subband frames, 2D wavelet transform is applied to decompose each frame into some spatial subbands. Third, coefficients in the ultimate spatio-temporal subbands are processed bit plane by bit plane to form an embedded compressed stream. The side information in Figure 6.1, including motion vectors, macroblock modes, and other auxiliary parameters to control the decoding process, is also entropy coded. Finally, the streams of subband coefficients and side information are assembled to form the final video stream packets.

The stream generated by the above scheme is highly scalable. The bit plane coding technique for subband coefficients provides fine granularity scalability in reconstruction quality. The hierarchical lifting structure of temporal transform is ready to provide scalability in the frame rate. The multi-resolution property of wavelet representation naturally provides scalability in resolution. When the bit rate, frame rate, and spatial resolution of a target video are specified, a sub-stream for reconstructing the video can be easily extracted by identifying relevant spatio-temporal subbands and retaining a partial or complete stream of them while discarding the others.

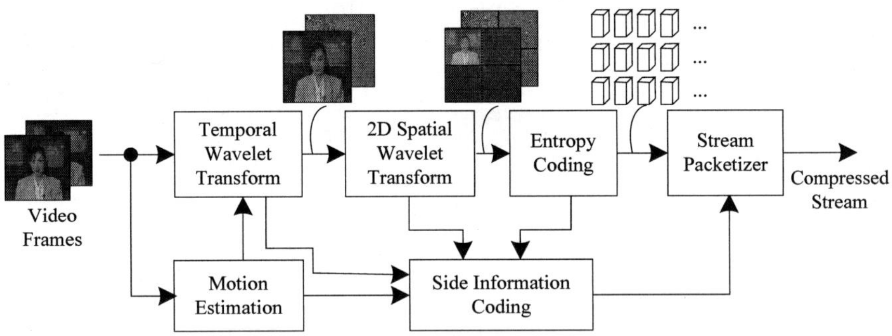

Figure 6.1 Block diagram of the Barbell-lifting coding scheme.

The following sections will discuss the core techniques employed in our proposed coding scheme, such as Barbell lifting, layered motion coding, 3D entropy coding, and base layer embedding. At the same time, we also cite several related techniques used in other schemes so as to give the audience a fuller picture.

6.2.1 Barbell Lifting

In many previous 3D wavelet coding schemes, the concept of lifting-based 1D wavelet transform is simply extended to the temporal direction as a transform along motion trajectories. In this case, the temporal lifting is actually performed as if in 1D signal space. This requests an invertible one-to-one pixel mapping between neighboring frames so as to guarantee that the prediction and update lifting steps operate on the same pixels. However, the motion trajectories within real-world video sequences are not always as regular as expected, and are sometimes even unavailable. For example, pixels with fractional-pixel motion vector are mapped to "virtual pixels" on reference, which cannot be directly updated. In the case of multiple pixels, mapping to one pixel on reference, the related motion trajectories will merge. For covered and uncovered regions, motion trajectories will disappear and appear. The direct adoption of 1D lifting in temporal transform cannot naturally handle these situations. It motivates us to develop a more general lifting scheme for 1D wavelet transform in a high-dimensional signal space, where multiple predicting and updating signals are supported explicitly through Barbell functions.

When the lifting scheme developed by Daubechies and Sweldens [80] is directly used in temporal direction, the basic lifting step can be illustrated in Figure 6.2a. A frame f_1 is replaced by superimposing two neighboring frames on it with a scalar

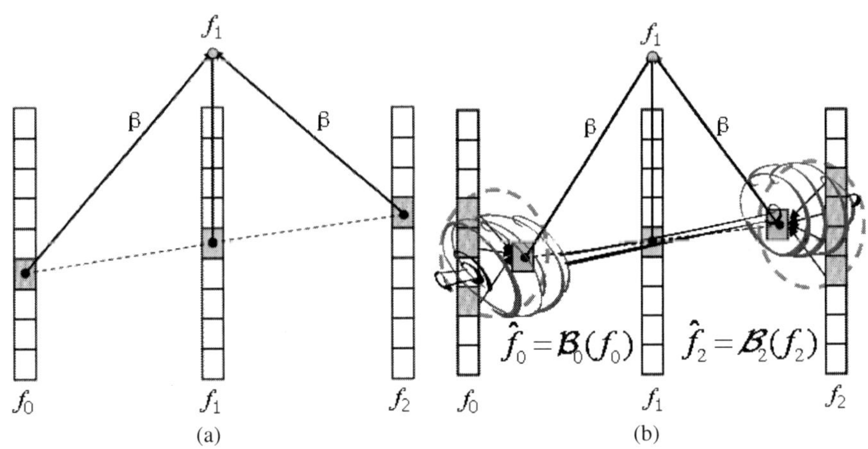

Figure 6.2 Basic lifting step. (a) Conventional lifting. (b) Proposed Barbell lifting.

factor β specified by the lifting representation of the temporal wavelet filter. Notice that only one pixel, of the signals f_0 and f_2 respectively, is involved in the lifting step. In the proposed Barbell lifting as shown in Figure 6.2b, instead of using a single pixel, we use a function of a set of nearby pixels as the input. The functions $B_0()$ and $B_2()$ are referred to as Barbell functions. They can be any linear or nonlinear functions that take any pixel values on the frame as variables. The Barbell function can also vary from pixel to pixel. Therefore, the basic Barbell-lifting step is formulated as

$$f_1 = \beta B_0(f_0) + f_1 + \beta B_2(f_2). \tag{6.1}$$

According to the definition of the basic Barbell-lifting step, we give a general formulation for M-level motion compensated temporal filtering (MCTF), where the mth MCTF ($0 \le m \le M-1$) consists of $\Lambda(m)$ lifting steps. Assume that $f^{(m,0)} = f_k^{(m,0)}$ denotes input frames of the mth MCTF and $f^{(m,\lambda)} = f_k^{(m,\lambda)}$ denotes the result of the λth lifting step $1 \le \lambda \le \Lambda(m)$ of the mth MCTF. k indicates the frame index.

For odd λ, the λth lifting step modifies odd-indexed frames based on the even-indexed frames, as formulated in Eq. (6.2). For even λ, the λth lifting step modifies even-indexed frames based on the odd-indexed frames, as formulated in Eq. (6.3). Here $\beta_0^{(m,\lambda)}$ and $\beta_1^{(m,\lambda)}$ are filter coefficients specified by the lifting representation of the mth level temporal wavelet filter. $B^O_{2i\to 2k+1}$ $(i = k, k+1)$ and $B^E_{2i+1\to 2k}$ $(i = k-1, k)$ are the Barbell function operators to generate lifting signal in odd and even steps, respectively. After all the lifting steps, we get the low-pass frames and high-pass frames, defined by $L_k^{(m+1)} = f_{2k}^{(m,\Lambda(m))}$ and $H_k^{(m+1)} = f_{2k+1}^{(m,\Lambda(m))}$, respectively. Theoretically, arbitrary discrete wavelet filter can be adopted in MCTF easily based on Eq. (6.2) and Eq. (6.3). But the biorthogonal 5/3 filter is the one that has already been verified as practical with good coding performance so far. It consists of two lifting steps: $\Lambda(m) = 2$, $\beta_0^{(m,1)} = \beta_1^{(m,1)} = -0.5$, and $\beta_0^{(m,2)} = \beta_1^{(m,2)} = 0.25$. In this case, $B^O_{2i\to 2k+1}$ and $B^E_{2i+1\to 2k}$ are commonly called prediction and update steps, respectively. In multilevel MCTF, the low-pass frames of a MCTF level are fed to the next MCTF level by $f_k^{(m+1)} = L_k^{(m+1)}$. Finally, the M-level MCTF outputs $M+1$ temporal subbands: high-pass subbands $H_k^{(1)}, H_k^{(2)}, \cdots, H_k^{(M)}$, and low-pass subband $L_k^{(M)}$,

$$f_{2k+1}^{(m,\lambda)} = f_{2k+1}^{(m,\lambda-1)} + \beta_0^{(m,\lambda)} B^O_{2k\to 2k+1}\left(f_{2k}^{(m,\lambda-1)}\right) + \beta_1^{(m,\lambda)} B^O_{2k+2\to 2k+1}\left(f_{2k+2}^{(m,\lambda-1)}\right). \tag{6.2}$$

$$f_{2k}^{(m,\lambda)} = f_{2k}^{(m,\lambda-1)} + \beta_0^{(m,\lambda)} B^E_{2k-1\to 2k}\left(f_{2k-1}^{(m,\lambda-1)}\right) + \beta_1^{(m,\lambda)} B^E_{2k+1\to 2k}\left(f_{2k+1}^{(m,\lambda-1)}\right). \tag{6.3}$$

(1) MC Prediction—We discuss the Barbell function of MC prediction. Assume there is a multiple-to-multiple mapping from frame $f_{2k+1}^{(m,0)}$ to frame $f_{2i}^{(m,0)}$, based on the motion between these frames and the correlation in related pixels. For any

pixel $a \in f_{2k+1}^{(m,0)}$, we define $M_{2k+1\to 2i}^{(m)}(a) \subset f_{2i}^{(m,0)}$ as the set of pixels in $f_{2i}^{(m,0)}$ that a is mapped to. For each pair of pixels $(a,b)(b \in M_{2k+1\to 2i}^{(m)})$, weighting parameter $w(a,b)$ is introduced for prediction, to indicate the correlation strength between pixels a and b. The operator $f_p = B_{2i\to 2k+1}^{O}(f_{2i}^{(m,\lambda-1)})$ based on Barbell lifting is defined as

$$f_p(x_a) = \sum_{b\in M_{2k+1\to 2i}^{(m)}(a)} w(a,b) f_{2i}^{(m,\lambda-1)}(x_b). \tag{6.4}$$

Here, x_a and x_b are coordinates of pixels a and b, in frames $f_{2k+1}^{(m,\lambda-1)}$ and $f_{2i}^{(m,\lambda-1)}$, respectively. The weighting parameters are subject to the constraint

$$\sum_{b\in M_{2k+1\to 2i(a)}^{(m)}} w(a,b) = 1.$$

There are two types of parameters in the Barbell function: the mapping from $f_{2k+1}^{(m,0)}$ to $f_{2i}^{(m,0)}$ and the weighting parameters $w(a,b)$. The mapping can be derived from motion vectors estimated based on the block-based motion model. In general, motion vector is measured up to fractional pixels for accurate prediction, such as 1/2-pel and 1/4-pel in H.264/AVC. Barbell lifting also supports fractional-pixel motion accuracy. In this case, each pixel in the current frame is mapped to multiple pixels in the neighboring reference frame, while the weighting parameters $w(a,b)$ are determined by the interpolation filter (the formulation is given by Xiong et al. [90]).

To achieve a proper trade-off between the efficiency of motion prediction and the coding cost of motion information, variable block-size partitioning is used for motion representation in the Barbell function. All macroblock partitions in H.264/AVC, such as 16×16, 16×8, 8×16, 8×8 and subpartitions in an 8×8 block, are supported in Barbell lifting. In addition, five motion coding modes are defined [88], including the *Bid*, *FwD*, *BwD*, *DirInv*, and *Skip* modes, to further reduce the cost of coding the mapping relationship. These modes jointly signal the connectivity in two directions and the motion vector assigned with a macroblock. In the *FwD* and *BwD* modes, the coding for motion on one side is skipped. Furthermore, the *Skip* and *DirInv* modes exploit the spatial and temporal correlations in motion fields, respectively, and therefore save coding bits.

Although the smaller block size allowed in the variable block-size partition in motion alignment can significantly reduce the average energy of predicted errors, it also increases the number of blocks used in motion compensation and causes more blocking boundaries in the prediction frame. This leads to many high-amplitude coefficients in spatial high-pass subbands of residue frame after spatial subband decomposition. The overlapped block motion compensation (OBMC) technique is adopted in Barbell lifting [88] to smooth the transition at block boundaries. In this case, the parameters $w(a,b)$ are determined by both the interpolation filter and the weighting window of OBMC (the formulation is given by Xiong et al. [90]). In addition to the OBMC technique, it is also possible to support any other multi-hypothesis techniques, for example, the multiple-reference MC prediction by the proposed Barbell-lifting model. These techniques can improve the compression efficiency of 3D wavelet video coding.

In the prediction step of our proposed coding scheme, we mainly borrow some mature MC prediction techniques from conventional video coding schemes and incorporate them into the Barbell-lifting framework. There are also several other techniques developed in other wavelet coding schemes for the MC prediction step. A typical one is hierarchical variable size block matching (HVSBM) [62, 67], which consists of constructing an initial full motion vector tree and pruning it subject to a given bit rate.

(2) Motion Compensated Update—The update step in Barbell lifting is performed according to the idea proposed by Feng et al. [89]. For a pair of pixels (a,b), where $a \in f_{2i+1}^{(m,0)}$, $b \in f_{2k}^{(m,0)}$, and $w(a,b) > 0$, since the pixel a is predicted from pixel b with a weighting parameter $w(a,b)$ in the prediction step, we propose to use the prediction error on the pixel a to update the pixel b, with the same weighting parameter. For any pixel $b \in f_{2k}^{(m,0)}$, we further define

$$M_{2k\to 2i+1}^{(m)}(b) = \left\{ a \in f_{2i+1}^{(m,0)} \mid b \in M_{2i+1\to 2k}^{(m)}(a) \right\}$$

as the set of pixels in $f_{2i+1}^{(m,0)}$ that b is mapped from. Therefore, the operator $f_u = B_{2i+1\to 2k}^{E}(f_{2i+1}^{(m,\lambda-1)})$ based on Barbell lifting is defined as

$$f_u(x_b) = \sum_{a \in M_{2k\to 2i+1}^{(m)}(b)} w(a,b) f_{2i+1}^{(m,\lambda-1)}(x_b). \tag{6.5}$$

Generally, the update step has an effect of temporal smoothing for regions with accurate motion alignment, and thus can improve coding performance. But when the motion in the video sequence is too complicated to be accurately represented by the employed motion model, temporal high-pass frames may contain large prediction residues. This makes the coding of temporal low-pass frames difficult when the prediction residue is superimposed on the even frames. It also results in ghost-like artifacts in the temporal low-pass frames, which is not desired for temporal scalability. To solve this problem, a threshold function $T_c()$ is applied to the updating signal

$$f_u(x_b) = T_c\left(\sum_{a \in M_{2k\to 2i+1}^{(m)}(b)} w(a,b) f_{2i+1}^{(m,\lambda-1)}(x_a) \right), \tag{6.6}$$

where

$$T_c(x) = \begin{cases} -c, & if\ \ x < -c \\ x, & if\ \ \|x\| \le c \\ c, & if\ \ x > -c \end{cases}.$$

By empirically setting $c = 2-5$, most visually noticeable artifacts can be removed from low-pass frames but the advantage of the update step in coding performance is still maintained.

In addition to the simple but effective threshold approach used in the Barbell-lifting coding scheme, Mehrseresht and Taubman [97] and Turaga and Van der Schaar [98] also introduce some techniques to adjust the update signal and, hence, reduce ghosting effects. Song et al. [99] propose regularizing the updating signal based on HVS. In another more interesting work, Girod et al. [100] investigate the update step as an optimal problem and derive a closed-form expression of the update step for a given linear prediction step. Chen et al. [101] further reveal the relationship between the energy distributed update [89] and the optimum update [100], and proposes a set of new update coefficients for improving coding efficiency and reducing quality fluctuations.

6.2.2 Layered Motion Coding

A stream generated by the Barbell-lifting coding scheme can be decoded at different bit rates, frame rates, and resolutions. In terms of rate-distortion optimization (RDO), a fixed set of motion vectors is not an optimum solution for different reconstructions. To achieve an optimum trade-off between motion and texture data, the Barbell-lifting coding scheme requests at least a layered motion coding.

If we use $D(R_T;MV)$ to denote the distortion of reconstructed video, R_T is the bit rate allocated to texture data, MV is motion data, R_{MV} is the bit rate for motion data, the optimization problem is to minimize $D(R_T;MV)$ subject to the total bit rate constraint $R = R_T + R_{MV} \leq R_{MAX}$. Using the Lagrange approach, it leads to the optimum problem as Eq. (6.7) and the solution is given in Eq. (6.8)

$$argmin_{R_T,R_{MV}} J = D(R_T;MV) + \lambda(R_T + R_{MV}). \tag{6.7}$$

$$\frac{\partial D(R_T;MV)}{\partial R_T} = \frac{\partial D(R_T;MV)}{\partial R_{MV}} = -\lambda. \tag{6.8}$$

It means that the texture and motion stream should achieve the equal R-D slope on its respective distortion-rate curve.

We propose a layered structure for the representation of motion data by Xiong et al. [91], which consists of multiple layers, L_1^{MV}, L_2^{MV}, $\cdots$, L_n^{MV}. These motion layers are selected with different lambda $\lambda_1 > \lambda_2 > \cdots > \lambda_n$. During the motion estimation and coding of the layer L_i^{MV}, its previous layer L_{i-1}^{MV} is used as prediction. The motion data can be refined in two ways: increasing the accuracy of the motion vector or splitting a block into smaller subblocks. When two adjacent motion layers have different resolutions, the motion vectors in the lower layer should be scaled and the macroblock partition modes should be converted before they are used as predictors.

The layered motion is applied to our coding scheme in the following way. Encoder uses the finest motion to perform temporal transform on video frames. This accurate motion provides efficient energy compaction and guarantees optimal coding performance with an increase in bit rate. But the decoder may receive only some of these motion layers for synthesis when the bit rate is low, giving a higher priority to textures. When the motion used in the encoder and decoder is different, perfect reconstruction cannot be obtained even if all texture coefficients are decoded without losses. From the observations by Xiong et al. [91], the distortion produced by motion mismatch is nearly constant in terms of mean square error (MSE) in a wide range of bit rates for texture. In other words, the distortion from motion mismatch is highly independent of texture quantization. This facilitates the estimation of the rate distortion property of the compressed texture and motion data. Let us use $D_i(R)$ to denote the distortion function when the decoder employs L_i^{MV}. Since a bit rate of $R_{L_i^{MV}}$ is allocated to motion and $R_T = R - R_{L_i^{MV}}$ is allocated to texture, it can be approximated by

$$D_i(R) = D\left(R_T; L_n^{MV}\right) + D_{mismatch}\left(L_n^{MV}, L_i^{MV}\right). \tag{6.9}$$

The first item on the right side of Eq. (6.9) is the quantized distortion with motion L_n^{MV}, and the second item is the distortion of motion mismatch. Based on Eq. (6.9), the R-D optimized motion layer selection can be performed during stream truncation at either the frame level or the sequence level.

In additon to the proposed layered motion coding used in the Barbell-lifting coding scheme, Secker et al. also propose a scalable motion representation by applying a subband transform and bit plane coding on motion field [72, 102]. Furthermore, the effect of motion parameter quantization on the reconstructed video distortion is estimated based on the power spectrum of the reference frame.

6.2.3 Entropy Coding in Brief

After temporal transform and 2D spatial wavelet transform, spatio-temporal subbands are available for entropy coding. Taking each spatio-temporal subband as a three-dimensional coefficient volume, we code it bit plane by bit plane using context-based adaptive arithmetic coding technique. The entropy coding is similar to the Embedded Block Coding with Optimized Truncation (EBCOT) algorithm [103, 104] employed in JPEG-2000. But unlike the coding in JPEG-2000, the coding of 3D wavelet coefficients involves exploiting correlations in all three dimensions. We propose a coding algorithm 3D-ESCOT [74] as an extension of EBCOT.

We divide each spatio-temporal subband into coding blocks and code each block separately bit plane by bit plane. For each bit plane, three coding passes are applied. The significance propagation pass codes the significance information for the coefficients, which are still insignificant but have significant neighbors. The magnitude refinement pass codes the refinement information for the coefficients that have already become significant. And the normalization pass codes the significance information for the remaining insignificant coefficients. In 3D-ESCOT, the formation of the contexts to code the significance information and the magnitude refinement information involves both temporal and spatial neighboring coefficients. Furthermore, the correlation of temporal coefficients is often stronger than that of spatial coefficients.

In addition to 3D-ESCOT being used in the Barbell-lifting coding scheme, 3D Embedded Zero Block Coding (EZBC) [105], 3D set partitioning in hierarchical trees (SPIHT) [63], and embedded morphological dilation coding (EMDC) [106] are other approaches that can be used to code 3D wavelet coefficients. 3D EZBC and 3D SPIHT use zero-tree algorithm to exploit the strong cross-subband dependency in the quadtree of subband coefficients. Furthermore, to code the significance of the quadtree nodes, the context-based arithmetic coding is used. The EMDC algorithm [106] predicts the clusters of significant coefficients by some form of morphological dilation.

6.2.4 Base Layer Embedding

As mentioned above, the MCTF decomposes original frames temporally in an open-loop manner. It means that the decomposition at the encoder does not take the reconstruction of video frames at the decoder into account. Open-loop decomposition makes the encoder simple because no reconstruction is needed. However, the weakness of no reconstruction at the encoder is that for a certain bit rate, the encoder does not know the mismatch between the encoder and the decoder so that the coding performance cannot be well optimized. For instance, at the encoder, original frames are used in motion compensation, motion estimation, and mode decision. While at the decoder, reconstructed frames will actually be used. The motion data and coding mode estimated on original frames may not be optimum for the motion compensation on reconstructed frames. The mismatch is large when the quality of the reconstructed frame is low, for example, at a low bit rate. That deteriorates the coding performance especially at a low bit rate. To improve the coding performance at a low bit rate, we incorporate a base layer into the Barbell-lifting coding scheme, which is coded using a close-loop standard codec. Another advantage in doing so is that such a base layer provides the compatibility to the standard coding scheme. The base layer can further exploit redundancy within the temporal low-pass subband $L_k^{(M)}$ without introducing further coding delay.

Suppose that a base layer is embedded into the Barbell-lifting coding scheme after the mth level $(0 \leq m \leq M-1)$ MCTF. The output low-pass video after MCTF is $f_k^{(m)}$, with a frame rate of $F/2^m$, where F is the frame rate of the original video. $D(f_k^{(m)})$, a down-sampled version of $f_k^{(m)}$, is fed to a standard video codec, for example, H.264/MPEG-AVC, where $D()$ is a down-sampling operator. Let $ENC()$ and $DEC()$ denote the base layer encoding and decoding processes. We can get the reconstructed frames at the low resolution by

$$\hat{f}_k^{(m)} = DEC(ENC(D(f_k^m))). \tag{6.10}$$

The reconstructed frames $\hat{f}_k^{(m)}$ are both available at the encoder and the decoder, which provide a low resolution and low frame-rate base layer at a certain bit rate. The coding of base layer can be fully optimized as done in H.264/AVC. Any standard compliant decoder can decode this base layer.

The up-sampled version of the reconstructed frames $U(\hat{f}_k^{(m)})$ is also used as a prediction candidate in the prediction step of the remaining MCTF. For example, in the m_1th MCTF $(m < m_1 \leq M-1)$, for those macroblocks which use the base layer as prediction, the prediction step is

$$f_{2k+1}^{(m_1,2)} = f_{2k+1}^{(m_1,1)} - U\left(\hat{f}_{(2k+1)*2^\wedge(m_1-m)}^{(m)}\right). \tag{6.11}$$

And for those macroblocks, no update step is performed. To make the base layer coding fit for the spatial transform of the Barbell-lifting coding scheme, the down-sampling operator extracts the low-pass subband after one or several levels of spatial wavelet transform. The up-sampling operation is the corresponding wavelet synthesis process.

6.3 Comparisons with SVC

The H.264/AVC scalable extension, or the SVC standard for short, is a new scalable video coding standard developed jointly by the ITU Telecommunication Standardization Sector (ITU-T) and the International Organization for Standardization (ISO). The SVC standard was originally developed from the Heinrich Hertz Institute (HHI) proposal [107], which extends the hybrid video coding approach of H.264/AVC toward MCTF [108]. Since both schemes are developed from MCTF, they have many commonalities, especially in the temporal decorrelation part. They also have many differences. In this section, we discuss the major commonalities and differences of the Barbell-lifting video coding to the SVC standard.

6.3.1 Coding Framework

The SVC standard uses a bottom-up layered structure to fulfill scalabilities, which is similar to the scalability supported in previous MPEG and ITU-T standards. A base layer is coded using H.264/AVC compliant encoder to provide a reconstruction at low resolution, low frame-rate and/or low bit rate. An enhancement layer that may be predicted from the base layer is coded to enhance the signal-to-noise ratio (SNR) quality for SNR scalability or to provide a higher resolution for spatial scalability. Multilevel scalability is supported by multiple enhancement layers. Once several lower layers are given, the coding of the current layer can be optimized, which leads to a one-by-one layer optimization. However, inefficient inter-layer prediction cannot totally remove redundancy in neighboring layers and will sacrifice performance in higher layers.

Unlike the SVC standard, the Barbell-lifting scheme uses a top-down coding structure. As mentioned above, video signal is decomposed temporally, horizontally, and vertically to spatio-temporal subbands. In a given frame-rate and resolution required, a certain set of spatio-temporal subbands corresponding to that spatio-temporal resolution are extracted and sent to the decoder. Decorrelation is achieved by temporal and 2D spatial wavelet transforms. The advantage is to represent the signal in a multi-resolution way so that scalability is inherently supported. The disadvantage of the top-down structure is that it may not favor coding performance at a low resolution or bit rate, since all decomposition is done with an open-loop structure and on the full resolution.

6.3.2 Temporal Decorrelation

Temporal decorrelation is one of the most important issues in video coding. Although MCTF can be supported at the encoder, the close-loop hierarchical B-structure [109], is the de facto decorrelation process in SVC. Suppose a M-level hierarchical B-prediction is performed on video sequence f_i, the mth level prediction can be expressed as

$$\begin{aligned} r_{(2k+1)*2\wedge(M-m)} =& f_{(2k+1)*2\wedge(M-m)} \\ & - \frac{1}{2} B^{O}_{2k*2\wedge(M-m)\to(2k+1)*2\wedge(M-m)} \left(\tilde{f}_{2k*2\wedge(M-m)}\right) \\ & - \frac{1}{2} B^{O}_{(2k+2)*2\wedge(M-m)\to(2k+1)*2\wedge(M-m)} \left(\tilde{f}_{(2k+2)*2\wedge(M-m)}\right), \end{aligned} \tag{6.12}$$

where r_i is the residual frame after prediction and $\tilde{f}_i$ is the reconstructed image of f_i, which is available at both the encoder and the decoder. $B^{O}_{i\to j}$ is defined as in Section 6.2.1. Basically, it corresponds to the prediction stage at the $(M-m+1)$th MCTF. The difference is that the prediction in the close-loop hierarchical B-structure is generated from the reconstructed images, while the MCTF is performed on the original images. Since in the case of lossy coding the decoder cannot get the original images, mismatch exists at the prediction stage of MCTF between the encoder and the decoder. It may cause coding performance degradation, especially at a low bit rate, where the mismatch between the original images and the reconstructed images is large.

The other difference is that there is no update step in the hierarchical B-prediction while there is in MCTF. The update step in MCTF, together with the prediction step, constructs a low-pass filter that makes the output low-pass frames smooth so that they can be better coded. It has been observed that the update step is effective in improving coding performance in 3D wavelet video coding. However, in most cases, the update step in SVC does not show much difference in terms of coding performance. Even for some cases in which the update step improves the coding performance in SVC, the gain can be similarly achieved using prefiltering. A possible reason for the different coding performance improvement of the update step in SVC and in 3D wavelet coding is that in SVC, integer approximation is applied to both temporal decorrelation and spatial transform. This may absorb most signals at the update step, which are often of low energy.

Despite these differences, the close-loop hierarchical B-prediction and MCTF have similar prediction structures. Actually, if high-pass temporal frames are skipped, the close-loop hierarchical B-prediction and MCTF are the same at the decoder because both prediction steps are performed on reconstructed images. The hierarchical prediction structure can both exploit the short-term correlation and the long-term one. Most frames are predicted bidirectionally in both structures. This accounts for why both schemes have shown significant coding performance gains over the H.264/AVC codec in traditional I-B-P prediction structure. Schwarz, Marpe, and Wiegand [110] give an analysis on hierarchical B-frames and MCTF.

6.3.3 Spatial Scalability

In the layered coding scheme of SVC, spatial scalability is supported by coding multiple resolution layers. The original full-resolution input video is down-sampled to provide input at lower resolution layers. To exploit cross-layer redundancy, the reconstructed images at a lower resolution can be used as prediction for some macroblocks when the prediction within the current resolution is not effective. The advantage is that the different resolution input can be flexibly chosen, which enables arbitrary down-sampling to generate low resolution video and non-dyadic spatial scalability. However, in the coding of a higher-resolution video, many macroblocks do not use the lower resolution as prediction, which means the corresponding bits at lower resolution do not contribute to the coding of higher resolutions. This affects coding performance in spatial scalability scenarios.

In the Barbell-lifting coding scheme, any lower resolution video is always embedded in higher-resolution video. The spatial low-pass subbands are used to reconstruct the low resolution video. Because of the critical sampling of the wavelet transform, the number of transform coefficients to be coded is the same as the number of pixels, even when multiple spatial scalability levels are supported. Bits of the low-pass subband contribute both to the lower resolution layer and the higher resolution layer. However, the constraint is that the low-resolution video is corresponding to the wavelet low-pass filter, which may not fit for all applications. It is also difficult to support an arbitrary ratio of spatial scalability, since dyadic wavelet transform is generally used.

6.3.4 Intra Prediction

As an extension of H.264/AVC, SVC still uses block-based DCT transform for spatial decorrelation. Such a block-based transform enables that each macroblock can be reconstructed instantly after the encoding to assist the coding of neighboring blocks. Intra prediction in H.264/AVC and SVC are such examples of this technology. By further introducing several directional intra-prediction modes, H.264 can efficiently exploit the directional correlation within images. It can significantly improve the coding performance of intra-frame and intra-macroblocks in P- or B-frames. However, similar technology is relatively difficult to use in 3D wavelet video coding since the spatial transform of each macroblock is not independent.

6.4 Advances in 3D Wavelet Video Coding

There are still several challenges in scalable video coding. One is how to achieve efficient spatial scalability. Both the Barbell-lifting coding scheme and SVC suffer from considerable performance degradation when spatial scalability is enabled. The other challenge is how to further improve the performance of temporal decorrelation. Two techniques, in-scale MCTF and subband adaptive MCTF, are developed, although the integrated Barbell-lifting coding scheme is not available yet in MPEG.

6.4.1 In-Scale MCTF

As shown in Figure 6.3, the temporal transform is performed prior to the 2D spatial transform in the encoder of the Barbell-lifting coding scheme. When a low-resolution video is requested at the decoder, spatial high-pass subbands higher than the target resolution are dropped. The other subbands are decoded by inverse wavelet transform and inverse MCTF at low resolution to reconstruct the target video.

Two kinds of mismatches exist between the encoder and the decoder in Figure 6.3. First, MCTF at the encoder and the decoder is performed at different resolutions, which results in artifacts at those regions with complex motion [111]. Second, as reported [97, 111, 112], all spatial subbands of video signals are coupled during the MCTF process due to motion alignment. The dropped spatial high-pass subbands are effectively referenced during the temporal transform at the encoder. But they become unavailable at the decoder, and thus result in an extra reconstruction error. To remove the first kind of mismatch, several modified decoding schemes are investigated by Xiong et al. [111]. To solve the second kind of mismatch, a new rate allocation scheme is proposed by Xiong et al. [113, 114], which allocates part of the bit budget to spatial high-pass subbands based on their importance in reconstruction.

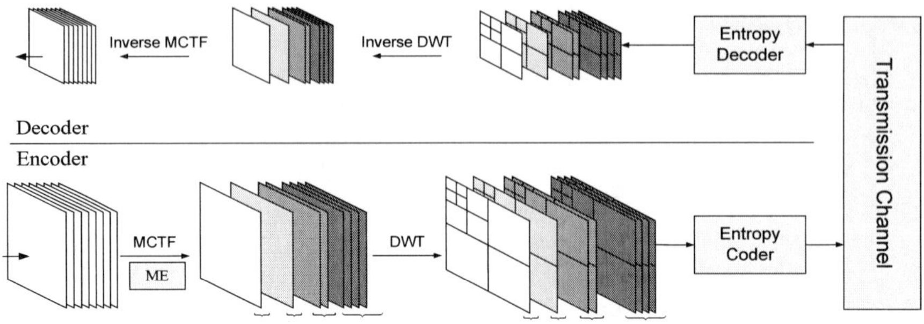

Figure 6.3 Encoding and decoding process for spatial scalability in the Barbell-lifting coding scheme.

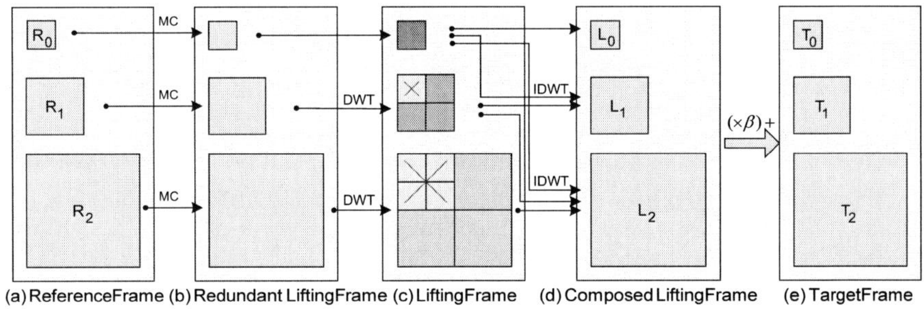

Figure 6.4 Lifting steps of in-scale MC temporal filtering. (a) ReferenceFrame, (b) Redundant LiftingFrame, (c) LiftingFrame, (d) Composed LiftingFrame, (e) TargetFrame.

The better way to solve the problem of spatial scalability is from the aspect of coding structure. Thus, an elegant in-scale MCTF is first proposed by Xiong et al. [115, 116], as shown in Figure 6.4. Assume there are three resolutions to be supported. In addition to the input resolution of frames denoted by subscript 2, two low-resolution versions, denoted by subscript 1 and 0, respectively, are generated by the wavelet filter that is also used in spatial transform. These frames constitute a redundant pyramid representation of original frames. But the multi-resolution temporal transform is designed as a whole so that coded coefficients are not redundant. The multi-resolution temporal transform is depicted in Figure 6.4. First, from Figure 6.4a to 6.4b, each independent motion compensation is performed on reference frame R_i $(i=0,1,2)$ to generate corresponding prediction. Second, from Figure 6.4b to 6.4c, a one-level wavelet transform, which has filters identical to those in spatial transform, is performed on each prediction except for that of the lowest resolution. The low-pass subband of each prediction is dropped in Figure 6.4c. Third, from Figure 6.4c to 6.4d, a new prediction is generated by inverse transforming the remaining high-pass subbands and all information available in all low-resolution layers. Finally, the signal L_i $(i=0,1,2)$ is used in the temporal lifting transform. In this way, the signal at a lower resolution layer is always exactly the wavelet low-pass subband of the signal at the next higher resolution layer. Thus, redundancy can be removed.

The proposed in-scale transform can also be described in the Barbell-lifting framework. We define $A_{L,n}$, $A_{H,n}$, $S_{L,n}$, and $S_{H,n}$ as analysis and synthesis operators of n-level discrete wavelet transform (DWT). After n-level DWT, $SI = \{(L,n),(H,n),(H,n-1),...,(H,1)\}$ is the set of subband index, and $f_s(s \in SI)$ denotes the subband s of any frame f. For example, $f_{L,n}$ is the coarsest scale of f and $f_{H,j}(j = n,n-1,...,2,1)$ is finer scales containing high-frequency details at high resolutions. With these notations, the in-scale lifting steps are formulated as follows.

For odd λ, the λth lifting step modifies odd-indexed frames based on the even-indexed frames. The lifting step for low-pass subband at the coarsest scale is performed according to Eq. (6.13), and the lifting steps for subbands at finer scales are performed according to Eq. (6.14),

$$\begin{aligned} f_{2k+1,L,n}^{(m,\lambda)} =& f_{2k+1,L,n}^{(m,\lambda-1)} + \beta_0^{(m,\lambda)} B_{2k\to 2k+1}^{O}\left(f_{2k,L,n}^{(m,\lambda-1)}\right) \\ &+ \beta_1^{(m,\lambda)} B_{2k+2\to 2k+1}^{O}\left(f_{2k+2,L,n}^{(m,\lambda-1)}\right), \end{aligned} \tag{6.13}$$

$$\begin{aligned} f_{2k+1,H,j}^{(m,\lambda)} =& f_{2k+1,H,j}^{(m,\lambda-1)} + \beta_0^{(m,\lambda)} A_{H,1}\left(B_{2k\to 2k+1}^{O}\left(f_{2k,L,j-1}^{(m,\lambda-1)}\right)\right) \\ &+ \beta_1^{(m,\lambda)} A_{H,1}\left(B_{2k+2\to 2k+1}^{O}\left(f_{2k+2,L,j-1}^{(m,\lambda-1)}\right)\right). \end{aligned} \tag{6.14}$$

For even λ, the λth lifting step modifies even-indexed frames based on the odd-indexed frames similarly, as formulated in Eq. (6.15) and Eq. (6.16),

$$\begin{aligned} f_{2k,L,n}^{(m,\lambda)} =& f_{2k,L,n}^{(m,\lambda-1)} + \beta_0^{(m,\lambda)} B_{2k-1\to 2k}^{E}\left(f_{2k-1,L,n}^{(m,\lambda-1)}\right) \\ &+ \beta_1^{(m,\lambda)} B_{2k+1\to 2k}^{E}\left(f_{2k+1,L,n}^{(m,\lambda-1)}\right), \end{aligned} \tag{6.15}$$

$$\begin{aligned} f_{2k,H,j}^{(m,\lambda)} =& f_{2k,H,j}^{(m,\lambda-1)} + \beta_0^{(m,\lambda)} A_{H,1}\left(B_{2k-1\to 2k}^{E}\left(f_{2k-1,L,j-1}^{(m,\lambda-1)}\right)\right) \\ &+ \beta_1^{(m,\lambda)} A_{H,1}\left(B_{2k+1\to 2k}^{E}\left(f_{2k+1,L,j-1}^{(m,\lambda-1)}\right)\right). \end{aligned} \tag{6.16}$$

The operator $A_{H,1}$ can be viewed as a part of $B()$. But, to easily understand Eq. (6.13) to Eq. (6.16), we keep it separate. The performance of the proposed technique in wavelet video coding can be found by Xiong et al. [115, 116]. Furthermore, the in-scale motion compensation technique is also applicable to current SVC because of the pyramidal multi-resolution coding structure in SVC. We extend the in-scale technique to support arbitrary up- and down-sampling filters and applied it to SVC in both open-loop and close-loop form, with macroblock-level R-D optimized mode selection [117–119]. Experimental results show that the proposed techniques can significantly improve spatial scalability performance of SVC, especially when the bit rate ratio of lower resolution bitstream to higher resolution bitstream is considerable [120].

6.4.2 Subband Adaptive MCTF

In general, a frame to be coded highly correlates with previous frames. This correlation can be exploited by generating a prediction through motion compensation. The correlation strength is dependent on the distance between this frame and its references and the accuracy of estimated motion vectors. Additionally, for a pair of the current frame and its prediction, the correlation strength also varies in different spatial frequency components. As shown in Figure 6.5, there is a frame to be coded and its prediction. After packet wavelet transform, 16 subbands are generated. The correlation coefficients between them in different subbands are quite different. For example, the correlation of low-pass subband is 0.98 but that of the highest subband is only 0.08.

It motivates us to differentiate various spatial subbands during MCTF. The basic idea comes from the optimum prediction problem of random signals. Let X and Y be two correlated random signals. We predict Y from X by a linear model $Z = Y - \alpha X$. The optimum parameter to minimize the mean square prediction error can be solved as

$$\frac{\partial E[z^2]}{\partial \alpha} = \frac{\partial E[Y^2 - 2\alpha YX + \alpha^2 X^2]}{\partial \alpha}. \tag{6.17}$$

In the case $\sigma_X \approx \sigma_Y$, $m_X \ll \sigma_X$, and $m_Y \ll \sigma_Y$, Eq. (6.17) can be approximated to $\alpha \approx \rho_{X,Y}$. It means the best parameter to achieve the optimum prediction is mainly determined by the correlation of the two signals. Therefore, we similarly adjust the strength of temporal filtering for various spatial subbands. It is formulated as follows. For odd λ, the λth lifting step modifies odd-indexed frames based on the even-indexed frames. The lifting step is performed for each subband s as in Eq. (6.18). For even λ, the λth lifting step modifies even-indexed frames based on the odd-indexed

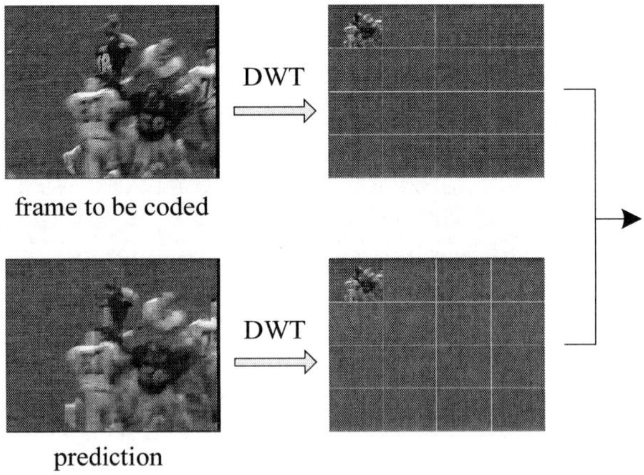

(0,0) 0.98	(0,1) 0.76	(0,2) 0.38	(0,3) 0.55
(1,0) 0.77	(1,1) 0.63	(1,2) 0.29	(1,3) 0.47
(2,0) 0.4	(2,1) 0.26	(2,2) 0.08	(2,3) 0.19
(3,0) 0.61	(3,1) 0.48	(3,2) 0.22	(3,3) 0.38

correlation

Figure 6.5 Correlation coefficients between a frame and its prediction in different subbands.

frames similarly, as in Eq. (6.19). The parameters $\alpha^{(m)(s)}_{2k\to 2k+1}$ and $\alpha^{(m)(s)}_{2k+2\to 2k+1}$ are determined by characteristic of subband-wise temporal correlation in the mth MCTF according to the method discussed by Xiong et al. [121]. Similarly, the operator A_s can be viewed as a part of $B()$. But, to easily understand Eq. (6.18) and Eq. (6.19), we keep it separate. The performance gain of this technique is reported by Xiong et al. [121].

$$\begin{aligned} f^{(m,\lambda)}_{2k+1,s} =& f^{(m,\lambda-1)}_{2k+1,s} \\ &+ \beta^{(m,\lambda)}_0 \alpha^{(m)(s)}_{2k\to 2k+1} A_s \left(B^O_{2k\to 2k+1}\left(f^{(m,\lambda-1)}_{2k}\right)\right) \\ &+ \beta^{(m,\lambda)}_1 \alpha^{(m)(s)}_{2k+2\to 2k+1} A_s \left(B^O_{2k+2\to 2k+1}\left(f^{(m,\lambda-1)}_{2k+2}\right)\right). \end{aligned} \tag{6.18}$$

$$\begin{aligned} f^{(m,\lambda)}_{2k,s} =& f^{(m,\lambda-1)}_{2k,s} \\ &+ \beta^{(m,\lambda)}_0 \alpha^{(m)(s)}_{2k\to 2k-1} A_s \left(B^E_{2k-1\to 2k}\left(f^{(m,\lambda-1)}_{2k-1}\right)\right) \\ &+ \beta^{(m,\lambda)}_1 \alpha^{(m)(s)}_{2k\to 2k+1} A_s \left(B^E_{2k+1\to 2k}\left(f^{(m,\lambda-1)}_{2k+1}\right)\right). \end{aligned} \tag{6.19}$$

6.5 Experimental Results

In this section, we conduct experiments to evaluate the coding performance of the proposed Barbell-lifting coding scheme. In Section 6.5.1, we compare our scheme to MC-EZBC [67, 69], a well-recognized scheme in the literature of 3D wavelet video coding. In Sections 6.5.2 and 6.5.3, we compare our scheme to SVC, the state-of-the-art scalable coding standard, from the aspect of SNR scalability and combined scalability, respectively.

6.5.1 Comparison with Motion Compensated Embedded Zero Block Coding (MC-EZBC)

We compare the Barbell-lifting scheme with MC-EZBC in this section. Only SNR scalability is considered here. Experiments are conducted with *Bus*, *Foreman*, *Coastguard*, *Mobile*, *Stefan*, and *Silence* CIF 30 Hz sequences, which represent different kinds of video. For MC-EZBC, two versions are investigated. One is the old scheme described in MPEG document m9034 [122], in which a comprehensive summary on its SNR scalability performance is provided. The other one is the latest improved MC-EZBC developed by Rensselaer Polytechnic Institute (RPI) [123]. Its performance is obtained based on the executables and configurations provided by Dr. Yongjun Wu and Professor John W. Woods.

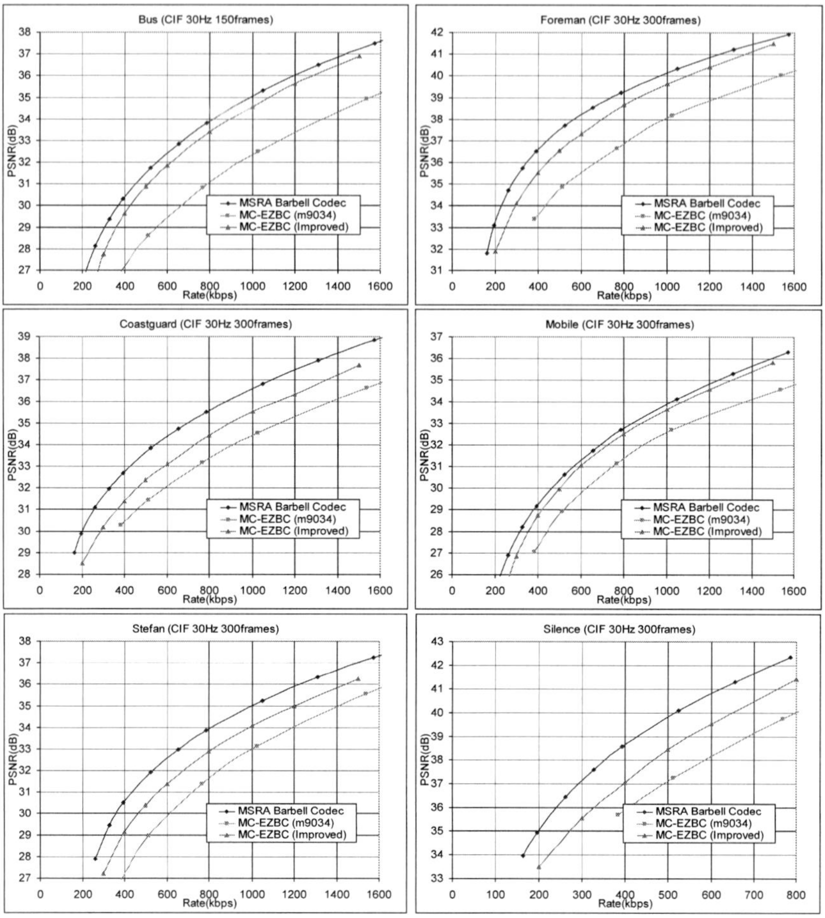

Figure 6.6 Coding performance comparison between the MSRA Barbell codec and MC-EZBC.

To obtain the performance of our proposed scheme, the bit rate ranges by Chen and Woods [122] are used. Four levels of MCTF are applied to all sequences. The resulting temporal subbands are spatially decomposed by a Spacl transform [83]. Base layer coding is not enabled in this experiment. The lambdas for motion estimation at all MCTF levels are set to 16 in our scheme. Figure 6.6 shows the results of our Barbell-lifting coding scheme, basic MC-EZBC (MPEG m9034), and the latest improved MC-EZBC. From Figure 6.6, one can see that for each sequence, the Barbell-lifting coding scheme and improved MC-EZBC scheme outperforms the basic MC-EZBC (m9034) over a wide bit-rate range. The PSNR gain can be about 1.3–3.2 dB. The Barbell-lifting coding scheme still performs better than the improved MC-EZBC scheme. Since many differences exist among the three schemes, it is difficult to analyze which part of the coding algorithm leads to performance difference and how much. But the main reasons accounting for the gain may be the following.

1. In basic MC-EZBC, Haar transform is used in MCTF, while in the Barbell-lifting coding scheme, 5/3 filter is used. The prediction step of wavelet transform using 5/3 filter is bidirectional, which is more effective than the uni-directional prediction in Haar transform. The difference between them is similar to the difference between B-picture coding and P-picture coding in video coding standards. A low-pass filter of 5/3 is better than that of Haar in terms of low-pass property, which makes the low-pass subband generated using 5/3 filter easier to be coded.
2. In the Barbell-lifting coding scheme, adaptively choosing the Barbell functions contributes to the performance gain. Variable block-size motion model similar to the one in H.264/AVC is used, with five motion coding modes, which has been shown to be effective. It makes a good trade-off between the prediction efficiency and the overhead of motion information. Moreover, overlapped block motion compensation and the update operator matching with the prediction step further improve the efficiency of temporal decomposition of the video signals.

6.5.2 Comparison with SVC for SNR Scalability

We also compare our proposed scheme with the latest SVC under the testing conditions defined by JVT [124]. First, we test the SNR scalability performance of both schemes. For SVC, its performance is quoted directly from JVT-T008 [125], which presents the results of the latest stable JSVM reference software, that is, JSVM6 [126]. To obtain the performance of our proposed scheme, the parameters for MCTF levels are set to 5, 5, 4, 4, 4, and 2, for *Mobile*, *Foreman*, *Bus*, *Harbour*, *Crew*, and *Football*, respectively. For each sequence, the spatio-temporal low-pass subband is coded as a base layer by a H.264/AVC codec. Temporal subbands are further spatially decomposed by a three-level Spacl DWT transform. The lambdas for motion estimation at all MCTF levels are set to 16. Figure 6.7 shows the performance of our scheme and the SVC (JVT-T008).

In general, the Barbell-lifting coding scheme works worse than SVC in the testing conditions of SNR scalability. In spite of SVC being developed and optimized extensively by JVT, there are still several possible reasons for the performance differences.

1. The close-loop prediction structure of SVC can reduce or remove the mismatch of the prediction between the encoder and the decoder. However, for the open-loop prediction structure in the Barbell-lifting scheme, the mismatch degrades coding performance. Although the base layer is used in the Barbell-lifting scheme, it only improves the efficiency of the spatio-temporal low-pass subband. It does not contribute to the coding of other subbands or reduce mismatch. However, for 4CIF sequences which are coded at comparatively high bit rates to lead to fewer mismatches, the performance gaps between these two schemes become small. In some cases, for example, with the Harbour sequence, the Barbell-lifting coding scheme can even outperform the SVC.

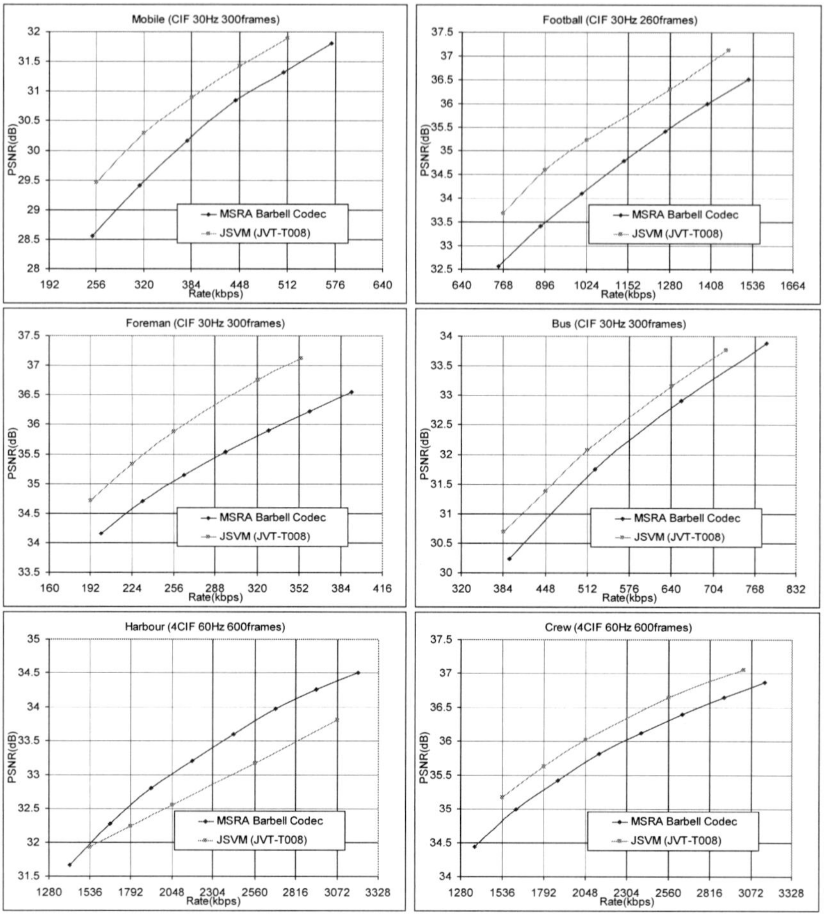

Figure 6.7 Coding performance comparison between the MSRA Barbell codec and SVC for SNR scalability.

2. In SVC, each macroblock is reconstructed instantly after its encoding. It enables effective intra prediction of the next macroblock. However, the absence of intra prediction prevents our scheme from efficiently coding the macroblocks, where motion compensation does not work well. That accounts for why the performance gap is large for the Football and Foreman sequences, which are either high motion sequences or have complex motion.

6.5.3 Comparison with SVC for Combined Scalability

The combined testing conditions defined by Wien and Schwarz [124] support both SNR scalability and spatial scalability. The stream to decode the low-resolution video is extracted from the high resolution one. In the Barbell-lifting coding scheme, a low-resolution video corresponds to the video down-sampled from a high-resolution one using a wavelet low-pass filter, specifically the 9/7 filter used in coding. Therefore, we also use the video down-sampled with a 9/7 wavelet filter as the low-resolution input in SVC, although it can support arbitrary down-sample filtering. Using the same down-sampling filter makes it possible to compare the low-resolution reconstruction qualities of the two schemes in PSNR.

For SVC, the configuration file in JVT-T008 [125] is reused except that the QP is adjusted slightly to support the lowest bit rate specified by Wien and Schwarz [124]. Bitstream is adapted to the given bit rate using quality layers. In the Barbell-lifting scheme, the number of MCTF levels are 5, 4, 3, and 2 for Mobile, Foreman, Bus, and Football, respectively. The lambdas are set to comparatively large values to favor performance at low resolution.

Figure 6.8 shows the comparison results for the Foreman, Football, Mobile, and Bus sequences in CIF format. For the performance at low-resolution, the Barbell-lifting coding scheme is still worse than SVC for the same reason addressed in Section 6.5.2 and the mismatch of MCTF between the encoder and the decoder. But for performance at high resolution, the Barbell-lifting coding scheme shows a coding performance close to SVC. For the Football sequence, the Barbell-lifting scheme even outperforms SVC by up to 0.6 dB, in spite of the higher bit rate at the low resolution. The reason may come from the different structures to support spatial

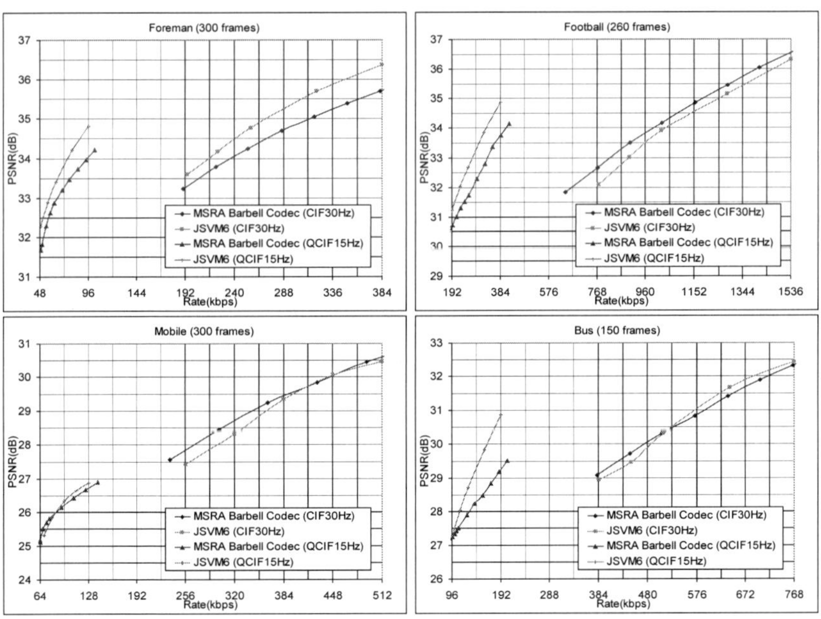

Figure 6.8 Coding performance comparison between the MSRA Barbell codec and SVC for combined scalability.

scalability, as described in Section 6.3.3. The inter-layer redundancy in SVC may lead to coding performance degradation at a high resolution, especially when the bit rate of the low resolution is high. However, for the Barbell-lifting coding scheme, the embedded structure of spatio-temporal decomposition aligns the coding of the low resolution and the high resolution.

6.6 Summary

This chapter first overviews the Barbell-lifting coding scheme. The commonalities and differences between the Barbell-lifting coding scheme and SVC are then exhibited for readers to better understand modern scalable video coding technologies. We then discuss two new techniques to further improve the performance of the wavelet-based scalable coding schemes that are also suitable for SVC. From the comparisons with SVC in terms of technique and performance, there is still a long way to go in perfecting wavelet-based video coding. For example, intra blocks in a high-pass frame are difficult to code efficiently because of global spatial transform; up-sampling and down-sampling filters are constrained to those used in spatial transform, which may result in aliasing visual artifacts in low-resolution video. Furthermore, an R-D optimized result is difficult to achieve because of the open-loop prediction structure used in 3D wavelet video coding.

Part III
Directional Transforms

This part studies directional transforms, another important component in image and video compression. Natural images inherently contain rich directional structures, beyond simple horizontal or vertical lines, as in the contemporary digital images we perceived today. It is true that conventional rectilinear transforms we have been using to represent images are unable to fully capture such directional correlations. For any transform to become superior in coding of natural images, not only does it need to properly capture the directional correlations, it also needs to generate the transform coefficients in such a way that they can be optimally encoded by subsequent entropy coding. It is based on such challenging principle that we propose an innovative mechanism to design directional transforms through incorporating directional operators in the lifting structure.

Chapter 7 presents a novel 2D wavelet transform scheme of adaptive directional lifting (ADL) in image coding. Instead of alternately applying horizontal and vertical lifting, as in present practice, ADL performs lifting-based prediction in local windows in the direction of high pixel correlation. Hence, it adapts far better to the image orientation features in local windows. The ADL transform is achieved by existing 1D wavelets and is seamlessly integrated into the global wavelet transform. The predicting and updating signals of ADL can be derived even at the fractional pixel precision level to achieve high directional resolution, while still maintaining perfect reconstruction. To enhance the ADL performance, a rate-distortion optimized directional segmentation scheme is also proposed to form and code a hierarchical image partition adapting to local features. Experimental results show that the proposed ADL-based image coding technique outperforms JPEG 2000 in both PSNR and visual quality, with the improvement up to 2.0 dB on images with rich orientation features.

In Chapter 8, we introduce directional primary operators to the lifting-based DCT and thereby derive a new directional DCT-like transform, whose transform matrix is dependent on the directional angle and interpolation used there. Furthermore, the proposed transform is compared with the straightforward one, first rotated and then transformed. A JPEG-wise image coding scheme is also proposed to evaluate the performance of the proposed directional DCT-like transform. The first 1D transform is performed according to image orientation features, and the second 1D transform still in the horizontal or vertical direction. At the same time, an approach is proposed to optimally select transform direction of each block because selected directions of neighboring blocks will influence each other. The experimental results show that the performance of the proposed directional DCT-like transform can dramatically outperform the conventional DCT up to 2 dB even without modifying entropy coding.

While directional adaptation is introduced into traditional transforms, different orders of two 1D transforms will result in different results of one 2D transform. Based upon an anisotropic image model, Chapter 9 analyzes the effect of transform orders in terms of theoretical coding gain. Our results reveal that the transform orders have little effect on the coding gain with full decomposition, good directional modes, and good interpolation. However, in practical compression schemes, since high-pass bands are not decomposed fully because of the consideration on complexity, different transform orders have different coding performances, which can be utilized by an adaptive transform order. Motivated by our analyzed results, a directional filtering transform (dFT) (dFT, in order to distinguish from the common usage on DFT) is proposed to better exploit correlations among pixels in H.264 intra-frame coding. It provides an evenly distributed set of prediction modes with an adaptive transform order. Both inter-block and intra-block correlations are exploited in this scheme. Experimental results in H.264 intra-frame coding demonstrate its superiority both objectively and subjectively.

Chapter 7
Directional Wavelet Transform

7.1 Introduction

The past decade has seen increased sophistication and maturity of wavelet-based image compression technologies. Within the family of mathematical transforms for image coding, discrete wavelet transform has unseated discrete cosine transform (DCT) [127–129] as the transform of choice. The wavelet-based JPEG 2000 international standard for still image compression [130] not only obtains superior compression performance over the DCT-based old JPEG standard [129], but also offers scalability advantages in reconstruction quality and spatial resolution that are desirable features for many consumer and network applications.

However, the prevailing practice of 2D wavelet transform has a legacy from traditional 2D DCT transform in that it is implemented by separable 1D filtering in horizontal and vertical directions. This separable 2D wavelet transform is referred to as rectilinear 2D wavelet transform to distinguish it from another separable but adaptive 2D wavelet transform to be proposed, called adaptive directional lifting (ADL). A serious drawback of rectilinear wavelet transforms is that they are ill suited to approximate image features with arbitrary orientation that is neither vertical nor horizontal. In these cases, rectilinear wavelets transform results in large-magnitude high-frequency coefficients. At low bit rates, the quantization noise from these coefficients is clearly visible, in particular causing annoying Gibbs artifacts at image edges with arbitrary directions. This problem has been identified by numerous researchers [127, 128, 131, 132].

How to fully exploit the directional correlation in either the image or frequency domains has been a research topic for many years. Research on DCT-based image coding has converged at directional prediction. Feig et al. incorporate spatial prediction into a JPEG-like code in a manner similar to the fractal-based image compression [127]. Even though this method does not offer a better rate-distortion performance than pure DCT-based coding, it has far fewer block artifacts and markedly better visual quality at very low bit rates. Kondo and Oishi perform directional prediction on DCT blocks. Their prediction is based on one of four coded neighboring

DCT blocks [128]. The new video coding standard H.264 successfully applies the block-based spatial prediction technique to intra-frame coding. Significant coding gains are made over the version without spatial prediction [13].

Direction-aware wavelet/subband-based image coding methods fall into two major categories in their ways of using and coding directional information.

1. *Directional Filter and Transform* — Ikonomopoulos and Kunt first proposed a 2D filter bank that produces one low-pass component image and N directional component images containing high-frequency components in a given direction [131]. Li and He incorporate subband decomposition into the Ikonomopoulos scheme [133], where each rectangular subband contains a given direction. Bamberger and Smith propose a filter bank with critically sampled and wedge-shaped regions to describe directional information [134], which receives more attention for the virtues of maximal decimation and perfect reconstruction. Some new directional filter banks based on Bamberger's method are recently reported by Nguyen and Oraintara [135] and Lu and Do [136]. In addition, many new wavelet transforms have been proposed to preserve fine directional information in the wavelet domain, such as ridgelet [137, 138], curvelet [139], directional wavelet transform [140], contourlet [141], complex wavelet [142, 143], brushlet [144], and so on. These directional filters and transforms provide good presentations of directional data in the frequency domain. They are extensively applied in feature extraction, image enhancement, denoising, classification, and even retrieval. However, they are not suited for compression for lack of efficient entropy coding to exploit directional information in each wavelet region.
2. *Directional Prediction* — The key issue on applying directional prediction to wavelet decomposition is the conflict of global transform and local features. Natural images usually contain rich orientation features. Partitioning an image into many small regions according to correlation direction may cause severe boundary effects and hurt coding efficiency. Taubman et al. propose a technique to resample an image before conventional subband decomposition [132]. The resampling process rotates image edges into horizontal or vertical directions so that following subband decomposition can gain accurate predictions from neighboring horizontal or vertical pixels. Wang et al. use a similar idea but further propose an overlapped extension to prevent coding artifacts around the boundaries of different directional regions [145]. A similar idea on wavelet packets is also reported by Zhang and Wu [146] and Carre et al. [147]. An additional method for directional prediction is to separate images into two or more parts, in which pixels of one part can be directionally predicted from other parts during wavelet decomposition [148, 149]. Both of them are not appropriate to handle varying orientations of image features. Furthermore, the 2D transform used in this category usually is unable to be separated as two 1D transforms.

This chapter presents a new technique of seamlessly integrating directional prediction in arbitrary orientation into the familiar framework of lifting-based 2D wavelet transform. The lifting structure developed by W. Sweldens is an efficient and popular implementation of wavelet transform, in which each finite impulse response (FIR) wavelet filter can be factored into several lifting stages [150]. A local spatial prediction can be readily incorporated into each lifting stage because the lifting stage only involves a few neighboring pixels in the calculation of predicting and updating signals.

Works in this line of thinking have been reported recently. Taubman proposes an orientation adaptive lifting transform for image compression [151]. Boulgouris and Strintzis propose interpolative pyramids for lossless and progressive image coding [152] and an adaptive lifting scheme for lossless image coding [153]. Aiming to minimize the predicted error variance, this method derives four directional filters from the quincunx sampling scheme and selects one of them with a median operation. Claypoole et al. investigate the order of prediction and update in adaptive lifting transform [154]. Li et al. propose a variance-normalized autocorrelation function of the difference image to reconstruct a linear predictor [155]. Gerek and Cetin propose a 2D orientation adaptive prediction filter in lifting structure [156]. In all these schemes, feature directions are estimated from causal data of lower spatial resolution with limited accuracy. Moreover, most of them have only integer precision in directional lifting transform. These two drawbacks reduce coding efficiency.

In this chapter we focus on the ADL wavelet transform. The preliminary results of ADL were first reported by Ding et al. [157], in which simple direction estimation is used on variable-size blocks similar to those in H.264. Chang et al. subsequently proposed to use quincunx sampling in directional lifting transform [158]. ADL is a general framework that allows the use of any 1D wavelet filter, such as the popular Haar, 5/3 and 9/7 wavelets, to perform 2D decomposition on images. The proposed ADL-based image transform is not a true 2D wavelet transform and can be implemented by two 1D transforms, but it has the following advantages not shared by rectilinear 2D wavelet transforms.

- In each lifting stage, the predicting or updating operations are carried out in the direction of image edges and textures in a local window, and are not necessarily horizontal or vertical. This adaptation can significantly reduce the signal energy of high-pass subbands.
- High angular resolution in prediction is achieved by the use of fractional pixels in prediction and update operations. The fractional pixels can be calculated by any existing interpolation method.
- In order to guarantee perfect reconstruction, the predicted and updated pixels are always in integer pixel positions.
- When 2D transform is separated as two 1D transforms, the two 1D transforms are not necessarily performed in two perpendicular directions. However, the split of the low and high subbands is in the horizontal and vertical directions, still generating rectangular subbands for operational convenience.

Due to nonstationary orientation statistics of typical natural images, the proposed ADL image transform works best if coupled with an adaptive image segmentation that classifies the input image into regions of approximately uniform edge/texture orientations. In each of these homogeneous regions, ADL maximizes the compaction of signal energy into the low subband by adjusting the prediction and interpolation direction. To this end, we propose a quadtree-based segmentation technique to construct an adaptive image segmentation in a rate-distortion optimal sense. The rate-distortion optimization is done by the well-known Breiman, Friedman, Olshen, and Stone (BFOS) optimal tree pruning algorithm [159].

Other than the replacement of conventional rectilinear wavelet transform by ADL and the adaptive segmentation component, the other components of the proposed image-coding system resemble their counterparts of JPEG 2000. The coefficient quantization is done by embedded bit plane coding, which preserves the scalability of our code stream. The entropy coding of coefficients is done by the EBCOT technique [103].

7.2 2D Wavelet Transform via Adaptive Directional Lifting

This section presents a new adaptive directional lifting (ADL)-based 2D wavelet transform. The ADL-based transform can overcome the difficulty of rectilinear 2D wavelet transforms in approximating image signals of edges and textures in arbitrary directions. Since the prediction of ADL uses the fractional pixels, the interpolation methods are also described.

7.2.1 ADL Structure

The fundamental difference between the conventional lifting and the proposed ADL lies in the prediction. Instead of always making the predictions in the horizontal or vertical direction, the ADL analyzes the local spatial correlations in all directions, and then chooses a direction of prediction in which the prediction error is minimal.

Consider a 2D signal $x(m,n)_{m,n \in Z}$. Without loss of generality, we assume that this signal is first decomposed into high and low subbands by a 1D wavelet transform in the vertical direction and then in the horizontal direction. With the technique given by Kondi et al. [25], each 1D wavelet transform can be factored into one or multiple lifting stages. A typical lifting stage consists of three steps: split, prediction, and update.

First, all samples are split into two parts: the even polyphase samples x_e and the odd polyphase samples x_o

$$\begin{cases} x_e(m,n) = x(m,2n) \\ x_o(m,n) = x(m,2n+1) \end{cases}. \tag{7.1}$$

In the prediction step, the odd polyphase samples located at integer positions are predicted from the neighboring even polyphase samples. The resulting prediction residuals, or high subband coefficients, are

$$h(m,n) = x_o(m,n) - p_e(m,n). \tag{7.2}$$

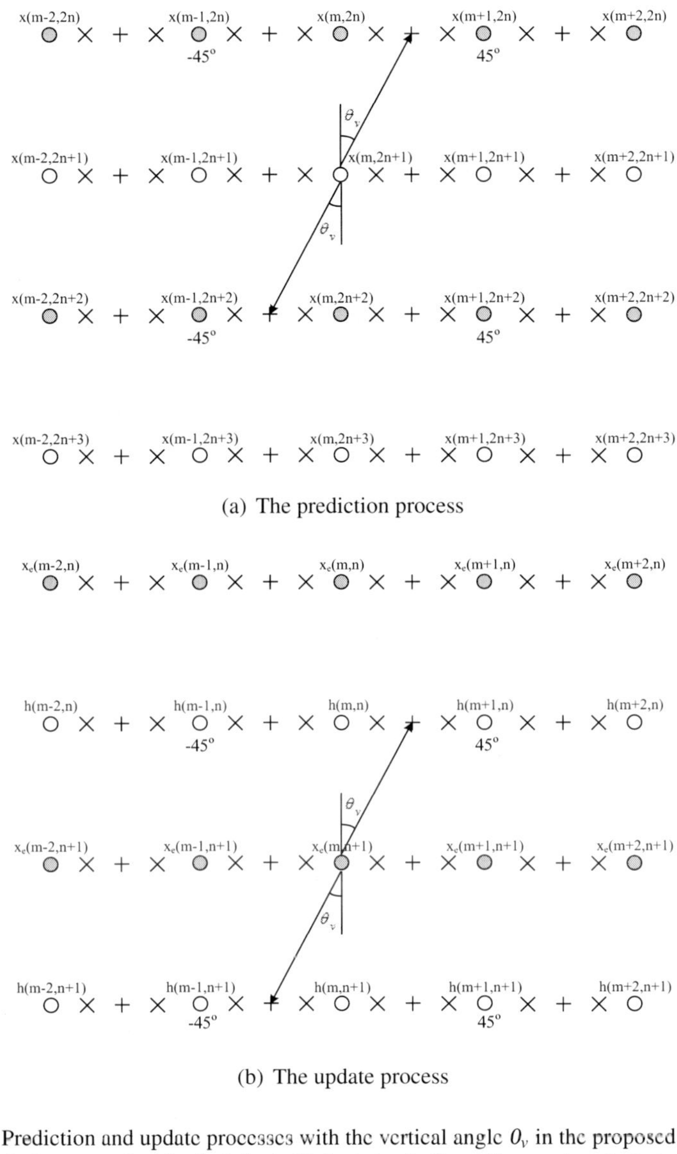

Figure 7.1 Prediction and update processes with the vertical angle θ_v in the proposed ADL, where the integer pixels are marked by "o," the half pixels by "+," and the quarter pixels by "x." (a) The prediction process. (b) The update process.

The prediction of each $x_o(m,n)$ is a linear combination of neighboring even coefficients with strong correlation. As shown in Figure 7.1, assume that the pixels have a strong correlation in the angle θ_v, where the integer pixels are marked by "o," the half pixels by "+," and the quarter pixels by "x." The prediction of $x(m,2n+1)$ is taken as a linear combination of the even polyphase samples identified by the arrows in Figure 7.1, specifically

$$P_e(m,n) = \sum_i \alpha_i x_e(m + sign(i-1) tan\theta_v, n+i), \tag{7.3}$$

where $sign(x)$ is 1 for $x \geq 0$ and -1 otherwise. The weight α_i are given by the filter taps. Note that $x_e(m+sign(i-1)tan\theta_v, n+i)$ is not necessarily sampled at an integer position. The corresponding finite impulse response function in z domain is

$$P(z_1, z_2) = \sum_{i=a}^{b} \alpha_i z_i^{sign(i-1)\tan\theta_v} z_2^i. \tag{7.4}$$

Here, the indexes a and b delimit the finite support of the FIR wavelet filter. Since the prediction is still calculated from the even polyphase samples, if the angle is known, the ADL can still perfectly reconstruct the odd polyphase samples with Eq. (7.2).

In the updating step, the even polyphase samples are replaced with

$$l(m,n) = x_e(m,n) + u_h(m,n). \tag{7.5}$$

Note that $l(m,n)$ always locates at an integer position. The update step of the proposed ADL scheme is performed in the same angle as that in the prediction step. We stress that the ADL framework is very general, and it does not have any restriction on the update angle. We keep the prediction and update angles the same to save the side information of coding the angles, also for the fact that the optimal update angle is consistent with the prediction angle for most images. Consequently, in the update step of the proposed ADL, the even polyphase samples are predicted as

$$u_h(m,n) = \sum_j \beta_j h(m + sign(j)\tan\theta_v, n+j). \tag{7.6}$$

The weights β_j are also given by the filter taps. $h(m+sign(j)\tan\theta_v, n+j)$ may not be integer high-pass coefficient due to $\tan\theta_v$. The corresponding finite impulse response function in z domain is

$$U(z_1, z_2) = \sum_{j=c}^{d} \beta_j z_1^{sign(j)\tan\theta_v} z_2^j. \tag{7.7}$$

Here, the indexes d and c delimit the finite support of the FIR wavelet filter. This step is also trivially invertible. Given $l(m,n)$ and $h(m,n)$, one can perfectly reconstruct the even polyphase samples. In addition, in order to achieve perfect reconstruction of the original 2D signal, we impose the predicted and updated samples to be at integer pixel positions in Eq. (7.2) and Eq. (7.5).

In summary, in Figure 7.2 we present a schematic diagram of the proposed ADL-based wavelet transform. The proposed FIR functions of the extended Haar, 5/3 and 9/7 filters are given as follows, respectively

$$Haar: \begin{cases} P_0(z_1,z_2) = -z_1^{-\tan\theta_v} \\ U_0(z_1,z_2) = z_1^{\tan\theta_v}/2 \\ s_0 = s_1 = 1 \end{cases}, \tag{7.8}$$

$$5/3: \begin{cases} P_0(z_1,z_2) = -\left(z_1^{-\tan\theta_v} + z_1^{\tan\theta_v} z_2\right)/2 \\ U_0(z_1,z_2) = \left(z_1^{\tan\theta_v} + z_1^{-\tan\theta_v} z_2^{-1}\right)/4 \\ s_0 = s_1 = 1 \end{cases}, \tag{7.9}$$

$$9/7: \begin{cases} -1.586134\left(z_1^{-\tan\theta_v} + z_1^{\tan\theta_v} z_2\right) \\ -0.05298\left(z_1^{\tan\theta_v} + z_1^{-\tan\theta_v} z_2^{-1}\right) \\ 0.882911\left(z_1^{-\tan\theta_v} + z_1^{\tan\theta_v} z_2\right) \\ 0.443506\left(z_1^{\tan\theta_v} + z_1^{-\tan\theta_v} z_2^{-1}\right) \\ s_0 = 1.230174 \\ s_1 = 1/s_0 \end{cases}. \tag{7.10}$$

We can view the conventional lifting as a special case of the ADL, when $\theta_v = 0$.

Upon completion of the 1D ADL wavelet transform running on index n, which can be viewed as a generalized vertical transform, the generalized horizontal transform is performed in the same way running on index m. Note that the prediction angle θ_h of generalized horizontal transform is not required to be perpendicular to the vertical counterpart θ_v. In other words, the generalized horizontal transform can optimize the prediction direction for its lifting decomposition. We would like to emphasize the flexibility of the ADL scheme. When a 1D wavelet transform is factored into more than one lifting stage (e.g., the 9/7 filter), except for the first lifting stage, the spatial prediction may be disabled in subsequent stages by setting θ_v to zero if the previous lifting decomposition has removed the directional correlations.

Just like in the conventional rectilinear 2D wavelet transform, the 2D ADL wavelet transform can decompose an image into multiple levels of different scales.

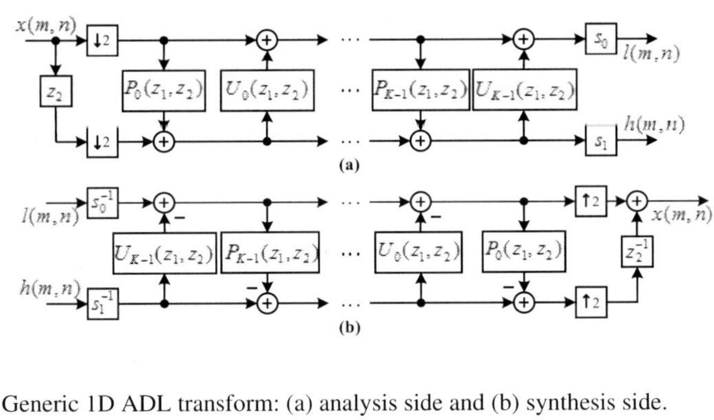

Figure 7.2 Generic 1D ADL transform: (a) analysis side and (b) synthesis side.

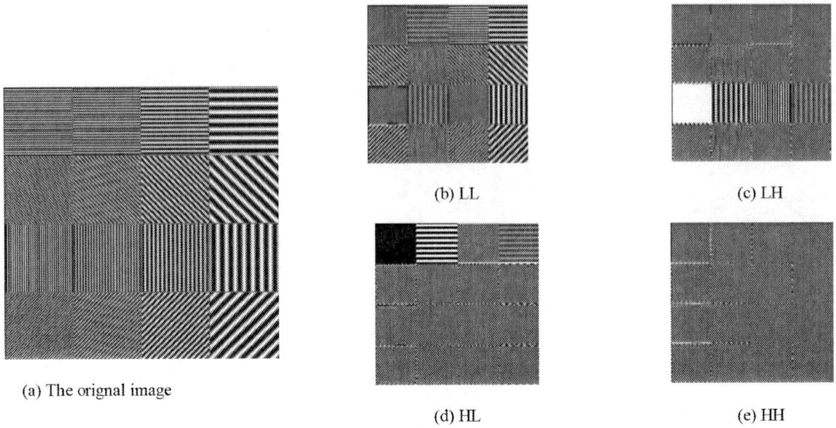

Figure 7.3 Exemplified image and the resulting four subbands with the proposed ADL decomposition. (a) The original image. (b) LL. (c) LH. (d) HL. (e) HH.

Although the generalized vertical and horizontal transforms of the ADL technique do not necessarily have their lifting directions to be perpendicular to each other, they still generate a subband structure that is identical to that of the rectilinear 2D wavelet transform. This is because the 2D ADL wavelet transform splits the low and high subbands horizontally and vertically in turn in exactly the same way as the conventional wavelet transform, creating LL, LH, HL, and HH subbands in one level of decomposition.

To visualize the effect of the 2D ADL wavelet transform, in Figure 7.3 we present the four subbands of the test image of Figure 7.3a, which are the results of one level of ADL decomposition. In the HL subband, only the top row of horizontal stripes contains significant amount of energy after the generalized vertical transform. The adaptive directional prediction successfully removes the statistical redundancy in all other directional patterns, which is exhibited by uniformly small prediction errors in the bottom three rows of the HL subband. The HH subband has even less amount of energy with no recognizable signal structures remaining, after the generalized vertical and horizontal transforms. In the LH subband, the energy compaction is somewhat less effective than in HL and HH subbands. In addition to the vertical stripes that still remain, some high-frequency diagonal textures also exist in the LH subband. This is because the down-sampling process of the generalized vertical transform makes the spatial resolution insufficient for the ADL scheme to find an accurate prediction direction. However, even in this case, the signal energy left in the LH subband is far lower than that of the conventional lifting scheme. To see the advantage of an ADL transform over rectilinear wavelet transform on real images, the reader is referred to Figure 7.6 for a quick preview.

7.2.2 Subpixel Interpolation

In order to perform directional prediction in an arbitrary angle θ_v, the proposed ADL scheme needs to know the intensity values at fractional pixel locations. In other words, $\tan\theta_v$ used in Eq. (7.3) and Eq. (7.6) is generally not an integer. Hence, the interpolation of subpixels becomes an issue.

We present an interpolation technique using Eq. (7.3) as an example. For perfect reconstruction, the integer pixels used to interpolate the fractional pixel at angle θ_v have to be even polyphase samples $x_e(m,n)$. No odd polyphase samples $x_o(m,n)$ can participate in the prediction in this case. The interpolation can be generally described as

$$x_e\left(m+sign(i-1)\tan\theta_v, n+i\right) = \sum_k a_k x_e(m+k, n+i). \tag{7.11}$$

Here, the subscript k indexes the integers around $sign(i-1)\tan\theta_v$, and a_k is the interpolation filtering parameters. Based on Eq. (7.3) and the z transform of Eq. (7.11), we have

$$z_1^{sign(i-1)\tan\theta_v} = \sum_k a_k z_1^k. \tag{7.12}$$

We adopt the popular Sinc interpolation technique [160]. The interpolation of Eq. (7.6) is performed the same way.

The subpixel interpolation problem can also be cast into an optimal filter design problem. Since the interpolation filter can be generalized to any FIR filter, one can design the filter to minimize the energy of the high subband

$$D = \sum_{m,n} h^2(m,n) = \sum_{m,n} \left| x_o(m,n) - \sum_k a_k x(m, n+k) \right|^2. \tag{7.13}$$

The minimization problem

$$\min_{\cdots, a_{k-1}, a_k, a_{k+1}, \cdots} \sum_{m,n} \left[x_o(m,n) - \sum_k a_k x(m, n+k) \right]^2 \tag{7.14}$$

can be solved by the standard least-square method.

The filter coefficients $\cdots, a_{k-1}, a_k, a_{k+1}, \cdots$ can be optimized for a given input image or a set of training images. In the former case, the optimal coefficients have to be sent as side information, whereas in the latter case the training set should have the same statistics of the input image.

In principle, the prediction angle θ_v can be a continuous variable. However, in practice, we found that nine uniformly quantized discrete angles, θ_i, $i = 0, \pm1, \pm2, \pm3, \pm4$, suffice to reap all the coding gains of ADL. Equivalently, the interpolation is done at the spatial resolution of the quarter pixel.

7.3 R-D Optimized Segmentation for ADL

In order for an image codec to benefit from the adaptability of ADL to the edge or texture orientation, it has to segment an image into regions of textures and edges of clear directional bias. This poses the following rate-distortion optimization problem. For ease of implementation, we partition the image recursively into blocks of variable sizes by quad-tree. All pixels in a quad-tree block will be subject to the same ADL transform. The finer the segmentation, the greater degree of gradient uniformity in the resulting blocks. This leads to better directional prediction of the image signal, hence lower distortion. However, the improved signal approximation of ADL is achieved at the expense of increased side information to describe the segmentation tree and the lifting directions of individual blocks. To find the quad-tree of optimal balance between the cost of coding the segmentation tree and the cost of coding ADL transform coefficients, we apply the well-known BFOS algorithm for optimal tree pruning [159].

First, we build a full quad-tree T by recursively partitioning each block into four subblocks until reaching a prespecified minimum block size. For a subtree $S \subset T$ we define its distortion

$$D(S) = \sum_{\tau \in S} \sum_{m,n} \mid h_{\tau,\vartheta}(m,n) \mid, \tag{7.15}$$

where $h_{\tau,\vartheta}(m,n)$ is the high subband coefficient of the ADL decomposition in the optimal direction ϑ, and the subscript τ denotes a terminal (leaf) node of the subtree S. Also, we define the rate of the subtree S as

$$R(S) = \sum_{\tau \in S} r_C(\tau) + \sum_{v \in S} r_T(v), \tag{7.16}$$

where $r_C(\tau)$ is the rate of entropy-coding all high subband coefficients in terminal node τ, and $r_T(v)$ is the rate of coding the side information of the tree node $v \in S$.

Recall from the previous section that ADL chooses the optimal prediction direction ϑ out of nine discrete angles $\theta_i, i = 0, \pm1, \pm2, \pm3, \pm4$. We adopt a predictive coding scheme to code ϑ based on the observation that the image gradient changes smoothly. Referring to Figure 7.4, in a general location of a terminal quad-tree block

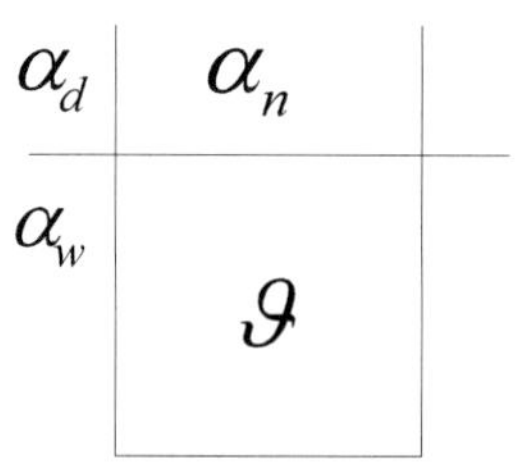

Figure 7.4 Prediction of lifting direction.

related to its neighboring blocks of lifting angles α_n, α_w, and α_d, we predict the optimal lifting direction ϑ to be

$$\hat{\vartheta} = \begin{cases} \alpha_w & |\alpha_d - \alpha_w| > |\alpha_d - \alpha_n| \\ \alpha_n & |\alpha_d - \alpha_w| \leq |\alpha_d - \alpha_n| \end{cases} . \tag{7.17}$$

Then, the prediction error $e_\vartheta = \vartheta - \hat{\vartheta}$ is arithmetic coded. The coded segment structure and predicted angle error are embedded in the user data part of JPEG 2000.

If $v \in S$ is an internal node, then $r_T(v) = 1$, that is, the bit to indicate the event of splitting the corresponding block; otherwise, for each leaf node $\tau \in S, r_T(\tau)$ is the number of bits to code the coefficients $h_{\tau,\vartheta}(m,n)$ in node τ and the optimal direction ϑ of the ADL, plus one bit to signal the terminal node. The rate of coding is estimated by the entropy of ADL coefficients $h_{\tau,\vartheta}(m,n)$ in node τ, and the rate of coding ϑ is estimated by the entropy of prediction residual $e_\vartheta = \vartheta - \hat{\vartheta}$.

Among all the pruned subtrees of T, an R-D optimal segmentation tree S^* is the one that minimizes the cost function

$$J = D(S) + \lambda R(S), \tag{7.18}$$

where λ is a Lagrangian multiplier that determines the total rate of coding the subtree S^* and the coefficients $h_{\tau,\vartheta}(m,n)$ of all the terminal nodes $\tau \in S$. Given a λ, we apply the BFOS optimal tree pruning algorithm to compute S^*. In order to meet a target rate R_0 of coding an image, one can apply the BFOS algorithm iteratively in a binary search of the corresponding value of λ_0. However, this process is expensive. Alternatively, we find a simple fast trick to approximate the value of λ_0, using the JPEG 2000 image code as a reference. It is well known that the EBCOT technique of JPEG 2000 generates an almost convex operational R-D curve. The proposed ADL image coding method also adopts the EBCOT technique for entropy coding of $h_{\tau,\vartheta}(m,n)$ and, hence, has an almost convex operational R-D curve as well. The only difference is that the ADL scheme adapts to the image signal better than the rectilinear wavelet transform. Everything else being equal, the former method approximately translates the operational R-D curve of the latter. Based on this observation, we run the JPEG 2000 code first to obtain its R-D slope $\lambda_J(R_0)$ at the target rate R_0. Then we estimate the Lagrangian multiplier of our method for the target rate R_0 to be $\lambda_0 = \lambda_J(R_0) - \delta$, where δ is an offset. Figure 7.5 shows the segmentation of the test image Barbara generated by the above algorithm and the optimal prediction directions of ADL in the quad-tree terminal nodes.

7.4 Experimental Results and Observations

The proposed ADL wavelet transform is implemented. In order to evaluate the ADL performance only as objectively as possible, we simply replace the wavelet transform module of JPEG 2000 by the ADL transform and use the same bit-plane coding (quantization module) and EBCOT technique (entropy coding module) as in JPEG

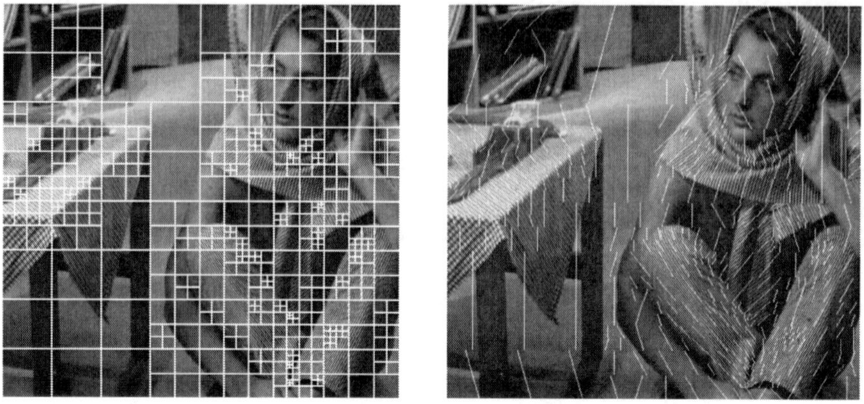

Figure 7.5 Partition of Barbara and the directions for each block.

2000. On a side note, the high modularity of the proposed ADL wavelet transform and its identical subband decomposition structure to other separable 2D transforms make our experiments very easy to conduct within the JPEG 2000 framework. This also means that the proposed ADL-based image codec can be made nearly compatible to the JPEG 2000 standard.

We report the experimental results of six common testing images: Barbara (512×512), Lena (512×512), Baboon (512×512), Bike (2560×2048), Woman (2048×2560), and Cafe (2560×2048), plus the first frame of the Foreman (352×288) video sequence from the MPEG testing set. The reference software VM 9.0 of JPEG 2000 is used in our comparison study. We compare the abilities of the proposed ADL wavelet transform and that of JPEG 2000 in energy packing, or spatial de-correlation. The same 5/3 filter is used in two methods.

Table 7.1 tabulates the average coefficient magnitudes of the LH, HL, and HH subbands. The ADL technique has a significant advantage over the conventional rectilinear wavelet transform on Barbara. This should be expected because this testing image contains strong directional textures. The reduction in average coefficient

Table 7.1 Average coefficient magnitudes in the LH, HL, and HH subbands; the numbers in brackets represent the percentage of reduction.

Subbands	Methods	Barbara	Bike	Cafe	Foreman
LH	JPEG 2000	5.31	3.01	5.01	1.3
	ADL	2.35(55.8%)	2.25(27.3%)	4.04(19.2%)	0.92(24.3%)
HL	JPEG 2000	2.26	3.19	5.51	2.07
	ADL	1.44(36.0%)	2.39(25.1%)	4.4(20.1%)	1.43(31.5%)
HH	JPEG 2000	1.48	1.05	1.56	0.6
	ADL	0.97(34.6%)	0.88(15.5%)	1.5(4.5%)	0.51(16.0%)

magnitude is 55.8% in the LH subband, 36% in the HL subband, and 34.6% in the HH subband. On other test images, the ADL technique also outperforms JPEG 2000.

Parallel to Table 7.1, Figure 7.6 shows one level of wavelet decomposition of testing image Barbara by JPEG 2000 and the proposed ADL. For clear visualization, the magnitudes of coefficients in the LH, HL, and HH subbands are scaled into the range [0, 255] in the plot. We bring the readers' attention to the fact that the ADL technique not only has an appreciably lower signal energy in the high-frequency subbands in comparison with JPEG 2000, but also its residual signals in the high-frequency subbands are much less correlated to the original spatial features of the image. This property contributes to superior visual quality of the reconstructed images by ADL, as we will see in Figure 7.7.

To compare the coding performance of the two methods, we present in Table 7.2 the peak signal-to-noise ratio (PSNR) results of all the test images at 0.125, 0.25, 0.5, and 1.0 bpp. For either method, all the images are decomposed by the three-level 2D transform. Both the popular biorthogonal 9/7 and 5/3 filters are tested. The Sinc interpolation is used by the ADL to generate subpixels. The coding gain can be up to 2.0 dB for Barbara coded with the 5/3 filter. On the test set the gain of ADL ranged from 0.21 dB to 1.36 dB, depending on the presence or lack of orientation features in the image. However, even on relatively smooth images like Lena, the ADL method still enjoys an advantage over the rectilinear 2D wavelet transform. The results with different prediction precisions are also given in Table 7.2. The results of subpixel prediction are significantly better than those of the integer prediction with both the 5/3 and 9/7 filters. However, the results of quarter-pixel precision are only marginally better than those of half-pixel prediction on images of strong edge features such as Barbara and Women. On relatively smooth images, the angular prediction of quarter-pixel precision can be even slightly worse than that of half-pixel precision (see the case of Foreman at 1.0 bpp).

Table 7.3 shows the average number of side information bits for coding segmentation tree and associated directions, which depends on overall bit rates and image contents. One can see that the side information bits constitute only a small portion of total code length. The percentage of overhead bits at low bit rates is less than that at high bit rates because of the proposed R-D optimized segmentation and direction estimation.

Due to its adaptability to directional signal features, the ADL wavelet transform tends to reconstruct image edges better than the rectilinear 2D wavelet transforms at the same bit rates. Since the human visual system is highly sensitive to edges, the former method should have superior visual quality than the latter method, which is indeed corroborated by our experimental results. As an example, Figure 7.7 presents the decoded Barbara images by the two methods, both at the rate 0.3 bpp and using the 5/3 filter. The image decoded by JPEG 2000 has severe moir patterns on the scarf and pants, which even change the texture orientations of the original image. Also, edge ringing effects are clearly visible in the JPEG 2000 decoded image. In contrast, the ADL-based image code eliminates most of the above artifacts. It reproduces the high-frequency scarf pattern almost perfectly, and greatly reduces the ringing effects around the edges.

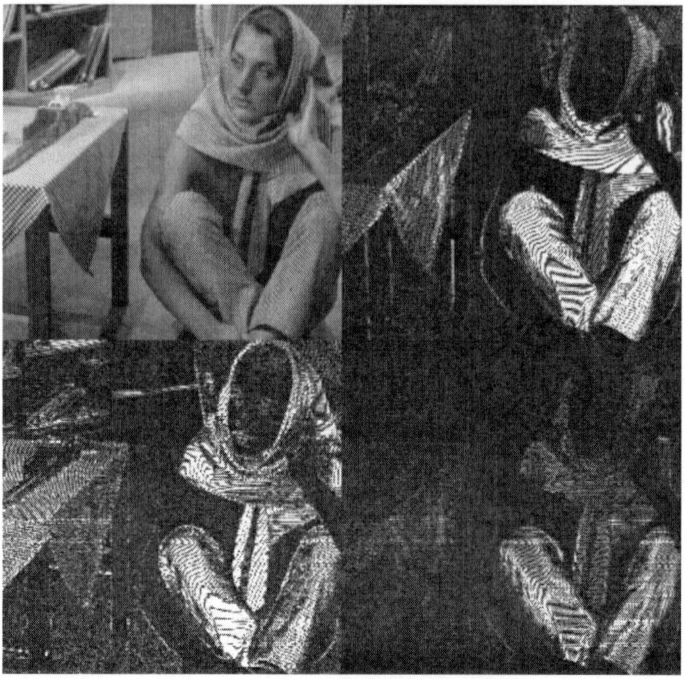

(a)

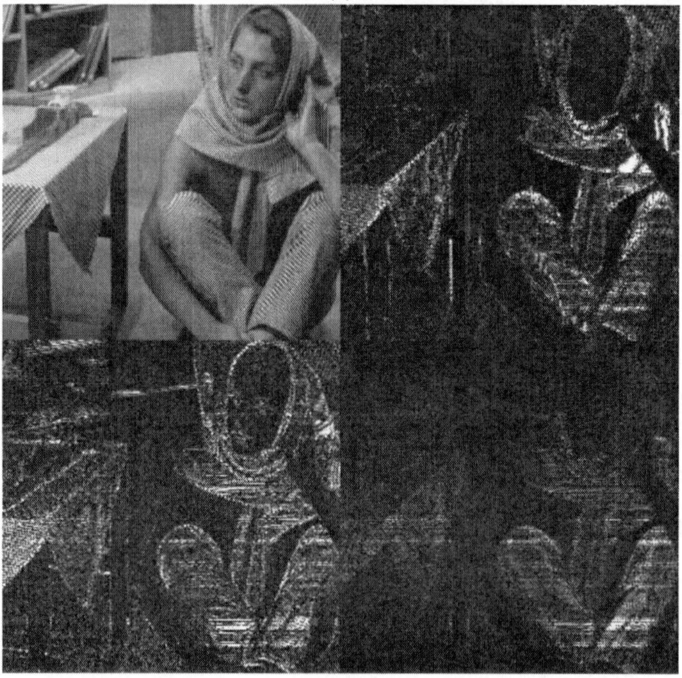

(b)

Figure 7.6 Coefficient magnitudes of (a) JPEG 2000 and (b) ADL after one level of 2D decomposition.

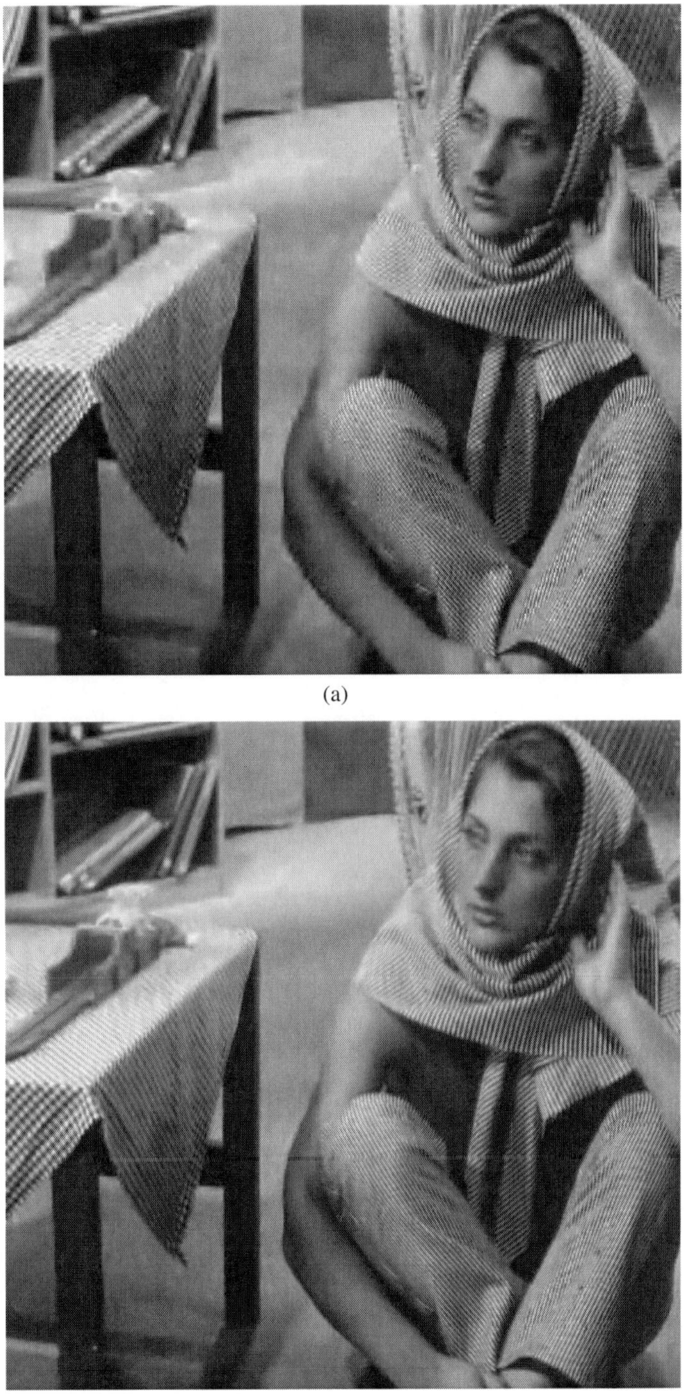

(a)

(b)

Figure 7.7 Visual quality comparison of decoded Barbara at 0.3 bpp: (a) JPEG 2000 versus (b) ADL.

Table 7.2 Comparisons of coding performance (in decibels) between JPEG 2000 and the proposed ADL-based coding scheme with the 5/3 and 9/7 filters.

Images	bpp	5/3 filter				9/7 filter			
		J2K	ADL 1-pel	ADL 1/2-pel	ADL 1/4-pel	J2K	ADL 1-pel	ADL 1/2-pel	ADL 1/4-pel
Barbara	0.125	24.59	25.69	26.07	25.95	25.02	26.16	26.32	26.45
	0.25	27.38	28.79	29.13	29.22	28.27	29.37	29.70	29.78
	0.5	30.95	32.35	32.87	32.95	32.15	33.10	33.50	22.58
	1	36.04	36.89	37.16	37.24	37.11	37.58	37.84	37.88
Lena	0.125	30.11	30.66	30.75	30.68	30.41	30.88	31.00	31.02
	0.25	33.22	33.76	33.90	33.88	33.78	34.13	34.18	34.25
	0.5	36.45	36.80	36.91	36.94	37.02	37.17	37.23	37.30
	1	39.51	39.68	39.74	39.73	40.06	40.09	40.17	40.20
Baboon	0.125	21.40	21.49	21.44	21.39	21.50	21.50	21.70	21.69
	0.25	22.87	23.03	23.05	23.04	23.10	23.10	23.25	23.23
	0.5	25.17	25.35	25.39	25.34	25.52	25.52	25.73	25.73
	1	28.62	28.79	28.83	28.88	29.02	29.02	29.29	29.32
Bike	0.125	25.74	26.62	26.66	26.61	25.93	26.77	26.78	26.78
	0.25	29.06	29.96	29.96	29.97	29.36	30.09	30.11	30.12
	0.5	33.09	33.80	33.79	33.76	33.38	33.98	33.99	34.00
	1	37.73	38.27	38.29	38.28	38.04	38.46	38.47	38.48
Cafe	0.125	20.42	20.42	20.57	20.55	20.70	20.80	20.83	20.84
	0.25	22.74	22.74	23.00	22.99	23.09	23.25	23.29	23.32
	0.5	26.42	26.42	26.66	26.67	26.78	26.93	27.02	27.06
	1	31.72	31.72	31.87	31.90	32.02	32.08	32.13	32.18
Women	0.125	26.59	26.77	26.96	27.00	26.94	27.16	27.33	27.35
	0.25	29.24	29.35	29.57	29.65	29.67	29.82	29.98	30.06
	0.5	33.00	33.03	33.14	33.24	33.42	33.51	33.60	33.72
	1	37.96	37.96	37.97	38.07	38.31	38.33	38.36	38.47
Foreman	0.125	28.71	29.82	29.85	29.94	29.15	30.06	30.11	30.16
	0.25	32.32	33.48	33.50	33.53	32.81	33.68	33.70	33.77
	0.5	35.89	37.19	37.19	37.25	36.33	37.27	37.44	37.43
	1	40.60	41.34	41.38	41.30	40.85	41.55	41.68	41.63

The ADL wavelet transform retains the PSNR and spatial scalability of JPEG 2000. Although the segmentation tree for ADL is optimized for a given target rate as we discussed in Section 7.3, the rate-distortion performance of the proposed ADL-based image coding still outperforms JPEG 2000 in a wide range of bit rates when operating progressively. In Figure 7.8, we compare the rate-distortion performances of the scalable ADL coded stream optimized for the mid bit rate of 0.6 bpp versus JPEG 2000 of four quality layers. Only when the bit rate is extremely low (at or below 0.1 bpp), the ADL-based image coding has inferior performance to that of JPEG 2000 due to the fact that the side information for the segmentation tree has to be placed at the beginning of the coded stream, and, hence, this part of the coded stream is not scalable.

We have incorporated the ADL wavelet transform into the set partitioning in hierarchical tree (SPIHT) scheme to evaluate its performance when coupled with a

Table 7.3 Bits for coding the segmentation tree and associated directions (bits per pixel).

Total bpp	Barbara	Lena	Baboon	Bike	Cafe	Women	Foreman
ADL 5/3							
0.125	0.015	0.009	0.008	0.007	0.006	0.004	0.017
0.25	0.019	0.020	0.012	0.013	0.009	0.007	0.021
0.5	0.034	0.040	0.021	0.031	0.021	0.021	0.033
1	0.034	0.040	0.021	0.031	0.021	0.021	0.033
ADL 9/7							
0.125	0.008	0.005	0.004	0.003	0.002	0.002	0.011
0.25	0.013	0.009	0.007	0.006	0.003	0.003	0.013
0.5	0.018	0.018	0.008	0.013	0.007	0.005	0.021
1	0.018	0.018	0.008	0.013	0.007	0.005	0.021

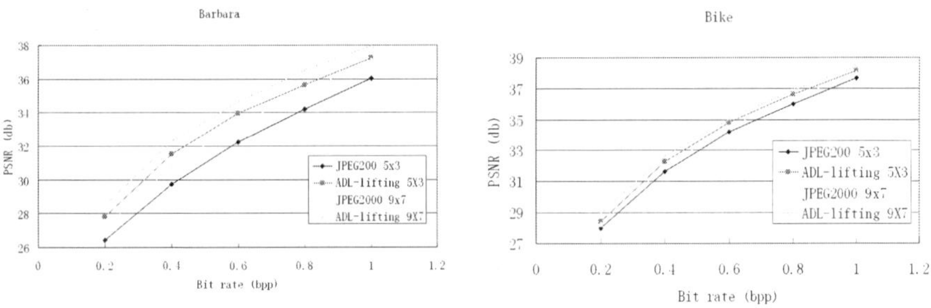

Figure 7.8 Scalable-rate PSNR curves of ADL-based code optimized for 0.6 bpp versus those of JPEG 2000.

Table 7.4 Comparison of the coding performance between SPIHT and ADL-based SPIHT in PSNR.

Total bpp	Barbara	Lena	Baboon	Bike	Cafe	Women	Foreman
SPIHT							
0.125	24.83	21.69	31.04	25.89	20.67	27.33	29.34
0.25	27.56	23.24	34.03	29.13	23.03	29.95	32.76
0.5	31.38	25.61	37.07	33.02	26.49	33.59	36.33
1	36.40	29.16	40.18	37.71	31.73	38.28	40.95
ADL-Based SPIHT							
0.125	26.38	21.79	31.13	26.41	20.78	27.60	30.19
0.25	29.28	23.37	34.18	29.72	23.19	30.20	33.76
0.5	33.14	25.76	37.22	33.54	26.69	33.74	37.34
1	37.63	29.29	40.25	38.06	31.86	38.33	41.60

different entropy coding scheme. Table 7.4 tabulates the coding results of the ADL-based SPIHT at quarter-pixel precision and the original SPIHT. The parameters used are the same as those in the original SPIHT. Only the wavelet transform is replaced with the proposed ADL wavelet transform. One can see that the ADL approach makes similar gains with the zero-tree entropy coding technique of SPIHT as with the EBCOT entropy coding technique of JPEG 2000.

Finally, let us discuss the computational complexity of the proposed ADL-based image codec in comparison with that of JPEG 2000. Clearly, the ADL-based image encoder has considerably higher complexity since it has to compute the image dependent segmentation tree T, and then searches for the optimal prediction direction in each terminal node of T. However, the decoder complexity of an ADL-based image compression method can be made comparable to that of JPEG 2000. This is because the inverse ADL transform is only slightly more expensive than inverse rectilinear transform, once the optimal lifting direction is given as side information. The asymmetric design of the ADL technique makes it suitable for some Internet and wireless applications that aim to achieve the highest possible rate-distortion performance by heavily optimizing the image coding offline and once and for all, while maintaining a reasonably low decoding complexity.

7.5 Summary

The rigidity of the existing rectilinear 2D wavelet transforms makes them ill suited for approximating image features of arbitrary orientations. This weakness can be overcome by the proposed new ADL technique. It can be seamlessly integrated into the conventional, global, and separable 2D wavelet transform, and can be implemented by any wavelet filter. In each ADL lifting stage, the prediction step can be performed in the direction of the strongest pixel correlation rather than stay mechanically fixed in the horizontal or vertical direction. Even though its prediction and update operations are based on fractional pixels, the ADL wavelet transform can guarantee perfect reconstruction without imposing any constraint on the interpolation method. A rate-distortion, optimized image-segmentation method is also developed so that ADL can efficiently approximate directional image features in local regions. Empirical results demonstrate the superior objective and perceived quality of the ADL-based image codec.

There are still several open issues to be investigated in the proposed ADL-based coding schemes. First, the interpolation used in the ADL wavelet transform is always performed in either the horizontal or vertical direction. It may blur the orientation property existing in raw images. Second, image blocks with different directions are continuously processed, which may cause boundary effects in the block boundaries. Third, the entropy coding does not take directional information into account, which should be used in the context model of arithmetic coding.

Chapter 8
Directional DCT Transform

8.1 Introduction

Most of the image compression schemes are constructed upon the identical architecture of the 2D transform followed by entropy coding, for example, JPEG uses the 2D discrete cosine transform (DCT) and variable length coding (VLC) [129], and JPEG 2000 uses the 2D wavelet transform and Embedded Block Coding with Optimized Truncation (EBCOT) [103, 130]. Either 2D DCT or 2D wavelet transform is implemented by separable 1D transform in horizontal and vertical directions. A serious drawback of these transforms is that they are ill-suited to approximate image features with arbitrary orientation that is neither vertical nor horizontal. In these cases, they result in large-magnitude high-frequency coefficients. At low bit rates, the quantization noise from these coefficients is clearly visible, in particular causing annoying Gibbs artifacts at image edges with arbitrary directions.

How to incorporate directional information into transform is a challenging problem in both image and video coding. Some work has been reported in wavelet and subband transforms. Taubman et al. first proposed a technique to resample an image before conventional subband decomposition [132]. The resampling process actually rotates image edges into horizontal or vertical direction so that following subband decomposition can gain accurate predictions from neighboring horizontal or vertical pixels. The lifting structure developed by Sweldens provides a good way to incorporate directional information into wavelet transform, in which each finite impulse response (FIR) wavelet filter can be factorized into several lifting stages [150]. Consequently, some adaptive lifting transforms have been proposed for the purpose [151, 153–158, 161]. In particular, the proposed adaptive directional lifting (ADL) by Ding et al. [157, 161] can outperform JPEG 2000 by up to 2 dB on images with rich orientation features by introducing directional data coding and subpixel directional prediction.

The efforts in DCT are to first introduce spatial prediction before transform. Feig et al. incorporate the idea of spatial prediction into a JPEG-like code in a manner similar to the fractal-based image compression [127]. Even though this method does not offer a better rate-distortion (R-D) performance than pure DCT-based coding, it has far fewer block artifacts and markedly better visual quality at very low bit rates. Kondo and Oishi perform directional prediction on DCT blocks. Their prediction is based on one of four coded neighboring DCT blocks [128]. The new video coding standard H.264 successfully applies the block-based spatial prediction technique to intra-frame coding. Significant coding gains are made over the version without spatial prediction [13].

To the best of our knowledge, the paper from Zeng and Fu is the first effort on how to incorporate directional information into DCT [162]. Their directional DCT is motivated by SA-DCT (shape-adaptive DCT) [163]. For an N-by-N block, the first 1D DCT is performed along with the direction selected, where each column is of a different length. Then DCT coefficients are reorganized according to their frequencies from DC to AC. Coefficients at the same frequency viewed as a row are transformed by the second 1D DCT, where each row is also of a different length. Since the directional transform is constrained within one block, direction selection is relatively easy in Zeng's scheme. But this also potentially hurts the performance of directional DCT because the correlation among neighboring blocks with the similar direction is not exploited. In addition, directions at fractional pixels are not considered yet by Zeng and Fu [162].

Inspired by the lifting-based transform [150] and our previous work on ADL [157, 161], a lifting-based directional DCT-like transform is presented in this chapter, which can be performed along arbitrary direction in theory. The lifting scheme factorizes 1D DCT into a series of so-called primary operations. We propose the corresponding directional form of each primary operation because introducing directional information in primary operation is much easier than in a whole transform. All of the directional primary operations construct the proposed directional DCT-like transform. Its matrix is dependent on selected direction angle and interpolation used there. The perfect reconstruction property is guaranteed by the lifting scheme when transform coefficients are not quantized at all.

8.2 Lifting-Based Directional DCT-Like Transform

In this section, the lifting structure of 1D DCT is first introduced in a self-contained way. And then the proposed 1D directional DCT-like transform is developed from the lifting structure. Since the 8-point DCT is extensively used in image coding, only the 8-point directional transform is taken as an example to be discussed here. The proposed ideas can be easily applied to a DCT of arbitrary size.

8.2.1 Lifting Structure of DCT

There are many approaches reported to factorize any N-point 1D DCT into a series of primary operations through plane rotation and butterfly for the fast implementation

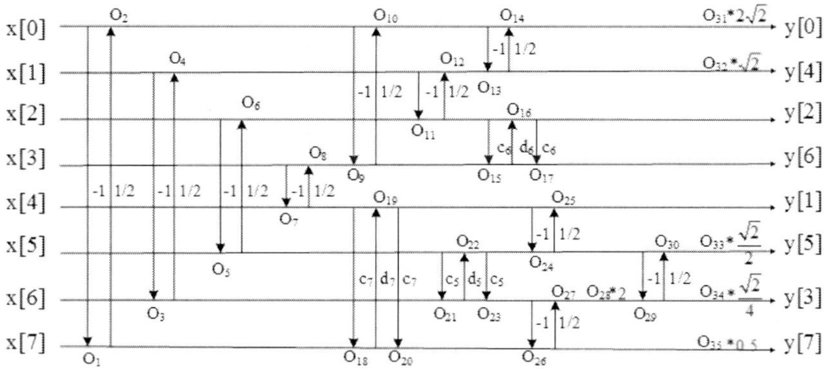

$$c_n = \frac{\cos(n\pi/16) - 1}{sin(n\pi/16)} \qquad d_n = \sin(n\pi/16)$$

Figure 8.1 Factorizing 8-point DCT into primary operations. $x[n]$ and $y[n]$ $(n = 0; 1; ...; 7)$ are the input signal and output DCT coefficient, respectively. $O(i = 1; 2; ...; 35)$ is the primary operation.

in either software or hardware [164–167]. With the lifting structure emerging, several lifting-based fast multiplierless approximations of DCT have been reported by Tran [168] and Liang and Tran [169]. It has been proven that any orthogonal filterbank including DCT can be decomposed into delay elements and plane rotations by lattice factorizations [170]. Since the lifting structure enables flexible implementation of a transform with perfect reconstruction, we take the work from Liang and Tran [169] as the basis to develop directional DCT-like transform.

Similar to the work of Liang and Tran [169], 8-point 1D DCT is implemented in the form of lifting structure as shown in Figure 8.1, where $O_{21} - O_{30}$ are slightly modified according to Loefflers factorization [166] so that the pixel distances in these operations are as close as possible, thus fully exploiting pixel correlation. Eight horizontal lines of Figure 8.1 represent 8 pixels to be transformed. The arrow lines and scales denote primary operations, where the DCT transform is factorized to totally 35 steps indicated by $O_1 - O_{35}$ in Figure 8.1. $O_9 - O_{17}$ and $O_{18} - O_{30}$ form two independent paths. The order of primary operations in different paths has no dependency.

To make the terminologies and discussions of this chapter consistent and clear, we will discuss both the lifting-based DCT given in Figure 8.1 and the proposed directional DCT-like transform in Z domain because the proposed 1D directional transform of a row or a column may influence each other with the transforms of other neighboring rows or columns through interpolation. Assume that x is a 2D image of N rows and M columns. 1D lifting DCT is performed in the vertical direction. Pixels in each row are granted as a 1D signal. The Z transform of this signal is defined as

$$X[n] = \sum_{m=1}^{M} x[n,m] z^{-m}, \quad n = 1, \cdots, N. \tag{8.1}$$

1D lifting DCT is performed on $X[n]$ according to the same lifting structure and primary operations in Figure 8.1. It actually contains M of 1D vertical lifting DCT although they are independent.

All primary operations in Figure 8.1 can be categorized into two types. If there is an arrow line from a pixel to another with a parameter α, the operation can be represented in Z domain as

$$O\left(X\left[n_i\right], X\left[n_j\right], \alpha\right) = \left\{ X \longleftarrow \begin{bmatrix} 1 & & & & & & & \\ & 1 & & & & & & \\ & & 1 & & & & & \\ & & & 1 & & & & \\ & & & & 1 & & & \\ & \alpha & & & & 1 & & \\ & & & & & & 1 & \\ & & & & & & & 1 \end{bmatrix} X \right\}. \tag{8.2}$$

X is a row vector, which consists of 8 continuous $X[n]$. For the convenience of later discussion, we call $X[n_i]$ and $X[n_j]$ as the target and source of a primary operation, respectively. α locates at the n_i-th row and the n_j-th column of operation matrix. If there is a scale parameter α beside a pixel, the second type of primary operation is defined as

$$O\left(X\left[n_i\right], \alpha\right) = \left\{ X \longleftarrow \begin{bmatrix} 1 & & & & & & & \\ & 1 & & & & & & \\ & & 1 & & & & & \\ & & & \alpha & & & & \\ & & & & 1 & & & \\ & & & & & 1 & & \\ & & & & & & 1 & \\ & & & & & & & 1 \end{bmatrix} X \right\}. \tag{8.3}$$

α is the diagonal element at the n_i-th row of operation matrix. According to Figure 8.1, the 8-point 1D DCT is factorized to 35 primary operations from left to right

$$Y = DCT_8\left(X\right) = O_{35} \circ O_{34} \circ \cdots O_2 \circ O_1\left(X\right). \tag{8.4}$$

Y is obtained DCT coefficient vector in Z domain. It contains all coefficients of 1D vertical DCT in all columns. Each column has a unique z^{-m}. $DCT_8()$ denotes the 8-point 1D DCT. O_k is the kth primary operation. Its parameter and operated pixels are depicted in Figure 8.1.

The inverse primary operations of $O(X[n_i], X[n_j], \alpha)$ and $O(X[n_i], \alpha)$ are defined as $O^{-1}(X[n_i], X[n_j], -\alpha)$ and $O^{-1}(X[n_i], 1/\alpha)$, respectively. Let O_k^{-1} denote the inverse operation of O_k. Then, the inverse 8-point DCT transform is represented as

$$X = DCT_8^{-1}\left(Y\right) = O_1^{-1} \circ O_2^{-1} \circ \cdots O_{34}^{-1} \circ O_3^{-1}5\left(Y\right). \tag{8.5}$$

Since each of these primary operations is perfectly reversible, the DCT given in Eq. (8.4) is also perfectly reversible.

8.2.2 Directional DCT-Like Transform

Introducing directional information in primary operations is much easier than in a whole DCT because a primary operation only involves a few pixels. The first step of developing directional transform is to define the directional primary operation in the lifting structure. As shown in Figure 8.2a, if a primary operation does not involve directional information, 1D vertical DCT is performed on each column and does not use any pixels in other columns. But, if a primary operation is performed along a direction rather than the vertical one, the 1D DCT transform may have to involve pixels of multiple columns. As shown in Figure 8.2b, if the source of a primary operation locates at a fractional pixel illustrated by a gray circle, its value is interpolated from neighboring integer pixels in a row. If the target of a primary operation locates at a fractional pixel, the resulting value is only an intermediate one. Similar to the energy distributed update (EDU) method proposed by Lee et al. [26], the intermediate value is immediately distributed to neighboring integer pixels according to their weighting factors of interpolation. Note that the target at a fractional pixel only happens on the upward primary operations. The downward and upward primary operations are illustrated by downward and upward arrows in Figure 8.1, respectively.

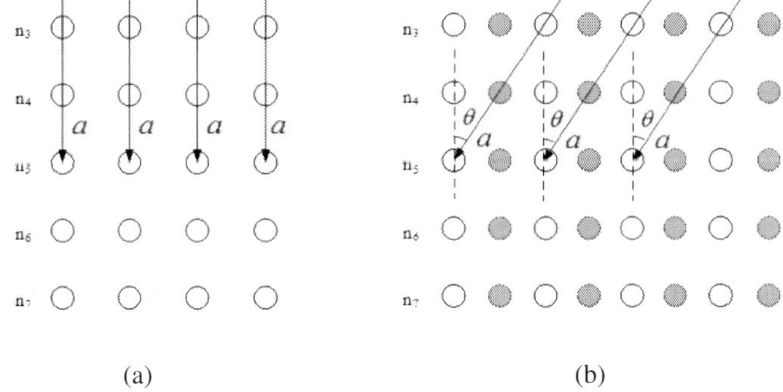

Figure 8.2 Exemplified primary operations: (a) Nondirectional and (b) directional, where the white circles denote integer pixels and the gray circles denote half pixels.

For a given direction angle θ unequal to zero as shown in Figure 8.2b, the first type of directional primary operation with a parameter α from one pixel to another is defined in Z domain as

$$O(X[n_i], X[n_j], \theta, \alpha) = \left\{ X \longleftarrow \begin{bmatrix} 1 & & & & & & & \\ & 1 & & & & & & \\ & & 1 & & & & & \\ & & & 1 & & & & \\ & & & & 1 & & & \\ & \alpha F(\theta) & & & & 1 & & \\ & & & & & & 1 & \\ & & & & & & & 1 \end{bmatrix} X \right\}, \quad (8.6)$$

where $n_i = n_5$ and $n_j = n_2$ in the case illustrated in Figure 8.2b. $F(\theta)$ is the function of shifting and interpolation in Z domain and locates at the n_i-th row and the n_j-th column of operation matrix. It is defined as

$$F(\theta) = Z^{-|n_i - n_j| tan(\theta)}. \quad (8.7)$$

If $|n_i - n_j| tan(\theta)$ is an integer, $F(\theta)$ only means a left shifting of $|n_i - n_j| tan(\theta)$ pixels; otherwise the interpolation is needed and $F(\theta)$ is redefined as

$$F(\theta) = \sum_k w_k z^{-k}. \quad (8.8)$$

Here, k indexes the integers around $|n_i - n_j| tan(\theta)$, and w_k's are the interpolation filtering parameters. Therefore, the directional primary operation Eq. (8.6) actually involves more than one pixel. Similarly, the inverse operation of Eq. (8.6) should be $O^{-1}(X[n_i], X[n_j], \theta, -\alpha)$. For the scaling primary operation, there is no difference between the directional version and the nondirectional version because it involves only one pixel at one time.

After the directional primary operations have been defined, it is not difficult to derive a new directional transform by assembling directional primary operations according to the order and the structure depicted in Figure 8.1. Similar to Eq. (8.4), the directional transform can be generally described as follows:

$$Y = DCT_{dir8}(X, \theta) = O_{35} \circ O_{34} \circ \cdots \circ O_2 \circ O_1 (X, \theta). \quad (8.9)$$

DCT_{dir8} indicates 1D 8-point directional DCT-like transform. The directional angle of every primary operation is θ. The first type of primary operation used in Eq. (8.9) is defined in Eq. (8.6) and the second type of primary operation in Eq. (8.3). The corresponding inverse 1D directional DCT-like transform is given as

$$X = DCT_{dir8}^{-1}(Y, \theta) = O_1^{-1} \circ O_2^{-1} \circ \cdots \circ O_{34}^{-1} \circ O_{35}^{-1} (Y, \theta). \quad (8.10)$$

The proposed directional DCT-like transform can guarantee the perfect reconstruction property if transform coefficients are not quantized at all, namely,

$$X = DCT_{dir8}^{-1}\left(DCT_{dir8}\left(X,\theta\right),\theta\right), \tag{8.11}$$

because the product of operation matrix of each primary operation and its inverse operation matrix is always equal to a unit matrix.

What is the exact transform matrix of the proposed directional DCT-like transform? From Eq. (8.6), it is dependent on directional angle and interpolation used there. The transform matrix in the case of $\theta = argtan(1/2)$ and linear interpolation is discussed here. Numerical values of the transform matrix are calculated from Eq. (8.9) and illustrated in Figure 8.3a. The transiting result of a numerical matrix multiplying a row vector $[z^6, z^5, \cdots, z^{-3}]^T$ in the first row is a column vector. It is the first row of an 8×8 transform matrix. For example, the first element of the first row is equal to

$$-0.006z^{-3} + 0.099z^{-2} - 0.613z^{-1} + 1.392 - 0.613z + 0.099z^2 - 0.006z^3.$$

It is a low-pass filter with 7 nonzero coefficients. It means that the 1D vertical transform will involve 6 neighboring columns. The longer the interpolation filter is, the more neighboring columns that are involved in the transform. Other elements are similar to the first element. Although the transform matrix looks a bit complicated, the computation of transform is still very easy thanks to the lifting structure.

8.2.3 Comparison with Rotated DCT

The straightforward way to make DCT adaptive to different orientations is to first rotate the image to a horizontal or vertical direction and then perform DCT on the rotated image, just like that done by Taubman and Marcellin [132]. For the same example of $\theta = argtan(1/2)$ and linear interpolation, numerical values of the transform matrix of the straightforward way are calculated and depicted in Figure 8.3b in the similar form. For the given $\theta = argtan(1/2)$, pixels at the first, third, fifth, and seventh row locate at a half pixel, thereby corresponding elements in the transform matrix contains two items in Z domain due to linear interpolation. Other elements only contain one item because they correspond to integer pixels.

One can observe the significant differences on the transform matrixes between the proposed directional DCT-like transform and the rotated DCT. It is not difficult to be explained. In the proposed transform, the interpolation is performed in each primary operation of lifting structure. Just as we have mentioned, if the source of a primary operation is a fractional pixel, its value is interpolated from neighboring integer pixel; and if the target of a primary operation is a fractional pixel, the result of primary operation is propagated to neighboring integer pixels. However, in the rotated DCT, the interpolation is performed before DCT and there is no interpolation anymore during the transform.

In addition to the differences on transform matrixes, the results of inverse transform are also different. The result of inverse transform in the proposed method

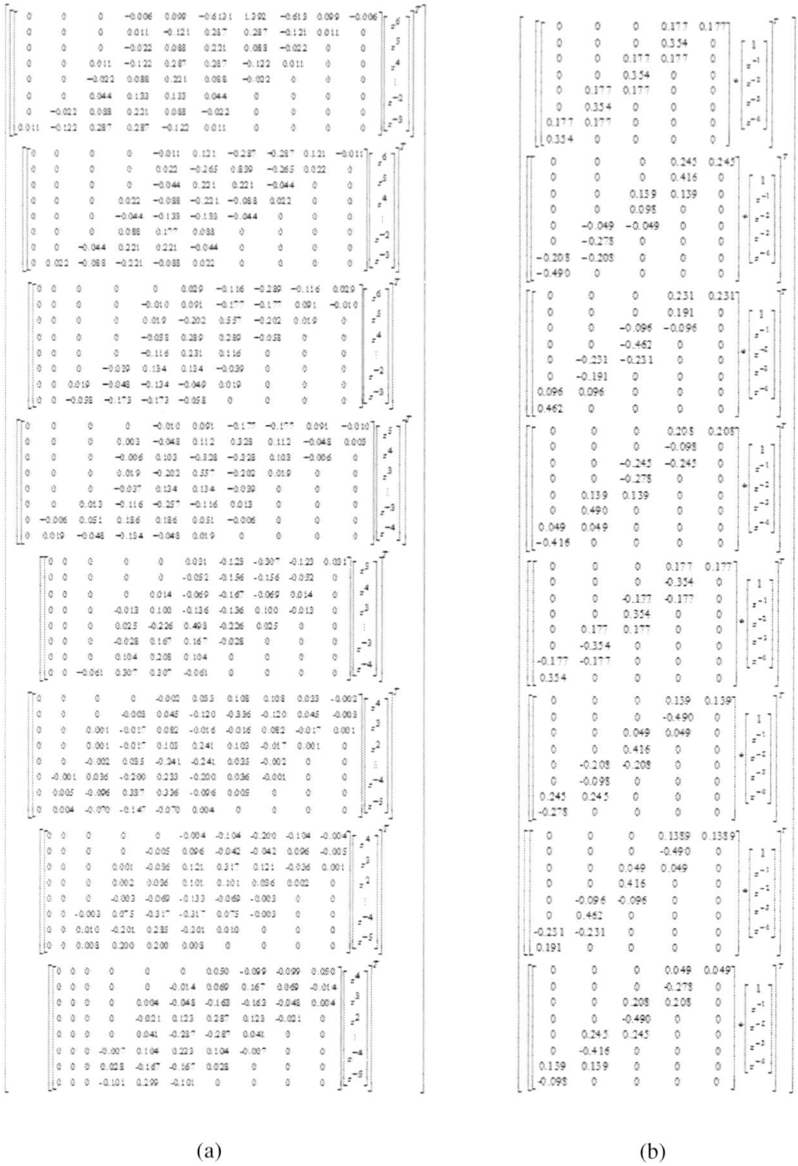

(a) (b)

Figure 8.3 The transform matrixes: (a) the proposed directional DCT-like transform; (b) the straightforward directional DCT transform after rotation.

directly is the image to be reconstructed. The perfect reconstruction can be guaranteed by the lifting structure. However, in the rotated DCT, the result of inverse transform is the image after rotation and interpolation. The final reconstructed image is obtained by rotating the result back. Without the special process as proposed by Taubman and Zakhor [132], the perfect reconstruction cannot be guaranteed. The serious problem in the rotated DCT is that it is only suitable to the adaptation on large image regions. In other words, the whole image or some rows of image has the same direction of correlation. It is difficult to be applied on small size blocks. One can imagine how difficult it is rotating each small block (e.g., 8×8 size) with different directions and still assembling rotated blocks into an image for later rectangular block-based compression. It is the reason to prevent this technique from being applied in practical coding schemes.

8.3 Image Coding with Proposed Directional Transform

In this section, we apply the proposed directional DCT-like transform into a JPEG-wise image coding scheme as shown in Figure 8.4. Except for those modules related to the directional transform, quantization denoted by "Q" in Figure 8.4, dequantization denoted by "Q^{-1}," entropy coding and entropy decoding are the same as those in JPEG.

In the encoder side, input image is first analyzed block by block to decide their transform directions. Then 2D transform is performed in each block along with the selected directional angle. Note that only 1D directional DCT-like transform is discussed in Section 8.2. But in a practical image and video coding scheme, 2D transform is needed. Therefore, the proposed directional DCT-like transform is applied in a block as the first 1D transform (e.g., vertical direction). After the 1D transform, generated coefficients do not own orientation property anymore. They are organized in different rows, saying DC, AC_1, $\cdots$ and AC_7, from low frequency to high frequency. Then, the second 1D transform is a normal one and performed on each row. For simplification, the transforms are vertically aligned, so that they are well filled in the 8×8 blocks. When the 2D transform is completed, the coefficient at the upper-left corner is the DC of this 8×8 block, just the same as the conventional DCT. They are quantized after ZigZag scan and coded by VLC to generate compressed stream. In the decoder side, received data is decoded to recover coefficients. Coefficients are first transformed by the normal 1D inverse DCT on each row and then are

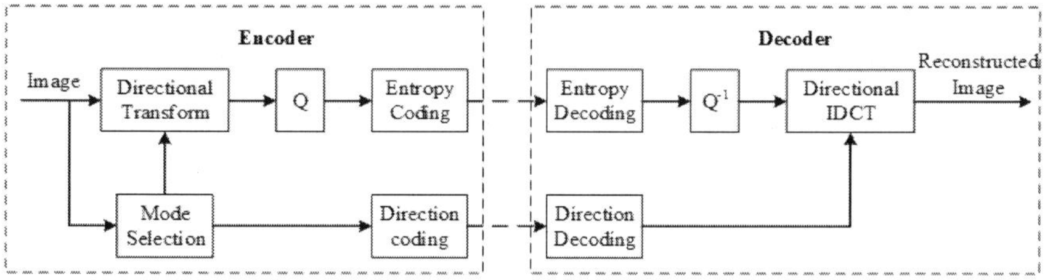

Figure 8.4 The coding scheme using a directional DCT-like transform.

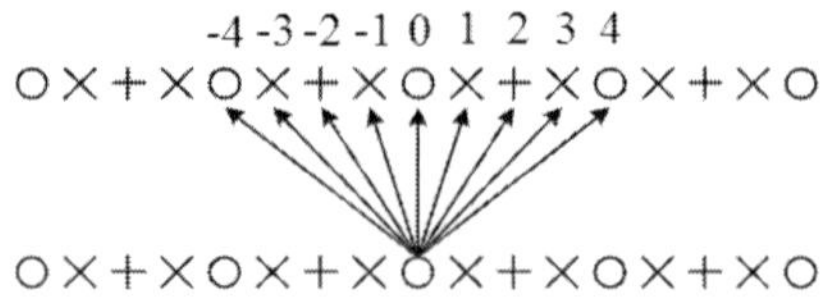

Figure 8.5 Available direction modes that are predefined in this chapter.

transformed by the proposed directional inverse transform along with the direction of each block to reconstruct the image.

In order to reduce the cost of such directional data, 9 modes are predefined as available directions, depicted in Figure 8.5 as the set $\{-4,\cdots,0,\cdots,4\}$, where integer pixels are marked by "o," half pixels by "+," and quarter pixels by "x." The angle is constrained to $\pm 45^\circ$ with quarter-pixel precision. If image features mainly orient 45° to 135°, the whole image can be rotated 90° before compression. When the precision is only a half pixel, there are 5 directional modes only. Each 8×8 block has to be assigned with a direction mode. The directional data is coded by predicted VLC method in the module of directional coding and sent to the decoder before coded coefficients.

In the mathematical analysis of the proposed directional DCT-like transform in Section 8.2, we assume that all blocks in a row have the same direction. It is impractical in coding a real image because of local orientation properties of the image. Therefore, in the proposed coding scheme in Figure 8.4, each block can have a unique direction. When neighboring blocks have different directions, a question is how to deal with direction transition in the proposed directional transform. In addition, the direction selection is also discussed here based on the dynamic programming algorithm.

8.3.1 Direction Transition on Block Boundary

To adapt to local orientation properties of the image, each block has its own transform direction. If two blocks have different transform directions, the direction transition will take place on the boundary of these two blocks. As shown in Figure 8.6, the direction of each block is indicated by dashed lines. The dashed-dotted line indicates the boundary of one block. The region that consists of fractional pixels between two dashed-dotted lines is the transition one. The solid line in the transition region illustrates each individual transition. The transition is assumed to always connect to the first column of integer pixels of the right block.

There are three different cases on transition according to our predefined modes. In the first case, both the directions of the two blocks are positive or negative modes, and the direction of the left block is not mode 3. The example of the direction of the

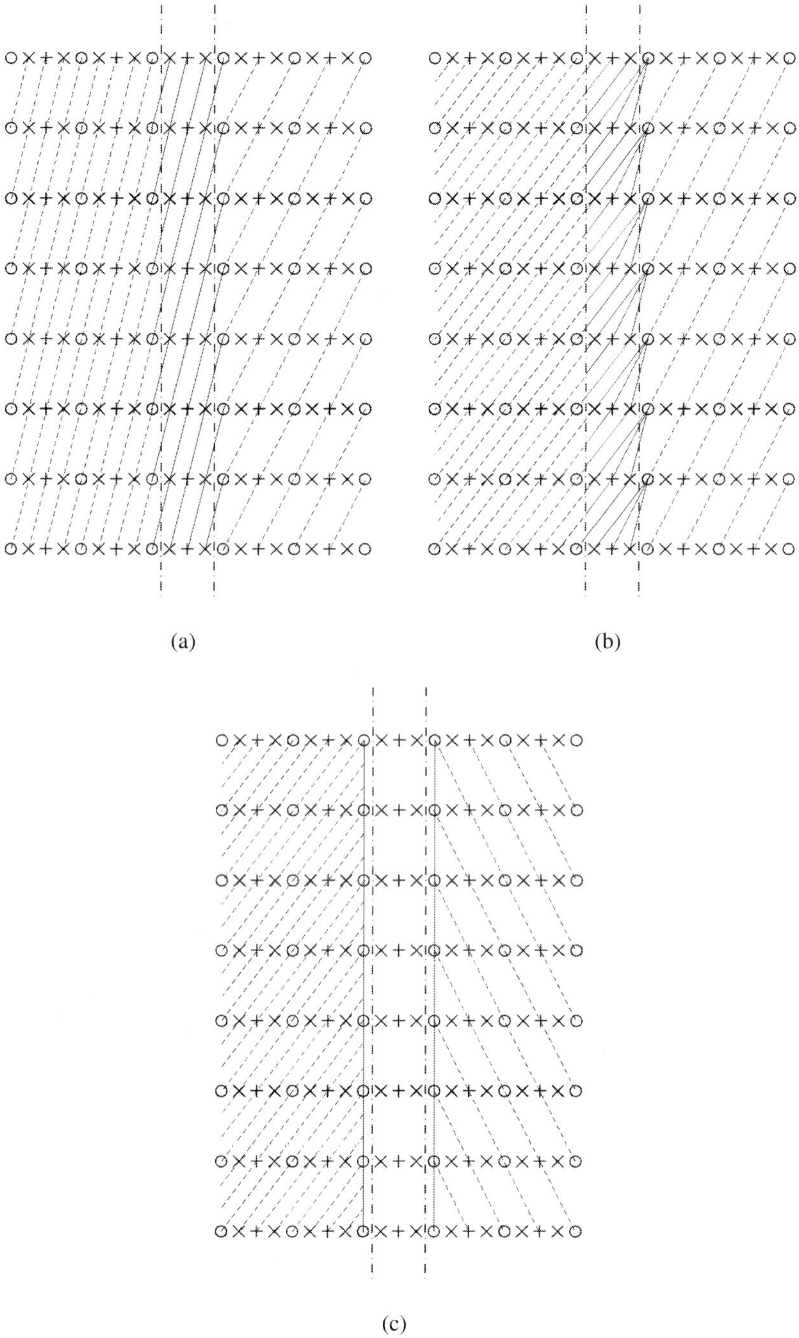

Figure 8.6 Three different transition cases and their processing.

left block equal to mode 1 and that of the right block equal to mode 2 is shown in Figure 8.6a. Along with the direction of the left block, each connection will be linked to one integer pixel on the first column of the right block. Just as the motion threading proposed by Xu et al. [74], we call the connection a directional thread. In this case, there only exists the one-to-one mapping during the transition. It is easy to perform the proposed directional DCT-like transform along with each directional thread.

In the second case, both the directions of two blocks are positive modes or negative modes, and the direction of the left block is mode 3. An example of the direction of the left block equal to mode 3 and that of the right block equal to mode 2 is shown in Figure 8.6b. Each directional thread of the left block is extended along with its direction to one of the fractional pixels in the transition region. Before it goes through the dashed-dotted line of the right block, it is forced to link to one integer pixel on the first column of the right block. In this case, there exists multi-to-one mapping during the transition. In the downward primary operation, the unique source can be found along with directional thread, no matter whether multi-to-one mapping exists or not. But in the upward primary operation, multiple sources may be used to modify a target according to the directional threads. Fortunately, this problem has been solved by the EDU proposed by Feng et al. [89]. We adopt the same approach in this chapter.

In the third case, the sign of the direction mode in the left block is reverse to that of the direction mode in the right block, or one of the direction modes in these two blocks is zero. The example of the direction of the left block equal to mode 3 and that of right block equal to mode –2 is shown in Figure 8.6c. In this case, these two blocks are granted as independent. When direction threads in the left block move to the last column of this block, they turn to the vertical direction along with the last column. Similarly, direction threads in the right block also turn to the vertical direction when they move to the first column of this block. Multi-to-one mapping also exists in the third case. They are also dealt with by the EDU approach.

8.3.2 Direction Selection

As we have discussed in Section 8.2, different rows of blocks of the proposed directional DCT-like transform are independent in mode selection. Assume that the mode set is $M = \{-4,\cdots,0,\cdots,4\}$, with the total number of K candidates, which means the predefined modes as shown in Figure 8.5. The task of the mode selection is, subject to a given bit budget, to select direction modes of blocks so that the minimum R-D cost of all these blocks can be achieved by applying the directional transform along the select directions. This optimum problem can be formulated as

$$argmin_{m_0,\cdots,m_{N-1}\in M} \sum_{n=0}^{N-1} P(n|m_0,\cdots,m_{N-1}). \tag{8.12}$$

N is the number of blocks in a row of blocks. m_n is the selected mode of the n-th block and is also used to denote the n-th block sometimes. $P()$ denotes the R-D cost of a block. Since $P()$ of a block is affected by not only the direction of this block but also the directions of neighboring blocks in the same row, the optimum problem Eq. (8.12) is not easy to be solved.

According to the predefined modes in this section, the direction of a block is constrained to $\pm 45°$. At the same time, if the direction mode signs of two neighboring blocks are inverse to each other, these two blocks will be granted as independent. Assume that there are three blocks m_{n-1}, m_n, and m_{n+1}. Only when all signs of these three blocks are positive or negative, the block m_{n-1} may influence the coding performance of the block m_{n+1}. Without loss of generality, all signs of these three blocks are positive, namely, their directions are within $(0°, 45°]$. Because the maximum direction angle is 45°, the block m_{n-1} will only use the pixels in the upper-left triangle of the block m_n, and the block m_{n+1} in the right-bottom triangle of the block m_n. Therefore, the block m_{n-1} does not directly use the pixels that are used by the block m_{n+1}. Of course, this discussion does not take the interpolation into account. But at least it indicates the effects between m_{n-1} and m_{n+1} is not as strong as expected.

To simplify the optimum problem Eq. (8.12), we consider the R-D cost of a block only to be influenced by neighboring left and right blocks, namely, the R-D cost of the block m_n is decided by the directions of m_{n-1}, m_n, and m_{n+1}. With this simplification, P() is defined as

$$P(n|m_{n-1}, m_n, m_{n+1}) = D(n|m_{n-1}, m_n, m_{n+1}) + \lambda R(n|m_{n-1}, m_n, m_{n+1}). \quad (8.13)$$

Given the three directions, $D(n|m_{n-1}, m_n, m_{n+1})$ is the reconstructed distortion of the n-th block, and $R(n|m_{n-1}, m_n, m_{n+1})$ are the bits for coding coefficients after quantization and direction data of the n-th block. λ is the Lagrange multiplier depending on the quantization step.

A solution using the dynamic programming algorithm [171] is proposed to select directions of blocks. Before we start to discuss the algorithm, some terminologies are defined here.

- m_n^k indicates that the k-th candidate mode is selected by the n-th block with $0 \le n < N$ and $0 \le k < K$.
- $P(n|m_{n-1}^{k_{n-1}}, m_n^{k_n}, m_{n+1}^{k_{n+1}})$ is the R-D cost of the n-th block when $m_{n-1}^{k_{n-1}}$, $m_n^{k_n}$, and $m_{n+1}^{k_{n+1}}$ are given.
- $\tilde{P}(n|m_n^{k_n}, m_{n+1}^{k_{n+1}})$ is the minimum R-D cost of previous $n+1$ blocks, i.e., b_0, b_1, $\cdots$, b_n, where b_i is the i-th block, with the directions of the boundary block $m_{n+1}^{k_{n+1}}$ and the last block $m_n^{k_n}$ of this layer are given.
- $\tilde{M}(n|m_n^{k_n}, m_{n+1}^{k_{n+1}})$ denotes the modes assigned to the previous n blocks, i.e., $b_0, b_1, \cdots, b_{n-1}$, to get $\tilde{P}(n|m_n^{k_n}, m_{n+1}^{k_{n+1}})$.

The dynamic programming algorithm is conducted layer by layer. The basic idea is, in the n-th layer, if the directions of the boundary block (i.e., the $[n+1]$-th block) and the n-th block are given, we find the best one from all possible mode combinations of previous n blocks, that is, $b_0, b_1, \cdots, b_{n-1}$, to get $\tilde{P}(n|m_n^{k_n}, m_{n+1}^{k_{n+1}})$. Then the triple of $\tilde{P}(n|m_n^{k_n}, m_{n+1}^{k_{n+1}})$, $\tilde{M}(n|m_n^{k_n}, m_{n+1}^{k_{n+1}})$, and $(m_n^{k_n}, m_{n+1}^{k_{n+1}})$ is recorded. For all K^2 combinations of $(m_n^{k_n}, m_{n+1}^{k_{n+1}})$, we have K^2 such triples available in the nth layer. In the $(n+1)$-th layer, for a given boundary $(m_{n+1}^{k_{n+1}}, m_{n+2}^{k_{n+2}})$, only the recorded K^2 triples are checked instead of all combinations of the previous $n+1$ blocks. The proposed direction selection algorithm is described as follows.

1. For $n = 0$, $\tilde{P}(0|m_0^{k_0}, m_1^{k_1})$ and $\tilde{M}(0|m_0^{k_0}, m_1^{k_1})$ with $0 \leq k_0 < K$ and $0 \leq k_1 < K$ are obtained by checking all K^2 combinations. The K^2 triples of $\tilde{P}(0|m_0^{k_0}, m_1^{k_1})$, $\tilde{M}(0|m_0^{k_0}, m_1^{k_1})$, and $(m_0^{k_0}, m_1^{k_1})$ are recorded (here $\tilde{M}(0|m_0^{k_0}, m_1^{k_1}) = \phi$ for any $(m_0^{k_0}, m_1^{k_1})$).
2. Given the $(n-1)$-th layer, for each $(m_n^{k_n}, m_{n+1}^{k_{n+1}})$ with $0 \leq k_n < K$ and $0 \leq k_{n+1} < K$, calculate

$$\tilde{P}(n|m_n^{k_n}, m_{n+1}^{k_{n+1}}) = argmin_{0 \leq k_{n-1} < K} \left(\tilde{P}\left(n-1|m_{n-1}^{k_{n-1}}, m_n^{k_n}\right) + P\left(n|m_{n-1}^{k_{n-1}}, m_n^{k_n}, m_{n+1}^{k_{n+1}}\right)\right). \tag{8.14}$$

   Based on the above conditional costs, merge the best $\tilde{M}(n-1|m_{n-1}^{k_{n-1}}, m_n^{k_n})$ under $m_{n+1}^{k_{n+1}}$ to generate $\tilde{M}(n|m_n^{k_n}, m_{n+1}^{k_{n+1}})$, that is,

$$\tilde{M}(n|m_n^{k_n}, m_{n+1}^{k_{n+1}}) = \left\{\tilde{M}(n-1|m_{n-1}^{k_{n-1}}, m_n^{k_n}) \cup m_{n+1}^{k_{n+1}}\right\}. \tag{8.15}$$

   Thus, K^2 triples of $\tilde{P}$ $(n|m_n^{k_n}, m_{n+1}^{k_{n+1}})$, $\tilde{M}$ $(n|m_n^{k_n}, m_{n+1}^{k_{n+1}})$, and $(m_n^{k_n}, m_{n+1}^{k_{n+1}})$ are recorded at the n-th layer.
3. In the last layer, where $n = N-1$, the last K^2 candidates are checked to get the modes for all blocks.

Since the direction selection is a forward recursion processing, the proposed directional DCT-like transform has much higher computation than normal DCT especially at the encoder side. But for the compression scenarios, the decoding computation is more important. The decoding computation of the proposed transform is also higher than that of normal DCT because of additional interpolation. The quantified comparison is given here for 1D 8-point transform. Normal DCT needs 29 additions, 12 shifts, and 13 multiplications. The proposed transform needs $29 + 29 \times (T-1)$ additions, 12 shifts, and $13 + 29 \times T/2$ multiplications for half-pel interpolation. T is the number of taps of the interpolation filter.

8.4 Experimental Results

Extensive experiments have been conducted to evaluate the performance of the proposed directional DCT-like transform. The proposed coding scheme is developed

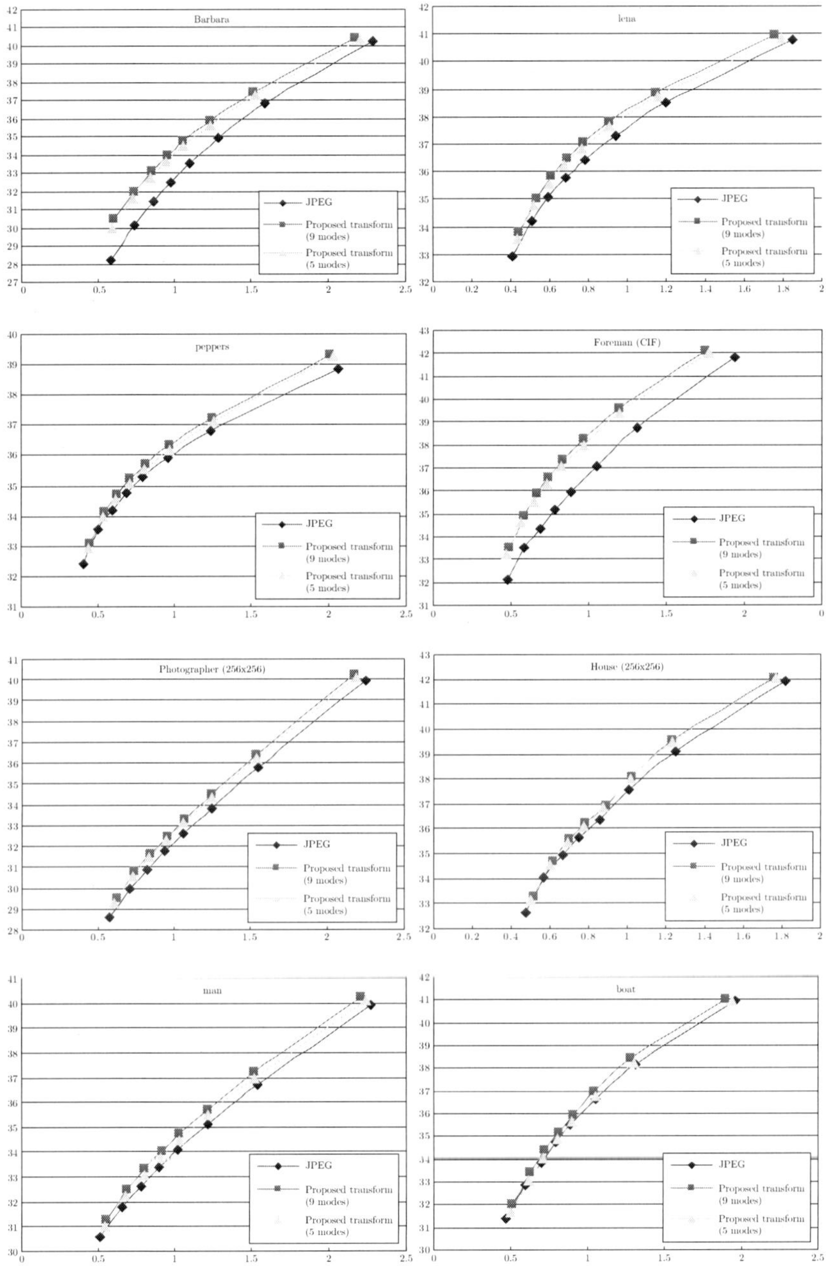

Figure 8.7 The PSNR versus bpp curves of JPEG and the proposed scheme.

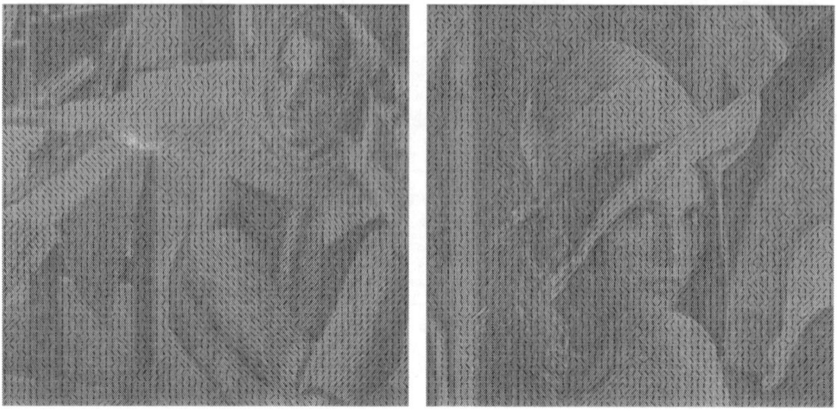

(a) (b)

Figure 8.8 Direction modes selected in Barbara with 1.07 bpp and in Lena with 0.69 bpp.

from JPEG, but the first 1D transform is replaced by the proposed directional DCT-like transform. The values of fractional pixels needed are calculated by the 8-tap Sinc filter. The direction of a block is first predicted from the coded directions of neighboring blocks and then is coded by the Huffman coding losslessly. Therefore, JPEG is selected as an anchor for comparisons.

We report the experimental results of eight common images: Barbara (512×512), Lena (512×512), Peppers (512×512), Photographer (256×256), House (256×256), Man (512×512), Boat (512×512), and the first frame of the Foreman (352×288) video sequence from the MPEG testing set. The Lagrange multiplier λ is set to 7 empirically. Two cases of 9 modes (quarter-pixel precision) and 5 modes (half-pixel precision) are tested in this experiment. The PSNR versus bit per pixel (bpp) curves are depicted in Figure 8.7, while the coding quality is set from 20 to 90 with an incremental step 10.

The coding gain can be up to 2.0 dB for Barbara. On the testing set the gain of the proposed directional DCT-like transform ranges from 0.2 dB to 2.0 dB, depending on the presence or lack of orientation features in the image. But even on relatively smooth images like Lena, the proposed method still enjoys an advantage over the normal DCT transform. The results with different prediction precisions are also given in Figure 8.7. The results of quarter-pixel precision can gain about 0.1 dB to 0.3 dB better than those of half-pixel prediction on images of strong edge features such as Barbara and Man. On relatively smooth images, the angular prediction of quarter-pixel precision can be similar to that of half-pixel precision. The similar results are observed in the adaptive directional lifting-based wavelet [161]. For the proposed scheme, the quarter-pixel precision does not give us much performance gain.

The modes selected for each 8×8 block in Barbara with 1.0 bpp and Lena with 0.69 bpp are depicted in Figure 8.8. The contrast of the two images is adjusted to highlight the direction modes drawn on the images, where the direction of the short

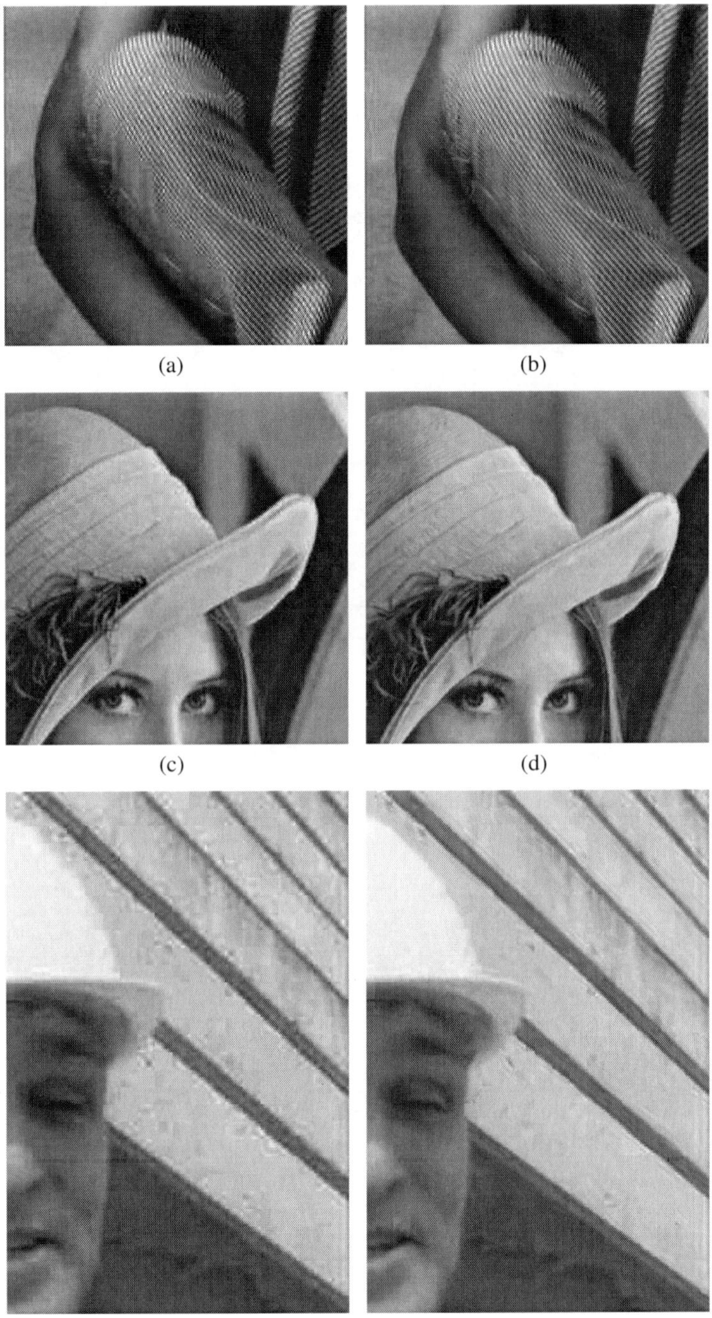

(a) (b)

(c) (d)

(e) (f)

Figure 8.9 Visual quality comparison: (a) Barbara, (c) Lena, and (e) Foreman are coded by JPEG at 0.58, 0.46, and 0.48 bpp, respectively; (b), (d), and (f) are coded by our scheme at 0.60, 0.44, and 0.49 bpp, respectively.

line is equal to the transform direction. Just as we expect, the areas with strong directional texture are mostly covered with the well matched modes, such as the clothes of Barbara and the hat of Lena.

Due to its adaptability to directional signal features, the proposed directional DCT-like transform tends to reconstruct image edges better than the normal DCT transform at the same bit rates. Since the human visual system is highly sensitive to edges, the former method should have superior visual quality than the latter one, which is indeed corroborated by our experimental results. As an example, Figure 8.9 presents the decoded Barbara, Lena, and Foreman images by the two methods. The image decoded by JPEG has severe moir patterns on the scarf and pants, which even change the texture orientations of the original image. Also, edge ringing effects are clearly visible in the JPEG decoded images. In contrast, the proposed directional transform coding eliminates most of the above artifacts. It reproduces the high frequency scarf pattern almost perfectly, and greatly reduces the ringing effects around the edges.

Tables 8.1 and 8.2 show the average number of side information bits for coding directions, which depends on overall bit rates and image contents. One can see that the side information bits constitute only a small portion of total coded size. The percentage of overhead bits at low qualities is more than that at high qualities because the constant λ is used in this experiment.

Table 8.1 Overhead bits (in bpp and percentage) for Barbara, Lena, Peppers, Man, and Boat.

	Quality	Barbara		Lena		Peppers		Man		Boat	
9 modes	20	0.040	6.64%	0.047	10.59%	0.047	10.58%	0.050	8.98%	0.049	9.49%
	30	0.038	5.15%	0.045	8.48%	0.044	8.21%	0.048	6.97%	0.048	7.67%
	40	0.035	4.17%	0.043	7.07%	0.043	6.91%	0.046	5.72%	0.047	6.50%
	50	0.034	3.57%	0.040	5.88%	0.042	5.90%	0.044	4.77%	0.045	5.55%
	60	0.033	3.11%	0.039	5.03%	0.040	4.95%	0.043	4.11%	0.044	4.83%
	70	0.032	2.55%	0.038	4.12%	0.039	4.02%	0.041	3.40%	0.040	3.83%
	80	0.031	2.00%	0.037	3.19%	0.039	3.07%	0.040	2.64%	0.037	2.90%
	90	0.030	1.37%	0.036	2.01%	0.040	1.95%	0.039	1.75%	0.036	1.89%
5 modes	20	0.031	5.29%	0.036	8.45%	0.037	8.59%	0.036	6.59%	0.032	6.39%
	30	0.029	4.01%	0.034	6.63%	0.035	6.69%	0.033	4.90%	0.031	4.98%
	40	0.028	3.30%	0.033	5.46%	0.034	5.74%	0.031	3.88%	0.030	4.26%
	50	0.027	2.79%	0.031	4.56%	0.032	4.59%	0.030	3.34%	0.029	3.61%
	60	0.026	2.42%	0.030	3.88%	0.031	3.89%	0.029	2.81%	0.029	3.17%
	70	0.025	1.99%	0.029	3.15%	0.030	3.06%	0.028	2.28%	0.027	2.53%
	80	0.023	1.52%	0.027	2.37%	0.030	2.35%	0.027	1.75%	0.025	1.92%
	90	0.023	1.04%	0.027	1.50%	0.030	1.49%	0.026	1.16%	0.024	1.24%

Table 8.2 Overhead bits (in bpp and percentage) for the Mandrill, Photographer, House, and Foreman.

	Quality	Mandrill		Photographer		House		Foreman	
9 modes	20	0.046	5.39%	0.055	8.88%	0.048	9.22%	0.045	9.59%
	30	0.043	4.00%	0.054	7.31%	0.051	8.28%	0.043	7.70%
	40	0.043	3.34%	0.053	6.31%	0.048	6.86%	0.042	6.46%
	50	0.042	2.87%	0.052	5.43%	0.047	6.02%	0.041	5.62%
	60	0.042	2.52%	0.051	4.75%	0.047	5.34%	0.040	4.92%
	70	0.041	2.09%	0.048	3.87%	0.045	4.40%	0.039	4.06%
	80	0.041	1.68%	0.046	3.01%	0.042	3.42%	0.039	3.21%
	90	0.041	1.16%	0.047	2.17%	0.041	2.33%	0.039	2.18%
5 modes	20	0.030	3.52%	0.041	6.67%	0.034	6.72%	0.036	7.67%
	30	0.028	2.60%	0.040	5.42%	0.035	5.90%	0.034	6.10%
	40	0.028	2.18%	0.038	4.57%	0.033	4.85%	0.032	4.98%
	50	0.027	1.86%	0.037	3.87%	0.033	4.32%	0.031	4.32%
	60	0.027	1.64%	0.035	3.34%	0.034	3.84%	0.030	3.70%
	70	0.026	1.32%	0.034	2.71%	0.031	3.07%	0.030	3.09%
	80	0.027	1.11%	0.033	2.13%	0.031	2.49%	0.028	2.30%
	90	0.027	0.77%	0.032	1.47%	0.029	1.62%	0.029	1.62%

8.5 Summary

In this chapter, we develop a directional DCT-like transform, which can be done along arbitrary direction in theory. The directional transform is based on the lifting structure of DCT by introducing directional primary operations. At the same time, a JPEG-wise coding scheme is also developed by using the proposed directional DCT-like transform. In the coding scheme, each 8×8 block can have its own transform direction in order to better adapt to local orientation properties of the image. In this scheme, the direction transition between block boundaries is investigated and dealt with carefully. Furthermore, a direction selection approach is proposed based on the dynamic programming algorithm. Experimental results show that the lifting-based directional DCT-like transform significantly outperforms the normal DCT transform in terms of PSNR and visual quality.

Chapter 9
Directional Filtering Transform

9.1 Introduction

In natural images, there usually exist many textures, lines, and edges. Along the edges, sample intensities change smoothly; but across the edges they will change dramatically even with a small shift from the edges. This kind of correlation shows a strong anisotropic feature because it depends upon not only the distance of samples but also their link orientation. Furthermore, the link orientation often goes in all directions rather than only the vertical or the horizontal. Traditional transforms such as the discrete cosine transform (DCT) and discrete wavelet transform (DWT) cannot handle this directional correlation well and energy compaction is not fully achieved. Therefore, many methods have been proposed to exploit the directionality in images/intra frames in terms of transforms and predictions.

In this chapter, we mainly focus on those methods for image and intra-frame compression. First, some transforms are proposed from a filter bank to well represent the multidirectional information of images in the frequency domain [134, 138, 141, 172–177]. Some attempts on applying these transforms to image compression have been reported [175–178]. The main challenge for them is that coefficients after these transforms hardly use existing mature quantization and entropy coding methods. On the other hand, some methods are proposed to introduce directionality into the traditional transforms. All these directional transforms have demonstrated a consistent and stable performance gain over their nondirectional counterparts [157, 158, 161, 179–186], no matter which compression schemes are used. These methods can be categorized into two classes, according to which traditional transform is used.

One category is the directional wavelet transform. Taubman and Zakhor first propose to adapt the filtering direction to local image content by resampling each rectangular block along its dominant edge orientation and then perform conventional subband transforms in the resampled domain [132]. Different from this work, some researchers propose to use a lifting structure for local directional filtering [151, 153, 154, 156, 187], where all subbands reside on the rectilinear grid. Since they do not transmit filtering directions to the decoder, instead using certain synchronization between the encoder and the decoder, their performances are constrained. No significant objective improvements are reported.

However, with direction information estimated at the encoder and explicitly transmitted to the decoder, both the curved wavelet transform [179, 180] and the adaptive directional lifting (ADL) [157, 161] report good objective and subjective performance over JPEG 2000. To improve the interpolation accuracy in ADL, Liu and Ngan [181] and Dong et al. [182] propose adaptive interpolation filters. Further, Chang and Girod [183] propose a direction-adaptive DWT with long-distance prediction to avoid interpolation and also demonstrate its advantages over ADL. Without being limited by rectilinear subsampling, Chang et al. also propose direction-adaptive lifting by quincunx subsampling [158].

Another category considers the directionality in DCT transform. Zeng and Fu [184] propose a block-independent directional DCT from the idea of shape-adaptive DCT. Xu et al. [185] introduce directional primary operations into a lifting-based DCT, which exploits inter-block correlations, too, but it has to estimate prediction directions of all blocks jointly by dynamic programming. A more recent work by Chang et al. incorporate the idea of directional DCT into H.264 intra-frame coding [13] and propose the direction-adaptive partitioned block transform [186].

In addition to directional transforms, H.264 intra-frame coding utilizes directional prediction to exploit directional correlation, where a block is unidirectionally predicted from reconstructed boundary samples in its neighboring blocks along a certain direction. However, the correlation among samples is not fully exploited yet since samples close to neighboring blocks tend to have better prediction than the others. Many schemes have been proposed recently to improve the performance, such as bidirectional prediction [188] and template matching [189].

In these 2D directional transforms, since every 1D transform is often performed neither horizontally nor vertically, it is difficult to prove the equivalence on the orders of two 1D transforms first horizontal and then vertical, and vice versa. Chang et al. [158] report that the order of two 1D transforms results in different coding performances for a 2D transform. Chang and Girod [183] also point out this problem in the coding gain analysis. However, it is still not clear what exact roles the 1D transform order plays in 2D directional transforms if the best nonseparable prediction filters are chosen.

In this chapter, the mathematical analysis based upon an anisotropic image model is first presented for this problem. The analysis results show that, if the neighboring samples used to predict one pixel can angularly cover the whole plane and the best weights based upon the mean square error (MSE) criterion are provided in each prediction, lifting in either order brings no difference. But in practice, the conditions cannot be satisfied because high-pass bands are often simply decomposed by conventional lifting or are not decomposed at all. Also the prediction modes cannot cover as wide an angular scope as supposed. Therefore, the lifting order problem for images with different orientations and sharpness exists.

Subsequently, we propose the directional filtering transform (dFT, in order to distinguish from the common usage on DFT) to better exploit image and intra-frame correlations in H.264 intra-frame coding. Some preliminary results are reported by

Peng et al. [190]. dFT consists of directional filtering and an optional DCT. In the directional filtering, there are two different modes. One is the unidirectional filtering (UDF), similar to H.264 intra prediction. Another is the bidirectional filtering (BDF), which is inspired by ADL but without an update. According to our analyses, an important feature in dFT is that every block can adaptively select its lifting order, which is difficult to achieve in directional wavelet transforms. Since dFT not only enables closer neighboring samples for prediction but also well exploits the directional correlation, it can achieve better performance than the current intra-frame coding in H.264.

9.2 Adaptive Directional Lifting-Based 2D Wavelet Transform

The ADL locally adapts the filtering direction to image orientations within a lifting-based DWT framework and still preserves the rectangular shape of subbands. Like traditional DWT, it consists of 2D lifting processes in one-level decomposition. Without loss of generality, we suppose that lifting in the vertical dimension is performed first. By lifting in the first dimension along a certain direction, the L and H bands are generated. Afterward, another lifting in the horizontal dimension is performed on the L and H bands, respectively, resulting in the final LL, LH, HL, and HH subbands. Different from conventional DWT, the filtering directions in the two dimensions may not be orthogonal and are not limited to the vertical or horizontal directions only.

Let $x[t]$, $t = [t_1, t_2]$ denote the original image signal. In the first dimension lifting, odd-row pixels are predicted from even rows along an angle d. Then even-row samples are updated from the odd-row residues along the same direction d. Though any biorthogonal wavelet transform can be factored into one or several lifting stages, we only consider one-step lifting here, which means one prediction step followed by one update step and scaling. Then the lifting process can be expressed as

$$y_1[t] = g_H \cdot \left(x_1[t] - P_{x_0}^d[t] \right), \tag{9.1}$$

$$y_0[t] = g_L \cdot \left(x_0[t] + g_H^{-1} \cdot U_{y_1}^d[t] \right), \tag{9.2}$$

where $x_0[t]$ is the even-row samples and $x_1[t]$ the odd-row samples. $P_{x_0}^d[t]$ is the prediction for $x_1[t]$ and $U_{y_1}^d[t]$ is the update for $x_0[t]$. $y_0[t]$ and $y_1[t]$ denote the L and H bands, respectively. g_H and g_L are scaling factors.

Since the filtering can adapt to image orientations, samples across multiple columns may be used in the prediction step, but only even-row samples are available to predict odd-row samples. Thus, the prediction and update function for the 5/3 wavelet, which is used throughout this chapter, can be expressed as

$$P_{x_0}^d[t] = \frac{1}{2} \sum_{i=0}^{1} \sum_{k=0}^{L-1} \omega_{i,k} x_0 \left[t_1 + \triangle t_{1\,i,k}, t_2 + \triangle t_{2\,i,k} \right], \tag{9.3}$$

$$U_{y_1}^d[t] = \frac{1}{4} \sum_{j=-1}^{0} \sum_{l=0}^{L-1} \omega_{j,l} y_1 \left[t_1 + \triangle t_{1\,j,l}, t_2 + \triangle t_{2\,j,l} \right], \tag{9.4}$$

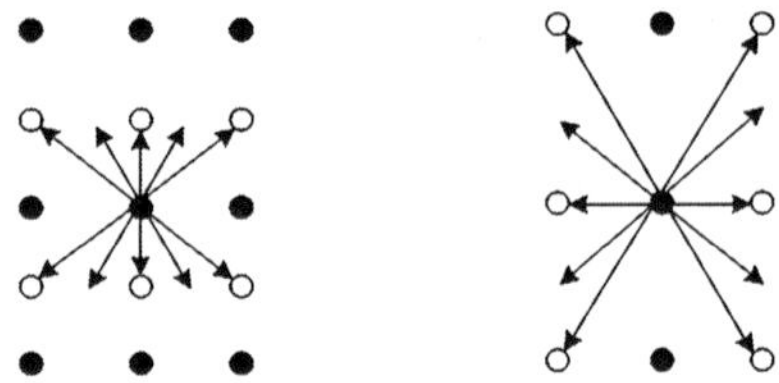

Figure 9.1 Modes in two dimensions for ADL with lifting in the vertical dimension performed first: (a) the first dimension; (b) the second dimension. The black circles denote the pixel to be predicted and the white circles denote samples used for prediction. Interpolation is needed where there is no sample at all.

where $[\triangle t_{1_{i,k}}, \triangle t_{2_{i,k}}]$ is the displacement vector between the pixel used for prediction and $x_0[t]$. $[\triangle t_{1_{j,l}}, \triangle t_{2_{j,l}}]$ is for the update stage. L is the number of samples used for the prediction of each i. $\omega_{i,k}$ and $\omega_{j,l}$ are weighting factors.

After the first dimension lifting, the second is similarly performed on the L and H bands, producing four subband signals: $z_{00}[t]$, $z_{01}[t]$, $z_{10}[t]$, and $z_{11}[t]$. They correspond to the LL, LH, HL, and HH subbands, respectively. Until now, one-level decomposition is completed. The multilayer wavelet structure can be obtained by iteratively performing 2D ADL on the LL subband.

Since ADL is operated on a global basis, a fixed lifting order, vertical lifting first followed by horizontal lifting, is chosen to guarantee the rectangular shape of each subband and the convenience for cross-block processing. However, this will hurt the performance. If a block has an orientation near horizontal, the vertical lifting cannot be just performed along the block orientation (seeing all possible prediction modes for two dimensions in Figure 9.1) and consequently energy is spread into the high-pass band. In this case, it may favor a horizontal lifting first. Considering the noneven distribution of modes in ADL, Chang et al. introduce a more evenly distributed set of modes in each dimension by using samples far away [183]. Since these modes can angularly cover almost the whole plane, insensitiveness to image transposition is achieved despite a fixed lifting order. However, the long prediction distance decreases the correlation and deteriorates the prediction accuracy. Thus, it is only useful for images with sharp edges. Moreover, there is no exact mode to handle the right horizontal direction.

So the question that arises here is, what is the effect of the order of two 1D transforms in a one directional 2D transform? In the next section, we try to answer this question by analyzing the coding gain of the directional 2D transform on an anisotropic image model.

9.3 Mathematical Analysis

In this section, the coding gain for an adaptive directional lifting-based wavelet similar to that by Chang and Girod [183] is first presented in terms of power spectral density (PSD). Afterward, numerical results will be given to analyze the effect of the lifting order.

9.3.1 Coding Gain of ADL

First, we analyze the coding gain of lifting in the vertical dimension. It consists of split, prediction, update, and scaling steps [see Eq. (9.1) and Eq. (9.2)]. The process is rewritten as

$$y_1[t] = g_H \cdot \left(x_1[t] - h_P^d[t] * x_0[t]\right), \tag{9.5}$$

$$y_0[t] = g_L \cdot \left(x_0[t] + g_H^{-1} \cdot \left(h_U^d[t] * y_1[t]\right)\right), \tag{9.6}$$

where $h_P^d[t]$ and $h_U^d[t]$ are prediction and update filters, respectively, on the down-sampled grid of $x[t]$.

To simplify the computation, we extend related signals onto the original image sampling grid $x[t]$. Let $\hat{x}_0[t]$ denote the up-sampled even-row samples and $\hat{h}_P^d[t]$ the filters applied to $\hat{x}_0[t]$. According to Eq. (9.3) and Eq. (9.4), we get

$$\hat{h}_P^d[t] = \begin{cases} \frac{\omega_{i,k}}{2}, t = \left[\triangle t_{1_{i,k}}, 2\triangle t_{2_{i,k}} + 1\right], k = 0, 1, ..., L-1 \\ 0 \end{cases}. \tag{9.7}$$

Its Fourier transform is

$$\hat{H}_P^d\left(e^{j\omega}\right) = \frac{1}{2}\sum_{i=0}^{1}\sum_{k=0}^{L-1}\omega_{i,k}e^{-j\left(\triangle t_{1_{i,k}}\omega_1 + \left(2\triangle t_{2_{i,k}}+1\right)\omega_2\right)}. \tag{9.8}$$

Since $h_P^d[t] * x_0[t]$ can be obtained by convoluting $\hat{x}_0[t]$ with $\hat{h}_P^d[t]$, shifting the resulting signal by -1 along the vertical dimension and then down-sampling it, we get the transfer function of $h_P^d[t]$ as

$$H_P^d\left(e^{j\omega}\right) = e^{j\omega_2/2}\hat{H}_P^d\left(e^{j\omega_1}, e^{j\omega_2/2}\right). \tag{9.9}$$

Similarly

$$H_U^d\left(e^{j\omega}\right) = e^{-j\omega_2/2}\hat{H}_U^d\left(e^{j\omega_1}, e^{j\omega_2/2}\right) = \frac{1}{2}e^{-j\omega_2/2}\hat{H}_P^d\left(e^{j\omega_1}, e^{j\omega_2/2}\right). \tag{9.10}$$

Assume that the image $x[t]$ comes from a 2D stationary continuous random field $\tilde{x}$ with zero mean and a variance σ^2. Similar to the work of Chang and Girod [183], we model the anisotropic auto-correlation as

$$R_{\tilde{x}\tilde{x}}(\tau) = E\{\tilde{x}_t\tilde{x}_{t-\tau}\} = \sigma^2 e^{-\sqrt{\tau^T\Gamma\tau}}, \tag{9.11}$$

where $\Gamma = \begin{pmatrix} cos\varphi & sin\varphi \\ -sin\varphi & cos\varphi \end{pmatrix}\begin{pmatrix} \Omega_\alpha^2 & 0 \\ 0 & \Omega_\beta^2 \end{pmatrix}\begin{pmatrix} cos\varphi & sin\varphi \\ -sin\varphi & cos\varphi \end{pmatrix}^T$, $\Omega_\alpha > \Omega_\beta > 0$. In this model, contours of equal correlation form an ellipse with the major axis aligned to the edge direction. φ denotes the orientation of the minor axis. Ω_α and Ω_β are parameters related to the minor and major axes, respectively. The larger Ω_α is, the faster the correlation orthogonal to the edge direction decays; while the smaller Ω_β is, the slower the correlation along the edge direction declines. By Fourier transform, we get the PSD of $\tilde{x}$ as

$$\tilde{X}_{\tilde{x}\tilde{x}}(\Omega) = \frac{2\pi\sigma^2}{\Omega_\alpha \Omega_\beta}\left(1+\Omega^T \bar{\Gamma}\Omega\right)^{-\frac{3}{2}}, \tag{9.12}$$

where $\Omega = (\Omega_1, \Omega_2)^T$ and $\bar{\Gamma} = \begin{pmatrix} cos\varphi & sin\varphi \\ -sin\varphi & cos\varphi \end{pmatrix} \begin{pmatrix} \Omega_\alpha^{-2} & 0 \\ 0 & \Omega_\beta^{-2} \end{pmatrix} \begin{pmatrix} cos\varphi & sin\varphi \\ -sin\varphi & cos\varphi \end{pmatrix}^T$.
For the discrete signal $x[t]$, its PSD is

$$X_{xx}\left(e^{j\omega}\right) = \sum_{i=-B}^{B}\sum_{j=-B}^{B} \tilde{X}_{\tilde{x}\tilde{x}}\left((\omega_1 - 2i\pi) f_1, (\omega_2 - 2j\pi) f_2\right), \tag{9.13}$$

where B is a band-limit parameter. f_1 and f_2 are the sampling frequencies in the horizontal and vertical directions, respectively.

Based upon this image model, we can deduce the PSD of $y_1[t]$ and $y_0[t]$ according to their relation to $x[t]$ in Eq. (9.5) and Eq.(9.6). That is

$$\begin{aligned} Y_{y_1y_1}^d\left(e^{j\omega}\right) &= g_H^2\Big(\left(1+|H_P^d\left(e^{j\omega}\right)|^2\right) X_{x_0x_0}\left(e^{j\omega}\right) \\ &\quad -2H_P^d\left(e^{j\omega}\right) X_{x_0x_1}\left(e^{j\omega}\right)\Big), \end{aligned} \tag{9.14}$$

$$\begin{aligned} Y_{y_0y_0}^d\left(e^{j\omega}\right) &= g_L^2\Big(\left(|1-H_U^d\left(e^{j\omega}\right) H_P^d\left(e^{j\omega}\right)|^2 + |H_U^d\left(e^{j\omega}\right)|^2\right) X_{x_0x_0}\left(e^{j\omega}\right) \\ &\quad +2\left(1-H_U^d\left(e^{j\omega}\right) H_P^d\left(e^{j\omega}\right)\right) H_U^{d*}(e^{j\omega}) X_{x_0x_1}\left(e^{j\omega}\right)\Big), \end{aligned} \tag{9.15}$$

where

$$X_{x_0x_0}\left(e^{j\omega}\right) = \frac{1}{2}\sum_{k=0}^{1} X_{xx}\left(e^{j\omega_1}, e^{j((\omega_2/2)+k\pi)}\right),$$

$$X_{x_0x_1}\left(e^{j\omega}\right) = \frac{1}{2}\sum_{k=0}^{1} (-1)^k e^{-j(\omega_2/2)} X_{xx}\left(e^{j\omega_1}, e^{j((\omega_2/2)+k\pi)}\right).$$

Substituting Eq. (9.9) and Eq. (9.10) into Eq. (9.14) and Eq. (9.15), we get

$$\begin{aligned} Y_{y_1y_1}^d(e^{j\omega}) &= \frac{1}{2} g_H^2 \sum_{k=0}^{1}\left[1-(-1)^k \hat{H}_P^d(e^{j\omega_1}, e^{j\omega_2/2})\right]^2 \\ &\quad \cdot X_{xx}(e^{j\omega_1}, e^{j(\omega_2/2+k\pi)}), \end{aligned} \tag{9.16}$$

$$Y^d_{y_0y_0}(e^{j\omega}) = \frac{1}{2}g_L^2\sum_{k=0}^{1}\left[1+(-1)^k\frac{1}{2}\hat{H}^d_P(e^{j\omega_1},e^{j\omega_2/2})-\frac{1}{2}\left(\hat{H}^d_P(e^{j\omega_1},e^{j\omega_2/2})\right)^2\right]^2 \cdot X_{xx}(e^{j\omega_1},e^{j(\omega_2/2+k\pi)}), \tag{9.17}$$

where the scaling factors normalizing the energy of the synthesis filters are computed as

$$g_L^2 = Aver\left(\left[1+\hat{H}^d_P\left(e^{j\omega}\right)\right]^2\right), \tag{9.18}$$

$$g_H^2 = Aver\left(\left[1-\frac{1}{2}\hat{H}^d_P\left(e^{j\omega}\right)-\frac{1}{2}\left(\hat{H}^d_P\left(e^{j\omega}\right)\right)^2\right]^2\right). \tag{9.19}$$

$Aver(\cdot) = (1/4\pi^2)\int\int_{-\pi}^{\pi}(\cdot)d\omega_1 d\omega_2$ is an integral function.

Thus, the variances of $y_1[t]$ and $y_0[t]$ are

$$\begin{aligned}\sigma^2_{y_1} &= Aver\left(Y^d_{y_1y_1}\left(e^{j\omega}\right)\right)\\ &= g_H^2 Aver\left(\left[1-\hat{H}^d_P\left(e^{j\omega}\right)\right]^2 X_{xx}\left(e^{j\omega}\right)\right),\end{aligned} \tag{9.20}$$

$$\begin{aligned}\sigma^2_{y_0} &= Aver\left(Y^d_{y_0y_0}\left(e^{j\omega}\right)\right)\\ &= g_L^2 Aver\left(\left[1+\frac{1}{2}\hat{H}^d_P\left(e^{j\omega}\right)-\frac{1}{2}\left(\hat{H}^d_P\left(e^{j\omega}\right)\right)^2\right]^2 X_{xx}\left(e^{j\omega}\right)\right).\end{aligned} \tag{9.21}$$

Assuming high-rate scalar quantization and the independence of quantization errors in two bands, the coding gain of a one-dimensional ADL can be defined as that by Taubman and Marcellin [130]

$$G_{ADL} = \frac{\sigma^2}{\sqrt{\sigma^2_{y_0}\sigma^2_{y_1}}}. \tag{9.22}$$

Although ADL cannot be orthonormal, this definition can still be applied due to the normalization of synthesis filter energy.

The process can be easily extended to the 2D case, where a second-dimension lifting is again performed on $y_0[t]$ and $y_1[t]$ before scaling, generating four subband signals $z_{00}[t]$, $z_{01}[t]$, $z_{10}[t]$, and $z_{11}[t]$. It can be expressed as

$$y_1[t] = x_1[t] - h_P^{d_1}[t] * x_0[t], \tag{9.23}$$

$$y_0[t] = x_0[t] + h_U^{d_1}[t] * y_1[t], \tag{9.24}$$

$$z_{01}[t] = g_{01}\left(y_{01}[t] - h_P^{d_2}[t] * y_{00}[t]\right), \tag{9.25}$$

$$z_{00}[t] = g_{00}\left(y_{00}[t] + g_{01}^{-1}\left(h_U^{d_2}[t] * z_{01}[t]\right)\right), \tag{9.26}$$

$$z_{11}[t] = g_{11}\left(y_{11}[t] - h_P^{d_3}[t] * y_{10}[t]\right), \tag{9.27}$$

$$z_{10}[t] = g_{10}\left(y_{10}[t] + g_{11}^{-1}\left(h_U^{d_3}[t] * z_{11}[t]\right)\right). \tag{9.28}$$

Above d_1 is the prediction direction vector for the first dimension lifting and d_2 and d_3 for the second dimension lifting performed on the low and high pass bands, respectively. $y_{00}[t]$ and $y_{01}[t]$ are even and odd columns of $y_0[t]$; $y_{10}[t]$ and $y_{11}[t]$ are that of $y_1[t]$. $h_P^{d_2}[t]$ and $h_U^{d_2}[t]$ are prediction and update filters on the low-pass signal in the second dimension; and are that on the high-pass signal. g_{00}, g_{01}, g_{10}, and g_{11} are scaling factors which normalize the energy of four synthesis filters.

Similarly, we can compute the four subband variances as

$$\begin{aligned}\sigma^2_{z_{01}z_{01}} &= Aver\left(G_1^{d_2}\left(e^{j\omega_1}, e^{j2\omega_2}\right) G_0^{d_1}\left(e^{j\omega}\right)\right) \\ &\quad \cdot Aver\left(H_1^{d_2}\left(e^{j\omega_1}, e^{j2\omega_2}\right) H_0^{d_1}\left(e^{j\omega}\right) X_{xx}\left(e^{j\omega}\right)\right),\end{aligned} \tag{9.29}$$

$$\begin{aligned}\sigma^2_{z_{00}z_{00}} &= Aver\left(G_0^{d_2}\left(e^{j\omega_1}, e^{j2\omega_2}\right) G_0^{d_1}\left(e^{j\omega}\right)\right) \\ &\quad \cdot Aver\left(H_0^{d_2}\left(e^{j\omega_1}, e^{j2\omega_2}\right) H_0^{d_1}\left(e^{j\omega}\right) X_{xx}\left(e^{j\omega}\right)\right),\end{aligned} \tag{9.30}$$

$$\begin{aligned}\sigma^2_{z_{11}z_{11}} &= Aver\left(G_1^{d_3}\left(e^{j\omega_1}, e^{j2\omega_2}\right) G_0^{d_1}\left(e^{j\omega}\right)\right) \\ &\quad \cdot Aver\left(H_1^{d_3}\left(e^{j\omega_1}, e^{j2\omega_2}\right) H_1^{d_1}\left(e^{j\omega}\right) X_{xx}\left(e^{j\omega}\right)\right),\end{aligned} \tag{9.31}$$

$$\begin{aligned}\sigma^2_{z_{10}z_{10}} &= Aver\left(G_0^{d_3}\left(e^{j\omega_1}, e^{j2\omega_2}\right) G_1^{d_1}\left(e^{j\omega}\right)\right) \\ &\quad \cdot Aver\left(H_0^{d_3}\left(e^{j\omega_1}, e^{j2\omega_2}\right) H_0^{d_1}\left(e^{j\omega}\right) X_{xx}\left(e^{j\omega}\right)\right),\end{aligned} \tag{9.32}$$

where

$$\begin{aligned}H_0^d\left(e^{j\omega}\right) &= \left[1 + \frac{1}{2}\hat{H}_P^d\left(e^{j\omega}\right) - \frac{1}{2}\left(\hat{H}_P^d\left(e^{j\omega}\right)\right)^2\right]^2, \\ H_1^d\left(e^{j\omega}\right) &= \left[1 - \hat{H}_P^d\left(e^{j\omega}\right)\right]^2, \\ G_0^d\left(e^{j\omega}\right) &= \left[1 + \hat{H}_P^d\left(e^{j\omega}\right)\right]^2, \\ G_1^d\left(e^{j\omega}\right) &= \left[1 - \frac{1}{2}\hat{H}_P^d\left(e^{j\omega}\right) - \frac{1}{2}\left(\hat{H}_P^d\left(e^{j\omega}\right)\right)^2\right]^2.\end{aligned}$$

The coding gain then is defined as

$$G_{ADL} = \frac{\sigma^2}{\sqrt[4]{\sigma^2_{z_{01}z_{01}}\sigma^2_{z_{00}z_{00}}\sigma^2_{z_{11}z_{11}}\sigma^2_{z_{10}z_{10}}}}. \tag{9.33}$$

9.3.2 Numerical Analysis

In this section, all schemes are analyzed from their coding gains when one-level decomposition is considered. If vertical lifting is performed first, $x[t]$ is first decomposed into L and H bands in the vertical dimension and then a horizontal dimension lifting is performed on the L and H bands, and vice versa. Hereafter, we represent the two cases as VF and HF, respectively.

In order to examine the impact of the lifting order independent of the angular precision and concrete mode settings, we choose the adaptive weighting based approach for analysis. It finds the best weighting factors of samples used for prediction by minimizing the prediction MSE under high rate condition. From another viewpoint, it is equal to maximizing the coding gain since the high-pass variance is minimized and the update brings little change to the low-pass variance. According to Eq. (9.1) and Eq. (9.2), this problem can be formulated as

$$\begin{cases} D = E\left[\left(x_1[t] - \frac{1}{2}\sum_{i=0}^{1}\sum_{k=0}^{L-1}\omega_{i,k}x_0\left[t_1 + \triangle t_{1_{i,k}}, t_2 + \triangle t_{2_{i,k}}\right]\right)^2\right], \\ \omega_{i,k} = argminD, \omega_{0,k} = \omega 1, k. \end{cases} \tag{9.34}$$

It can be solved by the Wiener-Hopf equation based upon the correlation model in Eq. (9.11).

As mentioned in Section 9.2, different lifting orders can affect the performance of ADL for its nonevenly distributed modes, which cover only part of the whole plane. Lifting performed first in the vertical dimension will be beneficial for images with a near-vertical orientation rather than ones near-horizontal. To simulate this condition, samples denoted by black circles in Figure 9.2 are used to predict the pixel denoted by the gray circles in the VF case. For the HF case, similar samples are adopted. Obviously, the prediction direction covers the same range as ADL in two dimensions. We have drawn the coding gain curves in the two cases for x with different orientations and sharpness in Figure 9.3, based upon the coding gain definition in Eq. (9.33) and the image model. The horizontal axis represents the orientation of the signal x with φ varying from $-45°$ to $135°$ and the vertical axis denotes the maximized coding gain. At each φ, the prediction weighting factors in both dimensions are recomputed according to Eq. (9.34) for adaptation. According to the statistical results on practical images by Chang and Girod [183], we set Ω_β/Ω_0 to 0.4, where Ω_0 is a normalization parameter related to (f_1, f_2). With fixed Ω_β/Ω_0, the sharpness of images is determined only by Ω_α/Ω_0. The larger Ω_α/Ω_0 is, the sharper the image is. In our analysis, Ω_α/Ω_0 is set to 1, 2, 4, and 8 for comparison.

From Figure 9.3, we can see when the angular scope of one-dimension lifting is confined to that between the two dashed lines, ADL with different lifting orders performs quite differently. For VF, it is superior to HF ranging from $-45°$ to $45°$ while it is worse than HF at other orientations. This is consistent with our supposition, since φ corresponds to the direction with sharp intensity transitions. With the decrease of

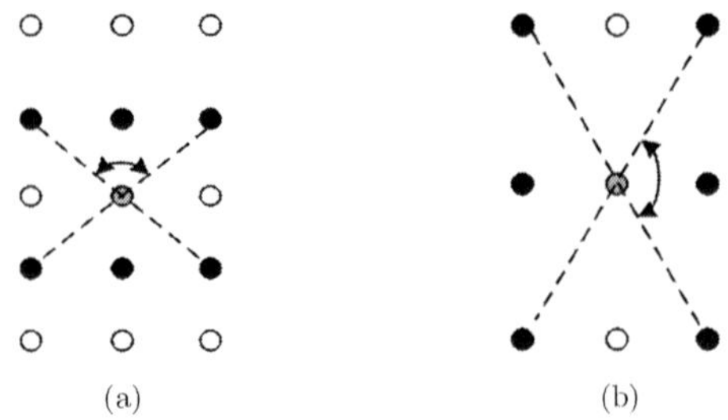

Figure 9.2 Pixels used in the prediction step of the VF case for lifting order analysis with adaptive weighting; (a) the first dimension, (b) the second dimension. The gray circles denote the pixel to be predicted on odd rows. The black circles residing on the even rows are samples used for prediction.

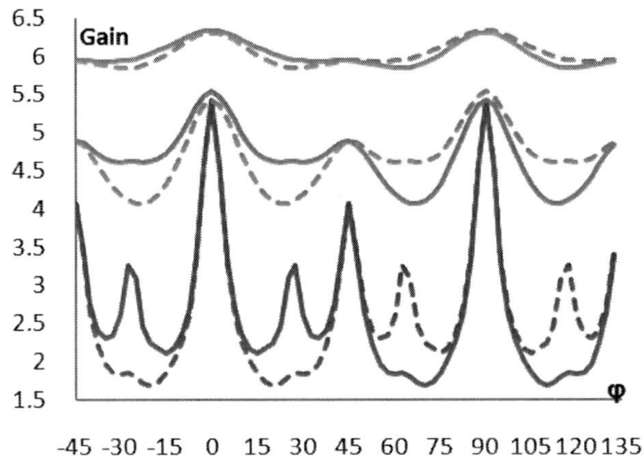

Figure 9.3 Lifting order analysis of ADL. The three solid lines correspond to the VF case with $\Omega_\alpha/\Omega_0 = 8, 2, 1$, respectively, from the bottom to top and the three dashed lines denote the corresponding coding gain curves for the HF case.

Ω_α/Ω_0, the image becomes smooth. Therefore, coding gain at different orientations varies little and the difference between VF and HF vanishes gradually.

As shown previously, we consider the case that the directional scope of each lifting is limited and show that it is sensitive to image transposition. Then a natural question will be: how do they perform if lifting can cover almost the whole plane? Hence, a lifting order analysis based upon more samples in prediction and a wider angular scope (see Figure 9.4) is presented. In Figure 9.4, only first dimension lifting is depicted for VF. The similar samples are used for the second dimension and the HF case. The coding gain curves are depicted in Figure 9.5. It can be observed that VF and HF perform almost the same at different Ω_α/Ω_0. When it equals 8, there are two peaks between 0° and 45° in VF, which correspond to d_1' and d_2' in Figure 9.4 and there are also two peaks between 45° and 90° along d_3' and d_4', respectively. In other words, the coding gain reaches its local maximum when there is an integer pixel residing at the even rows of the orthogonal sampling grid along the image orientation.

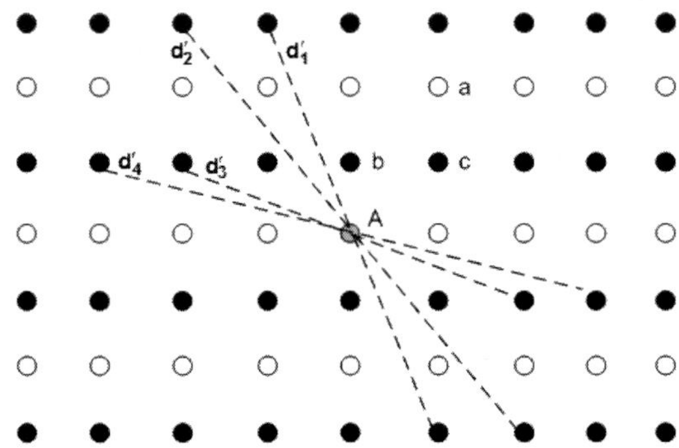

Figure 9.4 Samples used in the prediction step of the VF case for lifting order analysis with more samples and a wider angular scope. The gray circle denotes the pixel to be predicted on odd rows. The black circles residing on the even rows are samples used for prediction.

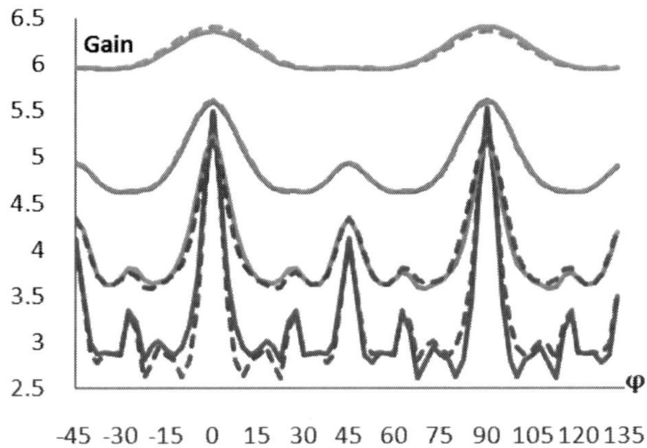

Figure 9.5 Lifting order analysis of ADL with more pixels for prediction. The four solid lines correspond to the VF case with $\Omega_\alpha/\Omega_0 = 8, 2, 1$, respectively, from the bottom to top and the four dashed lines denote the corresponding coding gain curves for the HF case.

This validates the effectiveness of the mode setting by Chang and Girod [183], where d'_2 does not exist. However, this is only the case when the edge is extremely sharp. When it becomes smooth, weighting factors fall on several nearer pixels at 26.6° in the first dimension of the VF case and adaptive interpolation [181, 182] takes effect. Anyhow, this experiment indeed illustrates, if the lifting in each dimension can cover almost the whole plane, the lifting order will make little sense.

All the previously mentioned analyses consider that the high-pass bands are decomposed as fully as the low-pass bands. However, in practical compression systems the high-pass residues are usually no longer decomposed if the transform complexity is taken into account. Without further decomposition on high-pass bands, there

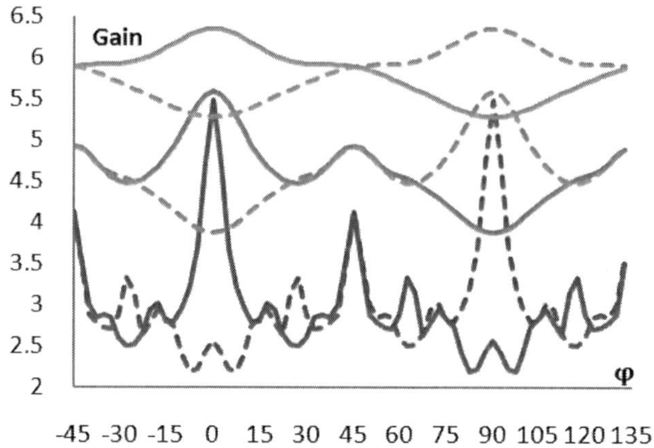

Figure 9.6 Lifting order analysis of ADL without decomposing the high-pass bands. The three solid lines correspond to the VF case with $\Omega_\alpha/\Omega_0 = 8, 2, 1$, respectively, from the bottom to top and the three dashed lines denote the corresponding coding gain curves for the HF case.

sometimes exist some large coefficients in these bands, thus, hurting the performance. The coding gain comparisons for the case that high-pass bands are not decomposed are depicted in Figure 9.6, where lifting in the first dimension dominates the performance. Now the VF curves apparently differ from the HF curves, and which is better depends upon both the image orientation and its sharpness. For the vertical direction, lifting in the vertical dimension should be performed first because the correlation among rows is much stronger than that among columns. For diagonal directions, the two cases perform the same since they are symmetrical relative to the diagonal lines. But, at half-pel directions, horizontal lifting performs better when the edge is sharp because there is an integer pixel “a” on the right column of the predicted pixel “A” with the strongest correlation, as shown in Figure 9.4. With the decrease of Ω_α/Ω_0, the correlation of “b” and “c” with “A” becomes stronger and dominates the directional row correlation. Therefore, we see that VF becomes superior to HF when Ω_α/Ω_0 equals 1. From these, we can conclude that directionality shows more importance when the edge is sharp; while the distance is more important when it is smooth. The coding gain curves for the cases with or without decomposing the high-pass bands are also depicted in Figure 9.7. Here the lifting order is adaptively selected. It demonstrates, without decomposing the high-pass bands, that almost the same performance can be achieved with an adaptive lifting order. In practice, it can save both bits and computations since there is no need to transmit its modes to the decoder.

According to the analyses in this section, it can be concluded that the adaptive lifting order can compensate the shortcomings brought by nonevenly covered or long-distance modes and bring us better performance. Although the analyses are theoretical and the absolutely adapted weighting factors for prediction cannot be achieved in practice, it indeed gives us a qualitative study of the compression performance

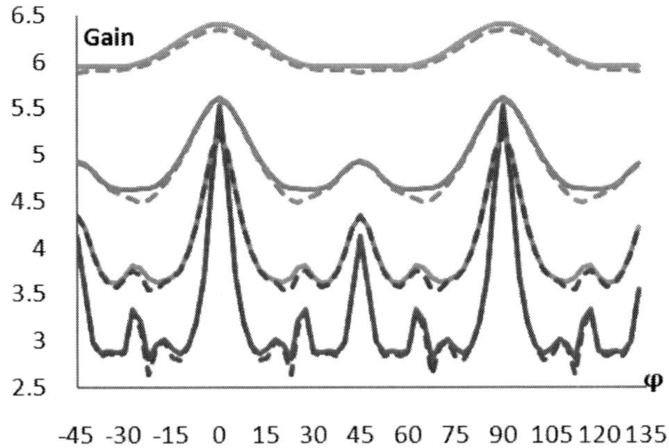

Figure 9.7 Comparison of adaptive lifting order based ADL with or without decomposing the high-pass band. The four solid lines correspond to the case that the high-pass bands are decomposed with adaptive prediction filters where $\Omega_\alpha/\Omega_0 = 8, 2, 1$, respectively, from the bottom to top; the four dashed lines denote the corresponding coding gain curves for the case that the high-pass bands are not decomposed at all.

influenced by different lifting orders. These analyses motivate us to develop our dFT compression scheme, which chooses an evenly distributed set of modes with an adaptive lifting order and short-distance prediction.

9.4 Directional Filtering Transform

In this section, the directional filtering transform (dFT) allowing an adaptive lifting order is proposed to exploit intra-frame correlations more fully. It overcomes the disadvantage of directional intra prediction in H.264 by allowing bidirectional prediction from the neighborhood, which may be samples from not only neighboring blocks but also the current block.

9.4.1 Proposed Intra-Coding Scheme

The proposed intra-coding scheme is depicted in Figure 9.8. There are two ways to exploit intra-frame correlations. One is that the input image is predicted by UDF from neighboring block boundaries. After that, residues are decorrelated by a DCT transform and then quantized with a uniform quantizer. Another is to predict the input image by BDF, which possesses evenly distributed filtering modes by which the lifting order can be adaptively selected. After the filtering, two choices are presented. If the filtered residues exhibit an isotropic correlation, a DCT-like transform is performed on them to further decorrelate them and the resulting coefficients are then

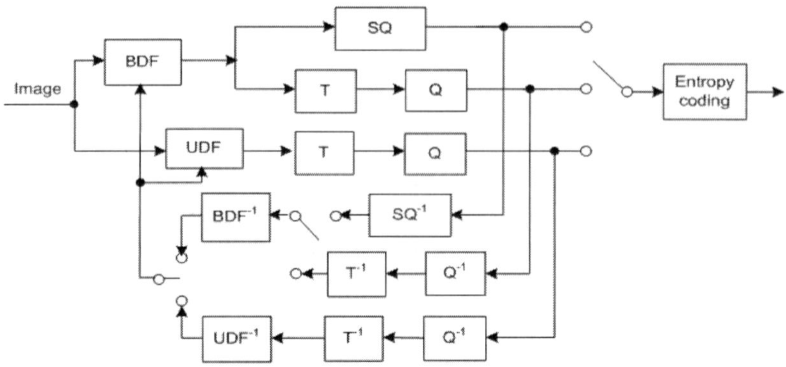

Figure 9.8 Proposed intra-coding scheme. T denotes the DCT-like integer transform in H.264 and Q denotes scalar quantization.

quantized and coded; otherwise the transform is skipped and residues are directly quantized and entropy coded. Rate distortion optimization is adopted to make this decision. The BDF, UDF, and optional transform are generically referred to as dFT.

9.4.2 Directional Filtering

x is a zero-mean 2D random signal representing an image block $(N+M)\times(N+M)$. $\{x_t \mid t=(t_1,t_2), t_1=M,\cdots,M+N-1, t_2=M,\cdots,M+N-1\}$ is the current block $N\times N$ to be coded. The samples in the first M columns and rows are the reconstructed boundaries of neighboring blocks. $\{y_t \mid t=(t_1,t_2), t_1=M,\cdots, M+N-1, t_2= M,\cdots,M+N-1\}$ is the residue block after filtering. In the following discussion, the case of $M=1$ and $N=4$ is considered.

The BDF is discussed first. To locally adapt to the orientations of images/intra frames, eight directional modes are adopted, which are the same as that in H.264 intra-frame coding (see Figures 9.9 and 9.10). Along each direction, there are three filtering steps within one-level decomposition to explore both intra-block and inter-block correlations.

For modes V_0-V_4, a vertical filtering is performed first on odd rows denoted by black circles in Figure 9.9a, leaving even rows unprocessed and denoted as $\hat{y}_1[t]$. This process corresponds to the vertical lifting without an update in ADL and, thus, it can be expressed as

$$\begin{cases} y_1[t] = x_1[t] - h_P^d[t] * x_0[t]. \\ \hat{y}_1[t] = x_0[t]. \end{cases} \tag{9.35}$$

The prediction filter $h_P^d[t]$ is a two-point average filter along d for integer-precision modes such as V_0, V_3, and V_4 in Figure 9.9a. For half-precision modes V_1 and V_2,

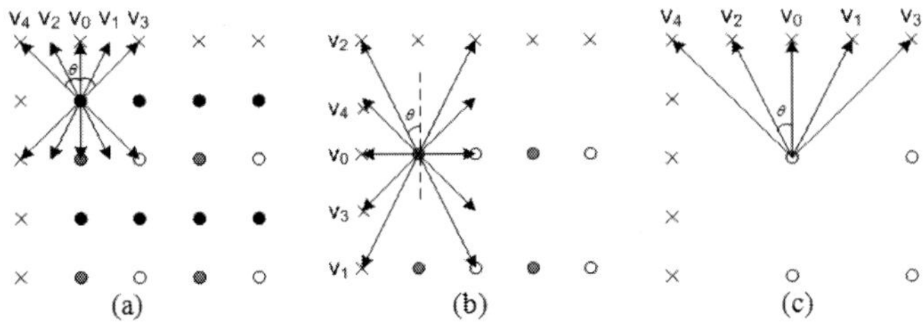

Figure 9.9 Three-step filtering in dFT with vertical filtering first. (a) First step. (b) Second step. (c) Third step.

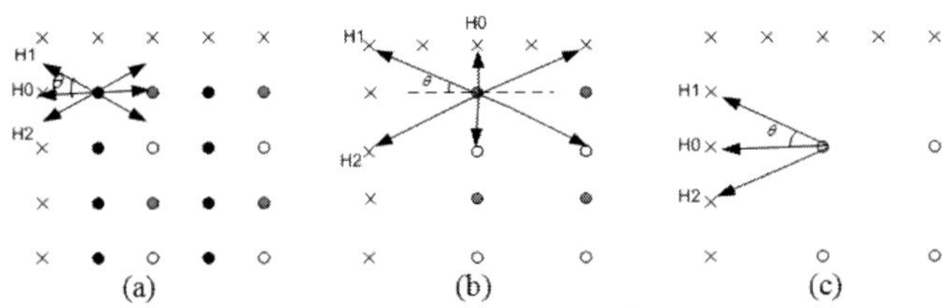

Figure 9.10 Three-step filtering in dFT with horizontal filtering first. (a) First step. (b) Second step. (c) Third step.

a 4-tap interpolation filter is used before prediction. At block boundaries, reconstructed pixels of neighboring up, left or left-up blocks are also used for prediction, which are denoted as crosses in Figure 9.9.

After the first step, horizontal filtering is performed on the odd columns of $\hat{y}_1[t]$, which are denoted as gray circles in Figure 9.9b. It is expressed as

$$\begin{cases} y_2[t] = y_{11}[t] - h_P^{d'}[t] * y_{10}[t]. \\ \hat{y}_2[t] = \hat{y}_{10}[t]. \end{cases} \tag{9.36}$$

$\hat{y}_{10}[t]$ is the down-sampled block of $\hat{y}_1[t]$ by keeping even columns and $\hat{y}_{11}[t]$ the odd columns. $y_2[t]$ is the filtered residues and $\hat{y}_2[t]$ the even columns of $\hat{y}_1[t]$. d' is the prediction direction for this filtering. Orientation is considered as unchanged in up-sampled $\hat{y}_1[t]$ and therefore d' should be along the same direction as d. At V_3 and V_4, this rule holds and it also needs to interpolate within even columns. But at V_0, no sample exists along the vertical direction. The prediction is then operated along the horizontal direction at V_0 for closer distance and corresponding strong correlation.

For the third step, $\hat{y}_2[t]$ is unidirectionally predicted from the reconstructed boundaries of neighboring blocks along the same orientation as d. The predicted residual signal $y_3[t]$ then is expressed as

$$y_3(t_1,t_2) = \begin{cases} \hat{y}_2(t_1,t_2) - x(2(t_1 - t_2\tan\theta),0), if\ t_1 \geq t_2\tan\theta, \\ \hat{y}_2(t_1,t_2) - x(0,2(t_2 - t_1/\tan\theta)), if\ t_1 < t_2\tan\theta, \end{cases} \tag{9.37}$$

$t_1, t_2 = 1, 2, \cdots, N/2$, where θ is the angle of prediction direction and the vertical axis in Figure 9.9.

For the modes $H_0 - H_2$, the horizontal filtering is performed first, followed by the vertical lifting and unidirectional prediction (see Figure 9.10). With the adaptive lifting order, BDF provides us with an evenly distributed set of modes covering the whole plane. If the image block exhibits a strong vertical correlation, lifting in the vertical dimension is performed first, and vice versa. Therefore, the directional correlation can be fully exploited.

As previously mentioned, we take one-level lifting as an example. The hierarchical structure can be obtained by iteratively performing two dimensional lifting on the remaining samples, such as $\hat{y}_2[t]$, before the third prediction step like that in ADL. Moreover, all residues are also scaled to preserve the energy of quantization noise in the reconstructed image. In our experiment, these scaling factors are computed numerically to normalize the synthesis filters.

The other filtering method is UDF, which can be similarly expressed as

$$y(t_1,t_2) = \begin{cases} x(t_1,t_2) - x(t_1 - t_2 \tan\theta, 0), if\ t_1 \geq t_2 \tan\theta, \\ x(t_1,t_2) - x(0, t_2 - t_1/\tan\theta), if\ t_1 < t_2 \tan\theta, \end{cases} \tag{9.38}$$

$t_1, t_2 = 1, 2, \cdots, N/2$. Here we only give a simplified expression. The weighted prediction and DC prediction are also applied to ensure prediction accuracy and accommodate smooth blocks.

With these two filtering modes, the directional correlation can be effectively exploited. Rate-distortion optimization (RDO) is adopted to choose between them and 1-bit flag is used to indicate which mode is selected. At the entropy coding stage, these flags are arithmetically coded by using neighboring ones as the context.

9.4.3 Optional Transform

After the filtering, there are two ways to process the residues. If there is still correlation among them, a 2D transform is performed to further decorrelate them; otherwise the transform is skipped and residues are directly quantized and entropy coded.

Similar to that by Ye and Karczewicz [188], a nonseparable directional transform derived from the Karhunen-Loève transform (KLT) is adopted for blocks at each orientation. Each residue block is first divided into 4×4 blocks and then optionally transformed with its corresponding transform matrix. Let y denote the 1×16 vector of one 4×4 residue block, the transform is given as

$$z = T'y. \tag{9.39}$$

T is the 4×4 transform matrix with each column representing the basis vector. Residue blocks of a data set with the same orientations and block types are gathered to train the matrix T. After the transform, the coefficients z are scalar quantized and then entropy coded. Separate context models are used for different types of blocks.

If there is no need to perform the transform, the quantization method that has been integrated into the Key Technical Area (KTA) software [191] is adopted to quantize the residues. Quantized coefficients are then scanned row by row for simplicity and entropy coded. Since the residuals still preserve some spatial features, the significant map is coded by taking the spatial correlations into account, where the left and upper significant bits are taken as the context for the current bit.

In order to determine whether the transform is performed or not, the rate-distortion optimization in H.264 is modified and used as the criterion. For the 4×4 mode, a transform is always performed at each 4×4 block so there are no overhead bits. For the 8×8 and 16×16 modes, the transform can be selected for each 4×4 block and, thus, there is one bit for each 4×4 block. All these bits are coded by context-adaptive arithmetic coding.

9.5 Experimental Results

In our experiment, H.264 intra-frame coding (JM14.1 High Profile) is taken as the benchmark. Experiments for dFT are performed across 4×4, 8×8, and 16×16 blocks. At 16×16, a two-level filtering is performed for BDF with the same filtering directions, while at 4×4 and 8×8 one-level filtering is performed for each block. The eight filtering directions are explicitly coded as with that of H.264 4×4 intra mode. Context adaptive binary arithmetic coding (CABAC) is chosen at the entropy coding stage.

The comparisons between dFT and H.264 intra coding are depicted in Figure 9.11. "H.264 Intra" denotes the intra coding of H.264 and "dFT" denotes the proposed scheme with two filtering modes: UDF and BDF. One can see that dFT shows its superiority especially on images with rich edges. There is up to a 1 dB gain for Foreman, 0.4 dB for Barbara, 1.8 dB for Spoke, 1.2 dB for Straw, and 0.1 dB for Lena and Monarch. In order to examine the roles that two filtering modes play, experiments on fixed block sizes are also carried out.

The performances of dFT on 16×16 blocks only are depicted in Figure 9.12. "H.264_16" denotes the intra coding of H.264 on 16×16 blocks, which is equivalent to the UDF mode of dFT. "BDF_16" denotes the dFT scheme with only the BDF mode being enabled. And "dFT_16" denotes the dFT scheme with the two filtering modes BDF and UDF both being enabled. We can see, when the block size is large, dFT shows a strong superiority to H.264 intra prediction by combination of the two filtering modes. Moreover, BDF performs much better than H.264 intra prediction on 16×16 blocks since it decreases the prediction distance by predicting from the neighborhood rather than the boundaries of neighboring blocks. On the other hand, the intra prediction in H.264 can compensate the disadvantages of BDF too, such as poor interpolation on blocks with sharp edges, weak prediction at V_0 in the second-step filtering (see Figure 9.9b) on blocks with sharp vertical edges, and lack of inter-band correlation exploration. Also the DC prediction in H.264 intra coding followed by a DCT-like integer transform and a subsequent Hadamard transform can give us

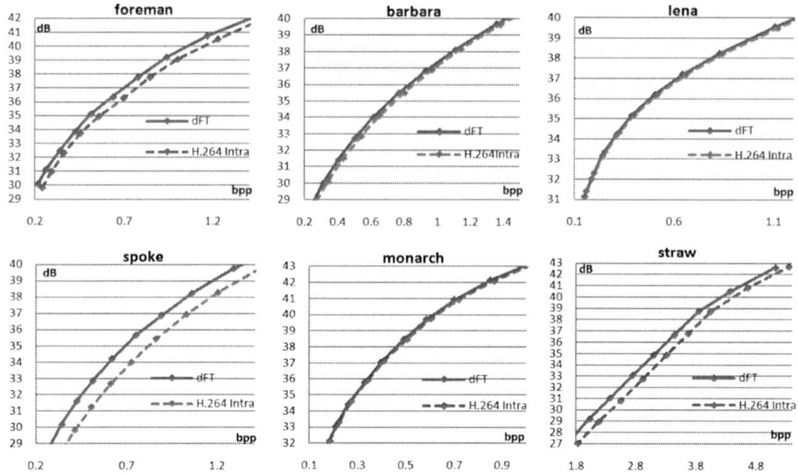

Figure 9.11 Experimental results for Foreman (176 × 144), Barbara (512 × 512), Lena (512 × 512), Spoke (512 × 512), Monarch (768 × 512), and Straw (512 × 512). The dashed curves denote the performance of H.264 intra coding and the real ones denote that of the proposed dFT.

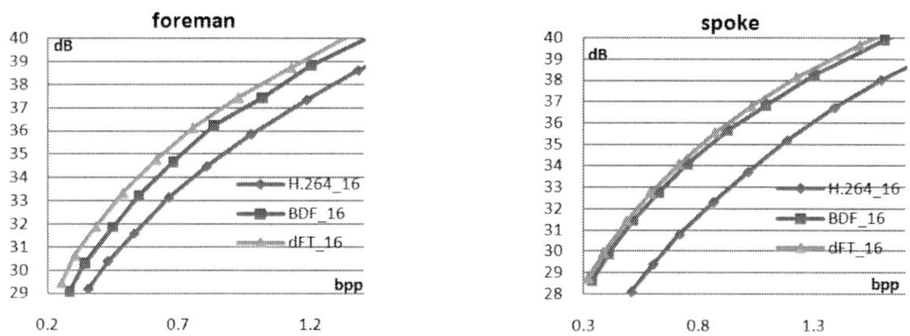

Figure 9.12 Experimental results for Foreman (176 × 144) and Spoke (512 × 512) on 16 × 16 blocks.

a compact representation of smooth blocks with little directionality, which further brings benefits to H.264 intra prediction. Therefore, both of the two filtering modes have their advantages and the combination of BDF and UDF can well exploit intra frame/image correlations. There is up to a 2.4 dB gain for Foreman and a 3.8 dB gain for Spoke with rich edges.

The evaluation of H.264 and dFT on 8×8 blocks only is also depicted in Figure 9.13. Now the performance gap between "H.264_8" and "BDF_8" diminishes and the two filtering modes show comparable performance on Foreman, since the block size is smaller. But by combination of BDF and UDF, dFT can still outperform H.264 intra prediction by up to 0.8 dB on Foreman and 2.2 dB for Spoke. While for 4 × 4 blocks (see Figure 9.14), the H.264 intra prediction can perform well enough because the 4 × 4 size prediction can describe the local characteristics of images

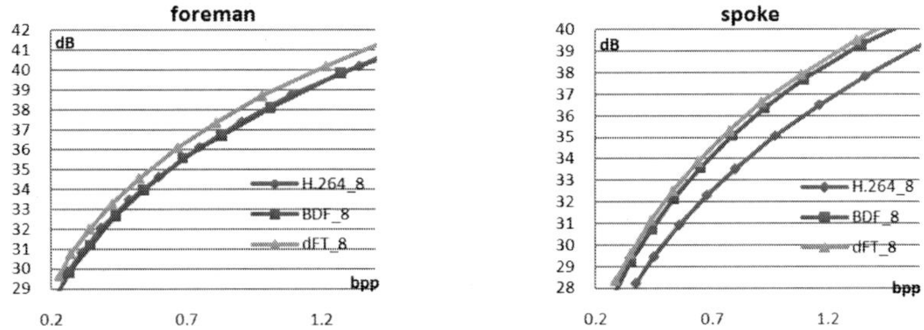

Figure 9.13 Experimental results for Foreman (176 × 144) and Spoke (512 × 512) on 8 × 8 blocks.

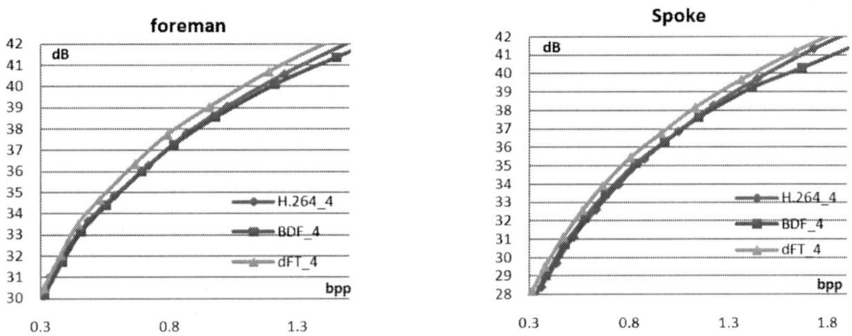

Figure 9.14 Experimental results for Foreman (176 × 144) and Spoke (512 × 512) on 4 × 4 blocks.

more precisely. Consequently, only a comparable performance is achieved by BDF on Foreman and Spoke at low bit rates. At high bit rates, there is even a loss on both images since there is no DC mode for BDF. When combining BDF with UDF, dFT can achieve up to a 0.5 dB gain on Foreman and 0.8 dB on Spoke, which are smaller than those for 8 × 8 and 16 × 16 blocks. Also we can see from Figure 9.11 that the superiority of dFT over H.264 intra coding is weakened by allowing variable block size predictions, which can partially adapt to the local characteristics of images.

In addition to objective evaluations, reconstructed frames of Foreman and Spoke at 0.3 bpp are also depicted in Figure 9.15. Apparently, the quality of dFT is much better than H.264 intra prediction. For Foreman, there is a clearer outline of the right eye and ringing artifacts around sharp edges disappear with dFT. For Spoke, apparent ringing artifacts are seen with H.264 intra prediction, while the edges with dFT look more sharp and smooth. Similar results are also observed in Straw. For Barbara, Lena, and Monarch, the subjective performances of H.264 intra coding are preserved by dFT. In conclusion, dFT not only presents a significant objective performance gain but also shows a much better subjective performance over H.264 intra coding.

As for computational complexity, BDF modes clearly have higher complexity than UDF modes at the encoder. To be specific, the prediction of BDF modes has similar computations as that of UDF modes. The increased computation from BDF modes lies in the nonseparable transforms and the RDO selection on each block

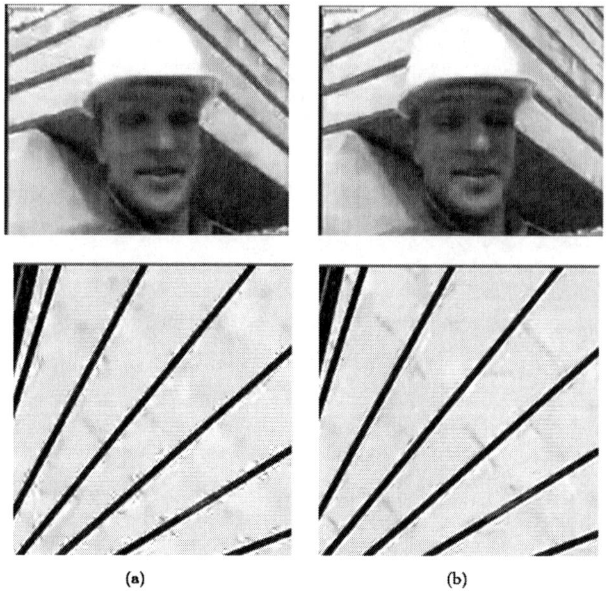

Figure 9.15 Reconstructed frame of Foreman and part of Spoke at 0.3 bpp. (a) H.264 Intra; (b) dFT.

deciding whether a transform is performed or not. According to current implementation without any speed optimization, the encoding time of dFT is about 7 to 8 times that of H.264 intra coding. From the experiences on H.264 speed optimization, the encoding computation can be significantly reduced and still maintain a similar performance. At the decoder side, since all prediction modes and transform decisions are known, there is hardly any increase in complexity.

9.6 Summary

We present a mathematical analysis of the lifting order problem of ADL in this chapter and show the advantages of lifting with an adaptive order. Also the directional filtering transform (dFT) is proposed according to the theoretical analysis to better exploit the directional correlations in images/intra frames. The band-based feature of BDF makes up the shortcomings of directional intra prediction from neighboring block boundaries in H.264. Also with an adaptive lifting order, an evenly distributed set of filtering modes occupying the whole plane offers it the possibility to predict from the most correlated samples. Both inter-block and intra-block correlations are exploited by dFT. Experiments in H.264 intra coding demonstrate its effectiveness especially in images with sharp edges. Subjective results also show its superiority in preserving the geometric flow of edges.

Part IV
Vision-Based Compression

Conventional visual data compression technologies base on Shannon's information theory. In most of existing image compression schemes and standards, the statistical correlation among pixels is considered as a dominant factor on designing prediction, transform, quantization, and entropy coding. During the past three decades, the coding technologies have evolved significantly. However, it has been felt difficult to further improve the coding technologies. In order to improve the coding performance, the computational complexity has been and still will be increased significantly. Nevertheless, it is more and more difficult to get a decent improvement even with very complicated schemes. It is time to reconsider if the current coding framework can have a bright future and if there are new and promising frameworks. In this part, we are trying to extend the current coding framework from different aspects.

In Chapter 10, image compression utilizing visual redundancy is investigated. Inspired by recent advancements in image inpainting techniques, we present an image compression framework toward visual quality rather than pixel-wise fidelity. In this framework, an original image is analyzed at the encoder side so that portions of the image are intentionally and automatically skipped. Instead, some information is extracted from these skipped regions and delivered to the decoder as assistant information in the compressed fashion. The delivered assistant information plays a key role in the proposed framework because it guides image inpainting to accurately restore these regions at the decoder side. Moreover, to fully take advantage of the assistant information, a compression-oriented edge-based inpainting algorithm is proposed for image restoration, integrating pixel-wise structure propagation and patch-wise texture synthesis. We also construct a practical system to verify the effectiveness of the compression approach in which edge map serves as assistant information and the edge extraction and region removal approaches are developed accordingly. Evaluations have been made in comparison with baseline JPEG and standard MPEG-4 AVC/H.264 intra-picture coding. Experimental results show that our system achieves up to 44% and 33% bits-savings, respectively, at similar visual quality levels. Our proposed framework is a promising exploration toward future image and video compression.

Current image coding schemes make them hard to utilize external images for compression even if highly correlated images can be found in the cloud. To solve this problem, in Chapter 11 we present a method of cloud-based image coding that is different from current image coding even on the ground. It no longer compresses images pixel by pixel and instead tries to describe images and reconstruct them from a large-scale image database via the descriptions. First, we describe an input image based on its down-sampled version and local feature descriptors. The descriptors are used to retrieve highly correlated images in the cloud and identify corresponding patches. The down-sampled image serves as a target to stitch retrieved image patches together. Second, the down-sampled image is compressed using current image coding. The feature vectors of local descriptors are predicted by the corresponding vectors extracted in the decoded down-sampled image. The predicted residual vectors are compressed by transform, quantization, and entropy coding. The experimental results show that the visual quality of reconstructed images is significantly better than that of intra-frame coding in HEVC and JPEG at thousands to one compression.

In Chapter 12, we present a novel feature-based coding scheme by using inter-image correlation for cloud photo storage. Different from other schemes on image set compression, our key technical contribution is to use local feature descriptors on analyzing the correlation among images. This content-based matching is invariant to scale and rotation, and is less sensitive to illumination changes. According to the analysis result, we can reorganize correlated images as a pseudo sequence for compression by minimizing predictive cost. We also observe that the correlation in pseudo sequences is much more complicated than that in natural video. By using locally matched regions, a multi-model approach is presented characterizing local geometric deformation and photometric transformation between images, which significantly outperforms global transformation. For one reference image, multiple predictions can be generated according to the multi-model approach. Finally, we further exploit the correlation between current image and multiple predictions by block-based motion compensation, similar to video compression. Experimental results show that our scheme coding of correlated images as a pseudo sequence is 10 times more efficient than individual JPEG compression.

Chapter 10
Edge-Based Inpainting

10.1 Introduction

Over the last two decades, great improvements have been made in image and video compression techniques driven by a growing demand for storage and transmission of visual information. State-of-the-art JPEG 2000 and MPEG-4 AVC/H.264 are two examples that significantly outperform their previous rivals in terms of coding efficiency. However, these mainstream signal-processing-based compression schemes share a common architecture, namely, transform followed by entropy coding, where only the statistical redundancy among pixels is considered as the adversary of coding. Through two decades of development, it has been becoming difficult to continuously improve coding performance under such architecture. Specifically, to achieve high compression performance, more and more modes are introduced to deal with regions of different properties in image and video coding. Consequently, intensive computational efforts are required to perform mode selection subject to the principle of rate-distortion optimization. At the same time, more and more memory-cost context models are utilized in entropy coding to adapt to different kinds of correlations. As a result, small improvements in coding efficiency are accomplished with great pain of increased complexity in both encoder and decoder.

In addition to statistical redundancy, visual redundancy in videos and images has also been considered in several works. They are motivated by the generally accepted fact that minimizing overall pixel-wise distortion, such as mean square error (MSE), is not able to guarantee good perceptual quality of reconstructed visual objects, especially in low bit-rate scenarios. Thus, the human visual system (HVS) has been incorporated into compression schemes by Jayant et al. [192] and Hontsch and Karam [193], trying to remove some visual redundancy and to improve coding efficiency as well as visual quality. Moreover, attempts have been made to develop compression techniques by identifying and utilizing features within images to achieve high coding efficiency. These kinds of coding approaches are categorized as second-generation techniques by Reid et al. [194], and have raised a lot of interest due to the potential of high compression performance. Nevertheless, taking the segmentation-based coding method as an example, the development of these coding schemes is greatly influenced by the availability as well as effectiveness of appropriate image analysis algorithms, such as edge detection, segmentation, and texture modelling tools.

Recently, technologies in computer vision as well as computer graphics have shown remarkable progress in hallucinating pictures of good perceptual quality. Indeed, advancements in structure/texture analysis [195, 196] and synthesis are leading to promising efforts to exploit visual redundancy. So far, attractive results have been achieved by newly presented texture synthesis techniques to generate regions of homogeneous textures from their surroundings [197–205]. Furthermore, various image inpainting methods have been presented, aiming to fill-in missing data in more general regions of an image in a visually plausible way. In fact, the word inpainting was initially invented by museum or art restoration workers. It is first introduced into digital image processing by Bertalmio et al. [206], where a third order partial differential equation (PDE) model is used to recover missing regions by smoothly propagating information from the surrounding areas in isophote directions. Subsequently, more models are introduced and investigated in image inpainting, for example, total variation (TV) model [207], coupled second order PDE model taking into account the gradient orientations [208], curvature driven diffusion (CDD) model [209], and so on. All these approaches work at pixel level and are good at recovering small flaws and thin structures. Additionally, exemplar-based approaches have been proposed to generate textural coarseness; by augmenting texture synthesis with certain automatic guidance, edge sharpness and structure continuity can also be preserved [210–212]. Combining PDE diffusion and exemplar-based synthesis presents more encouraging inpainting results [213–215]. Moreover, inpainting capability is further improved by simple human interactions when human knowledge is borrowed to imagine what unknown regions should be, so that the restoration results look natural to viewers [216, 217].

Due to its potential in image recovery, image inpainting likewise provides current transform-based coding schemes another way to utilize visual redundancy in addition to those that have been done [192–194]. This inference has been successfully exemplified in error concealment when compressed visual data is transmitted over error-prone channels [215, 218]. Moreover, it has been reported that improvement is achieved by employing image inpainting techniques in image compression even though in a straightforward fashion [215]. In addition, image compression also brings new opportunities to image inpainting, as we have pointed out [219]. Since the complete source images are available, many kinds of assistant information can be extracted to help inpainting deal with complex regions that contain structures or other features and which are unable to be properly inferred from the surroundings. Thus, inpainting here becomes a guided optimization for visual quality instead of a blind optimization for image restoration. Accordingly, new inpainting techniques may be developed to better serve image compression.

When image inpainting and image compression are jointly considered in an integrated coding system, two main problems need to be addressed. The first: what should be extracted from a source image as assistant information to represent important visual information. The second: how to reconstruct an image with this assistant information. On the one hand, it has been reported that using different image analyzers, various kinds of assistant information can be extracted, including edge, object, sketch [196], epitome [220, 221], and so on, to represent an image or portion of an image. Then, given a specific kind of assistant information, the corresponding restoration method should be developed to complete a desired reconstruction by making full use of it. On the other hand, from the compression point of view, the effectiveness of restoration methods as well as the efficiency of the compression of assistant information would also influence the choice of assistant information. Such dependency makes the problems more complicated.

In this chapter, we present an image coding framework in which currently developed vision techniques are incorporated with traditional transform-based coding methods to exploit visual redundancy in images. In this scheme, some regions are intentionally and automatically removed at the encoder and are restored naturally by image inpainting at the decoder. In addition, binary edge information consisting of lines of one-pixel width is extracted at the encoder and delivered to the decoder to help restoration. Techniques, including edge thinning and exemplar selection, are proposed and an edge-based inpainting method is presented in which distance-related structure propagation is proposed to recover salient structures, followed by texture synthesis. The basic idea has been discussed in our conference papers [219, 222]. However, some problems have not been investigated carefully in those papers, including questions such as why the edges of an image are selected as assistant information, or how to select the exemplar blocks automatically, and so on.

10.2 The Proposed Framework

As the basic idea of "encoder removes whereas decoder restores" has been mentioned in literature for image compression [215, 223], we would like to point out the novelties of our proposed method here. First, in our approach, the original image is not simply partitioned into two parts: one is coded by conventional transform-based approach, and the other is skipped during encoding and restored during decoding. Instead, techniques for image partition, block removal, and restoration in our proposed scheme are carefully designed toward compression rather than straightforward adoption. Furthermore, skipped regions will not be completely dropped at the encoder side if they contain a portion of information that is difficult to be properly recovered by conventional image inpainting methods. In fact, assistant information is extracted from the skipped regions to guide the restoration process and further induce new inpainting techniques.

The framework of our proposed compression scheme is depicted in Figure 10.1. In this scheme, an original image is first analyzed at the encoder side. The image analysis module automatically preserves partial image regions as exemplars and sends them to the exemplar encoder module for compression using conventional approaches. Meanwhile, it extracts designated information from skipped regions as assistant information and sends it to the assistant info encoder module. Then, the

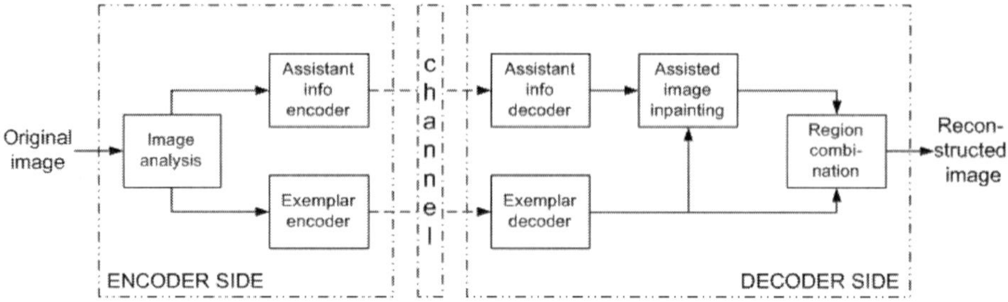

Figure 10.1 The framework of our proposed image compression scheme.

coded exemplars and coded assistant information are banded together to form final compressed data of this image. Correspondingly, at the decoder side, exemplars and assistant information are first decoded and reconstructed. Then, the regions skipped at the encoder are restored by image inpainting based on the twofold information. At the end, the restored regions are combined with the decoded exemplar regions to present the entire reconstructed image.

Figure 10.1 shows a general framework of the proposed compression scheme that does not constrain which kind of assistant information should be used there. Since source image is always available at the encoder side, there are many choices of assistant information extracted from the skipped regions, for example, semantic object, visual pattern, complete structure, simple edges, and so on. Here we start from the mathematical models in image inpainting to discuss what on earth the assistant information should be.

As shown in Figure 10.2, suppose that we are given an image function $f(x), x \in I$, where I is a square region in R^2. Ω, depicted as the gray region in Figure 10.2, is an open bounded subset of I with Lipschitz continuous boundary. It is just the region to be restored by image compression, image inpainting, or a combination of them. This restoration problem can be generalized as

$$argmin\left(\int_{\Omega} D\left(f_{\Omega}(x) - \hat{f}_{\Omega}(x)\right)dx + \lambda R\right). \tag{10.1}$$

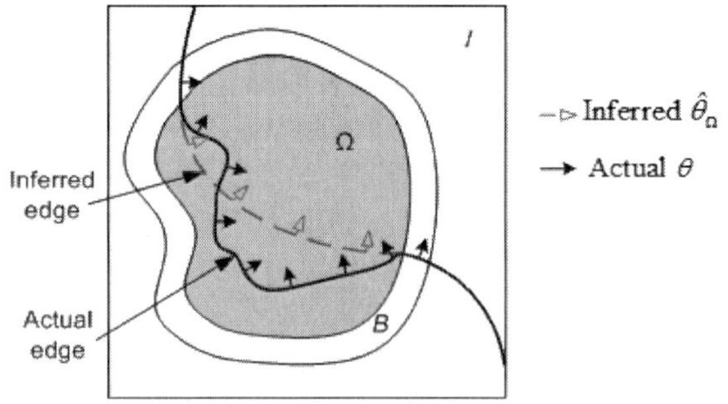

Figure 10.2 Illustration of image inpainting, where the gray region is to be restored.

Here, $f_{\Omega}(x)$ is the original image function in Ω, where it should satisfy $f_{\Omega}(x) = f(x)$ for any $x \in \Omega$. $\hat{f}_{\Omega}(x)$ is a reconstruction of $f_{\Omega}(x)$ at decoder. λ is a Lagrange factor. Clearly, Eq. (10.1) is to find the optimal function $\hat{f}_{\Omega}(x)$ by minimizing the joint cost consisting of reconstructed distortion $D()$ and coding bits R for Ω. Thus, image compression and image inpainting can be viewed as two extreme cases of Eq. 10.1. Specifically, in traditional image compression, $f_{\Omega}(x)$ is directly coded and sent to the decoder, where many bits may be needed to represent $f_{\Omega}(x)$; whereas in image inpainting, there is no bit to represent $f_{\Omega}(x)$ since $\hat{f}_{\Omega}(x)$ is inferred from $f_{I\setminus\Omega}(x)$. However, our proposed method, which is quite different from compression or inpainting, can be granted as a combination of them.

In typical inpainting scenarios, the restoration of $f_{\Omega}(x)$ is usually an ill-posed problem because information in Ω is totally unknown. Fortunately, an image is a 2D projection of the 3D real world. The lost region often has similar statistic, geometric, and surface reflectivity regularities as those in the surroundings. It makes the above ill-posed problem possible to be solved. Therefore, some models are introduced in image inpainting to characterize statistic, geometric, and surface regularities. These models should employ generic regularities, rather than rely on a specific class of images, so that model-based inpainting can be applied in generic images.

One such model, the TV model, is presented by Chan and Shen [207] for image inpainting, in which the variation regularity is first introduced. Since local statistical correlation usually plays a more important role than the global one, as shown in Figure 10.2, B instead of $I \setminus \Omega$ is used to infer the regularities in Ω, where B is a band around Ω. Then, the TV model is to find a function $\hat{f}_{\Omega}(x)$ on the extended inpainting region $B \cup \Omega$ such that it minimizes the following energy function:

$$argmin\left(\int_{B\cup\Omega} |\nabla \hat{f}_{\Omega}(x)|dx + \lambda \int_{B} |\hat{f}_{\Omega}(x) - f(x)|^2 dx\right). \tag{10.2}$$

The first term in Eq. (10.2) is to measure local homogeneity of image function in the region $B \cup \Omega$, and the second term, called the fidelity term, is the sum of the squared difference (SSD) between the reconstructed and the original B in $\hat{f}_{\Omega}(x)$ and B the original in $f(x)$. Eq. (10.2) can be solved by the Euler-Lagrange method described by Chan and Shen [207]. Accordingly, TV inpainting is good at restoring homogenous regions. But, if the lost region contains rich structures, it does not work well, especially when structures are separated far apart by the lost region.

To solve it, another parameter θ is introduced in the inpainting model [208]. Let θ be the vector field of normalized gradient of $f(x)$. $\hat{\theta}_{\Omega}$ is the corresponding parameter to be restored on Ω. With the new parameter of gradient directions, the inpainting problem is posed as extending the pair of functions (f, θ) on B to a pair of functions $(\hat{f}_{\Omega}, \hat{\theta}_{\Omega})$ on Ω. It is completed by minimizing the following function:

$$\begin{aligned} argmin(&\int_{B\cup\Omega} |div\,(\hat{\theta}_{\Omega}(x))\,|^p\,(a + b|\nabla k * \hat{f}_{\Omega}(x)|)\,dx \\ &+ \alpha \int_{B\cup\Omega} (|\nabla \hat{f}_{\Omega}(x)| - \hat{\theta}_{\Omega}(x)\cdot\nabla\hat{f}_{\Omega}(x))\,dx) \end{aligned}. \tag{10.3}$$

The first term presents smooth continuation demand on $\hat{\theta}_{\Omega}$, where a and b are positive constants, and k is a smoothing kernel. It is the integral of the divergence (in L^p function space) of the vector field $\hat{\theta}_{\Omega}$, with respect to the gradients of the smoothed $\hat{f}_{\Omega}(x)$. The second term is a constraint between $\hat{\theta}_{\Omega}$ and $\hat{f}_{\Omega}(x)$, where α is a positive weighting factor. $\hat{\theta}_{\Omega}$ should be related to $\hat{f}_{\Omega}(x)$ by trying to impose $\hat{\theta}_{\Omega} \cdot \nabla \hat{f}_{\Omega} = |\nabla \hat{f}_{\Omega}|$. The use of the vector field θ is the main point of the model given in Eq. (10.3). Thus, it enables image inpainting to restore missing regions by continuing both the geometric and photometric regularities of images.

However, the model in Eq. (10.3) assumes that the parameter $\hat{\theta}_{\Omega}$ can be inferred from θ under a certain smooth constraint. But this assumption is not always true for nature images. Taking Figure 10.2 as an example, the area to be restored consists of two homogenous regions divided by an edge denoted by the solid curve. The dashed curve is the inferred edge in Ω according to Eq. (10.3), which is quite different from the actual one. This problem is hard to be solved in conventional inpainting scenarios even using human intelligence as proposed by Sun et al. [217]. Therefore, in our proposed coding framework, assistant information should be used to correctly infer $\hat{\theta}_{\Omega}$ on Ω. As we have discussed, $\hat{\theta}_{\Omega}$ is the vector field of normalized gradient and is independent from the absolute magnitudes of gradients. It contains two parts of information: where $\hat{\theta}_{\Omega}$ exists and what its direction is. Commonly, it can be simply represented by binary edges of one-pixel width for the purpose of efficient compression.

Consequently, edge information is selected as assistant information for image inpainting. With assistant information, we could remove more regions in an image. Thus, it greatly enhances the compression power of our method. Since edges are low-level features in image, there are some mature tools available to automatically track them in an image. Moreover, edge information is concise and easy to describe in compressed fashion. Therefore, the employment of edge information can, on the one hand, help preserve good visual quality of the reconstructed image. On the other hand, it enables high compression performance by removing some structural regions and efficiently coding the edge information.

Accordingly, an overview of our approach is exemplified in Figure 10.3. Figure 10.3a is the input of the original image of Lena. After image analysis, an edge map denoted by curves in Figure 10.3b is generated, based on which the exemplars denoted by the non-black blocks in Figure 10.3c are selected. Consequently, the exemplars and the needed edge information shown as curves in Figure 10.3c will be coded into bitstream. Then, at the decoder side, the edge information is utilized to guide the structure propagation for the recovery of edge-related regions. The corresponding result is given in Figure 10.3d. The remainder of unknown regions will be restored by texture synthesis. The final reconstructed image after region combination is given in Figure 10.3e.

In the following two sections, we will explain the modules in our framework in detail, especially on the two most important modules, namely image analysis and assisted image inpainting. Here, we would like to emphasize that the introduction of assistant edge information raises different demands on both the encoder and decoder. We deal with them comprehensively.

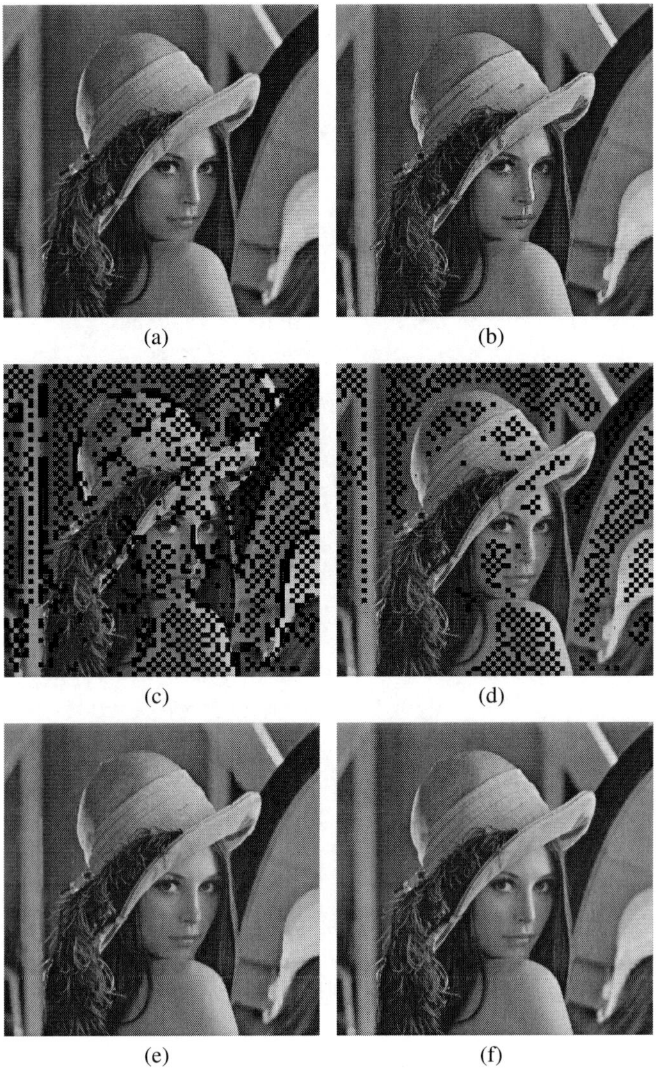

Figure 10.3 Comparison with baseline JPEG on test image Lena. (a) Original image. (b) Edge map. (c) Removed blocks (black blocks) and assistant edge information, note that the assistant edge information is a subset of the entire edge map. (d) Reconstructed image after structure propagation. (e) Reconstructed image after texture synthesis. (f) Reconstructed image by baseline JPEG.

10.3 Edge Extraction and Exemplar Selection

The image analysis module at the encoder side consists of two sub-modules: The first is to extract edge information from image and the second is to select exemplar and skipped regions at block level according to available edge information. They are discussed in the following two sections.

10.3.1 Edge Extraction

As discussed in Section 10.2, edge information plays an important role in the proposed coding scheme. It assists the encoder to select exemplar and skipped regions and the decoder to restore skipped regions with our proposed edge-based inpainting. Extracted edges do not need to represent complete and continuous topological properties of an image because our purpose is not to segment or restore an object. Discontinuous edges can likewise play the role of assistant information in the proposed scheme. But taking the topological properties into account in edge extraction will make edges more meaningful in terms of low-level vision.

Therefore, though there are many mature tools available to extract edges from images, the topology-based algorithm presented by Rothwell et al. [224] is adopted in our system to extract assistant information. The algorithm presents good results especially on extracting intersection edges. According to this method, an input image is first smoothed by a two-dimensional isotropic Gaussian filter so as to avoid noise. Second, $|\bigtriangledown f(x)|$ and θ are calculated on the filtered image for each pixel x. If $|\bigtriangledown f(x)|$ is the local maximum gradient along the direction θ and larger than a threshold, then pixel x belongs to an edge. At last, the pixels with non-maximum gradients are checked by spatially adapted thresholds to prevent missing edges caused by the unreliable estimation of θ.

As shown in Figure 10.4a by the curves, edges extracted with the above algorithm (or most of the existing methods as well) are often of more than one-pixel width. This causes ambiguous directions in guiding the restoration at the decoder side and also increases the number of bits to code the edge information. Although [224] also proposes a thinning method, it does not satisfy the special requirement in our proposed edge-based inpainting. It is because pixel values on edges are not coded but rather inferred from connected surrounding edges in our proposed scheme. Thus, a new thinning method is proposed here by taking into account the consistence of pixel values on edges as well as the smoothness of edges.

Here, we present the details of our proposed thinning method. Given the detected edge pixels, we first group them into eight connective links and each edge-link (also known as a connected component in the graph that is made up by edge pixels) is thinned independently. Complying with the terminologies defined in Section 10.2, our goal is to find a one-pixel-width line which contains N pixels, that is, $f(x_n)$ for $n = 1, 2, \cdots, N$, yielding the minimal energy

$$J = \alpha \sum_{n=1}^{N} |\triangle f(x_n)| + \beta \sum_{n=1}^{N} \sum_{k=1}^{N} d\left(f(x_n), f(x_k)\right) + \gamma \sum_{n=1}^{N} |\kappa(x_n)|^p, \qquad (10.4)$$

where α, β, and γ are positive weighting factors. The energy function Eq. (10.4) consists of three terms. The first term is the Laplacian of each edge pixel. The second

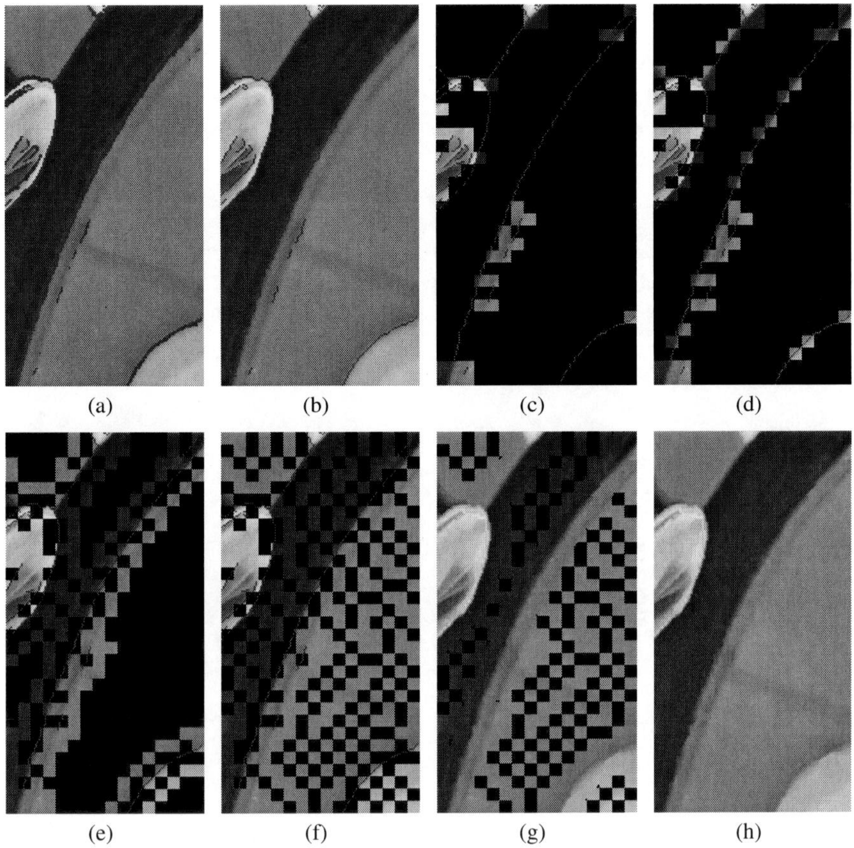

Figure 10.4 Step-wise results of our scheme on test image Lena, zoomed-in partition. (a) Original image with detected edge pixels. (b) Thinned edges. (c) Chosen necessary structural blocks. (d) Chosen additional structural blocks. (e) Chosen necessary textural blocks. (f) Chosen additional textural blocks. (g) Structure propagation result. (h) Texture synthesis result.

term is the constraint on pixel values of all edge pixels. After thinning, remaining edge pixels should have similar values. To make this constraint as simple as possible, only the difference among eight neighboring pixels are considered, and the function $d()$ is defined as

$$d\left(f(x_n), f(x_k)\right) = \begin{cases} |f(x_n) - f(x_k)|, & if x_k \in \mu_8(x_n) \\ 0, & otherwise \end{cases}. \tag{10.5}$$

$\mu_8(x_n)$ denotes the 8-neighbor of x_n. The last term of Eq. (10.4) evaluates the curvature of the edge at each pixel. Similar to [208, 209], is defined as

$$\kappa(x_n) = div\left(\frac{\nabla f(x_n)}{|\nabla f(x_n)|}\right). \tag{10.6}$$

In addition, we want to emphasize that the thinning process should not shorten the edge, thus only redundant pixels on the edge can be removed.

The optimal thinning solution for each edge-link is obtained through dynamic programming algorithm. Given a start point of each edge-link, the energies of all possible paths, linked in eight connective manner, are calculated according to Eq. (10.4). Referring to the width of the initial edge-link, several paths with smaller energies are recorded in the dynamic programming. Then, each recorded path is extended consequently by adding one neighbor pixel, which results in the minimal energy. Note that the thinning algorithm can be performed in parallel manner for all edge-links in an image, because they are independent in terms of thinning process. Figure 10.4b presents the corresponding thinning results using our proposed method.

10.3.2 Exemplar Selection

After the edges are extracted, exemplar selection is performed based on these available edges. Here, for simplicity, the exemplar selection process is performed at block level. Specifically, an input image is first partitioned into non-overlapped 8×8 blocks, and each block is classified as structural or textural according to its distance from edges. In detail, if more than one-fourth of pixels in a block are within a short distance (e.g., five-pixel) from edges, it is regarded as a structural block, otherwise a textural one. Then, different mechanisms are used to select the exemplars for textural blocks and structural blocks. Blocks that are not selected as exemplars will be skipped during encoding. Moreover, exemplar blocks are further classified into two types, the necessary ones and the additional ones, based on their impacts on inpainting as well as on visual fidelity. Generally, one image can not be properly restored without necessary exemplar blocks, whereas additional blocks help to further improve visual quality.

(1) Textural Exemplar Selection: Figure 10.5a illustrates the process of exemplar selection for textural blocks. In this figure, edge information is denoted by thickened lines, based on which the image is separated into structural regions (indicated by gray blocks) and textural regions (indicated by white and black blocks).

It is generally accepted that pure textures can be satisfactorily generated even given a small sample. However, in practice, image regions are often not pure textures, but rather contain kinds of local variations, such as lighting, shading, and gradual changing. Furthermore, exemplar-based texture synthesis is sensitive to the chosen samples. In image inpainting, a common solution to unknown textural regions is to synthesize them from samples in their neighborhood.

In our scheme, the necessary textural exemplars are selected in the border of textural regions. That is, as shown in Figure 10.5a, denoted by white blocks, if a textural block is next to a structural one, along either horizontal or vertical direction, it is considered as necessary. Such blocks are selected because they contain the information of transitions between different textural regions, which are hard to be restored

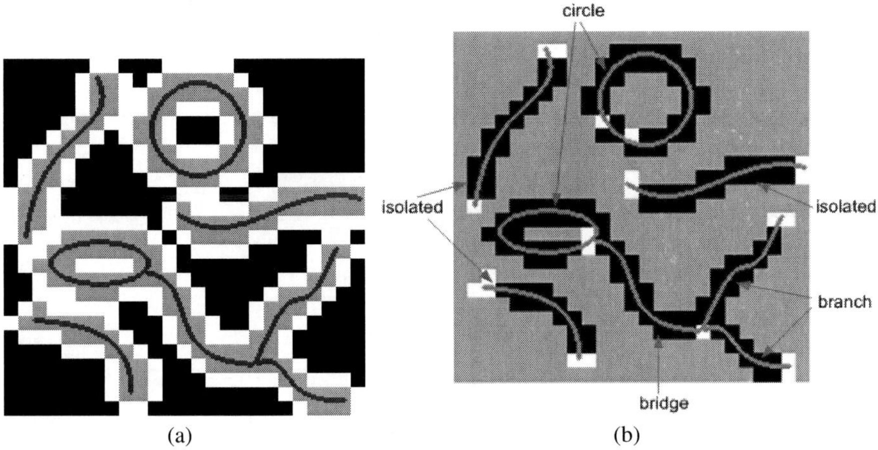

Figure 10.5 An example of necessary exemplar selection in which curves denote edges and black blocks denote skipped regions. (a) Textural exemplar selection in which white blocks are necessary textural exemplars; (b) structural exemplar selection in which white blocks are necessary structural exemplars, four types of edges are also distinguished in (b). Also see Figure 10.4 for a practical example.

by inner samples. In addition, propagation of these blocks, from outer to inner, can reconstruct the related textural regions.

To further improve visual quality of reconstructed images, additional blocks can be progressively selected to enrich exemplars. In this process, we consider additional blocks as representatives of local variations. On the one hand, if a block contains obvious variation, it should be preserved in advance. On the other hand, because the variation is a local feature, removing large-scale regions should be avoided in exemplar selection. Thus, each non-necessary textural block B_i is related to a variation parameter V_i defined as

$$V_i = w_i Var(B_i) + w_2 \sum_{B_j \in \mu_4(B_i)} |E(B_i) - E(B_j)|. \tag{10.7}$$

Here, w_1 and w_2 are positive weighting factors. $\mu_4()$ indicates 4-neighbor of a certain block. The functions $Var()$ and $E()$ are the variance and mean value of the pixel values in a block, respectively. In our system, according to an input ratio, the blocks with higher variation parameters will be selected, during which we also check the connective degree of each block so that the removed blocks do not constitute a large region.

(2) Structural Exemplar Selection: Figure 10.5b shows the exemplar selection method for structural blocks. In this figure, edges are represented by lines with indicated different types, and structural regions are indicated in white and black blocks, whereas all textural regions are in gray.

As we have discussed, besides many textural blocks, some structural blocks are also skipped at the encoder side and restored at the decoder side by the guidance of edge information. Therefore, necessary and additional structural exemplars are also selected based on available edges. To better introduce the method, edges are categorized into four types according to their topological properties, as indicated in Figure 10.5b: "isolated" edge traces from a free end (i.e., an edge pixel connected with only one other edge pixel) to another free end; "branch" edge traces from a free end to a conjunction (i.e., an edge pixel connected with more than three other edge pixels); "bridge" edge connects two conjunctions; and "circle" edge gives a loop trace.

Commonly, edge acts as the boundary of different region partitions. For the sake of visual quality, in image inpainting, two textural partitions along both sides of an edge should be restored independently. The tough problem here is how to restore the transition between two partitions. We may use a model to interpolate the transition from textures of two partitions, but usually the results look very artificial and unnatural. Therefore, the blocks containing the neighborhood of free ends should be selected as exemplar so that the transitions of textural partitions can be restored by propagating information in these blocks along the edges. Conjunction blocks of edges are also selected as exemplar for similar reason because there are transitions among more than three textural regions. For circle edges, a circle completely divides the image into two partitions — inner part and outer part — so we choose two blocks as necessary exemplars, which contain the most pixels belonging to inner region and outer region of a circle edge, respectively. In a few words, by necessary exemplars, we provide not only samples for different textures separated by an edge, but also the information of the transitions between these textures, and thus the decoder is able to restore the structural regions.

Additional structural blocks can also be selected as exemplars to further improve visual quality. Given an input ratio, the process is quite similar to that for textural blocks. Each non-necessary structural block is also related to a variation parameter, which can be calculated by Eq. (10.7). Here, the different partitions separated by the edges are independently considered in calculating the mean value as well as the variance, and resulting parameters of different partitions are summed up to get the total variation parameter of a block.

In Figure 10.4, we present the step-wise results of the exemplar selection method. Based on the edge information shown in Figure 10.4b, the selected necessary and additional structural exemplars are shown in Figure 10.4c,d gradually. Similarly, in Figure 10.4e,f we add the necessary and additional textural exemplars, as well.

10.4 Edge-Based Image Inpainting

Based on the received edges and exemplars, we propose an edge-based image inpainting method to recover the non-exemplar regions at the decoder side. Different from the encoder, the inpainting algorithm is not block-wise but rather designed to deal with arbitrary-shaped regions. Still, the non-exemplar regions are classified into structures and textures according to their distances to the edge as the encoder. Generally, structures are propagated first, followed by texture synthesis (as shown in Figure 10.3d,e). A confidence map, similar to that by Drori et al. [211] and Criminisi et al. [212], is constructed to guide the order of structure propagation as well as texture synthesis. Specifically, at the very beginning, known pixels (pixels in decoded exemplars) are marked with confidence 1 and unknown pixels (pixels in removed blocks) are marked with confidence 0. Afterward, each generated pixel is related with a confidence value between 0 and 1 during the inpainting process. In addition, known pixels as well as generated pixels are all called "available" ones in this section.

In the literature, exemplar-based inpainting methods can be roughly classified into two types, that is, pixel-wise schemes and patch-wise schemes. Pixel-wise methods are suitable for restoration of small gaps, but may introduce blurring effects or ruin texture pattern while dealing with large areas. Patch-wise methods, on the contrary, are good at keeping texture pattern, but may introduce seams between different patches, which are quite annoying. In our scheme, these two strategies are adapted for different circumstances.

10.4.1 Structure Propagation

A sketch map of structure propagation is shown in Figure 10.6. The gray block in Figure 10.6 indicates an unknown structural block; the black curve with circle points represents an edge piece and related pixels; and four dash-dot lines restrict a region, namely *influencing region*, including unknown pixels within a short distance (e.g., ten-pixel) from the edge. Notice that it is the edge piece together with the influencing region, rather than a structural block, that is treated as a basic unit in the structure propagation. Since the free ends and conjunctions of edges are all selected as exemplars, the textural regions along an edge can be readily divided and independently generated in inpainting process.

To recover a basic unit, the unknown pixels belonging to the edge piece are first generated. As shown in Figure 10.6a, the unknown pixels (denoted by black points) are generated from the known pixels (indicated by white points) using linear interpolation, that is,

$$\hat{f}(x_n) = \frac{\sum_{k=1}^{N} \lambda_{nk} \hat{f}(x_k)}{\sum_{k=1}^{N} \lambda_{nk}}, \tag{10.8}$$

$$\lambda_{nk} = \begin{cases} |n-k|^{-2}, & if\ x_k\ is\ known \\ 0, & otherwise \end{cases}.$$

where, similar to Eq. (10.4), N gives the number of pixels in this edge piece and n and k index different pixels.

After the edge restoration, the neighboring structure as well as texture within the influencing region will be filled-in with regard to the recovered edge. The inpainting method for completion of influencing region is designed concerning the

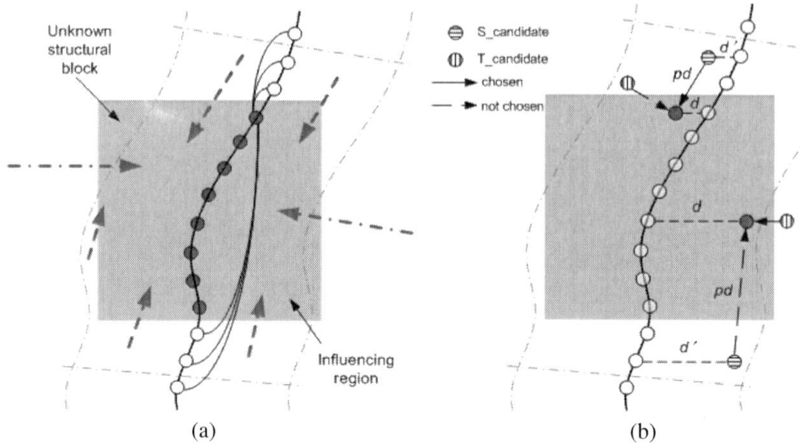

Figure 10.6 Pixel-wise structure propagation. (a) A piece of edge and its influencing region, with arrowed dash-dot lines and dashed lines showing the propagation directions. (b) Restoration of the influencing region in which each generated pixel is copied from one of two candidate pixels.

following facts. First, the pixel-wise approach is preferred since the narrow regions along the edge pieces are to be handled. Second, edges are expressed by one-pixel-width curves, which can be quite different in geometric shapes among exemplar and non-exemplar regions, so we have to wrap the edges to reconstruct the unknown structure. Finally, the widths of structures are local variant, which means that it is hard to tell the exact boundary between structure and texture in an influencing region. Therefore, in our scheme, each pixel in the influencing region will have two candidates: one is treated as a structural pixel to be propagated parallel along the edge; the other is regarded as a textural pixel to be generated from the neighboring available pixels. Then, the one that makes a smooth transition from structure to texture will be selected to fill-in the unknown pixel. Moreover, as the decision making on candidate pixels is highly relevant to its available neighbors, the order for pixel completion is another important issue that should be considered. Thus, we also construct a confidence map, as mentioned at the beginning of this section, to control the generation order. For the unknown pixel, the higher the neighboring confidence is, the earlier it will be generated.

Accordingly, the recovery of influencing region is performed as follows. Here, unknown pixels to be recovered in the influencing region are called target pixels. They are denoted by black points in Figure 10.6b. For each target pixel, two candidate pixels are searched out from the surrounding available pixels. The structural candidate (S-candidate) of the target pixel, which lies within the influencing region, is indicated by horizontal striped point in Figure 10.6b; whereas the textural candidate (T-candidate) of the target pixel is denoted by vertical striped point, which locates within a short distance from the target pixel despite whether it is within the influencing region or not.

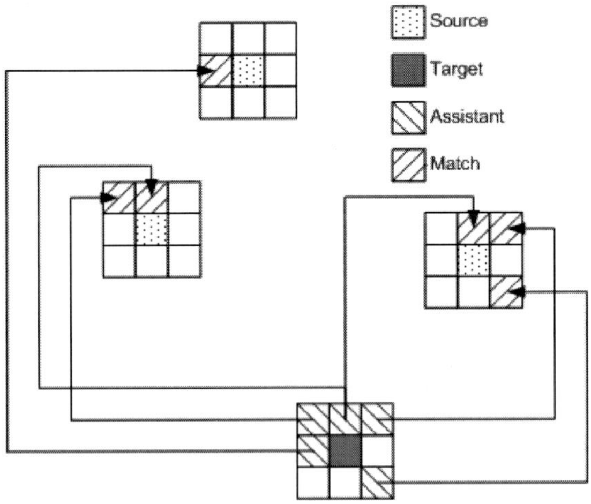

Figure 10.7 Pair matching in our structure propagation algorithm.

A pair matching method, similar to that by Ashikhmin [199], is utilized to generate both the S-candidate and the T-candidate. As illustrated in Figure 10.7, for each assistant pixel, also known as any available pixel belonging to the 8-adjacent neighborhood of the target pixel, we will search for its match pixel(s) with the most similar value to it. Then, complying with the spatial relation between the assistant pixel and the target one, a pixel adjacent to a match pixel in the same relative spatial position is selected as a source pixel. As indicated in Figure 10.7, an assistant pixel may correspond to several match pixels and gives several source pixels; meanwhile, several assistant pixels in 8-adjacent neighborhood may generate the same source pixel, as well. After obtaining several source pixels, we propose to use a weighted-SSD (sum of squared difference) criterion to choose the S-candidate, as given in

$$D_s = \sum_i \left(|d(x_S^i) - d(x_t^i)| + 1\right) \times |\hat{f}(x_S^i) - \hat{f}(x_t^i)|^2, \tag{10.9}$$

where x_S^i and x_t^i are corresponding, the ith pixel in the neighborhood of the S-candidate and the target pixel, respectively, and $d()$ indicates the distance from each pixel to the edge, $\hat{f}()$, as used before, is the reconstructed image. By minimizing Eq. (10.9), we can find the S-candidate from the obtained source pixels, which is situated in a similar relative position to that of the target pixel with respect to the edge, thus ensuring the parallel diffusion of structural information.

Differently, since no direction information is involved in the textural region, only the ordinary SSD between the neighborhood of source pixels and target pixel is considered as the criterion to choose the T-candidate,

$$D_T = \sum_i |\hat{f}(x_T^i) - \hat{f}(x_t^i)|^2. \tag{10.10}$$

Similar to that in Eq. (10.9), x_t^i here represents the ith pixel in the neighborhood of the T-candidate. Thus, the source pixel that has the most similar neighboring values to the target one will be selected as the T-candidate.

In fact, the two diffusions, or S-candidate selection and T-candidate selection, are simultaneous and competitive. These two candidates have to compete with each other and only one of them will be chosen to fill-in the target pixel. Normally, if target pixel nears edge, the choice will bias to the S-candidate. In addition, it can be observed that long-distant parallel diffusion of structural information often leads to blurring artifacts. Thus, the determination is made by comparing DT and D'_S, which are defined in Eq. (10.10) and Eq. (10.11), respectively

$$D'_S = \left(\frac{d}{d_0} + \frac{pd}{pd_0}\right) \times \sum_i |\hat{f}(x_S^i) - \hat{f}(x_t^i)|^2. \tag{10.11}$$

Here, d_0 and pd_0 are constants and stands for the distance from the target pixel to the edge and pd indicates the distance from the target pixel to the S-candidate, as shown in Figure 10.6b. If D_T is less than D'_S, then the T-candidate is chosen to fill-in the target pixel, otherwise the S-candidate is selected. In this way, all unknown pixels within the influencing region of an edge are generated.

10.4.2 Texture Synthesis

The edges as well as their influencing regions are readily restored by structure propagation. Then, in this section, the remainder of unknown regions are treated as textural regions, so texture synthesis is employed to fill in these holes.

For textural regions, we prefer patch-wise algorithms because they are good at preserving large-scale texture pattern. We choose square patches as the fundamental elements while a confidence map is introduced to guide the order of synthesis. Unknown textural regions are progressively restored during texture synthesis by first reconstructing the prior patches and then the others that remain. The priority of a patch is determined by calculation of confidence and the distance from the edge. As shown in Figure 10.8, for each patch centered at a marginal pixel of unknown regions (denoted by target patch), we calculate the average confidence value of all pixels in this patch, as well as the average distance of all pixels from the edge. Then the patch with the highest confidence rating and the greatest distance from the edge will be synthesized first.

Afterward, a source patch, which is most similar to the target patch, will be searched out from the neighborhood of the target patch. Here, the similarity of two patches is measured by the SSD of pixel values between overlapped available pixels of two patches. A patch that results in the least SSD will be chosen as the source patch. Note that the filling-in process is not as simple as copy-paste work, we have to deal with overlapped regions as well as seams. In our algorithm, the graph-cut method proposed by Kwatra et al. [203] is used to merge the source patch into the existing image, and Poisson editing [216] is utilized to erase the seams. After one

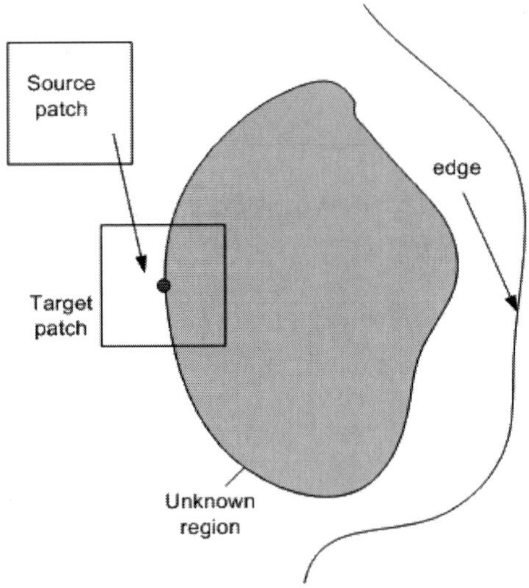

Figure 10.8 Pair matching in our structure propagation algorithm.

patch is restored, the confidence map is updated. All newly recovered pixels are treated as available pixels in the following synthesis steps. Then, the next target patch is searched and processed until no unknown pixel exists.

10.5 Experimental Results

10.5.1 Implementation

Our presented approach can be integrated with the state-of-the-art coding schemes to enhance compression performance. In our experiments, two compression standards, JPEG and MPEG-4 AVC/H.264 (referred to simply as H.264 hereafter), are adopted. Thus, two fully automatic image coding systems, based on JPEG and H.264 respectively, have been constructed to evaluate the effectiveness of our proposed compression approach. In this section, we would like to clarify several implementation details of the systems.

First, in both systems, the one-pixel-width edge information is coded using JBIG method. Note that the edge information coded into the final bitstream is only a subset of the entire edge map. In other words, the edges that are fully covered by the exemplar regions will not be coded (it can be observed by comparing the edges in Figure 10.4b and 10.4f). Second, in the JPEG-based system, the exemplar locations are denoted at block level by a binary map, in which 1 stands for a removed block and 0 for an exemplar block, and the map is coded by an arithmetic coder. The original image is then coded by JPEG coding method, during which the removed blocks will be skipped in encoding but to be filled with the DC values copied from previous blocks, so that the DC prediction in JPEG can still be performed in exemplar block compression. Third, in the H.264-based system, since the basic coding unit is macro-block 16×16, we consider two instances: if a macro-block is totally removed, then a new macroblock type I_SKIP is coded; otherwise, the macro-block has a new element called block removal pattern (BRP) for indicating which of the four 8×8 blocks is removed, and the BRP is later coded by the arithmetic coder, too. Similar to the JPEG-based method, the exemplar blocks are coded using H.264 scheme and DC values from previous blocks are filled to the removed blocks to enable the intra prediction of H.264 scheme.

In addition, there are some predefined parameters in both the encoder and decoder. To test the robustness of our system, we fix these parameters as follows for all test images. At the encoder side, the weighting factors are defined as $\alpha = 1.0$, $\beta = \gamma = 0.2$, and $p = 1$ (suggested by Ballester et al. [208]) for Eq. (10.4) in edge thinning, while $w_1 = w_2 = 1.0$ for Eq. (10.7) in exemplar selection. At the decoder side, structure propagation works on pixels that have less than 10-pixel distances from edges. The search range for T-candidate is 9×9, and the S-candidate is found in the entire influencing region. The search range and patch size for texture synthesis are 11×11 and 7×7, respectively. The parameters in Eq. (10.11) are set to 5. We would like to remark that the weighting factors for edge thinning have been carefully tuned using our test images, while the other parameters are just empirically selected with consulting the existing systems (e.g., [217]). However, it should be noted that the parameters can greatly influence the computational complexity of both the encoder and decoder, which will be further analyzed in the following section. At last, the only two flexible parameters in our experiments are the additional block ratios for the structural exemplar and textural exemplar; actually, they act as quality control parameters in our system.

10.5.2 Test Results

We test our compression systems on a number of standard color images from the USC-SIPI image database[1] and the Kodak image library.[2] Some results are presented here to evaluate the compression ratio as well as reconstructed quality of our scheme. In all tests, the quality parameter (QP) of JPEG coding method is set to 75, while the QP of H.264 intra coding is set to 24. Bit-rate savings are listed in Table 10.1.

Figure 10.3 shows test image Lena and corresponding results of our JPEG-based system. As mentioned before, the coded exemplars and the edge information are denoted in Figure 10.3c. In this test, 10% additional structural blocks as well as 50% additional textural blocks are preserved. Based on the preserved blocks and

[1] http://sipi.usc.edu/services/database/

[2] http://r0k.us/graphics/kodak/

Table 10.1 Bit-rate savings of our scheme compared to JPEG (QP is set to 75) and H.264 (QP is set to 24).

Images	Additional Block Ratio		JPEG-Based Comparison			H.264-Based Comparison		
	Structural	Textural	JPEG	Proposed	Bit Saving	H.264	Proposed	Bit Saving
Jet	0.1	0.3	1.156	0.919	20.6%	0.985	0.880	10.7%
Lena	0.1	0.5	1.112	0.888	20.1%	0.993	0.869	12.5%
Milk	0.1	0.3	0.913	0.512	44.0%	0.616	0.415	32.7%
Peppers	0.1	0.3	1.217	0.965	20.8%	1.311	1.080	17.6%
Kodim02	0.1	0.2	1.058	0.709	33.0%	0.948	0.701	26.0%
Kodim03	0.2	0.3	0.895	0.608	32.1%	0.710	0.562	20.9%
Kodim05	0.1	0.3	2.027	1.719	15.2%	2.086	1.845	11.6%
Kodim07	0.1	0.5	1.079	0.802	25.7%	0.876	0.751	14.3%
Kodim11	0.2	0.3	1.368	1.047	23.5%	1.354	1.098	18.9%
Kodim19	0.2	0.5	1.276	0.915	28.4%	1.246	0.956	23.3%
Kodim20	0.1	0.4	0.897	0.638	28.9%	0.823	0.636	22.6%
Kodim23	0.2	0.3	0.821	0.567	30.9%	0.622	0.504	18.9%

assistant edge information, our presented structure propagation gives inpainting results in Figure 10.3d, and the final reconstructed image after texture synthesis is shown in Figure 10.3e. Compared with the reconstructed image shown in Figure 10.3f by baseline JPEG, our scheme saves 20% bits (as given in Table 10.1) but presents similar visual quality. More comparison results in visual quality concerning standard images can be found in Figure 10.9. It shows that up to 44% bits-saving (shown in Table 10.1) is achieved by our scheme at similar visual quality levels, compared to baseline JPEG.

In Figure 10.10, our proposed structure propagation method is evaluated. In this test, we remove only structural blocks and use different approaches to recover them. The details of partial images together with the assistant edge information are given in the first column. Then, results generated by the PDE-based diffusion [225], which is the traditional solution to structural regions, are shown in the second column. This method works well only for smooth regions and certain simple structures. In addition, image completion method presented by Sun et al. [217] is also tested in the third column, only the user interaction is omitted. Since no edge information is utilized, these two methods result in annoying distortion in most of the structural regions. In the last column, we give our results that are accomplished by the help of edge information. It is clearly demonstrated that the assistant edges help greatly in structure restoration, and thus empower the encoder to remove more blocks.

Furthermore, our JPEG-based system is also tested using the images in the Kodak image library, which contains photographs of natural scenes at high resolution. The comparison results on visual quality are shown in Figure 10.11, in which the top row shows our results whereas the bottom row presents baseline JPEG results. It can be observed that the visual quality of our resulting image is very similar to that of the JPEG. The bits-saving of our JPEG-based system is indicated in Table 10.1. Our method on average saves 27% bits for the five images shown in Figure 10.11 at the similar visual quality levels.

Figure 10.9 Comparisons with baseline JPEG on test images Jet, Milk, and Peppers. From left to right: removed blocks (black blocks) and assistant edge information; reconstructed image by our scheme; reconstructed image by baseline JPEG.

To investigate the detailed bit-rate cost in our scheme, we list the percentage of different coded elements in Table 10.2, from which we note that even different images will lead to different allocations of coded elements, the exemplar location information as well as the edge information still costs only a little in overhead. Commonly, the bits used to code the exemplar blocks occupy more than 90% of total bits cost. However, it is still possible to further reduce the bits cost on edge information, taking into account the exemplar locations, or skipping those edges that can be inferred from the exemplar blocks.

In Figures 10.12 and 10.13 we show the reconstructed images by our H.264-based system in comparison with standard H.264 intra coding. Both results show similar visual quality to each other, as in the JPEG comparisons. The bit-rate saving is also noticeable, shown in Table 10.1, but not as much as the comparison with JPEG. This is caused by two reasons. On the one hand, the H.264 intra coding is more

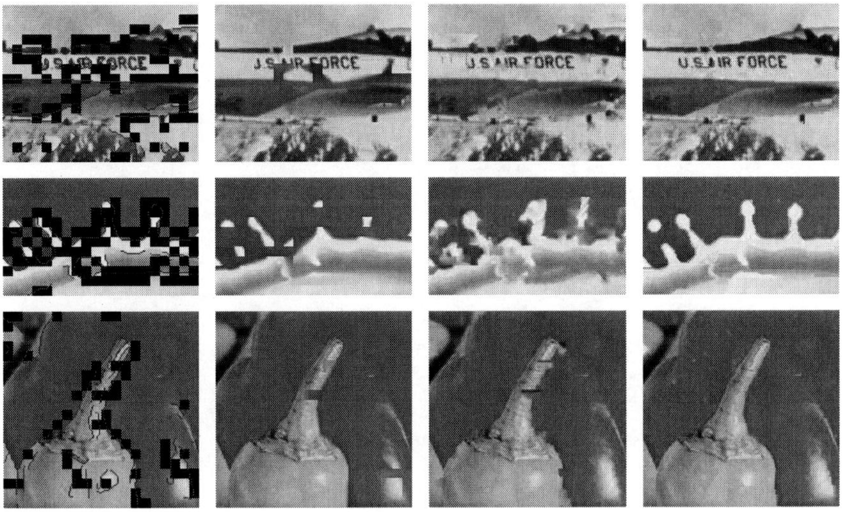

Figure 10.10 Comparisons of different structure propagation approaches, zoomed-in partitions. From left to right: removed blocks (black blocks) with assistant edge information; inpainting result by PDE-diffusion [225]; inpainting result by patch-wise synthesis [217]; inpainting result by our scheme. Note that only our scheme takes advantage of the edges.

Table 10.2 Percentage of different coded elements in our JPEG-based system (QP is set to 75).

Images	Exemplar locations (arithmetic coder)	Assistant edges (JBIG)	Exemplar blocks (JPEG)
Jet	1.4%	6.4%	92.3%
Lena	1.4%	6.6%	92.0%
Milk	2.7%	5.2%	92.1%
Peppers	1.4%	6.7%	91.9%

efficient than JPEG in coding performance, so the nonexemplar blocks, especially the textural ones, will cost fewer bits in standard H.264 than in baseline JPEG, but in our scheme the edge information still cost the same bits in either realization. On the other hand, due to the complicated spatial predictions performed in the H.264 intra coding, the filling of only DC values for removed blocks is proved not good enough, since it breaks the original spatial relations between neighboring blocks, but for JPEG this filling of DC values seems enough since JPEG only conduct DC prediction. Nevertheless, our scheme can still acquire up to 33% bit-rate savings compared to the state-of-the-art H.264 intra coding.

10.5.3 Discussions

It can be observed that the ratio of additional textural exemplar has a big effect on visual quality of the reconstructed images. As given in Table 10.1, for homogeneous

(a)

(b)

(c)

(d)

(e)

Figure 10.11 Comparisons with baseline JPEG on the Kodak Image Library. (a) kodim02; (b) kodim03; (c) kodim05; (d) kodim07; (e) kodim19. The top row shows the reconstructed images by our scheme and the bottom row shows the reconstructed images by baseline JPEG.

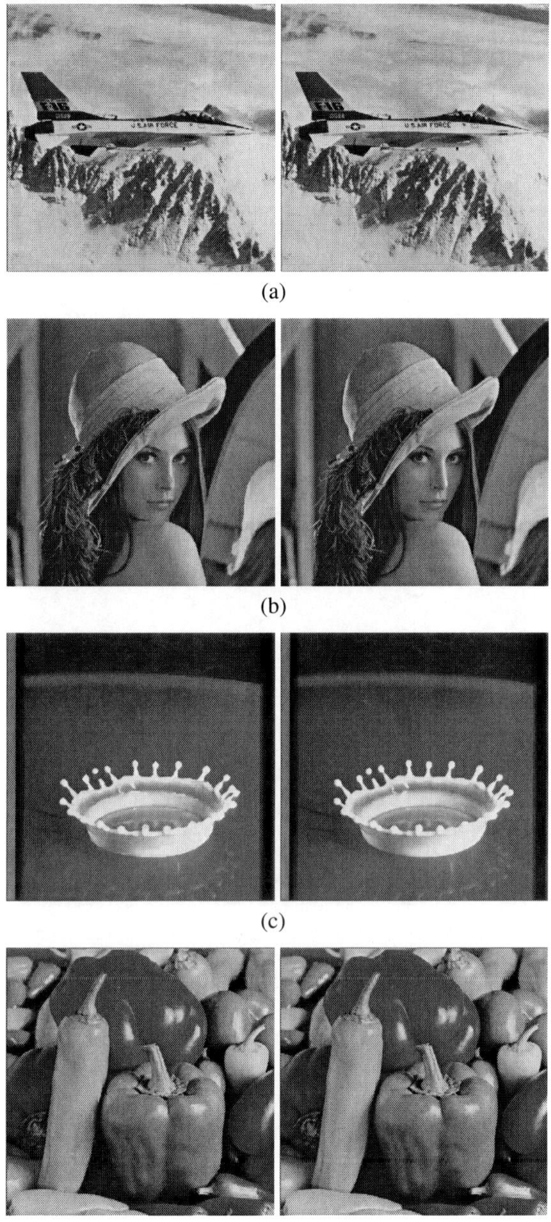

Figure 10.12 Comparisons with standard H.264 intra-picture coding. (a) Jet. (b) Lena. (c) Milk. (d) Peppers. The top row shows the reconstructed images by our scheme and the bottom row shows the reconstructed images by standard H.264 intra coding.

Figure 10.13 Comparisons with standard H.264 intra-picture coding. (a) kodim11. (b) kodim20. (c) kodim23. The top row shows the reconstructed images by our scheme and the bottom row shows the reconstructed images by standard H.264 intra coding.

textural regions, such as the wooden door in kodim02 (Figure 10.11a), low exemplar ratio is used to pursue high compression ratio; whereas for complex and irregular textural regions, for example, flowers and leaves in kodim07 (Figure 10.11d), high ratio is preferred to ensure good visual quality. However, thanks to the given edge information, the reconstructed quality of structural regions is less sensitive to the additional structural exemplar ratio.

In addition, the improvement of our scheme in terms of compression ratio is various for different images. Commonly, the more complicated the image is, the less gain we can provide. It is because when coding images with lots of details (such as kodim05, Figure 10.11c), the extracted edge map usually contains miscellaneous edges which makes many blocks as necessary exemplars. Thus, only a limited number of regions can be removed at encoder. However, in this case, 15% bits-saving is still provided by our JPEG-based system without noticeable visual loss, as shown in Table 10.1.

The computational complexity of our scheme is relatively higher than that of the traditional coding schemes, since at the encoder side we perform extra edge extraction and exemplar selection, and at the decoder side we add the inpainting process. In particular, the computation of the decoder is greatly related with the parameters used in the inpainting, such as search range and patch size, which determine the necessary number of SSD calculations. There are several previous works proposed to reduce the computations of SSD, so as to accelerate the image synthesis [198,202,211], and those methods can be adopted in our system as well.

The visual quality assessment is highly related to our work, that is, if we have a good metric used to measure visual quality, we are able to not only better evaluate our scheme, but also further improve the performance by rate-"distortion" optimization, where "distortion" measures the perceptual quality in addition to the statistical fidelity. Unfortunately, we have not yet found such a good metric for our purposes. Thus, for visual quality comparisons in our experiments, we always set the same quality parameters for both the standard compression scheme and our inpainting-based scheme. Thus, the exemplar regions will have the same quality (both subjectively and objectively). Additionally, the restored regions still have acceptable visual quality. Accordingly, the "comparable quality" or "similar visual quality levels" indicates visually similar qualities, which are examined by human observations.

10.6 Summary

In this chapter, we present an image compression framework that adopts inpainting techniques to remove visual redundancy inherent in natural images. In this framework, some kinds of distinctive features are extracted from images at the encoder side. Based on the obtained features, some regions of an image are skipped during encoding, only to be recovered by the assisted inpainting method at the decoder side. Due to the delivered assistant information, our presented framework is able to remove enough regions so that the compression ratio can be greatly increased. Our presented inpainting method is capable in effectively restoring the removed regions for good visual quality, as well.

Moreover, we present an automatic image compression system, in which edge information is selected as the assistant information because of its importance in preserving good visual quality. The main techniques we proposed for this compression system, that is, edge thinning, exemplar selection, and edge-based inpainting, are also addressed here. Experimental results using many standard color images validate the ability of our proposed scheme in achieving higher compression ratio while preserving good visual quality. Compared to JPEG and H.264, at the similar visual quality levels, up to 44% and 33% bits-saving can be acquired by our approach, respectively.

Further improvements of the current scheme are still promising. First, the assistant information as well as the selected exemplars can be described and compressed into bitstream in a more compact fashion. Second, extraction of the distinctive features can be more flexible and adaptable. In addition to edge information, there are other candidates, such as sketch [196] and epitome [220, 221], which could be derived from source images to assist the vision technologies and the compression methods as well. Furthermore, image inpainting is still a challenging problem when some kinds of assistant information are provided, into which we need to put more effort in the future.

Chapter 11
Cloud-Based Image Compression

11.1 Introduction

The cloud is characterized by a large amount of computing resources, storage, and data [226]. Imagining a cloud that collects a huge number of images, for example, Google street view images [227], when you randomly take a picture with your phone on the street, you can often find some highly correlated images in the cloud that were taken at the same location at different viewpoints and angles, focal lengths, and illuminations. If you try to share the photo with friends through the cloud, it is problematic to use conventional image coding (e.g., JPEG) that usually provides only 8:1 compression ratio [228]. It will consume a lot of precious power and network bandwidth to transmit such a high-resolution and high-quality JPEG image. It would be more convenient to take advantage of the cloud for compression and transmission if there is a high probability of finding very similar images in the cloud.

However, state-of-the-art image coding, consisting of directional intra prediction and transform [13, 229], makes it difficult to take advantage of highly correlated images in the cloud. Intra prediction uses decoded neighboring pixels from the same image to generate predictions and then the pixels are coded by subtracting the predictions. It requires that the predictions used at the encoder and decoder must be identical. The idea of intra prediction cannot be extended to external images. First, when mobile devices compress images, it is difficult to know which highly correlated images can be found in the cloud. Second, the number of images in the cloud is huge, so it is impossible to store them in mobile devices even partially. Third, the images in the cloud are changing dynamically. The result is that it will be quite costly to maintain image consistency between the cloud and all mobile devices.

Image search has been demonstrated to be a successful application on the Internet [230]. By submitting the description of one image, including semantic content [231], outline [232, 233], and local feature descriptors [234, 235], one can easily retrieve many similar images. Near and partial duplicate image detection is a hot research topic in this field [234, 236, 237]. However, the purpose of image search is not to generate an image from search results. In fact, reconstructing a given image from similar images is tougher than the image search itself.

Recent efforts have shed light on using a large-scale image database to recover, compose, and even reconstruct an image [238–243]. In particular, Weinzaepfel et al. are the first to reconstruct an image by using local feature descriptors SIFT (Scale Invariant Feature Transform) [244]. The follow-up work presented by Daneshi and Guo [245] tries to reconstruct an image using SIFT and SURF (Speed Up Robust Features) descriptors. However, it is a very challenging problem to reconstruct a visually pleasing image using local feature descriptors only.

To solve the above problem, we propose describing input images by SIFT descriptors and their down-sampled images. SIFT descriptors are extracted from the original images and are used to retrieve near and partial duplicate images in the cloud and identify corresponding patches. The down-sampled images play an important role in making the reconstructed images visually pleasing. They are used to verify every retrieved image patch and guide how to stitch image patches like the given images.

The down-sampled image is compressed by conventional image coding or intra-frame coding in video. Using the correlation between the down-sampled image and SIFT descriptors to compress SIFT feature vectors is another important technical contribution. We first compress locations, scales, and orientations of SIFT descriptors and then use them to extract prediction vectors from the decoded down-sampled image so that high-dimension SIFT vectors can be efficiently compressed by prediction and transform coding.

The reconstruction of images is similar to that by Weinzaepfel et al. [244]. In addition to using the down-sampled image, there are two other technical differences. First, we adopt the approach proposed by Wu et al. [236] and Zhou et al. [237] for partial duplicate image retrieval. It uses bundling features instead of a single feature that significantly increases the retrieval accuracy. Second, we use all SIFT descriptors in a patch to estimate patch transformation with the RANdom SAmple Consensus (RANSAC) algorithm [246] and a perspective projection model is used.

11.2 Related Work

11.2.1 Visual Content Generation

Image inpainting is the task of filling in or replacing a region of an image [206]. Many approaches have been proposed to learn from the known regions of an image and then recover an unknown region from what has been learned. Hays and Efros are the first to propose image completion from a large-scale image database [238]. The proposed approach uses GIST descriptors to retrieve images of similar scenes that are applied to recover unknown regions of an image. Whyte et al. [239] later propose retrieving images of the same scene from the Internet via a viewpoint invariant search and replacing a user specified region.

Image composition is a long-standing research topic. Here, we focus on the papers that deal with composing an image from a sketch based on a large-scale image database. Eitz et al. propose the first sketch-to-photo scheme [240, 241]. It takes a sketch drawn by a user as input and retrieves correlated images from a sketch-based search. Finally, it composes an image by graph cut and possion blending. Chen et al. improve the scheme by taking both sketch and text labels as input [242].

Image composition can be viewed as the recovery of all objects in an image. The objects are only represented by input sketch. It is a much tougher problem than image inpainting and completion. Human interactions are required in current sketch-to-photo schemes. Although it requires only a small number of bits to represent a sketch and text labels, it is unacceptable to compress an image using interactions. From the compression viewpoint, composed images must look like input images in detail. But for an input sketch, image composition may generate quite different results in color and texture. Taking CG (computer graphic) images as input is an improvement over sketch [243].

Recently, Weinzaepfel et al. propose reconstructing an image from SIFT [244]. Although the purpose of this work is to study privacy in image search, it is actually the first to reconstruct an image from local feature descriptors. Daneshi and Guo further use SURF descriptors and study the role of scale data in reconstruction [245]. Although content can be observed from reconstructed images, which is consistent with the goal of image compression, reconstructed visual quality is terrible because SIFT descriptors only provide local information.

To make image description and reconstruction meet the goals of image compression, it is important to select descriptions with a better trade-off between reconstruction distortion and data size. Sketch is a highly abstracted description that does not contain any details. Local feature descriptors only provide local details. Neither of them can guarantee that reconstructed images look like input images. To solve this problem, we propose describing input images by local feature descriptors and their down-sampled images.

11.2.2 Local Feature Compression

SIFT descriptors, proposed by Lowe [247], present distinctive invariant features of images that consist of location, scale, orientation, and feature vectors. The scale and location of SIFT descriptors are determined by maxima and minima of difference-of-Gaussian images. One orientation is assigned to each SIFT descriptor according to the dominant direction of the local gradient histogram. The feature vector is a 128-dimension vector that characterizes a local region by gradient histogram in different directions. Since SIFT descriptors have a good interpretation of the response properties of complex neurons in the visual cortex [248] and an excellent practical performance, they have been extensively applied to object recognition, image retrieval, 3D reconstruction, annotation, watermarking, and so on.

Compression of SIFT descriptors has recently become a requirement of mobile-based applications. One image usually has thousands of SIFT descriptors. Without any data reduction and compression, the total size of SIFT descriptors may be even larger than the image size. Ke and Sukthankar propose applying principal components analysis (PCA) to greatly reduce the dimension of the feature vector [249]. Hua et al. propose linear discriminant analysis to reduce the dimension of the feature vector [250]. Chandrasekhar et al. propose the use of transform coding of the feature vectors [251]. Yeo et al. propose using coarsely quantized random projections of descriptors to build binary hashes [252]. Jegou et al. decompose feature vectors into a Cartesian product of low-dimension vectors that are quantized separately [253].

Several approaches propose directly reducing generated dimension of feature vectors. SURF reduces the dimension to 64 [245] with a performance similar to the SIFT. Compressed histogram of gradient (CHoG) descriptors, proposed by Chandrasekhar et al. [254, 255], are designed for compression. It not only changes the generation of the gradient histogram but also compresses it by tree coding and entropy coding. It achieves a high compression ratio. However, from both practical adoptions and recent evaluations in MPEG [256], SIFT descriptors are still a good choice in many applications.

Compared with conventional image coding, the approaches for the compression of SIFT descriptors are far away from being mature. Thus, several papers propose compressing images first and then extracting SIFT descriptors from decompressed images. Makar et al. propose compressing image patches at different scales and then generating SIFT descriptors from decompressed patches [257]. It performs better than the direct compression of SIFT descriptors. However, if there are many patches to code in an image, they will overlap each other and thus lead to low performance. Chao and Steinbach analyze locations of SIFT descriptors in an image and assign more bits to the regions with SIFT descriptors in JPEG compression [258].

All of the above approaches target the compression of SIFT feature vectors only. Since we target reconstructing an image, what we need is not only SIFT feature vectors but also SIFT locations, scales, and orientations because the latter tells us where retrieved image patches are stitched. For the same reason, we also need the down-sampled image. Since SIFT feature vectors have a strong correlation with the image, the problem examined in this chapter is how to compress both the down-sampled image and all information of the SIFT descriptors efficiently. The basic idea presented in this chapter mimics inter-frame coding in video. The prediction of feature vectors is extracted from the up-sampled decoded image. Residual feature vectors are compressed after prediction by transform, quantization, and entropy coding.

11.2.3 Image Reconstruction

Many approaches are reported to use SIFT descriptors for image retrieval [259–263], where SIFT feature vectors are quantized into visual words [260]. Quantization of

SIFT feature vectors makes image retrieval applicable to a large-scale database. But it also reduces the discriminative power of SIFT feature vectors.

To solve this problem, Chum et al. propose expanding highly ranked images from original query as new queries [260]. Philbin et al. propose quantizing a SIFT descriptor to multiple visual words [262]. Jegou et al. introduce binary signatures to refine visual words [263]. Zhou et al. propose quantizing feature vectors to bit-vectors [264]. For near and partial duplicate image retrieval, the geometric relationship of visual words plays an important role. To utilize this information, Wu et al. propose bundling a maximally stable region and visual words together [236]. Zhou et al. propose using spatial coding to represent spatial relationships among SIFT descriptors in an image [237].

Image alignment is a historic research topic and most key papers have been surveyed by Talluri [265]. When the SIFT descriptors are available for two images that are to be aligned, the most popular approach for estimating the transformation between them is RANSAC [246]. Torr et al. improve the algorithm by using the maximum likelihood estimation instead of the number of inliers [266]. Torr and Zisserman further introduce a pyramid structure with ascending resolutions to improve the performance of RANSAC [267]. Chum and Matas greatly speed up the approach by introducing confident match [268].

11.3 The Proposed SIFT-Based Image Coding

The block diagram of the proposed cloud-based image encoder is shown in Figure 11.1. For the input image, a down-sampled image is first generated and compressed. SIFT descriptors are also extracted from the original image. The location, scale, and orientation of every extracted SIFT descriptor guide to extract a feature vector as

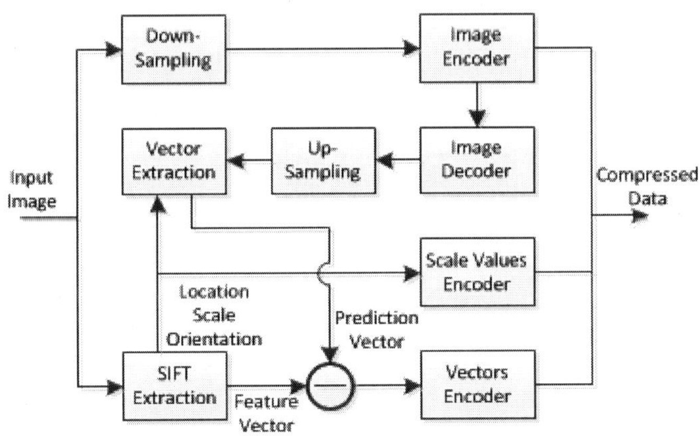

Figure 11.1 The block diagram of the proposed cloud-based image encoder.

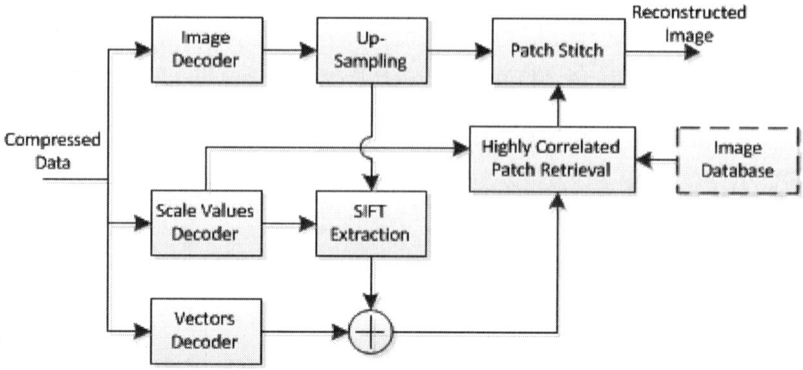

Figure 11.2 The block diagram of the proposed cloud-based image decoder.

prediction from the decompressed image after up-sampling. Finally, the prediction is subtracted from the feature vector. All components of SIFT descriptors are compressed and transmitted to the cloud with the compressed down-sampled image.

The block diagram of the proposed cloud-based image decoder is depicted in Figure 11.2. In the cloud, a server first decompresses the down-sampled image and SIFT data. By using the decompressed location, scale, and orientation of every SIFT descriptor again, one prediction vector, exactly the same as that in Figure 11.1, is extracted from the decompressed image after up-sampling. Then, the SIFT feature vector is reconstructed by adding the prediction. To reconstruct the input image, decompressed SIFT descriptors are used to retrieve highly correlated image patches. For every retrieved image patch, we estimate the transformation and then stitch it to the up-sampled decompressed image. Finally, a high-resolution and high-quality image is obtained.

The block diagrams of the encoder and decoder shown in Figures 11.1 and 11.2 look very different from those of conventional image coding. In the following sections, we will discuss the details of the proposed cloud-based image coding.

11.4 Extraction of Image Description

In this chapter, a down-sampled image is used to describe the target for reconstruction in the cloud because it carries enough information, including outline, shape, color, and objects. Furthermore, after an image is down-sampled, it can be efficiently compressed by conventional image coding. The down-sample process is described as follows

$$I_g(x,y) = \sum_{n=-T/2}^{T/2} \sum_{m=-T/2}^{T/2} I(x,y)G(x-m,y-n,\sigma). \qquad (11.1)$$

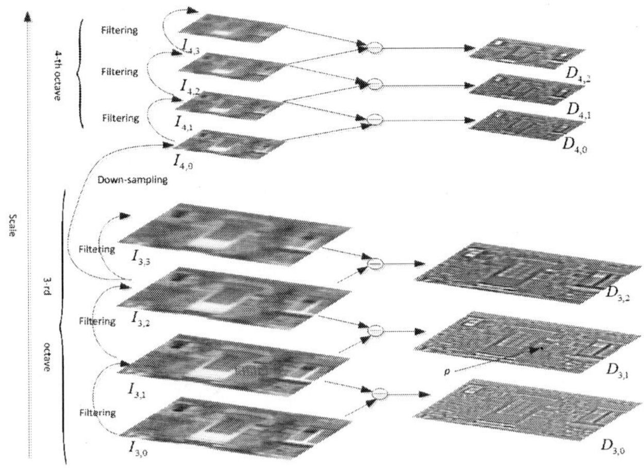

Figure 11.3 Part of the Gaussian scale space with $3rd$ and $4th$ octaves. p indicates an extreme point detected in $D_{3,1}$. The rectangle in $I_{3,1}$ indicates the corresponding SIFT descriptor.

$G()$ is the discrete low-pass filter with the support $T+1$. The kernel of the filter is a Gaussian function with variance σ. After the filtering, the down-sampled image is generated as

$$I_d(x,y) = I_g(rx,ry), r = 2^0, 2^1, \ldots, 2^n, \tag{11.2}$$

where n is an integer. Parameter r is the down-sampling ratio at one dimension. r is often set as 2 or 4 in most current applications. In this chapter, r is set as 8 or 16. Thus the down-sampling factor of images will be 64:1 or 256:1. It is much larger than that to be considered in super-resolution so that our targeted compression ratio can be achieved.

By repeatedly applying the filtering Eq. (11.1) and down-sampling Eq. (11.2) with $r = 2$ to the input image, a Gaussian scale space is generated, as shown on the left side of Figure 11.3. The image used here comes from the INRIA Holiday data set [269]. It consists of a set of images $\{(I_{0,0}, \ldots, I_{0,K}), \ldots, (I_{N,0}, \ldots, I_{N,K})\}$. n $(0 \leq n \leq N)$ is an octave index in the space. It indicates that the input image is down-sampled n times. k $(0 \leq k \leq K)$ is a filtering index in an octave. The differences of Gaussian images are generated by $D_{n,k} = I_{n,k+1} - I_{n,k}$. They are depicted on the right side of Figure 11.3.

In fact, all $D_{n,k}$ constitute a Laplacian scale-space $L(x)$ with $x = (x,y,s)^T$. s is a scale index and is defined as $s = n + (k-1)/(K-2)$ with $K \geq 3$ and $1 \leq k \leq K-2$. Feature points are detected by maxima and minima in $D_{n,k}$. If one sample is the largest or smallest of its 8 neighbors and 9 neighbors in $D_{n,k+1}$ and $D_{n,k-1}$ respectively, it will be a feature point in the image. As shown in Figure 11.3, p is such a point at (x_{f0}, y_{f0}) and s_{f0} is the scale index of $D_{n,k}$.

Let us define a vector $x_{f0} = (x_{f0}, y_{f0}, s_{f0})^T$. The subpixel and finer scale index to x_{f0} are derived by fitting a 3D quadratic to $L(x)$. The Taylor expansion of $L(x)$ at x_{f0} is described as follows

$$L(x) = L(x_{f0}) + \frac{\partial L^T(x)}{\partial x}|_{x_{f0}} \Delta x + \frac{1}{2} \Delta x^T \frac{\partial^2 L(x)}{\partial x^2}|_{x_{f0}} \Delta x, \tag{11.3}$$

where $\Delta x = x - x_{f0}$. The accurate extreme point can be calculated by

$$x_f = x_{f0} + (\frac{\partial^2 L(x)}{\partial x^2}|_{x_{f0}})^{-1} \frac{\partial L^T(x)}{\partial x}|_{x_{f0}}. \tag{11.4}$$

x_f is the location and scale index of a SIFT descriptor. Its scale value is $\sigma_f = 1.6 \times 2^{(s_f - n)}$. In the Gaussian scale space, only the limited number of $I_{n,k}$ is available. Assume the scale index of $I_{n,k'}$ is the closest one to s_f.

The orientation θ of the SIFT descriptor is calculated in a region of $I_{n,k'}$ around (x_f, y_f). Its size is set as $W_o \times W_o$, where $W_o = 9 \times \sigma_f$. Local image gradients d_x and d_y are calculated using a centered derivative mask [−1,0,1]. The orientation $\arctan(d_y/d_x)$ is evenly partitioned into 36 bins covering 360 degrees. A histogram with 36 bins can be calculated by

$$\theta(i) = \sum\nolimits_{(d_x,d_y)\in\Omega_\theta(i)} \omega_\theta(x,y) \sqrt[2]{d_x^2 + d_y^2}, 1 \le i \le 36, \tag{11.5}$$

with $\Omega_\theta(i) = \{(d_x, d_y) | \frac{\pi(i-1)}{18} \le \arctan(\frac{d_y}{d_x}) < \frac{\pi i}{18}\}$. $\omega_\theta(x,y)$ is a weighting factor and is defined as a Gaussian function $\sigma_\theta = 1.5\sigma_f$ centered at (x_f, y_f). The histogram is normalized and approximated to a polynomial function. The orientation at the highest peak of the function is θ. In addition to the highest peak, if other peaks exist with a value above 80% of the highest peak, multiple SIFT descriptors will be generated for these peak orientations with the same location and scale index.

The feature vector is also extracted in a region of $I_{n,k'}$ around (x_f, y_f). Its size is set as $W_v \times W_v$, where $W_v = \sqrt{2} \times B \times 3\sigma_f$.The region is first rotated to θ, which provides rotation invariance for the SIFT descriptors. The factor $\sqrt{2}$ guarantees a complete rectangle region after rotation. The rectangle region is further partitioned into uniform 4×4 sub-regions. Thus, $B = 4$ and each sub-region has $9\sigma_f^2$ samples. The gradient orientation is evenly partitioned into 8 bins. For a sub-region j, an 8-dimension vector v_j can be generated by

$$v_j(i) = \sum\nolimits_{(d_x,d_y)\in\Omega_v(i)} \omega_v(x,y) \sqrt[2]{d_x^2 + d_y^2}, 1 \le i \le 8, \tag{11.6}$$

with $\Omega_v(i) = \{(d_x, d_y) | \frac{\pi(i-1)}{4} \le \arctan(\frac{d_y}{d_x}) < \frac{\pi i}{4}\}$. $\omega_v(x,y)$ is a weighting factor and is defined as a Gaussian function $\sigma_v = B/2$ centered at (x_f, y_f). Combining all 16 sub-regions, we can get one 128-dimensional vector $v = (v_1, \ldots, v_{16})$. After normalization, it is the feature vector of a SIFT descriptor.

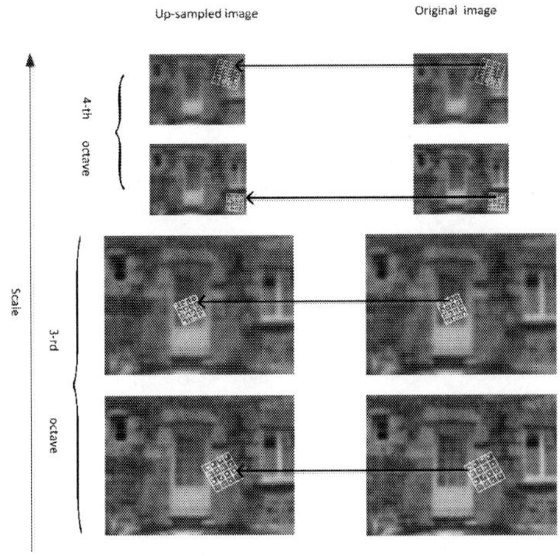

Figure 11.4 An exemplified prediction of SIFT descriptors. The right side is the Gaussian scale-space generated from original image and several extracted SIFT descriptors are drawn there. The left side is the Gaussian scale-space from the up-sampled decompressed image. The SIFT descriptors in this space are extracted by using the same locations and orientations as those in the right side.

11.5 Compression of Image Descriptors

The down-sampled image is directly compressed by the intra-frame coding of HEVC [14]. The core problem in our proposed scheme is how to efficiently compress SIFT descriptors. As shown in Figure 11.4, the key idea presented in this chapter is to use feature vectors extracted from a decompressed image as the prediction. It is similar to inter-frame coding in video. Two questions will be answered here. The first is whether the feature vectors extracted from the down-sampled image are good predictions. The second is how to efficiently use the predictions for the compression of SIFT descriptors.

11.5.1 Prediction Evaluation

First, we observe that SIFT descriptors extracted from a down-sampled image are different from those from the original image on location (x_f, y_f) and scale index s_f. This is reasonable because of the down-sampling process. Fortunately, we observe a strong correlation between feature vectors extracted from the down-sampled image and from the original image when scale index s_f is larger than a certain value. We will evaluate this correlation here.

A decompressed down-sampled image is first up-sampled to the original resolution. A Gaussian scale space is generated by the approach described in Section 11.4. For a SIFT descriptor $S = (x_f, y_f, s_f, \theta, v)$ extracted from the original image, we can

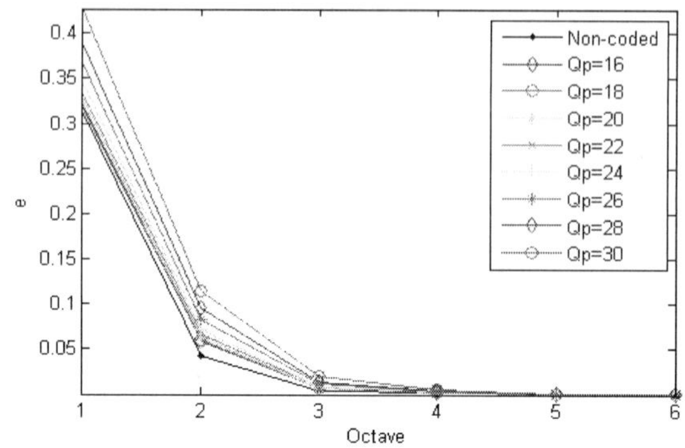

Figure 11.5 The predicted MSE with regard to v in different octaves.

generate a prediction vector $\bar{v}$ in the scale-space by using location (x_f, y_f), scale s_f, and orientation θ. We evaluate the prediction using the normalized mean square error (MSE)

$$e = \frac{\sum_{j=1}^{16}\sum_{i=1}^{8}(v_j(i) - \bar{v}_j(i))^2}{\sum_{j=1}^{16}\sum_{i=1}^{8}(v_j(i))^2}. \tag{11.7}$$

We take the same image used in Figures 11.3 and 11.4 as an example and draw the curves of the average e of all SIFT descriptors in an octave versus octave index as shown in Figure 11.5. We have tested other images in the data set and observed similar results.

In Figure 11.5, we use different approaches to generate $\bar{v}$. The curve "Non-coded" indicates that the down-sampled image is not coded. The curves "QP=x" indicate that the down-sampled image is coded by HEVC with a given QP. In this experiment, the image is down-sampled by $r = 2^4$, which corresponds to the scale index 4. We first observe that, when QP is equal to or less than 22, the curves are close to that without compression. We also observe that, if the scale index is equal to or larger than 4, the average e approaches zero. It indicates that feature vector $\bar{v}$ can predict v very well. When the scale index is 2 or 3, the prediction is good but has some errors. For the scale indices smaller than 2, the prediction becomes considerably poorer because the details have been removed by down-sampling.

11.5.2 Compression of SIFT Descriptors

Based on the above evaluation, we propose different strategies for compression of SIFT feature vectors at different octaves. For octaves with scale index $s_f \geq \log_2 r$, feature vector v is not coded and instead $\bar{v}$ is its reconstruction. For octaves with scale index $\log_2 r - s_0 \leq s_f < \log_2 r$, where s_0 is a constant and is usually set as

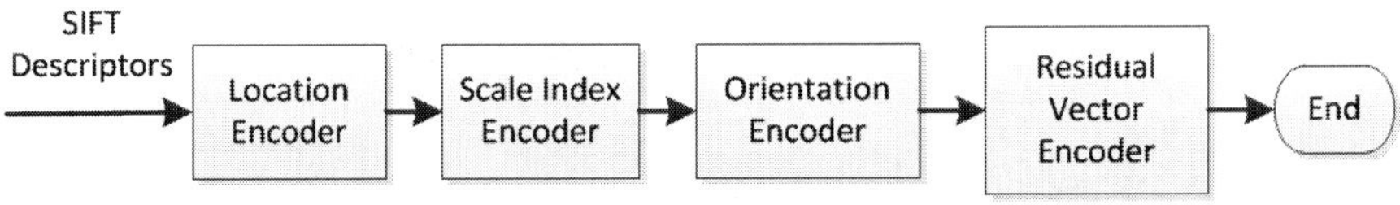

Figure 11.6 The coding process of SIFT descriptors in one image.

a small integer (e.g., 1 or 2), the residual vector $v - \bar{v}$ is coded by transformation, quantization, and entropy coding. For the rest of the octaves, since the scale indices are small, retrieved patches are too small to help image reconstruction. Those SIFT descriptors are discarded directly. Selecting part of the SIFT descriptors is a common method of mobile applications [254]. Although it may have a negative impact on search results and reconstructed quality, it is a cost-effective way for practical applications.

The coding process of SIFT descriptors is depicted in Figure 11.6. The locations of all SIFT descriptors are compressed first. We can determine the number of SIFT descriptors based on the number of locations. All scale indices and orientations are then compressed in raster scan order, respectively. According to the discussed coding strategies and coded scale indices, we know which residual feature vectors should be coded. Finally, they are compressed one by one.

Locations are important in the proposed scheme. First, they indicate where orientations and feature vectors are calculated. Second, they are used to calculate the transformation between retrieved image patches and the input image, which requires high precision. Thus, they are only quantized into integers in the original resolution. Assume that $\mathscr{L} = \{(x_f(1), y_f(1)), \ldots, (x_f(M), y_f(M))\}$ is the set of all SIFT locations, where M is the total number. We first generate a binary matrix B_l of the same size as the input image, for which the element is defined as

$$b_l(x,y) = \begin{cases} 1, if (x,y) \in \mathscr{L} \\ 0, if (x,y) \notin \mathscr{L}. \end{cases} \tag{11.8}$$

This matrix is compressed by binary arithmetic coding. For any $b_l(x,y) = 1$, an integer is used to indicate the number of SIFT descriptors at (x,y). It is most likely one but the maximum number allowed is four. It is compressed by small-alphabet arithmetic coding. This approach is similar to the coding of a block of discrete cosine transform (DCT) coefficients in H.264 [13].

s_f decides the region sizes to calculate θ and v, respectively. For every s_f, its octave index n is coded by 3 bits, and $s_f - n$ is quantized into 16 levels at precision of 1/16 that is coded by 4 bits. θ decides rotation invariance of SIFT descriptors. Every θ is quantized into 128 levels at precision of $45/16 \approx 3$ degrees and coded by 7 bits. Therefore, for a SIFT descriptor, s_f and θ will take 14 bits.

In the compression of the residual feature vector, a binary string is first generated, where each bit indicates if one residual feature vector is zero or not after transform and quantization. This string is compressed by binary arithmetic coding too. Given a residual vector to code, we organize it as two 8×8 matrices

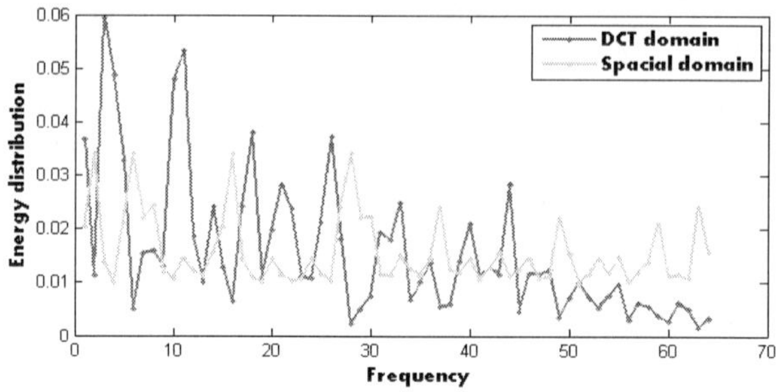

Figure 11.7 Energy compaction of residual feature vectors.

$$\begin{cases} B_1 = (\Delta v_1, \ldots, \Delta v_8)^T \\ B_2 = (\Delta v_9, \ldots, \Delta v_{16})^T \end{cases} \tag{11.9}$$

where $\Delta v_i = v_i - \bar{v}_i$ is an 8-dimension row vector. Each matrix is transformed by 8×8 DCT. After quantization, most of the coefficients are zero and nonzero coefficients, which are most likely small integers. Similar to approach, Eq. (11.8), DCT coefficients are converted to a binary matrix and a set of nonzero integers. One bit indicates if one coefficient is zero or not. An integer is the value of a nonzero coefficient.

Finally, we evaluate the DCT transformation in our scheme. The role of transformation is to compact signal energy. We randomly select 10,000 residual feature vectors in our scheme and calculate their average energy distribution in a pixel domain and DCT domain in Figure 11.7. We can observe that the energy after transformation concentrates at low and middle frequency and is reduced at high frequency. But obviously the energy compaction is not as good as in the residual image. Considering the computation of transform, it should be further studied if the transformation can be skipped in future.

11.6 Image Reconstruction

The decoding of SIFT descriptors is an inverse process to what we have described in Section 11.5. Thus we skip the details of decoding and assume that both the down-sampled image and SIFT descriptors have been decoded. This section will focus on image reconstruction.

11.6.1 Patch Retrieval

The first step for reconstruction is to retrieve highly correlated images. Assume that the SIFT descriptors of all images in the cloud have been extracted with $s_f \geq 2$.

Fifty million SIFT feature vectors are selected randomly and trained into 1 million visual words by approximate k-means [260]. Every SIFT feature vector in the cloud is quantized to a visual word if it has the minimum Euclidian distance to the visual word. In an image, the region of a SIFT descriptor with a large-scale index often partially or completely covers the regions of some SIFT descriptors with small scale indices. We bundle them as a group. One image often has tens to hundreds of groups. Every group is represented by a set of visual words and their geometric relationship in the image.

Decoded SIFT feature vectors are quantized to visual words and organized into groups too. Every group is matched with all groups in the cloud. The matching is scored by the number of matched visual words and their geometric relationship. This score is assigned to the image that contains the group. After all groups are matched, several images with high sum scores are selected as highly correlated images for reconstruction. More details can be found in Wu et al. [236] and Zhou et al. [237].

In order to guarantee matching precision, patch retrieval is operated on 128-dimension vectors instead of visual words. All decoded SIFT descriptors are denoted as $\tilde{\Omega}$, and all of the SIFT descriptors in the selected images as Ω. Every decoded SIFT descriptor $\tilde{S} = (\tilde{x}_f, \tilde{y}_f, \tilde{s}_f, \tilde{\theta}, \tilde{v}) \in \tilde{\Omega}$ is independently matched with every $S = (x_f, y_f, s_f, \theta, v) \in \Omega$. The matching criterion between $\tilde{S}$ and S is defined as

$$\mathscr{M}_{\tilde{\Omega},\Omega}(\tilde{S}, S) = \begin{cases} 1, \text{ if } S = S_{min}, D_{S,S'_{min}}/D_{S,S_{min}} > C \\ 0, \text{ otherwise.} \end{cases} \tag{11.10}$$

The feature vector of SIFT descriptor S_{min} is the closest one to that of $\tilde{S}$ at the square Euclidean distance, and the feature vector of S'_{min} is the second-closest one. $D_{S,S'_{min}}$ and $D_{S,S_{min}}$ are the Euclidean distances. C is a constant and usually set as 1.5. From a matched SIFT descriptor S, we can get a patch P_S from the retrieved images according to its location, scale, and orientation. Similarly, we can get another patch $P_{\tilde{S}}$ from the up-sampled decompressed image by $\tilde{S}$.

11.6.2 Patch Transformation

The flowchart of image reconstruction is similar to that by Weinzaepfel et al. [244]. In addition to the thumbnail being applied to verify every patch and guide the stitching, another key technical difference from that by Weinzaepfel et al. [244] is that we use all SIFT descriptors located on a patch to estimate the transformation of the patch. Therefore, the RANSAC algorithm is adopted [246] and a perspective projection model is used.

P_S is a high-resolution patch and $P_{\tilde{S}}$ is a low-resolution one. It is difficult to estimate an accurate transformation between them by pixel-based matching. Fortunately, SIFT locations of P_S and $P_{\tilde{S}}$ are extracted at high precision. Thus, feature-based matching is adopted in the proposed scheme. The corresponding feature points are detected by matching SIFT descriptors of P_S and $P_{\tilde{S}}$. The set of SIFT descriptors in P_S can be found by $\Omega_S = \{S | S \in \Omega \cap (x_f, y_f) \in X_S\}$, and the set in $P_{\tilde{S}}$ by $\Omega_{\tilde{S}} = \{\tilde{S} | \tilde{S} \in \tilde{\Omega} \cap (\tilde{x}_f, \tilde{y}_f) \in X_{\tilde{S}}\}$. X_S and $X_{\tilde{S}}$ are the sets of samples in P_S and $P_{\tilde{S}}$, respectively. Since the set size of Ω_S and $\Omega_{\tilde{S}}$ are much smaller than that of Ω and $\tilde{\Omega}$, more pairs of matched SIFT descriptors can be found by the same criterion as Eq. (11.10). They are written as

$$\Omega_{S \cap \tilde{S}} = \{(\tilde{S}, S) | \mathscr{M}_{\Omega_{\tilde{S}}, \Omega_S}(\tilde{S}, S) = 1, \tilde{S} \in \Omega_{\tilde{S}}, S \in \Omega_S\}.$$

In general cases, the transformation from P_S to $P_{\tilde{S}}$ is defined as a planar projection H with eight parameters. The parameters are estimated by RANSAC [246]. Several pairs of SIFT locations in $\Omega_{S \cap \tilde{S}}$ are randomly selected to calculate H. The remaining pairs of SIFT locations are used to verify H. For a pair $(\tilde{S}(i), S(i))$, if the Euclidean distance between $(\tilde{x}_f(i), \tilde{y}_f(i))$ and the point of $(x_f(i), y_f(i))$ after H transformation is smaller than a given threshold ε, it is called as an inlier to H; otherwise it is an outlier. H and the number of inliers are recorded. This process is repeated. If a new H has more inliers, H and the number of inliers are updated. Finally, we can get H with the maximum number of inliers.

Besides the estimated H, we can immediately write another H_0 from the location, scale index, and orientation of $\tilde{S}$ and S if the transformation between P_S and $P_{\tilde{S}}$ is a combination of translation, rotation, and scaling

$$H_0 = \left[\tilde{s}_f / s_f R \ T \right], \tag{11.11}$$

where $R = \begin{bmatrix} \cos(\tilde{\theta} - \theta) & -\sin(\tilde{\theta} - \theta) \\ \sin(\tilde{\theta} - \theta) & \cos(\tilde{\theta} - \theta) \end{bmatrix}$, $T = \begin{bmatrix} \tilde{x}_f - x_f \\ \tilde{y}_f - y_f \end{bmatrix}$. In the proposed scheme, H_0 is a rival transformation to H and we will select a better one during patch stitching. Although H_0 is usually not as accurate as H, it can be used to control error if H is not correct. Furthermore, it is easier to obtain the H_0 transformation because there is no need to run RANSAC.

11.6.3 Patch Stitching

Both transformations H and H_0 are applied to P_S to get two patches P_S^H and $P_S^{H_0}$. The up-sampled decompressed image $\tilde{I}_u$ is used to guide the patch stitching. Two patches P_S^H and $P_S^{H_0}$ are matched with a region of $\tilde{I}_u$ centered at $(\tilde{x}_f, \tilde{y}_f)$, respectively. The match at every location is scored by the MSE between the patch and a corresponding patch in the up-sampled decompressed image. The better patch and the best location are decided by the minimum score. Finally, we get the patch $\hat{P}_{\tilde{S}}$ that will be stitched. If the minimum score of $\hat{P}_{\tilde{S}}$ is larger than a threshold, we will discard this patch because it most likely is not correct for $\tilde{I}_u$. Since $\hat{P}_{\tilde{S}}$ may come from an image with different illumination, it is blended to the up-sampled image $\tilde{I}_u$ by Poisson editing [244, 270].

Figure 11.8 The selected images as inputs in our experiments, denoted as "a" to "j" in the raster scan order.

11.7 Experimental Results and Analyses

We use the INRIA Holiday data set [269], which has 1491 images in total, for our experiments in this chapter. Images in this data set have a resolution of up to 8M pixels. There are multiple images in the same scene captured at different viewpoints and focal lengths. As shown in Figure 11.8, 10 images are selected as input images and the rest are used as images in the cloud. For convenience, we denote them from "a" to "j" in the raster scan order. We select the intra-frame coding in HEVC and JPEG as the anchors. HEVC is an on-going video coding standard and is much better than H.264 [14]. JPEG is the most popular image coding standard on the Internet, although its coding efficiency is not good to date. Three schemes will be evaluated for both compression ratio and visual quality.

11.7.1 Compression Ratio

In the proposed scheme, all images are down-sampled by 256:1 and compressed by the intra-frame coding of HEVC. In order to get a high compression ratio, all coding tools in HEVC are enabled. The quantization step for all images is set as 22. All SIFT descriptors in the octaves with scale indices more than 2 are compressed. Since there are too many SIFT descriptors in the second octaves, only part of the SIFT descriptors, which can be predicted better, is selected for compression. The residual feature vectors are transformed and quantized. The quantization step is set as 30. It is a normal parameter for the inter-frame coding.

The experimental results are listed in Table 11.1. The average size of down-sampled images is 6.14 KB after HEVC compression. Each image has a different number of SIFT descriptors ranging from 227 to 722, which is dependent on image content. The average number of SIFT descriptors per image is 500. The average size of all SIFT descriptors per image is 1.82 KB after compression by the proposed scheme. It is equivalent to 30 bits per SIFT descriptor on average.

The total size and compression ratio of every image are listed in Table 11.2. For the proposed scheme, the total size is 7.96 KB on average and the corresponding compression ratio is 1885:1. The maximum compression ratio is as high as 4000:1.

Table 11.1 Image sizes, compressed HEVC sizes, and compressed SIFT sizes in the proposed scheme.

Image	Size	HEVC (KB)	No. SIFT	SIFT (KB)	Bits per SIFT
A	2816×2112	2.80	464	1.63	29
B	2560×1920	7.75	722	2.45	28
C	3264×2448	10.13	679	2.36	28
D	2048×1536	5.26	415	1.55	31
E	1536×2048	4.95	315	1.08	28
F	2048×1536	3.16	390	1.33	28
G	3264×2448	9.48	542	1.91	29
H	1536×1152	2.41	227	0.78	28
I	3264×2448	10.36	654	2.27	28
J	2048×1536	5.09	593	2.82	39
Ave.	4.69M	6.14	500	1.82	30

Table 11.2 Comparisons with intra-frame coding of HEVC and JPEG.

Image	JPEG		HEVC		Proposed	
	Size	Ratio	Size	Ratio	Size	Ratio
a	39.0	447	4.67	3731	4.43	3933
b	52.7	273	17.62	817	10.20	1411
c	66.1	354	22.07	1061	12.49	1874
d	27.0	341	8.65	1065	6.81	1353
e	35.4	260	13.13	702	6.03	1528
f	24.6	375	5.93	1554	4.49	2052
g	64.4	363	19.39	1207	11.39	2055
h	14.3	362	3.85	1347	3.19	1626
i	72.2	324	28.71	815	12.63	1853
j	32.7	282	10.36	890	7.91	1164
Ave.	42.84	338	13.44	1319	7.96	1885

JPEG results are generated by the software IrfanView. We try to make JPEG file sizes close to that of the proposed scheme. Even setting the lowest quality, the file size is still 42.84 KB on average and the corresponding compression ratio is only 338. HEVC results are generated using the reference software 4.0 by setting the quantization step as 51. The file size is 13.44 KB on average and the compression ratio is 1319. Even for HEVC, the most efficient image coding scheme to date, the file size is still larger than that of the proposed scheme by 70% on average.

11.7.2 Visual Quality

For the generated files listed in Table 11.2, we evaluate their visual qualities here. Since the coding artifacts are quite different even for JPEG and HEVC, it is hard to evaluate them by an objective criterion. We adopt the double-stimulus continuous

Table 11.3 Average scores and PSNR for JPEG, HEVC, and the proposed scheme.

Image	JPEG		HEVC		Proposed	
	Score	PSNR	Score	PSNR	Score	PSNR
a	1	25.06	1.79	31.26	4.42	29.65
b	1.16	20.62	3	21.55	2.47	18.29
c	1.16	23.28	3.42	26.70	4.32	17.50
d	1	22.58	2.42	24.93	4.21	20.74
e	1.05	20.62	1.63	21.24	3	19.27
f	1	24.19	2.16	26.65	2.58	22.78
g	1.11	22.04	2.26	25.01	3.11	20.77
h	1.16	22.66	2.32	26.54	4.05	23.35
i	1	20.28	2.26	21.89	4.21	18.51
j	1.37	22.61	2.95	25.00	3.74	22.36
Ave.	1.1	22.39	2.42	25.08	3.61	21.32

quality-scale (DSCQS) method [271] for subjective testing. Twenty undergraduate students without any coding experience served as assessors. The decoded images are displayed on an ultra HD monitor in random order. Image quality is scored from 1 to 5, indicating bad, poor, fair, good, and excellent.

The average scores for every image are listed in Table 11.3. The scores of JPEG and HEVC are only 1.1 and 2.42, respectively. These indicate that the quality of JPEG and HEVC are bad and poor. The average scores of the proposed scheme are 3.61. This indicates that the visual quality is between fair and good. We note that the scores of 50% for the images are more than 4 in the proposed scheme. In other words, their quality is close to the excellent grade even after the thousand-to-one compression. Images "b" and "f" have low scores for the proposed scheme because there are several reconstruction artifacts. The PSNR is also given in Table 11.3 as a reference.

Limited by the pages, only the reconstructed results of images "a," "e," and "f" at different scores are shown in Figure 11.9. Each group of results in Figure 11.9 consists of two rows. In the first row, images from left to right are the original image and the results of the proposed scheme, HEVC, and JPEG. We can observe that the results of the proposed scheme are very consistent with the input images. The second row shows the details in a corresponding region indicated by rectangles. The artifacts of HEVC and JPEG can be observed clearly. However, the results of the proposed scheme present the details clearly. As we have pointed out, the score of image "f" is low in the proposed scheme. Although the result looks like the original image, some people in front of the building are not constructed well as marked by the circles. This is the typical artifacts in the proposed scheme.

11.7.3 Highly Correlated Image

Highly correlated images in this chapter mainly indicate those images taken of the same scene with different viewpoints, focal lengths, and illuminations. In fact, the

Figure 11.9 The reconstructed results of images "a," "e," and "f." Each group of results consists of two rows. In the first row, images from left to right are original images and the results of the proposed scheme, HEVC, and JPEG. The second row is the details in a corresponding region indicated by the rectangles. The circles indicate the reconstructed artifacts in the proposed scheme.

proposed scheme can be extended to the cases of partial duplicate images [236,237], where scenes are different but some regions and objects are similar. Highly correlated images play an important role in the proposed scheme. We show four retrieved images for images "a," "e," and "f" in Figure 11.10. We can observe that the image

Figure 11.10 Highly correlated images retrieved for images "a," "e," and "f."

Figure 11.11 Images in the first row are the new correlated images for image "a." The second row is the result of the proposed scheme in this case. The rectangle indicates the stitched one patch.

retrieval performs well and highly correlated images can be found and ranked first. For images "a" and "e'," one and two images retrieved are non-correlated. But since their patches have big MSE values with the decompressed image, they are finally discarded and do not result in bad results.

What is the result of the proposed scheme if there are no highly correlated images retrieved? For image "a," we remove the first four correlated images in Figure 11.10 from the data set and then reconstruct it. Four new retrieved images are shown at the top of Figure 11.11 and obviously they are not correlated to image "a." The reconstructed result is also shown in Figure 11.11. We can observe that all patches except for one are removed because of the large MSE values of the image. The patch is stitched in the region marked by a rectangle. It does not introduce annoying artifacts. The final result is the up-sampled decompressed image that looks very blurry.

11.7.4 Complexity Analyses

The client needs to encode and decode a down-sampled image, extract SIFT descriptors, and encode SIFT descriptors. Since the size of the down-sampled image is small, its encoding and decoding are simple. As discussed in Section 11.5, the proposed SIFT compression has low computation without training, complicated prediction, and large size transform. The highest computation is needed for building a pyramid image space and extracting SIFT feature vectors. Since two octaves with small scales are skipped, the computation has been reduced greatly. In our current implementation, it usually takes $5-7$ seconds in a single-core phone at 1 GHz frequency. This time is expected to decrease to 1 second or less in a four-core phone that is now available on the market.

Computation in the cloud is a key problem. Correlated images are retrieved by visual words, and high correlation patches are retrieved only in several selected images. Their computations are not high. However, both the RANSAC [246] algorithm and Poisson editing have high computation. In our current implementation, it often takes a minute or more to reconstruct an image. The RANSAC algorithm of every patch is very suitable for parallel computation. We can take hundreds of cores in the cloud to calculate projection models of all patches simultaneously. Poisson editing can only be performed one by one. Recently, we have found that it can be approximated by the membrane interpolation [272]. In our parallel implementation, the reconstruction time has been reduced to several seconds [273].

11.7.5 Comparison with SIFT Feature Vector Coding

Finally, we evaluate the proposed compression approach in the image retrieval scenario by comparing it to CHoG [254, 255]. The Zurich Building Database (ZuBuD) is used [274]. It contains 1005 images of 201 buildings in Zurich of resolution 640×480. It also has 115 query images of resolution 320×240. The experiment is set up strictly according to the method by Chandrasekhar et al. [255].

For the proposed scheme, SIFT descriptors are extracted from the original query images. Thumbnails are generated by 2:1 down-sampling in each dimension and then compressed by HEVC intra-frame coding. SIFT locations, scales, orientations, and residual vectors are compressed by the approach described in Section 11.5. The compressed description size is adjusted by changing the quantization steps for the thumbnail from 20 to 50 at an interval 5. At the same time, the number of SIFT descriptors varies by changing scale indices and predicted MSE values. As shown in Figure 11.12, seven compressed descriptions have sizes from 0.33 kB to 8.62 kB, including the thumbnails. The vertical axis is the recall, which is defined as the percentage of query images correctly retrieved [255].

The SIFT curve shows the results without any compression. It should provide the best retrieval results by using SIFT descriptors. As shown in Figure 11.12, point D is the un-compressed version of points D1, D2, and D3, which have different

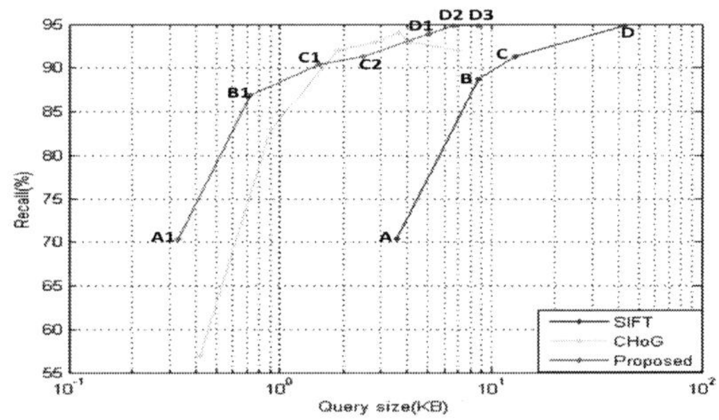

Figure 11.12 The recall rate versus query size of SIFT, CHoG, and the proposed compression approach.

quantization steps. We can observe that the proposed compression only causes up to 1.8% recall loss at point B1 as compared with point B. But the query size is reduced more than 90%. The curve of CHoG in the research by Chandrasekhar et al. [255] is drawn in Figure 11.12, too. We can observe that they have similar recall at similar query sizes. Both of them have a largely decreasing slope on recall when query sizes become too small. One thing that we have to mention here is, that for the proposed compression, the query includes the thumbnail that can be potentially applied to improve retrieval quality and re-rank searching results.

11.8 Further Discussion

11.8.1 Typical Applications

The proposed cloud-based image coding is not a replacement of the conventional image coding. The conventional image coding is mature on how to use the correlations of pixels in the same image for compression. Conventional image coding is more suitable and stable for image archives. Only when the large-scale database of images is available in the cloud, can the proposed coding scheme exploit the correlations of external images and significantly reduce the compressed data size. Therefore, the proposed scheme is more suitable for image sharing and browsing on the Internet. What are the typical applications for cloud-based image coding? Two are discussed here.

(1) Image Sharing — When you visit a sightseeing place, you take some images by your mobile phone and want to share them to your friends immediately. Each image is 4–5 MB in JPEG. It will take minutes to transmit an image over current wireless networks and also consume a lot of precious power on your mobile phone. One current solution is that the images are significantly down-sampled and sent to your friends. But these images look small and unimpressive because many details

are lost. If you select to transmit the down-sampled images to the cloud with an additional several kilobytes of compressed SIFT descriptors, the high-quality and high-resolution images can be reconstructed in the cloud and shared with your friends in JPEG over broadband networks.

(2) Image Search — Modern image search engines have stored thumbnails for result presentation and visual words for content-based matching. SIFT descriptors of crawled images can be stored efficiently by using the proposed compression. Search engines can definitely do more with SIFT descriptors. When users point to a thumbnail that they are interested in, a high-quality and high-resolution image can be generated and presented to them immediately without jumping to the host webpage.

Who will be the service providers? Many companies have owned large-scale databases of images, like Google Street View Image, Microsoft Bing, Yahoo Flickr, Apple iCloud, and so on. They are potential service providers. The network providers are also possible to provide this service. They have the advantage of being able to distribute large-scale databases of images in network nodes and wireless base stations.

11.8.2 Limitations

The large-scale database of images plays the most important role in the quality of reconstructed images. For an input image, if we cannot find highly correlated images in the cloud, the scheme cannot output a good quality reconstruction. Although mobile photo sharing is discussed as an exemplified application, it is not clear which practical applications in the cloud can always have highly correlated images available. It will limit the usage of the proposed scheme.

There are some technical limitations to the proposed scheme too. Images in some complicated scenes may be difficult to reconstruct. For example, there are often people in front of famous places. Although the background can be reconstructed from the correlated images, it is hard to accurately reconstruct a face from other images. One solution for this issue is to segment the foreground by several user interactions and then compress the foreground by image compression. In addition, an image in a dynamic scene (e.g., billboards, windmills, etc.) may also be difficult to reconstruct.

11.8.3 Future Work

The description of the input image needs to be further studied. If the dimension of feature vectors can be reduced as in SURF and CHoG, it will save more bits for residual feature vector coding. In addition, the optimization between distortion and rate should be carefully studied for coding SIFT descriptors. Some coding parameters are set empirically.

Last but not least, retrieved image patches are stitched into the up-sampled decompressed image one by one. This can be improved twofold. First, for complicated scenes, it is difficult to find a correct transformation model for a large patch. It can be split into multiple pieces and each piece is refined before stitching. Second, a patch replaces a corresponding region completely in the proposed scheme. Since the decompressed image is more reliable in low frequencies, the patch stitching can be improved in the frequency domain.

11.9 Summary

This chapter presents a cloud-based image coding scheme, where images are described by SIFT descriptors and down-sampled images. The SIFT descriptors are compressed by predicting from the corresponding SIFT descriptors extracted from the down-sampled images and transform coding. Finally, high-quality and high-resolution images are reconstructed from a set of images in the cloud by the proposed description. Experimental results show that the proposed scheme not only achieves 1885:1 compression on average but also achieves high subjective scores. The reconstructed quality in half of the testing images is close to that of the original images.

Chapter 12
Compression for Cloud Photo Storage

12.1 Introduction

During the last decade, there was an explosive growth of images and photos on the Web. Facebook announced in October 2012 that they have hosted 220 billion photos, and further, the amount was increasing at several hundred million every day [1]. Chinese Tencent announced in August 2012 that they have hosted over 150 billion photos [275]. Microsoft also announced in October 2012 that its cloud storage SkyDrive has hosted 11 billion photos [276]. Let us assume the size of every photo is 2.5 MB on the average, 200 billion photos needs 500,000 1T hard disks to store. For data safety, multiple copies (e.g., three copies) of every photo are needed and hard disks have to be replaced every three years. The current price of 1T hard disk is about \$60 and thus the storage cost will be about \$90 million per three years even without considering the cost of power supply and air conditioning. As a result, there is an urgent need for more efficient compression for cloud photo storage.

Images in cloud storage are currently stored as individual JPEG files. Although it is easy to access every image, it is inefficient for storage because the correlation among images is not used at all. To solve this problem, one class of approaches is proposed generating a representative signal (RS) (e.g., an average image) from correlated images and then compressing all correlated images by subtracting the RS [277–280]. But such one-to-multiple prediction structure is not efficient because other coded images cannot be used for prediction. The other class of approaches is proposed organizing correlated images as a pseudo sequence and then compressing the sequence like video [281–285]. This prediction structure looks good. But these approaches mainly use global transformation and block-based motion compensation, and ignore the fact that pseudo sequences are not natural video.

Figure 12.1a shows several frames in the Foreman video sequence, which present slow changes and strong correlation. However, the changes among images in photo album are large and complicated, as shown in Figure 12.1b. First, although they are taken in the same scene, the location, viewpoint, and focus length may be very different. Furthermore, the objects in the scene have different distances to camera lens.

(a) Video frames

(b) Images in photo album

Figure 12.1 Video frames and images in photo album.

Thus the geometric deformation between two images is difficult to be represented by one global model. Second, they may be taken at different time and thus there exist different illuminations and shadows. It is difficult to be compensated by block-based motion compensation. Third, images often contain foreground objects, which cover part of the background. In a word, although the images in Figure 12.1b look similar visually, the correlation among them is weak if we evaluate it pixel by pixel even with motion compensation. Obviously, it requires the advanced approaches and models to detect and exploit this kind of correlation.

In this chapter, we present a novel compression scheme based on local feature descriptors. By extracting local feature descriptors of images in photo album, we can match the corresponding regions among images, no matter what scale, rotation, and illumination are. It is one content-based matching similar to the classical content-based image retrieval [286, 287]. The correlated images and their correlation are then described by a directed graph. In order to decide the prediction structure, we further convert the graph to a minimum spanning tree (MST) by minimizing the predictive cost. According to the tree, the correlated images are organized into a pseudo sequence. One of the important contributions in this chapter is that the proposed scheme is based on the distance between local feature vectors instead of pixel values.

To characterize the complicated correlation between images, we further present a feature-based three-step prediction. First, we formulate the deformations of two images as a minimum energy problem based on the locations of matched local feature descriptors. By solving the minimum problem, we can automatically derive the number of deformations and the parameters of every deformation. Therefore, multiple prediction images can be generated from one reference image. It well solves the local region deformation problem between two images. Second, the feature-based linear photometric transformation is developed to compensate illumination changes from one prediction image to the current image. Finally, the advanced block-based motion compensation in HEVC [14] is used to handle the displacements of blocks. Experimental results show that our scheme can better exploit the inter-image correlation and its compression ratio is 10 times that of individual JPEG compression at the same quality.

12.2 Related Work

In this section, we will review the related work on image set compression and local feature descriptors.

12.2.1 Image Set Compression

The first class of approaches for image set compression is to generate a RS from all correlated images and then compress every correlated image by subtracting the RS. The research in this line focus on how to generate the RS and the proposed approaches, which include Karhunen-Loève transform (KLT) [277], the centroid method [278], the max-min differential method [279], the max-min predictive method [279], and the low-frequency template [280]. This class of approaches has two constraints. First, they are only applicable for highly correlated images (e.g., medical and satellite images) because good alignment among images is required to generate the RS. Second, the one-to-multiple prediction structure is not efficient because other coded images except for the RS cannot be used for prediction.

The second class of approaches for image set compression relaxes the alignment constraint. They assume that an image set contains well separated clusters of correlated images. Each cluster of correlated images is described by a minimum spanning tree (MST). Cheng et al. evaluated the hierarchical clustering methods for image set compression [281]. Chen et al. optimized the coding order of correlated images by minimizing the predictive cost [283]. Lu et al. adopted MPEG-like block-based motion compensation to exploit the correlation among images [282]. Au et al. introduced global motion compensation before the block-based motion compensation [284]. Zou et al. introduced the advanced block-based motion compensation in HEVC to image set compression [285].

The above approaches assume that correlated images in photo album have the similar correlation as video sequences. As the comparison shows in Figure 12.1, it is obviously not true. It is well known that global motion compensation only handles consistent motion and block-based motion compensation mainly handles 2D displacement. Therefore, these approaches cannot fully exploit the correlation of images in photo album. To well model local geometric deformation and photometric transformation, we present the feature-based matching for better aligning correlated images and the feature-based multi-model for compensating complicated local motion. It is an efficient way to exploit the correlation in photo album.

12.2.2 Local Feature Descriptors

Local feature descriptors are interesting points in images that can be extracted to identify and represent the images. The most famous local feature descriptors are scale-invariant feature transform (SIFT), proposed by Lowe [247]. SIFT descriptors present distinctive invariant features of images that consist of location, scale, orientation, and feature vector. The scale and location of SIFT descriptors are determined by maxima and minima of difference-of-Gaussian images. One orientation is assigned to each SIFT descriptor according to the dominant direction of the local gradient histogram. The feature vector is a 128-dimension vector that characterizes a local region by gradient histogram in different directions. Since SIFT descriptors have a good interpretation of the response properties of complex neurons in the visual cortex [248] and an excellent practical performance, they have been extensively applied to object recognition, image retrieval, 3D reconstruction, annotation, watermarking, and so on.

In addition to SIFT descriptors, some other local descriptors are proposed. SURF (Speed Up Robust Feature) descriptors reduce the dimension to feature vectors to 64 with a performance similar to the SIFT [288]. CHoG (Compressed Histogram of Gradient) descriptors, proposed by Chandrasekhar et al. [254, 255], are designed for compression. They not only change the generation of the gradient histogram but also compress them by tree coding and entropy coding. The successes on applying local feature descriptors to image retrieval have demonstrated the advantage of feature-based content matching in dealing with scale and rotation variations. Image retrieval can be regarded as an attempt to find inter-image correlation in a large-scale image set. Therefore, different from traditional pixel-based similarity measures, the proposed feature-based approach is more robust in evaluating the correlation among images in photo album.

Furthermore, with matched local feature vectors, the locations of corresponding points between two correlated images can be found. The deformation between them can be thus estimated by the RANdom SAmple Consensus (RANSAC) algorithm [246]. It is called the feature-based image alignment, which has been surveyed by Talluri [265]. Following this idea, we further take multiple local deformations into account and present the multi-model approach to estimate multiple geometric deformations and generate multiple predictions from one reference image. It can better exploit the correlation between images.

Before this chapter, feature-based image compression schemes have been proposed [273, 289]. Instead of compressing images pixel by pixel, they proposed describing images by thumbnails and local feature descriptors. Images are reconstructed from a large-scale image set via the compressed description. Although they have high compression ratio, they cannot achieve a good reconstruction under the MSE (mean square error) criteria. However, our scheme in this chapter uses local feature descriptors only to estimate geometric deformations and photometric transformations. The compression scheme still consists of prediction, transform, and entropy coding, like existing image and video coding schemes.

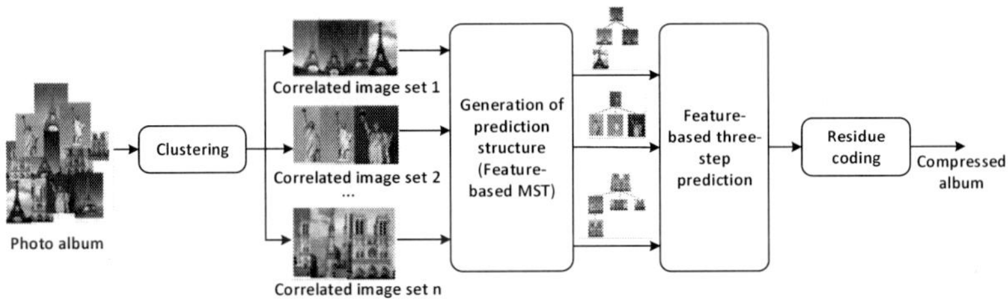

Figure 12.2 The proposed scheme. The solid boxes indicate the technologies studied in this chapter. The dashed boxes indicate using the existing technologies.

12.3 Proposed Scheme

Our scheme is shown in Figure 12.2. Photo albums often contain images taken in different scenes. They can be clustered by the approaches of content-based image retrieval especially for near and partial duplicated image retrieval [234, 236, 237]. Most of these approaches also use local feature descriptors. In this chapter, we do not discuss the clustering problem and assume that correlated image sets have been clustered in photo albums.

For a set of correlated images, we have to first determine an efficient prediction structure for compression. It can be formulated as a graph optimization problem. Each image is regarded as a node in a directed graph. The weighted edges are set by the predictive cost using one image to predict another. Therefore, the prediction structure with the minimum predictive cost can be found by searching the minimum spanning tree (MST) of the graph [290]. Since the feature-based measure is more robust to geometric deformations (e.g., zooming, rotation, and flection) and is also less sensitive to illumination changes and noise than the pixel-based measure, in our scheme edges are weighted by SIFT feature-based distances. Based on this graph, we search the SIFT feature-based MST (SFMST in short) by minimizing the overall SIFT feature-based distance.

According to the SFMST as shown in the middle of Figure 12.2, the inter-image prediction is performed from the root to each leaf node. Here, we present a feature-based three-step prediction to exploit inter-image correlation, which consists of SIFT feature-based geometric deformation, SIFT feature-based photometric transformation, and block-based motion compensation. Specifically, in the SIFT feature-based geometric deformation, we adopt multiple models to characterize the correlation of local corresponding regions between two images. With the models, multiple predictions are generated from the same reference image for coding the current image. In the SIFT feature-based photometric transformation, we estimate a linear model based on matched SIFT descriptors in corresponding regions to reduce intensity differences. Finally, the advanced block-based motion compensation in HEVC is adopted to further reduce the displacement of blocks and improve prediction accuracy.

After the proposed feature-based prediction, the residue images are further compressed by transform and entropy coding to generate the compressed data. In our implementation, these two modules come from HEVC. In the following sections, we will mainly discuss the proposed feature-based prediction structure and feature-based inter-image prediction in details.

12.4 Feature-Based Prediction Structure

Let us assume a correlated image set as $\mathscr{I} = \{I_1, I_2, \cdots, I_N\}$, where N is the number of images in the set. For every image I_n, there is a set of SIFT descriptors as $\mathscr{F}_n = \{f_n(1), f_n(2), \cdots, f_n(K_n)\}$. In other words, there are K_n SIFT local feature descriptors detected and extracted in the image I_n. Every SIFT local feature descriptor is expressed as $f_n(k) = \{\mathbf{p}_n(k), s_n(k), o_n(k), \mathbf{v}_n(k)\}$. $\mathbf{p}_n(k) = \{x_n(k), y_n(k)\}$ is the location of $f_n(k)$ in the image. $\mathbf{v}_n(k)$ is the 128 dimensional feature vector, which is a histogram representing the directional gradients in the local region around $\mathbf{p}_n(k)$. $s_n(k)$ is the scale of $f_n(k)$. It indicates the image area covered by $f_n(k)$. $o_n(k)$ is the dominant gradient orientation of $f_n(k)$.

12.4.1 Graph Building

For the correlated image set, we can build a directed graph $G = \{\mathscr{I}, \mathscr{E}\}$, as shown in Figure 12.3a. In the example, there are four correlated images $\mathscr{I} = \{I_1, \cdots, I_4\}$. Every node represents an image. The key problem in this graph is how to decide the edge weights $e_{i,j} \in \mathscr{E}$, which indicates the predictive cost using image I_i to predict image I_j. In general, the stronger the correlation between I_i and I_j, the smaller the predictive cost $e_{i,j}$ is. But it is decided by how well the correlation can be utilized. The measure for evaluating the correlation in pixel domain is often defined as

$$D_I(I_i, I_j) = \frac{1}{M} \sum \| I_j - H(I_i) \|^p . \tag{12.1}$$

$\| \cdot \|^p$ is the p norm and M is the number of pixels in I_j. $H(\cdot)$ is the transformation from image I_i toward image I_j. For highly correlated images, I_i and I_j are aligned well and thus Eq. (12.1) can be simplified as $\sum \| I_j - I_i \|^p / M$. For the complicated cases, some corresponding points in I_i and I_j have to be found and then the transformation $H(\cdot)$ is estimated based on the corresponding points. It is the so-called pixel-based alignment, which is readily affected by selected corresponding points and noise.

Therefore, we propose using the local feature descriptors to estimate the correlation among images. Let us first define the distance of any two SIFT local feature descriptors $f_i(k_i)$ and $f_j(k_j)$ as

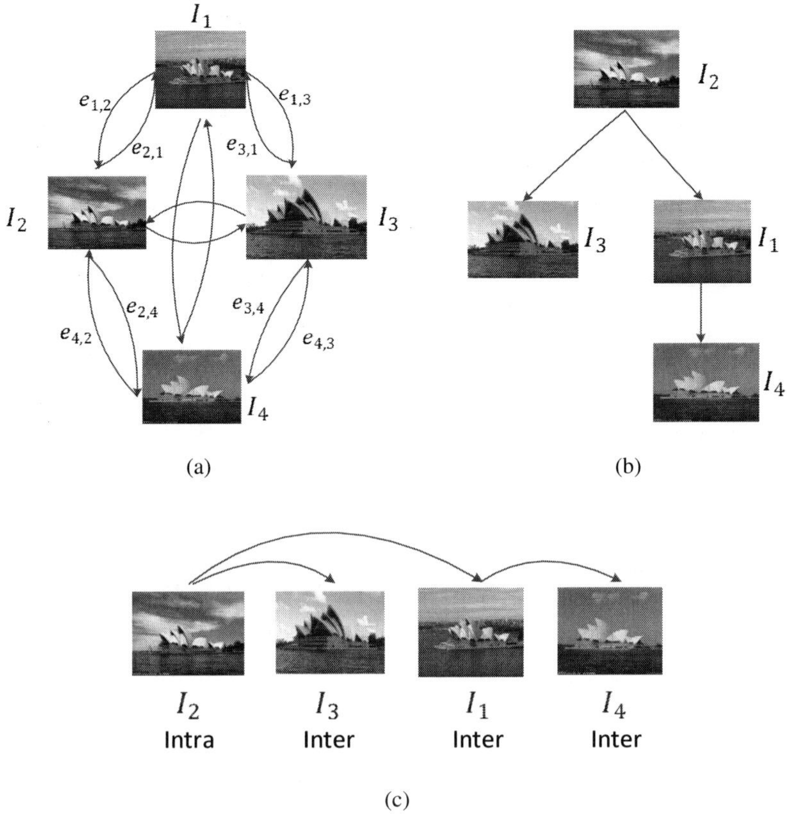

Figure 12.3 Feature-based prediction structure. (a) Directed graph; (b) feature-based minimum spanning tree; (c) prediction structure.

$$D_f(f_i(k_i), f_j(k_j)) = \frac{1}{|\mathbf{v}_j(k_j)|} \| \mathbf{v}_j(k_j) - \mathbf{v}_i(k_i) \|^2 . \quad (12.2)$$

$|\mathbf{v}_j(k_j)|$ is the dimension of $\mathbf{v}_j(k_j)$. For a given SIFT local feature descriptor $f_j(k_j)$ in I_j, assume that $f_i(k_i)$ and $f_i(k_m)$ are the closest and second closest SIFT local feature descriptors in I_i, namely,

$$\begin{aligned} f_i(k_i) &= argmin_{\forall f_i(k) \in \mathscr{F}_i} D_f\left(f_i(k), f_j(k_j)\right), \\ f_i(k_m) &= argmin_{\forall f_i(k) \in \mathscr{F}_i \& f_i(k) \neq f_i(k_i)} D_f\left(f_i(k), f_j(k_j)\right). \end{aligned}$$

We can say $f_i(k_i)$ and $f_j(k_j)$ as the matched SIFT local feature descriptors, if

$$D_f(f_i(k_i), f_j(k_j)) < \alpha D_f(f_i(k_m), f_j(k_j)). \quad (12.3)$$

α is a constant and is usually set as 0.6. In other words, the distance of $f_j(k_j)$ respect to the second closest SIFT local feature descriptor $f_i(k_m)$ in I_i should be larger than a certain ratio compared with the distance respect to the closest SIFT local feature descriptor $f_i(k_i)$.

By using the above matching approach, we can find all matched SIFT local feature descriptors in I_i and I_j, which is expressed as

$$\mathscr{F}_{i,j} = \{(f_i(1), f_j(1)), \cdots, (f_i(K_{i,j}), f_j(K_{i,j}))\}. \tag{12.4}$$

$K_{i,j}$ is the number of matched SIFT local feature descriptors, which should be equal to or less than $\min(K_i, K_j)$. The edge weight $e_{i,j}$ using image I_i to predict image I_j is estimated as

$$e_{i,j} = \frac{1}{\mid \mathscr{F}_{i,j} \mid} \sum_{\forall (f_i(k), f_j(k)) \in \mathscr{F}_{i,j}} D_f(f_i(k), f_j(k)). \tag{12.5}$$

$\mid \mathscr{F}_{i,j} \mid$ is the number of elements in $\mathscr{F}_{i,j}$. After computing all edge weights, we can get the directed graph as shown in Figure 12.3a.

12.4.2 Feature-Based Minimum Spanning Tree

A directed graph, as shown in Figure 12.3a, can be converted to different trees. We look for the tree which represents the correlation among images to facilitate compression. Since the edge weights are the predictive costs using one image to predict another, the prediction structure can be obtained by finding the tree with the minimum predictive cost. It can be described as the problem of seeking a MST in the graph.

Given a directed graph $G = \{\mathscr{I}, \mathscr{E}\}$, where $\mathscr{I} = \{I_1, \cdots, I_N\}$ and $\mathscr{E}$ is the edge set, we can calculate every edge weight $e_{i,j} \in \mathscr{E}$ according to Eq. (12.5). Then the total predictive cost of the graph G is defined as

$$C(G) = \sum_{e_{i,j} \in \mathscr{E}} e_{i,j}. \tag{12.6}$$

Our target is to generate a tree $T = \{\mathscr{I}, \mathscr{E}_T\}$ with $\mathscr{E}_T \subset \mathscr{E}$, which contains no cycle and has the minimal total predictive cost

$$T^* = argmin_{\forall T \subset G} C(T), \quad s.t. \;\; d(I_i) < d_{max} \;\; for \;\; \forall I_i \in \mathscr{I}. \tag{12.7}$$

$d(I_i)$ denotes the depth of image I_i in the tree, starting from the root node as 0. d_{max} is the maximum depth allowed in the tree. We introduce this constraint to control the maximum delay in accessing an image randomly. The larger d_{max}, the more images need to be processed for decoding images at leaf nodes.

We solve this minimization problem by two stages. In the first stage, we construct a MST according to the method described by Chu and Liu [291] without considering the depth constraint. To build the MST, we first generate a set of edges

$$E_0 = \{e_1, \cdots, e_N\}, \quad where \quad e_n = min_{\forall e_{i,n} \in \mathscr{E}} e_{i,n}. \tag{12.8}$$

e_n denotes the edge with the minimal predictive cost connecting to I_n. There are N edges in E_0. As we know, the MST with N nodes should have $N-1$ edges. So we remove the edge with the maximum predictive cost from the set E_0, resulting in E_1. If there is no circle in E_1, we have $\mathscr{E}_{T1} = E_1$; otherwise, all nodes of each circle are replaced by a virtual node and edge weights related to the virtual node are updated correspondingly. Then we repeatedly calculate E_0 and E_1 with reduced nodes in the graph until finding $\mathscr{E}_{T1}$ (for more details, please refer to [291]). At last, we have MST $T' = \{\mathscr{I}, \mathscr{E}_{T1}\}$.

In the second stage, we first check the depth of T'. The process stops if the depth of T' is less than d_{max}; otherwise, we modify T' with regard to the constraint on depth d_{max}. We first select one path from the root node to a leaf node with depth larger than d_{max} in T' and denote it as $T_{sub} = \{\mathscr{I}_{sub}, \mathscr{E}_{sub}\}$. And then, we modify the subtree by $\mathscr{E}'_{sub} = \mathscr{E}_{sub} \bigcup B$ and $B = \{e_{i,j} \mid d(I_i) < d(I_j), I_i \in \mathscr{I}_{sub}, I_j \in \mathscr{I}_{sub}\}$. It actually adds edges in the subtree. Assuming $I_r \in \mathscr{I}_{sub}$ as the root of T', we have subset nodes $\mathscr{I}'_{sub} = \mathscr{I}_{sub} - \{I_r\}$. Then we generate the optimal subtree T^*_{sub} in $T'_{sub} = \{\mathscr{I}'_{sub}, \mathscr{E}'_{sub}\}$ by the minimum spanning tree algorithm again [291]

$$T^*_{sub} = argmin(\sum_i (C(T'_{sub,i}) + e_{r,i})), \quad s.t. \;\; d(I_i) < d_{max} - 1 \;\; for \;\; \forall I_i \in \mathscr{I}'_{sub}. \tag{12.9}$$

$T'_{sub,i}$ presents one of the subtree of T'_{sub} with a children node $I_i \in \mathscr{I}'_{sub}$ as root. This process is repeated on all the subtree whose depths exceed $d_{max} - 1$, so that we get the final SFMST $T = \{\mathscr{I}, \mathscr{E}_T\}$, as exemplified in Figure 12.3b.

12.4.3 Prediction Structure

Once the SFMST $T = \{\mathscr{I}, \mathscr{E}_T\}$ is produced, the prediction structure of the images in $\mathscr{I}$ is determined by depth-first traversing the SFMST T. Accordingly, we organize the images into a sequence Q_I. The root image in T is the first frame in the sequence, which is intra coded without inter-image prediction. The other images are coded as inter frames predicted from their parents. The exemplified SFMST $T = \{\mathscr{I}, \mathscr{E}_T\}$, shown in Figure 12.3b, consists of the image set $\mathscr{I} = \{I_1, I_2, I_3, I_4\}$ and edges $\mathscr{E}_T = \{e_{2,3}, e_{2,1}, e_{1,4}\}$. We can generate the sequence Q_I shown in Figure 12.3c. The corresponding prediction structure is that both I_1 and I_3 are predicted from I_2 and I_4 is predicted from I_1 and I_2.

Note that this prediction structure can be readily supported by the state-of-the-art video coding standards (e.g., H.264 and HEVC), which enable multiple references and long-term prediction.

12.5 Feature-Based Inter-Image Prediction

After the prediction structure of images in a photo album is decided, the feature-based inter-image prediction is applied to better exploit the correlation among images. The block diagram of the feature-based inter-image prediction is shown in Figure 12.4. As we have discussed, since the correlation of images in photo album is very complicated, the proposed prediction consists of three steps. First, we estimate geometric deformations between current and reference images in the module "Deformation Estimation." Multiple deformations are allowed during this estimation. But, since High Efficient Video Coding (HEVC) video coding defines four reference buffers, as shown by $r_0, \cdots, r_3$ in Figure 12.4, we finally select at most four deformations. The modules "G" generate prediction images using the deformation models $\mathscr{H} = \{H_0, \cdots, H_3\}$. Second, the modules "P" estimate illumination changes between current image and prediction images and compensate them by a linear model. It has been supported by the HEVC standard, known as weighted prediction. Finally, the block-based motion estimation is applied to stored prediction images and current image. The one with the minimum rate-distortion cost is selected to predict the blocks of the current image.

12.5.1 Feature-Based Geometric Deformations

For a pair of images I_i and I_j, it is a hard problem to derive the number of deformations between them and the parameters of every deformation. We propose solving this problem based on the matched SIFT descriptors $\mathscr{F}_{i,j}$ instead of pixel values. In fact, any four pairs of matched SIFT descriptors can deduce a projective

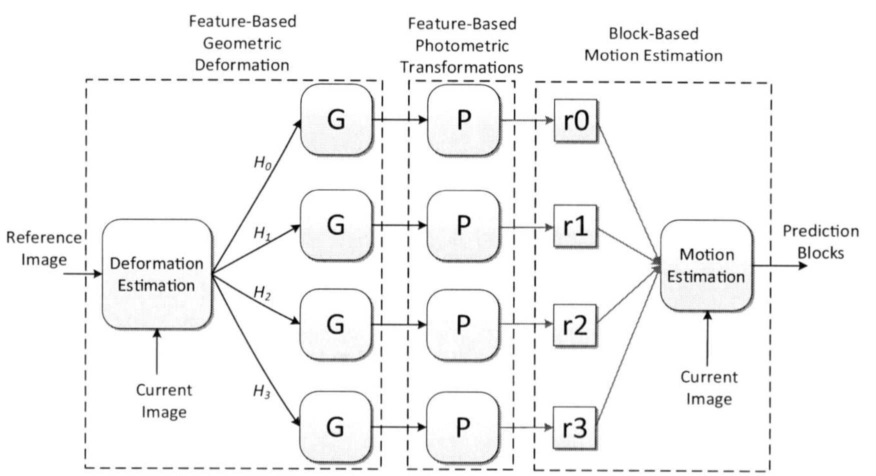

Figure 12.4 The block diagram of SIFT feature-based inter-image prediction.

transformation. However, modeling the deformations in this way will lead to a huge number of transformations. Fischler and Bolles proposed the RANSAC algorithm [246] to fit the deformations with a single model, which covers the largest set of matched features. This single model fitting works well when there is a unified deformation between two images. However, when there exist multiple regions/objects following different motions, the algorithm cannot get a good result.

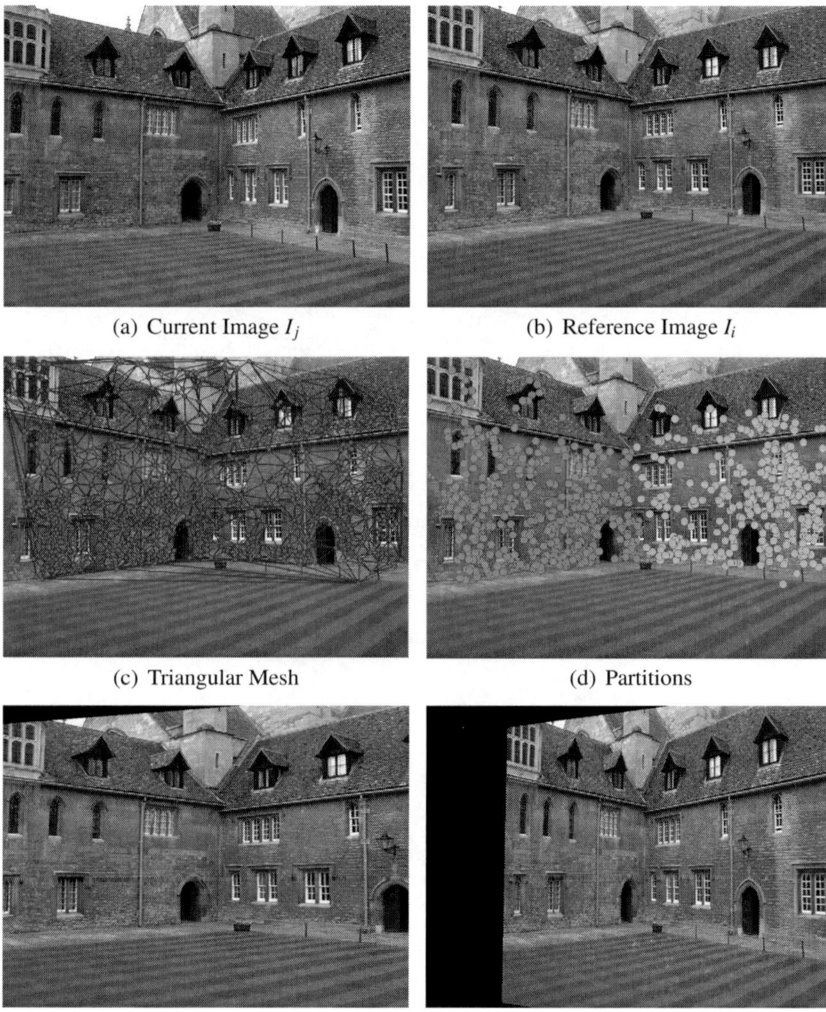

(a) Current Image I_j (b) Reference Image I_i

(c) Triangular Mesh (d) Partitions

(e) Prediction 1 (f) Prediction 2

Figure 12.5 Multi-model geometric deformations.

For two images I_j and I_i, as shown in Figure 12.5a,b, our multi-model deformation manages to automatically divide all the matched SIFT features between I_i and I_j into several groups and accordingly generate the deformation models $\mathscr{H}$ by minimizing the following energy function

$$E = D + \lambda \times S. \tag{12.10}$$

E is the energy of the estimated models, D is the data term, S is the smoothness term, and λ is a weighted parameter. The first term D measures the geometric distance of the matched feature points under a deformation model set $\mathscr{H} = \{H_l \mid l = 1, \cdots, L\}$

$$D = \sum_{l} \sum_{(f_i(k_{i,j}), f_j(k_{i,j})) \in \mathscr{F}^l_{i,j}} D_l(k_{i,j}). \tag{12.11}$$

$\mathscr{F}^l_{i,j}$ is the set of matched local feature descriptors in $\mathscr{F}_{i,j}$ fitting the model H_l. For each pair of matched features $(f_i(k_{i,j}), f_j(k_{i,j})) \in \mathscr{F}^l_{i,j}$, the deformation distance between their locations under the model H_l is defined as

$$D_l(k_{i,j}) = \parallel \mathbf{p}_j(k_{i,j}) - \mathbf{p}_i(k_{i,j}) \times H_l \parallel . \tag{12.12}$$

The smoothness term S in Eq. (12.10) evaluates the connectivity between features. We model the neighboring relationship among the feature points in image I_i by the Delaunay triangulation, as shown in Figure 12.5c. Then the smoothness between two neighboring feature points $\mathbf{p}_i(k_{i,j})$ and $\mathbf{p}_i(k_{m,n})$ is defined as

$$S_l(\mathbf{p}_i(k_{i,j}), \mathbf{p}_i(k_{m,n})) = \begin{cases} 0, \; if\ both\ \mathbf{p}_i(k_{i,j})\ and\ \mathbf{p}_i(k_{m,n})\ fit\ in\ H_l \\ 1, \qquad otherwise \end{cases} \tag{12.13}$$

Thus

$$S = \sum_{l} \sum_{k_{i,j}, k_{m,n} \in \mathfrak{N}(k_{i,j})} S_l(\mathbf{p}_i(k_{i,j}), \mathbf{p}_i(k_{m,n})). \tag{12.14}$$

$\mathfrak{N}(k_{i,j})$ denotes the set of neighboring feature points of $\mathbf{p}_i(k_{i,j})$ in the triangulation mesh. The weighted parameter λ is the discontinuity penalty factor.

According to Isack and Boykov [292], the approximate minimization of Eq. (12.10) can be solved using a graph-cut based approach. In the graph, each data node denotes a feature point $\mathbf{p}_i(k_{i,j})$ and each labelling node represents a model H_l. The edge weight between a data node and a labeling node is defined as the corresponding $D_l(k_{i,j})$, and the edge weight between two data nodes is defined as $S_l(\mathbf{p}_i(k_{i,j}), \mathbf{p}_i(k_{m,n}))$. Note that the edge weights change with the variation of assigned models. Based on the deductions discussed by Boykov et al. [293], the minimization of E is approximated by finding the minimum cuts on the graph. Then the matched features in image I_i is partitioned into L sets and each set has a deformation model H_l. As shown in Figure 12.5d, the matched features in image I_i are partitioned into two sets and two deformation models are produced accordingly. For more details on graph cut, please refer to Boykov et al. [293].

Once the deformation model set $\mathscr{H}$ is available, we can generate a deformed prediction set $\bar{\mathscr{I}}_i = \{\bar{I}_{1,i}, \cdots, \bar{I}_{L,i}\}$ from image I_i to image I_j, where each deformed prediction $\bar{I}_{l,i} \in \bar{\mathscr{I}}_i$ is obtained by $\bar{I}_{l,i} = I_i * H_l$. The number of prediction images from I_i to I_j is L. However, we note that the more the models, the higher the complexity is in the following block-based motion compensation. Meanwhile, we also observe that the models with limited inliers usually cover small regions and the corresponding deformed images may suffer from severe distortions outside the corresponding regions. This leads to the fact that these deformed images are not effective for prediction. According to the buffer size of H.264 and HEVC, we limit the model number L up to 4 in our scheme. In other words, we use at most four models to produce the deformed prediction images $\bar{\mathscr{I}}_i = \{\bar{I}_{1,i}, \cdots, \bar{I}_{L,i}\}$ with $L \leq 4$. The models with more inlier SIFT pairs have the higher propriety to be selected. Figure 12.5e,f show two deformed predictions.

12.5.2 Feature-Based Photometric Transformation

Our feature-based photometric transformation module is presented to reduce the intensity disparities between the geometrically deformed prediction images in $\bar{\mathscr{I}}_i$ and the current image I_j. For each prediction image $\bar{I}_{l,i}$, we define the photometric transformation as

$$\bar{\bar{I}}_{l,i} = \alpha_l \bar{I}_{l,i} + \beta_l . \tag{12.15}$$

α_l and β_l denote the scale and offset parameters, respectively. We continue to use the matched local feature descriptors to estimate photometric transform. The values of α_l and β_l are determined by minimizing the distance between $\bar{\bar{I}}_{l,i}$ and I_j based on the matched features set $\mathscr{F}^l_{i,j}$

$$z = \sum_{(f_i(k_{i,j}), f_j(k_{i,j})) \in \mathscr{F}^l_{i,j}} \| I_j(\mathbf{p}_j(k_{i,j})) - (\alpha_l \bar{I}_{l,i}(\bar{\mathbf{p}}_{l,i}(k_{i,j})) + \beta_l) \|^2 . \tag{12.16}$$

$\bar{\mathbf{p}}_{l,i}(k_{i,j})$ is the feature location after geometric deformation in image $I_{l,i}$. $I_j(\mathbf{p}_j(k_{i,j}))$ and $\bar{I}_{l,i}(\bar{\mathbf{p}}_{l,i}(k_{i,j}))$ are the gray value of pixels. For every pair of matched local feature descriptors, we only use the pixel values at the central point to estimate the parameters. But as shown in Figure 12.5d, every matched region has tens to hundreds of matched local descriptors. They are enough to estimate the parameters robustly. The parameters α_l and β_l are calculated by solving the partial differential functions of Eq. (12.16) as

$$\alpha_l = \frac{N \cdot A - B \cdot C}{N \cdot D - B^2}, \tag{12.17}$$

$$\beta_l = \frac{C \cdot D - A \cdot B}{N \cdot D - B^2}, \tag{12.18}$$

where

$$A = \sum_{(f_i(k_{i,j}),f_j(k_{i,j}))\in\mathscr{F}_{i,j}^l} \bar{I}_{l,i}(\bar{\mathbf{p}}_{l,i}(k_{i,j})) \times I_j(\mathbf{p}_j(k_{i,j})),$$
$$B = \sum_{(f_i(k_{i,j}),f_j(k_{i,j}))\in\mathscr{F}_{i,j}^l} \bar{I}_{l,i}(\bar{\mathbf{p}}_{l,i}(k_{i,j})),$$
$$C = \sum_{(f_i(k_{i,j}),f_j(k_{i,j}))\in\mathscr{F}_{i,j}^l} I_j(\mathbf{p}_j(k_{i,j})),$$
$$D = \sum_{(f_i(k_{i,j}),f_j(k_{i,j}))\in\mathscr{F}_{i,j}^l} \bar{I}_{l,i}^2(\bar{\mathbf{p}}_{l,i}(k_{i,j})),$$

and $N = \mid \mathscr{F}_{i,j}^l \mid$.

Having the values of α_l and β_l, we generate the photometric deformed prediction $\bar{\bar{I}}_{l,i}$ for each prediction image $\bar{I}_{l,i}$ by Eq. (12.15). Then all the deformed prediction images will be stored in the reference buffers $r_0, \cdots, r_3$ and used as prediction images in the following block-based motion compensation, as shown in Figure 12.4.

12.5.3 Block-Based Motion Compensation

After the geometric deformation and photometric transformation, the majority of geometric and illuminate disparities has been compensated efficiently. However, we note that the inter-image prediction still suffers from small local disparities (e.g., local shifting). To deal with the local disparities, we employ the block-based motion compensation, which follows the approach adopted in HEVC. Here, we would like to point out that the parameters of deformation models as well as those of photometric transformations are encoded and transmitted to the decoder side. Each parameter is scaled and rounded into integers before entropy coding.

12.6 Experimental Results

In this section, we evaluate the performance of our feature-based compression scheme in comparison with the state-of-the-art methods. We demonstrate the efficiency of our scheme using five different photo albums. "CastleEntry" is a set of images provided by Strecha. "RockBoat" is a personal album captured at Summer Palace. "WadhamCollege" is provided by VGG. "MailRoom" is a set of indoor images. "NotreDame" contains 45 public photos randomly crawled from Flickr. Figure 12.6 presents three examples of each image set for preview. The properties of each image set are summarized in Table. 12.1. All images in these sets are shot at different camera positions without duplicate images. In addition, RockBoat and NotreDame

Figure 12.6 Three example images of the five test sets. From top to bottom: CastleEntry, RockBoat, WadhamCollege, MailRoom, and NotreDame.

Table 12.1 Properties of the five image sets.

Image Set Name	Number of Images	Resolution	Location Variation	Time Variation	Camera Variation	Bright Variation
CastleEntry	10	3072×2048	Y	N	N	N
RockBoat	20	2592×1728	Y	Y	N	Y
WadhamCollege	5	1024×768	Y	N	N	N
MailRoom	7	1072×600	Y	N	N	N
NotreDame	45	Variable	Y	Y	Y	Y

sets are collected at different time under different lightings. Moreover, NotreDame set is shot by different people with different cameras at different resolutions.

In the following tests, we first evaluate the performance of our multi-model prediction and photometric transformation, respectively. Then we compare our scheme with the most popular image compression scheme JPEG. We also evaluate our scheme compared with the latest research work on the photo album compression, including Prof. Oscar's approach [284] and Zou's approach [285]. The HEVC standard software HM7.0 is used here as the coding platform [14]. At last, the complexity of our scheme is discussed.

In our tests, all these images in an album are randomly indexed before coding. For each album, the SFMST is first generated to determine the coding order, resulting in a sequence. The first image in the sequence is intra encoded and the rest of the images are inter coded. We use the nearest parent image in the SFMST as the reference image in the three-step inter-image prediction. Following HEVC coding, we generate the final compression results. All coding tools in HEVC are enabled and the QPs (quantization parameters) are set to 22, 27, and 32 for the images.

12.6.1 Efficiency of Multi-Model Prediction

We evaluate the efficiency of our multi-model prediction in Figure 12.7. The labels "multiple," "single," "inter," and "intra" denote the results of our multi-model, RANSAC-based single model, HEVC inter coding, and HEVC intra coding, respectively. Figure 12.7 shows the results of two image pairs and two image sets. In this test, the four prediction methods are applied to replace the prediction process in our scheme to generate the coded results, respectively. One can observe that our approach achieves 15% bits saving compared with the single-model prediction and 30% bits saving compared with the inter prediction of HEVC.

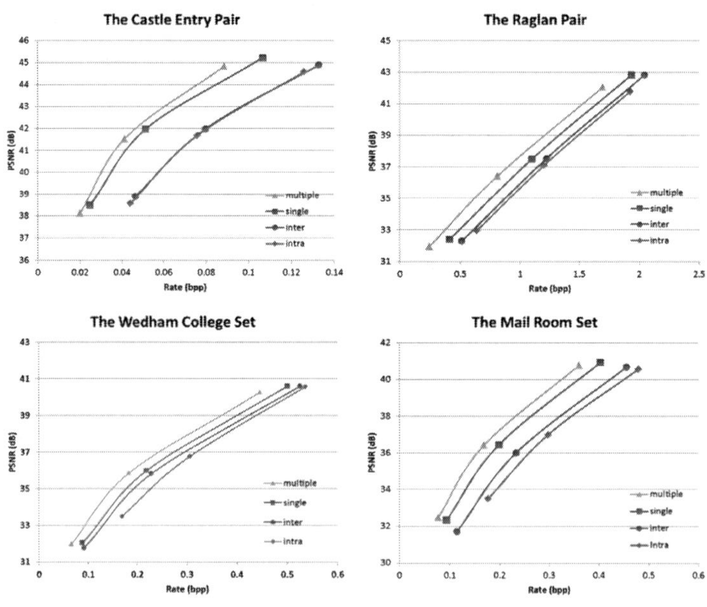

Figure 12.7 Performance evaluation of our multi-model prediction.

We also note that the improvement of our multi-model prediction differs for different sets. Our multi-model prediction prefers the case that the images contain multiple objects or regions, which follow different motions under different camera positions. For example, there are two walls with different motions and different camera locations in the scene of WedhamCollege. Our multi-model prediction generates two motion models and two predictions for the two walls, respectively. Thus, it can reduce the local redundancy much more efficiently than the single-model prediction, which fits only one of the walls. Similarly, MailRoom contains images with multiple objects shot in a limited space. The cabinet, the water canteen, and two bookshelves in the scene follow different motion models. In this case, both single and multiple model achieve good results. Our multi-model prediction performs the best.

12.6.2 *Efficiency of Photometric Transformation*

Different from indoor environments, outdoor scenes usually suffer luminance variances. In this test, our scheme is used to generate the results by enabling or disabling the photometric transformation (PT in the figures). The curves marked as PT on and PT off denote the results with PT enabled and disabled, respectively.

RockBoat contains 20 images, which are collected at different times under different lightings. Figure 12.8 shows the comparison results by rate-distortion curves. One can observe that the photometric transformation helps to reduce brightness variances and achieves more than a 5% bits saving.

NotreDame has 45 images collected at not only different lightings but also different resolutions. The columns in Figure 12.9 denote the bits savings compared with

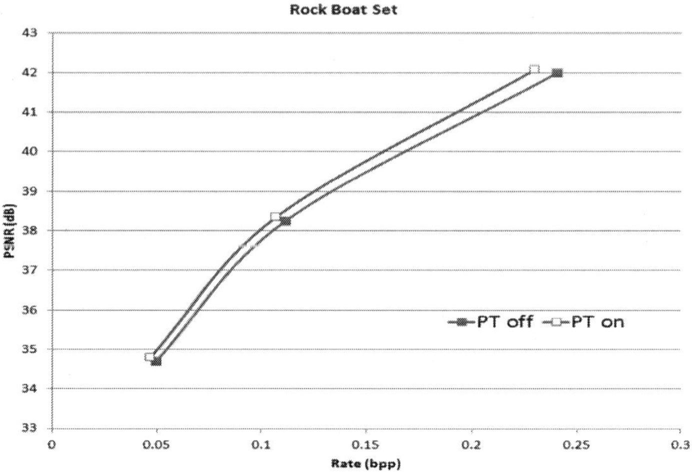

Figure 12.8 Efficiency of photometric transformation on RockBoat image set.

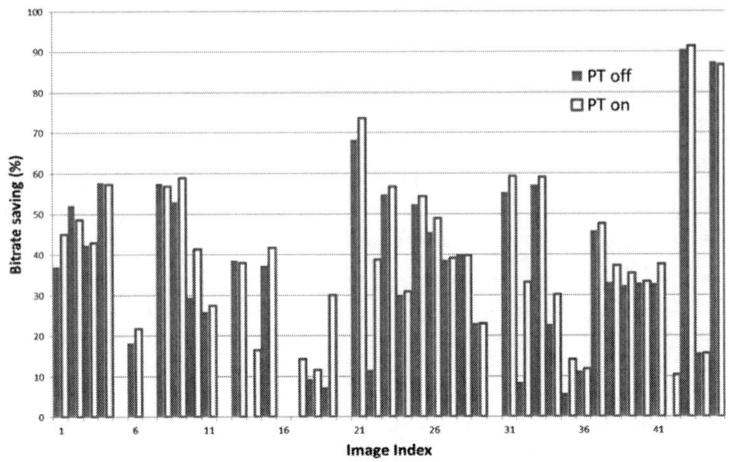

Figure 12.9 Efficiency of photometric transformation on NotreDame image set.

HEVC intra coding. The QP setting here is 27. The solid and hollow columns denote the performance by disabling and enabling the photometric transformation, respectively. The root image in this set is that marked as "Image 5." One can observe that the maximum bits savings is over 90% for "Image 43." The average bits savings for all the 45 images are 30.2% without PT and 34.6% with PT. One can also note that our scheme has no gain for several images (e.g., Images 7, 12, and 16). In fact, if the brightness variances of these images are too large (e.g., suffering overexposure), our photometric transformation cannot work well. Our scheme also loses efficiency when the noise is high enough to greatly reduce the accuracy of feature detection or matching.

12.6.3 Overall Performance

Figure 12.10 shows the coding performance of our feature-based compression method. In Figure 12.10a,b, the comparison results for our method, Oscar's approach, Zou's approach, and individual JPEG on CastleEntry and RockBoat are presented, respectively. The curve marked as "Oscar's" denotes the average performance of Oscar's approach. In this method, one center image is selected via the MSE-based selection and both global and local motion compensations are used in the inter-image prediction. The curve marked as "Zou's" shows the performance of Zou's approach, which utilizes the MSE-based MST to determine the coding structure followed by the block-based motion compensation in HEVC for the inter prediction. The results of JPEG are marked as "JPEG." One can observe that our scheme achieves the best coding performance. Our scheme outperforms JPEG with 10-times compression ratio at the same quality level. It also obtains more than 2 dB PSNR gains or 40%

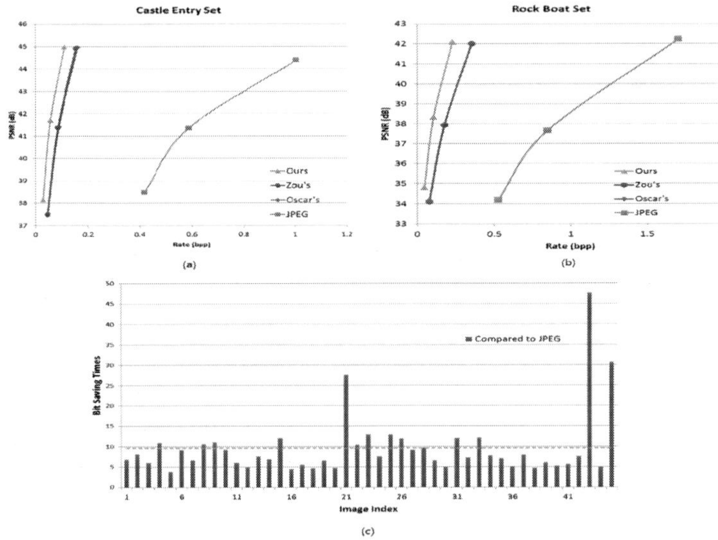

Figure 12.10 Performance comparisons: (a) CastleEntry image set; (b) RockBoat image set; (c) NotreDame image set.

bits saving compared with the other two solutions. One may note that Oscar's approach and Zou's approach have similar performance. It is because the complicated geometric distortions between the images cannot be reduced efficiently by either the pixel-based global motion compensation or the block-based motion compensation.

Figure 12.10c shows the results for NotreDame. Since the images in this set are with different resolutions, the traditional inter-prediction methods proposed for video coding cannot work directly in coding this set. Thus we evaluate the performance in comparison with JPEG only. The columns in this figure show how many times smaller our compressed files are than those of JPEG. For this set, our scheme constantly outperforms JPEG and achieves averagely 9.6 times compression ratio as shown by the dashed line in Figure 12.10.

12.6.4 Complexity

At the decoder side, we introduce the geometric deformations and photometric transformations to traditional video decoding. Since each geometric deformation can be regarded as an interpolation process, the complexity of the geometric deformation is $O(n)$. The complexity of the photometric transformation is $O(n)$, too. So the complexity increment in decoding is limited.

The encoder complexity of our scheme is higher than that of the corresponding video schemes. The complexity increment mainly comes from four aspects: SIFT detection, SIFT-based distance calculation, SFMST generation, and multi-model inter prediction. We can reduce the complexity of SIFT detection by computing at low resolution counterparts of the input images. The complexity of distance calculation can be accelerated by the k-dimensional tree (KD-tree) and approximate nearest neighbor (ANN) search. Moreover, the complexity of our scheme can be reduced by taking advantage of fast algorithms [294] and parallel mechanisms. The complexity of SFMST generation is proportional to the number of images in the graph. The multi-model inter prediction is based on the energy-based model fitting by Isack and Boykov [292], which has a much faster convergence speed compared with other multi-model fitting methods. Since our scheme targets the image storage in cloud environments, we believe that the computational resource should no longer be the bottleneck of the system.

12.7 Our Conjecture on Cloud Storage

If photos in cloud storage are stored as individual JPEG files, with the increasing number of photos, the storage needed also increases linearly, as shown by the dash-dot line in Figure 12.11. It indicates that the service cost in cloud storage will increase with the size of cloud storage. Users thus have to pay more money to increase their storage sizes in cloud.

However, if our scheme is adopted for image and photo storage in cloud, we guess that the service cost is not increasing linearly as the number of increased data. With more and more images and photos stored in cloud, if the correlation with stored data can be fully utilized, the actual size for an uploaded image after compression will become fewer and fewer because most of the contents in the image already exist. In other words, when the scale of cloud storage reaches a certain threshold, the

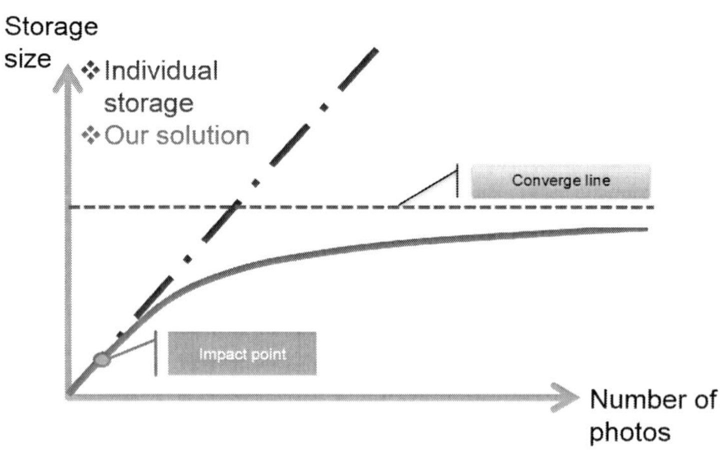

Figure 12.11 Our conjecture on the capacity of cloud storage.

increasing cost for storing a new image becomes very cheap. Our conjecture is that the cost for cloud storage will converge to a constant no matter how many images and photos are uploaded, as shown by the solid curve in Figure 12.11. It is important for the service providers of cloud storage and users because it indicates that one day the cloud storage paid by you is a fixed fee but the storage size is unlimited.

The same principle should also exist for video storage, although currently we do not have a scheme for coding video using the inter-video correlation, similar to that we present in this chapter. Since image and video are dominant data in cloud storage, if our conjecture is true, cloud storage will be a trend for personal data storage with low cost and high reliability.

12.8 Summary

We present a feature-based image compression scheme for photo album. It adopts SIFT local descriptors as measurements to analyze the correlation of images in photo album and generate the models for geometric deformation and photometric transform between images. The analysis is robust to scale and rotation variations and is less sensitive to illumination changes. With the presented prediction technologies, the images in photo album can be compressed with 10-times efficiency than individual JPEG compress. It points out a trend on compressing large-scale image and video in cloud storage, namely, the correlation across images and video.

In this chapter, we only consider the predictive coding for correlated images. Like H.264 and the HEVC video standard, the more efficient coding approach is the bipredictive approach. In the next step, we will further extend our scheme by using the bipredictive approach. It can better handle the complicated scenes with foreground and background objects. It can also handle the images belonging to two correlated image sets or more. Another important problem is that images in photo album are dynamic in cloud storage. Our current solution can deal with deleting and inserting images. If some new images are inserted into a compressed album, our scheme may have to rebuild the SFMST and re-encode the whole albums. Also, if some images need to be deleted from the compressed album, we need to revise the SFMST and may have to transcode the related images. Though useful, the current way in supporting dynamic album is straightforward. Advance approaches should be investigated in the future.

Part V
Compressive Communication

Compressive sensing (CS) is a theory for sampling. However, when it is described by a bipartite graph and is also constrained by binary input, it looks very similar to low-density parity-check (LDPC) code but with two differences. The first one is that the symbols in CS are generated by arithmetic sum instead of logical exclusive-OR in LDPC code. Therefore, we called binary input CS arithmetic code. In addition, when the operation is arithmetic, the bipartite graph is no longer necessary as binary weights. The second one is that the connections are assigned with weights. In this part, we will study how to apply compressive sensing to solving communication problems. We call this research compressive communication.

Chapter 13 presents the first complete design to apply compressive sampling theory to sensor data gathering for large-scale wireless sensor networks. The successful scheme developed in this research is expected to offer fresh frame of mind for research in both compressive sampling applications and large-scale wireless sensor networks. We consider the scenario in which a large number of sensor nodes are densely deployed and sensor readings are spatially correlated. The proposed compressive data gathering is able to reduce global scale communication cost without introducing intensive computation or complicated transmission control. The load balancing characteristic resulting from compressive sampling principles is capable of extending the lifetime of the entire sensor network as well as individual sensors. Furthermore, the proposed scheme can cope with abnormal sensor readings gracefully. We also carry out the analysis of the network capacity of the proposed compressive data gathering and validate the analysis through ns-2 simulations. Morc importantly, this novel compressive data gathering has been tested on real sensor data and the results show the efficiency and robustness of the proposed scheme.

Chapter 14 presents a novel compressive modulation (CM) scheme for fine-grained rate adaptation in time-varying wireless channel. Such compressive modulation concepts open up a new research avenue in modulation from symbol generation, mapping, to demodulation. Motivated by the emerging compressive sensing theory, we first propose random projection (RP) code that properly converts binary source to multilevel rateless symbols. Then, the generated symbols are sequentially and evenly mapped to a dense square quadrature amplitude modulation (QAM) constellation. At the receiver, the RP decoder is capable of recovering source bits directly from multilevel symbols so as to avoid that information used in the decoding mismatches to what is received from the channel. CM is a receiver adaptation scheme that allows the sender to transmit at fixed rate, but the achieved rate is decided by the actual channel condition. We carry out extensive simulations over random binary sources. CM consistently outperforms sender rate adaptation that is based on conventional modulation schemes. More importantly, we implement CM on OFDM physical layer and evaluate its performance on a software radio testbed. Results show that CM outperforms "Oracle" sender adaptation by 41%–100%, and achieves a 33% gain over the state-of-the-art receiver rate adaptation scheme.

Chapter 15 presents our proposed CM to simultaneously achieve joint source-channel coding and seamless rate adaptation. The embedding of source compression into modulation brings significant throughput gain when the physical layer data contains non-negligible redundancy. Two key design issues in the proposed CM are addressed in this chapter. First, we consider the RP code design for sources with different redundancies. Three principles are established and a concrete implementation is given. Second, we devise a linear-time decoding algorithm for the proposed RP code. In this belief propagation (BP) algorithm, we find that computing convolution in time domain is more efficient than that in frequency domain for binary variable nodes. Moreover, we invent a ZigZag deconvolution to further reduce the complexity. Analysis show that the proposed decoding algorithm is nearly 20 times faster than the state-of-the-art BP algorithm for CS called CS-BP. Emulations on traced data show that CM achieves significant throughput gain, up to 33% and 70%, respectively, over the Hybrid automatic repeat request (ARQ) with compression and bit interleaved coded modulation (BICM) with compression, under practical time-varying wireless channels.

Chapter 13
Compressive Data Gathering

13.1 Introduction

This chapter considers the data gathering problem in a large-scale wireless sensor network. Such a large-scale data gathering sensor network finds a variety of applications in habitat monitoring, traffic control, and surveillance. In general, large-scale sensor data gathering is accomplished through multi-hop routing from individual sensor nodes to the data sink. Successful deployment of such large-scale sensor networks faces two major challenges in effective global communication cost reduction and in energy consumption load balancing.

The need for global communication cost reduction is obvious because such sensor networks typically are composed of hundreds to thousands of sensors, generating a tremendous amount of sensor data to be delivered to a data sink. It is very much desired to take full advantage of the correlations among the sensor data to reduce the cost of communication. Existing approaches adopt in-network data compression, such as entropy coding or transform coding, to reduce global traffic. However, these approaches have introduced significant computation and control overheads that are often not suitable for sensor network applications.

The need for energy consumption load balancing is also clear because of the required multi-hop data transmission for such large-scale sensor networks. Figure 13.1 shows such a network where sensors are densely deployed in the region of interest and monitor the environment on a regular basis. A simple but typical example is the highlighted route in Figure 13.1. Suppose N sensors, denoted as s_1, s_2, ..., and s_N, form a multi-hop route to the sink. Let d_j denote the readings obtained by node s_j. The intuitive way to transmit d_j, $j = 1, 2, \ldots N$ to the sink is through multi-hop relay as depicted in Figure 13.2a. Node s_1 transmits its reading d_1 to s_2, and s_2 transmits both its reading d_2 and the relayed reading d_1 to s_3. At the end of the route, s_N transmits all N readings to the sink. It can be observed that the closer a sensor is to the sink, the more energy is consumed. Clearly, the sensor nodes closer to the data sink will soon run out of energy and the lifetime of the sensor network will be significantly shortened.

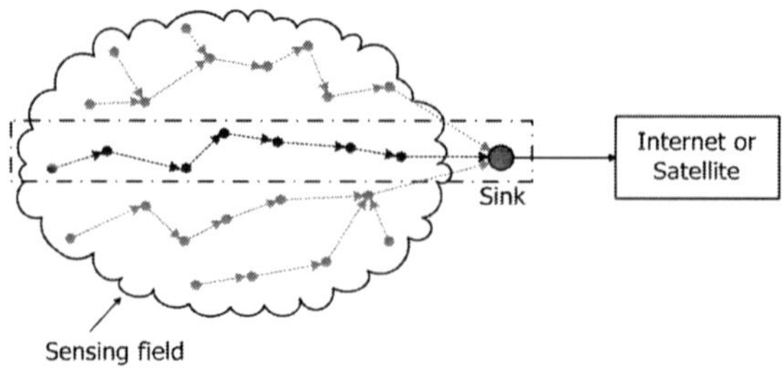

Figure 13.1 Data gathering sensor network.

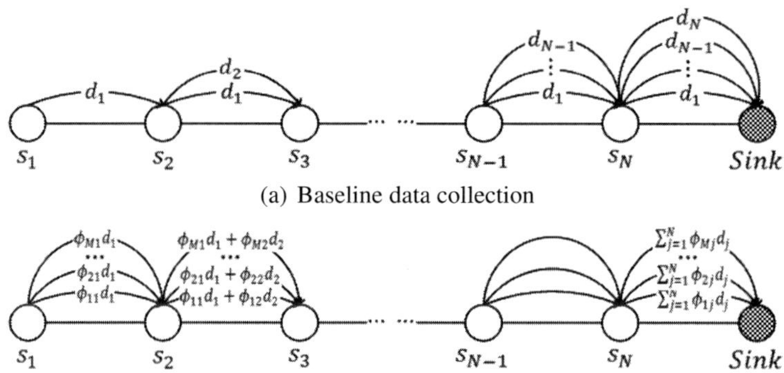

(a) Baseline data collection

(b) Compressive data gathering

Figure 13.2 Comparing baseline data collection and compressive data gathering in a multi-hop route.

This chapter presents the first complete design to apply compressive sensing theory [295–297] to sensor data gathering for large-scale wireless sensor networks, successfully addressing the two major challenges as outlined above. First, the proposed data gathering is able to achieve substantial sensor data compression without introducing excessive computation and control overheads. With elegant design, the proposed scheme is also able to disperse the communication costs to all sensor nodes along a given sensor data gathering route. This will result in a natural load balancing and extend the lifetime of the sensor network.

The basic idea of the proposed compressive data gathering (CDG) is depicted in Figure 13.2b. Instead of receiving individual sensor readings, the sink will be sent a few weighted sums of all the readings from which to restore the original data. To transmit the ith sum to the sink, s_1 multiplies its reading d_1 with a random coefficient ϕ_{i1} and sends the product to s_2. Upon receiving this message, s_2 multiplies its reading d_2 with a random coefficient ϕ_{i2} and then sends the sum $\phi_{i1}d_1 + \phi_{i2}d_2$ to s_3. Similarly, each node s_j contributes to the relayed message by adding its own product. Finally, the sink receives $\sum_{j=1}^{N} \phi_{ij}d_j$, a weighted sum of all the readings. This process is repeated using M sets of different weights so that the sink will receive M weighted sums.

With such design, all nodes transmit M messages and consume same the amount of energy. Each node only performs one addition and one multiplication in order to compute one weighted sum. Comparing Figure 13.2a,b, careful readers will observe that the first M nodes send more messages in CDG than in the baseline transmission, while the rest of the nodes send less messages in CDG. When N is large and M is much smaller than N, CDG can significantly reduce the total number of transmissions and save energy. The key problem now becomes whether the sink is able to restore N individual readings from M measurements, when M is far smaller than N. Fortunately, the compressive sensing theory has a positive answer to this question.

There are three main contributions in the proposed scheme. First, we extend the application of compressive sensing theory from one or a few sensors to large-scale multi-hop sensor networks. Beyond the basic idea, we propose a scheme which allows CDG to be practically applied to large sensor networks. Second, we carry out a theoretical analysis of the network capacity for CDG and validate the capacity gain of CDG through ns-2 simulations. Third and more importantly, we test CDG on two sets of real sensor data. The results show that CDG is practically applicable to various data gathering sensor networks. Even when sensor data exhibit little spatial correlations in which case conventional in-network compression approaches would fail, CDG is still able to reduce the traffic of the bottleneck node by two to three times and significantly prolong the network lifetime.

13.2 Related Work

The fundamental assumption of in-network data compression is that sensor nodes have spatial correlations in their readings. According to where the spatial correlation is utilized, we can classify existing in-network data compression techniques into two categories.

13.2.1 Conventional Compression

Conventional compression techniques utilize the correlation during the encoding process and require explicit data communication among sensors. Cristescu et al. [298] propose a joint entropy coding approach, where nodes use relayed data as side information to encode their readings. Again take the multi-hop route in Figure 13.2 as an example. First, node s_1 encodes its reading d_1 into message p_1 using $H(d_1)$ bits, where $H(d_1)$ is the entropy of d_1. Then, when s_2 receives p_1, it encodes its reading d_2 into message p_2 using $H(d_2|d_1)$ bits, where $H(d_2|d_1)$ is the conditional entropy. Since d_1 and d_2 are correlated, $H(d_2|d_1)$ is smaller than $H(d_2)$. Therefore, jointly encoded messages cost less bits than independently encoded messages.

The above approach utilizes data correlation only unidirectionally. If data are allowed to be communicated back and forth during encoding, nodes may cooperatively

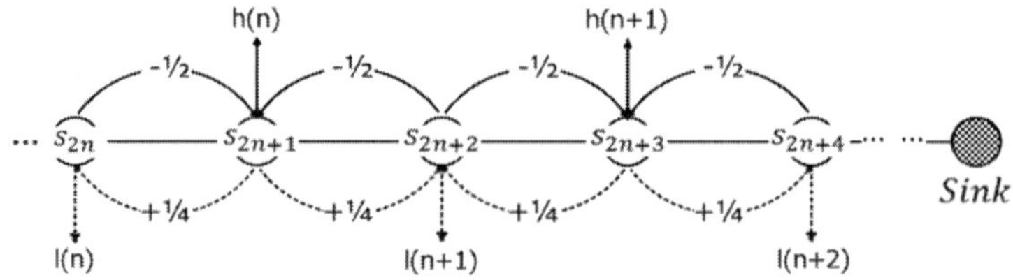

Figure 13.3 Cooperative wavelet compression.

perform transform to better utilize the correlation. Ciancio et al. [299] and Acimovic et al. [300] propose to compress piece-wise smooth data through a distributed wavelet transform. In doing so, even nodes first broadcast their readings. Upon receiving the readings from both sides, odd nodes compute the high pass coefficients $h(\cdot)$. Then, odd nodes transmit $h(\cdot)$ back and even nodes compute the low pass coefficients $l(\cdot)$. This process is illustrated in Figure 13.3. Although wavelet decorrelation can be performed for multiple levels, it is not suggested in distributed processing because of the communication overhead. After the transform, nodes transmit significant coefficients to the sink, usually in their raw form to avoid the complexity of entropy coding.

Quantization of a group of readings to one representative value is another form of conventional compression. The clustered aggregation (CAG) technique [301] forms clusters based on sensing values. By grouping sensors with similar readings, CAG only transmits one reading per group to achieve a predefined error threshold. Gupta et al. [302] exploit a similar idea. In each round of data gathering, it only involves a subset of nodes, which is sufficient to reconstruct data for the entire network.

There are two main problems with conventional compression techniques. First, the compression performance relies heavily on how the routes are organized. In order to achieve the highest compression ratio, compression and routing algorithms need to be jointly optimized. This has been proved to be a nondeterministic polynomial (NP)-hard problem [298]. Second, the efficiency of an in-network data compression scheme is not solely determined by the compression ratio, but also depends on the computational and communication overheads. However, joint entropy coding techniques perform complex computation in sensors, while transform-based techniques require a large amount of data exchanges.

13.2.2 Distributed Source Coding

Distributed source coding techniques [303–305] intend to reduce the complexity at the sensor nodes and utilize correlation at the sink. They are based on the Slepian-Wolf coding theory [306], which claims that compression of correlated readings, when separately encoded, can achieve the same efficiency as if they are jointly encoded, provided that the messages are jointly decoded. This important conclusion not only eliminates data exchanges, but decouples routing from compression. After encoding sensor readings independently, each node simply sends the compressed message along the shortest path to the sink to achieve the minimum energy consumption.

However, a prerequisite of Slepian-Wolf coding is that the global correlation structure needs to be known in order to allocate the appropriate number of bits to be used by each node. This is hard to fulfill in a large-scale wireless sensor network. In view of this, Yuen et al. [307] adopt a localized Slepian-Wolf coding scheme. Based on the assumption that sensors outside the immediate neighborhood have weak correlation in their readings, a node may only consider its data correlation with one-hop neighbors when determining the size of the encoded message. We will show that, for a set of real sensor data which do not satisfy this assumption, the localized coding scheme will fail to compress such data.

Distributed source coding techniques perform well for static correlation patterns. However, when the correlation pattern changes or abnormal events show up, the decoding accuracy will be greatly affected. Since detecting abnormal events is an important task of the sensor network, when an abnormal event is captured by a side node, the originally assigned number of bits will be inadequate to encode the reading, and cause decoding error at the sink. More seriously, when the abnormal reading appears at a main node, it will cause errors within a large range of reconstructed sensor readings.

13.2.3 Compressive Sensing

With the emergence of compressive sensing theory [295–297], we have seen a new avenue of research in the field of in-network data compression. The intuition behind compressive sensing is that a small number of linear projections (or so called measurements) of a sparse signal contains adequate information for its reconstruction. Mathematically, let $\mathbf{x} = [x_1 x_2 \ldots x_N]^T$ be a K-sparse signal, that is, there are only $K(K \ll N)$ nonzero elements in x_i's. Let us take M measurements of $\mathbf{x}$ through linear projection:

$$\mathbf{y} = \Phi \cdot \mathbf{x}, \tag{13.1}$$

where Φ is an $M \times N(M < N)$ random matrix, then $\mathbf{x}$ can be perfectly reconstructed from $\mathbf{y}$ under certain conditions.

Compressive wireless sensing (CWS) [308] appears to be able to reduce the latency of data gathering in a single hop network where every sensor can directly com municate with the sink. CWS delivers linear projections of sensor readings through synchronized amplitude-modulated analog transmissions. Due to the difficulties in analog synchronization, CWS is less practical for large-scale sensor networks. Rabbat et al. [309] leverages compressive sensing for data persistence, instead of data gathering, in a wireless sensor network. In an overview paper, Haupt et al. [310] also speculate the potential of using compressive sensing theory for data aggregation in a multi-hop wireless sensor network. However, no real scheme has been reported based on this initial idea.

When compressive sensing is applied to in-network data compression, it will bring a wealth of similar benefits as distributed source coding including the simple encoding process, saving of inter-node data exchange, and decoupling of compression from routing. Furthermore, compressive sensing has two additional advantages. First, it can deal with abnormal sensor readings gracefully. This advantage will be detailed in the next section. Second, data reconstruction is not sensitive to packet losses. In compressive sensing, all messages received by the sink are equally important. However, in distributed source coding, received data are predefined as main or side information. Losing main information will cause fatal errors to the decoder. All these desired merits make compressive sensing a promising solution to the data gathering problem in large-scale wireless sensor networks.

13.3 Compressive Data Gathering

The objective of compressive data gathering is two-fold: compress sensor readings to reduce global data traffic and distribute energy consumption evenly to prolong network lifetime. Similar to distributed source coding, the data correlation pattern shall be utilized on the decoder end. In addition, compression and routing are decoupled and therefore can be optimized separately.

13.3.1 Data Gathering

The intuition behind CDG is that higher efficiency can be achieved if correlated sensor readings are transmitted jointly rather than separately. We have given a simple example in Section 13.1, showing how sensor readings are combined while being relayed along a chain-type topology to the sink. In practice, sensors usually spread in a two-dimensional area, and the ensemble of routing paths presents a tree structure. Figure 13.4a shows a typical routing tree in a wireless sensor network. In this example, the sink has four children, each of them leading a subtree delimited by the dotted lines. Data gathering and reconstruction of CDG are performed on the subtree basis.

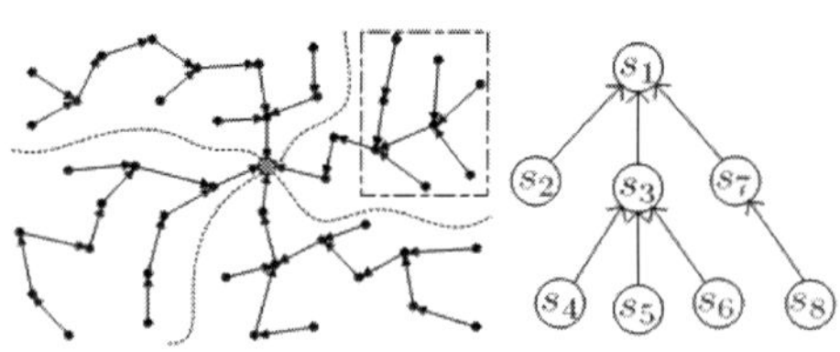

Figure 13.4 Data gathering in a typical routing tree.

In order to combine sensor readings while relaying them, every node needs to know its local routing structure. That is, whether or not a given node is a leaf node in the routing tree or how many children the node has if it is an inner node. To facilitate efficient aggregation, we have made a small modification to the standard ad hoc routing protocol: when a node chooses a parent node, it sends a "subscribe notification" to that node; when a node changes parent, it sends an "unsubscribe notification" to the old parent.

The data gathering process of CDG is illustrated through an example shown in Figure 13.4b. It is the detailed view of a small fraction of the routing tree marked in Figure 13.4a. After all nodes acquire their readings, leaf nodes initiate the transmission. In this example, s_2 generates a random number ϕ_{i2}, computes $\phi_{i2}d_2$, and transmits the value to s_1. The index i denotes the ith weighted sum ranging from 1 to M. Similarly, s_4, s_5, and s_6 transmits $\phi_{i4}d_4$, $\phi_{i5}d_5$, and $\phi_{i6}d_6$ to s_3. Once s_3 receives the three values from its children, it computes $\phi_{i3}d_3$, adds it to the sum of relayed values, and transmits $\sum_{j=3}^{6}\phi_{ij}d_j$ to s_1. Then, s_1 computes $\phi_{i1}d_1$ and transmits $\sum_{j=1}^{8}\phi_{1j}d_j$ to its parent. Finally, the message is forwarded to the sink, which contains the weighted sum of all readings in a subtree.

Assume that there are N nodes in a particular subtree, then the ith weighted sum can be represented by:

$$y_i = \sum_{j=1}^{N} \phi_{ij} d_j. \tag{13.2}$$

For this subtree, the sink obtains M weighted sums $\{y_i\}$, $i = 1,2,...M$. Mathematically, we have:

$$\begin{pmatrix} y_1 \\ y_2 \\ \vdots \\ y_M \end{pmatrix} = \begin{pmatrix} \phi_{11} & \phi_{12} & \cdots & \phi_{1N} \\ \phi_{21} & \phi_{22} & \cdots & \phi_{2N} \\ \vdots & & \vdots & \\ \phi_{M1} & \phi_{M2} & \cdots & \phi_{MN} \end{pmatrix} \begin{pmatrix} d_1 \\ d_2 \\ \vdots \\ d_N \end{pmatrix}. \tag{13.3}$$

In this equation, each column of $\{\phi_{ij}\}$ contains the series of random numbers generated at a corresponding node. In order to avoid transmitting this random matrix from sensors to the sink, we can adopt a simple strategy: before data transmission, the sink broadcasts a random seed to the entire network. Each node uses this seed and its unique identification to generate the series of pseudo-random numbers. This generation process can be reproduced by the sink to obtain the $\{\phi_{ij}\}$ matrix.

In Eq. (13.3), d_i $(i = 1,2,...N)$ is a scalar value. In a practical sensor network, each node is possibly attached with a few sensors of different type, for example, a temperature sensor and a humidity sensor. Thus sensor readings from each node become a multi-dimensional vector. To deal with this, we may separate readings of each dimension and process them respectively. Alternatively, since the random coefficients ϕ_{ij} are irrelevant to sensor readings, we may treat d_i as a vector. Then, the weighted sums y_i become vectors of the same dimension too.

When $M < N$, solving a set of M linear equations with N unknown variables is an ill-posed problem. However, sensor readings are not independent variables, rather, they are spatially correlated. Under this assumption, in the following section we will explain whether the set of linear equations are solvable, what requirements M should meet to solve them, and how these equations can be solved.

13.3.2 Data Recovery

13.3.2.1 Recovery of Spatially Correlated Data

According to the compressive sensing theory, a K-sparse signal can be reconstructed from a small number of measurements with a probability close to one. The weighted sums obtained in Eq. (13.3) are a typical type of measurements. Signal sparsity characterizes the correlations within a signal. An N-dimensional signal is considered as a K-sparse signal if there exists a domain in which this signal can be represented by K ($K \ll N$) nonzero coefficients. Figure 13.5a shows a 100-dimensional signal in its original time domain. Obviously, it is not sparse at all in this domain. Because of the signal correlation, it can be described more compactly in transform domains such as wavelet and discrete cosine transform (DCT). Figure 13.5b gives a representation of the same signal in DCT domain. We can see that there are only five nonzero DCT coefficients. Therefore, this signal is a 5-sparse signal in the DCT domain.

In densely deployed sensor networks, sensors have spatial correlations in their readings. Let N sensor readings form a vector $\mathbf{d} = [d_1\ d_2\ \dots\ d_N]^T$, then $\mathbf{d}$ is a K-sparse signal in a particular domain Ψ. Denote $\Psi = [\psi_1\ \psi_2\ \dots \psi_N]$ as the representation

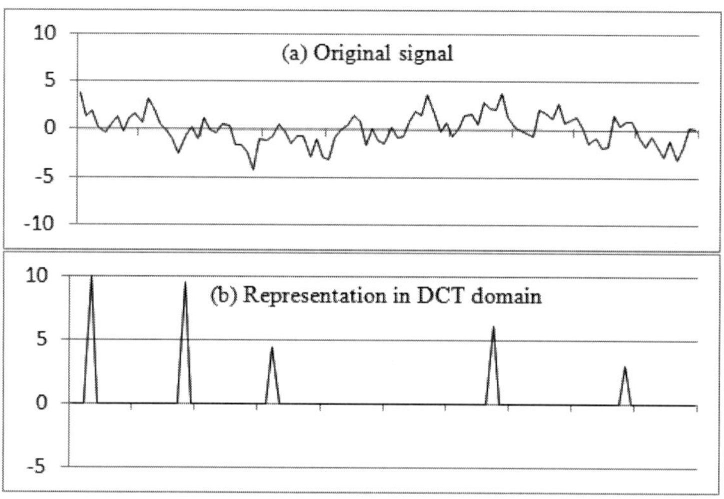

Figure 13.5 A 5-sparse signal in the DCT domain.

basis with vectors $\{\psi_i\}$ as columns, and $\mathbf{x} = [x_1, x_2, ... x_N]^T$ are the corresponding coefficients. Then, $\mathbf{d}$ can be represented in the Ψ domain as:

$$\mathbf{d} = \sum_{i=1}^{N} x_i \psi_i, \text{ or } \mathbf{d} = \Psi \mathbf{x}. \tag{13.4}$$

The compressive sensing theory tells that a K-sparse signal can be reconstructed from M measurements if M satisfies the following conditions [311]:

$$M \geq c \cdot \mu^2(\Phi, \Psi) \cdot K \cdot \log N, \tag{13.5}$$

where c is a positive constant, Φ is the sensing matrix as defined in Eq. (13.3), and $\mu(\Phi, \Psi)$ is the coherence between sampling basis Φ and representation basis Ψ. The coherence metric measures the largest correlation between any two elements of Φ and Ψ, and is defined as:

$$\mu(\Phi, \Psi) = \sqrt{N} \cdot \max_{1 \leq i, j \leq N} |\langle \phi_i, \psi_j \rangle|. \tag{13.6}$$

From Eq. (13.6), we can see that the smaller the coherence between Φ and Ψ is, the less measurements are needed to reconstruct the signal. In practice, using random measurement matrix is a convenient choice, since a random basis has been shown to be largely incoherent with any fixed basis, and $M = 3K \sim 4K$ is usually sufficient to satisfy Eq. (13.5).

With sufficient number of measurements, the sink is able to reconstruct sensor readings through solving an l_1-minimization problem:

$$\min_{\mathbf{x} \in \mathbb{R}^N} \|\mathbf{x}\|_{l_1} \quad s.t. \quad \mathbf{y} = \Phi \mathbf{d}, \ \mathbf{d} = \Psi \mathbf{x}. \tag{13.7}$$

In addition, for sparse signals whose random projections are contaminated with noise, reconstruction can be achieved through solving a relaxed l_1-minimization problem, where ε is a predefined error threshold:

$$\min_{\mathbf{x} \in \mathbb{R}^N} \|\mathbf{x}\|_{l_1} \quad s.t. \quad \|\mathbf{y} - \Phi \mathbf{d}\|_{l_2} < \varepsilon, \ \mathbf{d} = \Psi \mathbf{x}. \tag{13.8}$$

Suppose $\tilde{\mathbf{x}}$ is the solution to this convex optimization problem, then the proposed reconstruction of the original signal is $\tilde{\mathbf{d}} = \Psi \tilde{\mathbf{x}}$. Here, the Ψ matrix describes the correlation pattern among sensor readings. For example, if sensor readings are piece-wise smooth, we may use wavelet transform matrix as Ψ. As we have mentioned earlier, sensors do not use the correlation pattern during the data gathering process. This information is only used by the sink during data recovery.

13.3.2.2 Recovery of Data with Abnormal Readings

One of the main purposes of the sensor network is to monitor abnormal events. However, when abnormal events take place, the sparsity of sensor readings is

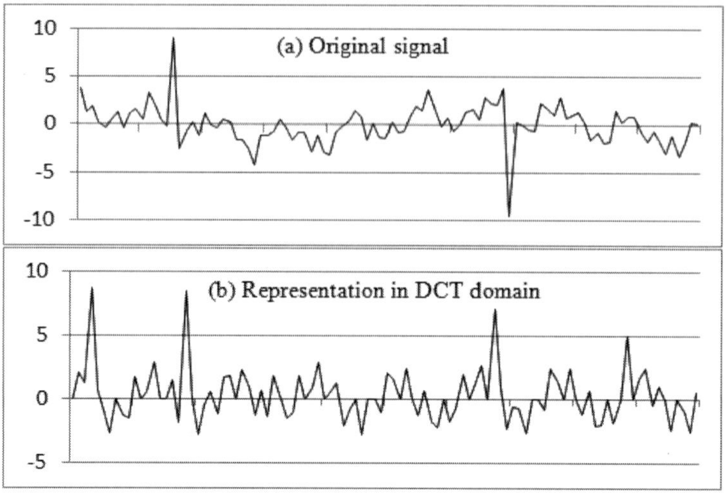

Figure 13.6 A signal with two abnormal readings.

compromised. As an example, Figure 13.6a differs from Figure 13.5a by only two abnormal readings. The corresponding DCT coefficients shown in Figure 13.6b are not sparse anymore. Therefore, the signal in Figure 13.6 is not sparse in either time domain or transform domain. In this situation, conventional compression techniques need to transmit significantly more data in order to reconstruct the original signal. Distributed source coding techniques will have a big degradation.

We have a better solution in compressive data gathering. Sensor data with abnormal readings can be decomposed into two vectors:

$$\mathbf{d} = \mathbf{d_0} + \mathbf{d_s}, \tag{13.9}$$

where $\mathbf{d_0}$ contains the normal readings, which are sparse in a certain transform domain, and $\mathbf{d_s}$ contains the deviated values of abnormal readings. Since abnormal readings are sporadic, $\mathbf{d_s}$ is a sparse signal in the time domain. Suppose the normal readings are sparse in Ψ domain, then Eq. (13.9) can be rewritten into:

$$\mathbf{d} = \Psi \mathbf{x_0} + I\mathbf{x_s}, \tag{13.10}$$

where I is the identical matrix, and both $\mathbf{x_0}$ and $\mathbf{x_s}$ are sparse. We can see that signal $\mathbf{d}$ is decomposed into two signals which are sparse in different domains. We can construct an overcomplete basis $\Psi' = [\Psi\ I]$, then $\mathbf{d}$ should be sparse in Ψ' domain:

$$\mathbf{d} = \Psi' \mathbf{x},\ \mathbf{x} = [\mathbf{x_0}^T \mathbf{x_s}^T]^T. \tag{13.11}$$

Incorporating Eq. (13.11) into Eq. (13.7) or Eq. (13.8), the signal recovery with abnormal readings can be solved similarly by the l1-norm optimization. Donoho et al. [312] showed the possibility of stable recovery under a combination of sufficient sparsity and favorable structure of the overcomplete system. Moreover, they also proved that stable recovery of the sparse signal in an overcomplete dictionary also works for noisy data, and the optimally sparse approximation to the noisy data, to within the noise level, differs from the optimally sparse decomposition of the ideal noiseless signal by at most a constant multiple of the noise level.

Suppose $\tilde{\mathbf{x}}$ is a vector of length $2N$, and is the solution to the l_1-minimization problem defined in Eq. (13.8) when an overcomplete dictionary is used. Similarly, the original sensor readings can be reconstructed by $\tilde{\mathbf{d}} = \Psi'\tilde{\mathbf{x}}$. Denote $\tilde{\mathbf{x_s}}$ as an N-dimensional vector composed of the last N elements of $\tilde{\mathbf{x}}$, then the nonzero values in $\tilde{\mathbf{x_s}}$ indicate the positions of abnormal readings.

13.4 Network Capacity of Compressive Data Gathering

The previous section illustrated how to gather and recover sensor readings acquired in one time instance. This section will investigate the benefit of CDG from the viewpoint of network capacity, that is, how frequent CDG allows sensors to acquire data while ensuring all readings can be transmitted to the sink. The capacity of a data gathering network is defined as follows.

Definition 8 (Network Capacity) *We shall define that a rate λ is achievable in a data gathering sensor network, if there exists a time instance t_0 and duration T such that during $[t_0, t_0 + T)$ the sink receives λT bits of data generated by each of the sensors s_i, $i = 1, 2, ...N$. Then, network capacity C is defined as the supremum of the achievable rate, or $C = sup\{\lambda\}$.*

Different from the pioneering work on network capacity analysis [313], the traffic pattern in our study is many-to-one. We let all sensors generate data at the same rate, and assume that sensor readings acquired at the same time instance are K-sparse.

13.4.1 Network Capacity Analysis

We assume a discal sensing area in which N sensor nodes are uniformly distributed, and the sink is located in the middle of the disk. All sensor nodes and the sink communicate over single frequency shared radio channel, accessed through time-division multiple access (TDMA) control. We denote W as the amount of data a node transmits in one time slot, and restrict that a node cannot transmit and receive at the same time.

Let $\{X_k,\ k \in V\}$ be the subset of nodes simultaneously transmitting over the shared channel in a specific time slot. Then a successful transmission from X_i, $i \in V$ to X_j can be defined under two interference models.

Definition 9 (Protocol Model) *Transmission from node X_i to X_j is successful under protocol model if and only if the following two conditions are satisfied:*

- $\|X_i - X_j\| \leq r$,
- $\|X_k - X_j\| > (1+\delta)r$, $\delta > 0$ *for* $k \in V - \{i\}$.

The first condition requires that the two communicating nodes are within a distance r. The second condition requires that the receiving node is at least $(1+\delta)r$ away from any other transmitting nodes.

Definition 10 (Physical Model) *Transmission from node X_i to X_j is successful under physical model if and only if:*

$$\frac{\frac{P_i}{\|X_i - X_j\|^{\alpha}}}{N_G + \sum_{k \in V, k \neq i} \frac{P_k}{\|X_k - X_j\|^{\alpha}}} \geq \beta,$$

where P_i is the transmission power for X_i, α is the fading parameter, and N_G is noise power level. The expression on the left is the signal to interference and noise ratio (SINR) at the receiving node. A successful transmission under physical model requires the SINR to be greater than a predefined threshold β.

13.4.1.1 Capacity under the Protocol Model

The capacity under protocol model can be analyzed in a similar way as Marco et al. [314]. Let us first recall the following lemma.

Lemma 13.1. *N nodes are uniformly distributed in a region of area A. When N is large, the number of nodes n within a sub-region R of area A_R can be bounded with high probability.*

$$Pr\left(\frac{NA_R}{A} - \sqrt{\alpha_N N} \leq n \leq \frac{NA_R}{A} + \sqrt{\alpha_N N}\right) \to 1,$$
$$as\ N \to \infty.$$

Sequence α_N is chosen such that $\alpha_N \to \infty$ as $N \to \infty$, and $\lim_{N\to\infty} \frac{\alpha_N}{N} = \varepsilon$, where ε is positive but arbitrarily small.

Proof. Each of the N nodes has the same probability A_R/A to fall in region R. Therefore, n follows binomial distribution with the mean being $\mu_n = \frac{NA_R}{A}$ and the variance being $\delta_n^2 = \frac{NA_R}{A}(1 - \frac{A_R}{A})$. According to Chebychev's inequality:

$$Pr(|n - \mu_n| \geq \sqrt{\alpha_N N}) \leq \frac{\delta_n^2}{\alpha_N N} = \frac{\frac{A_R}{A} \cdot (1 - \frac{A_R}{A})}{\alpha_N}. \tag{13.12}$$

The probability goes to 0 as $N \to \infty$ and $\alpha_N \to \infty$.

Theorem 13.1. *In a wireless sensor network with N uniformly distributed nodes, compressive data gathering can achieve network capacity of* $\lambda \geq \frac{W}{M}\frac{\pi r^2-\sqrt{\varepsilon}}{\pi(2+\delta)^2 r^2+\sqrt{\varepsilon}}$ *with a probability close to 1 as* $N \to \infty$*, where* ε *is arbitrarily close to 0, and M is the number of random measurements. Usually* $M = c_1 K$*, and* c_1 *is a constant in the range of* $[1,4]$.

Proof. Consider a node in transmission. According to Definition 9, the distance from any interfering source to this node is at most $(2+\delta)r$. In other words, all the interfering sources are contained in a disk of area $A_{R_1} = \pi(2+\delta)^2 r^2$. Based on Lemma 13.1, the number of nodes in this region, denoted by n_{itf}, is less than n_1 with high probability:

$$n_{itf} \leq n_1 = \frac{NA_{R_1}}{A} + \sqrt{\alpha_N N}. \tag{13.13}$$

Next we build the contention graph of the network by connecting interfering nodes. With high probability, the maximal node degree in the contention graph is $n_1 - 1$. According to the graph coloring theory, all nodes can be colored with at most n_1 different colors. If we associate each color with a transmission slot, every node gets one chance to transmit in n_1 slots. Therefore, the average transmission rate of each node is:

$$\gamma = \frac{W}{n_1}. \tag{13.14}$$

Then consider the one-hop neighbors of the sink. They are contained in a disk centered at the sink and with a radius of r. The area of the disk is $A_{R_2} = \pi r^2$. According to Lemma 1, the number of nodes in this region, denoted by n_2, can be bounded with high probability:

$$\frac{NA_{R_2}}{A} - \sqrt{\alpha_N N} \leq n_2 \leq \frac{NA_{R_2}}{A} + \sqrt{\alpha_N N}. \tag{13.15}$$

Recall that compressive data gathering is performed on a subtree basis. We shall adopt an appropriate routing protocol such that all subtrees are roughly of equal size. For simplicity, we consider the size of each subtree is $N_p = \frac{N}{n_2}$. Since the sensor data from the entire network is K-sparse, when $N \to \infty$, we can consider that each subset of the nodes are proportionally sparse, that is, K/n_2-sparse. The number of random measurements needed to reconstruct data is M/n_2 per subtree. To achieve the rate λ, the transmission rate of the subtree root should be $M\lambda/n_2$. Take Eq. (13.14) into account, we have:

$$\frac{W}{n_1} = \frac{M\lambda}{n_2}. \tag{13.16}$$

Substituting Eq. (13.13) and Eq. (13.15) into Eq. (13.16), we have:

$$\lambda = \frac{W}{M}\frac{n_2}{n_1} \geq \frac{W}{M}\frac{\frac{NA_{R_2}}{A} - \sqrt{\alpha_N N}}{\frac{NA_{R_1}}{A} + \sqrt{\alpha_N N}} = \frac{W}{M}\frac{\frac{A_{R_2}}{A} - \sqrt{\varepsilon}}{\frac{A_{R_1}}{A} + \sqrt{\varepsilon}}$$
$$= \frac{W}{M}\frac{\pi r^2 - \sqrt{\varepsilon}}{\pi(2+\delta)^2 r^2 + \sqrt{\varepsilon}}. \tag{13.17}$$

As $N \to \infty$, $\sqrt{\varepsilon} \to 0$, the lower bound of achievable capacity is arbitrarily close to $\frac{W}{M(2+\delta)^2}$.

13.4.1.2 Capacity under the Physical Model

Without loss of generality, we assume the following constraints for the physical model:

- All nodes transmit with equal and finite power P_0.
- All noises are of the same variance. Therefore, for a given small positive number η, there exist a noise level N_0 such that $prob(N_0 > N_G) < \eta$.
- Given α and β, P_0 is chosen such that the network is a connected graph when the noise level is N_0.

Theorem 13.2. *In a wireless sensor network with N uniformly distributed nodes, compressive data gathering can achieve network capacity of* $\lambda \geq \frac{W}{M}\frac{\pi r_0^2 - \sqrt{\varepsilon}}{\pi(2+\delta_0)^2 r_0^2 + \sqrt{\varepsilon}}$ *with a probability close to 1 as* $N \to \infty$, *given* $r_0 < \sqrt[\alpha]{\frac{P_0}{\beta N_0}}$ *and* $\delta_0 > \sqrt[\alpha-1]{\frac{2\pi\beta c_2}{1-\beta r^\alpha N_0/P_0}} - 1$.

Proof. Theorem 13.1 gives the network capacity under protocol model. We will prove that when $r = r_0$ and $\delta = \delta_0$, a feasible transmission schedule under protocol model is also feasible under physical model.

First, we restrict the communication to nodes within a distance of r_0. When node X_i transmits data to node X_j, the SINR at X_j is:

$$SINR_j = \frac{\frac{P_0}{|X_i - X_j|^\alpha}}{N_0 + \sum_{k \in V, k \neq i} \frac{P_0}{|X_k - X_j|^\alpha}}. \tag{13.18}$$

Denote P_s as the received signal strength and P_f as the interference strength in Eq. (13.18). Since $|X_i - X_j| < r_0$, we have:

$$P_s = \frac{P_0}{|X_i - X_j|^\alpha} > \frac{P_0}{r_0^\alpha}. \tag{13.19}$$

A necessary condition is that when $P_f = 0$, the selection of r_0 should ensure $SINR_j > \beta$. This can be satisfied if $r_0 < \sqrt[\alpha]{\frac{P_0}{\beta N_0}}$.

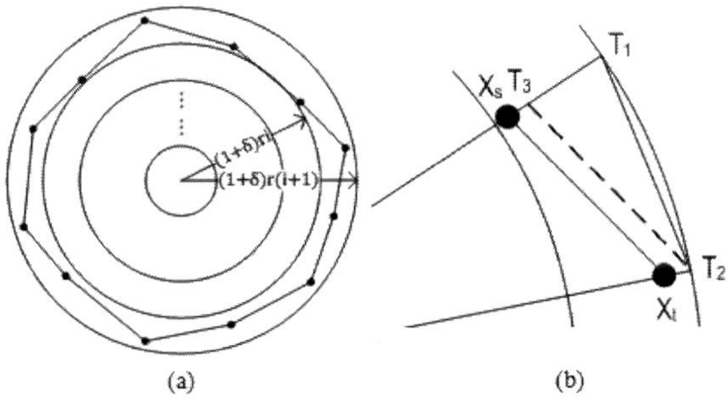

Figure 13.7 Connecting adjacent transmitting nodes in an annulus.

Next, let us look at the interference part P_f. A feasible schedule under protocol model ensures that there is no other simultaneous transmitter in the circular area centered at X_j and with a radius of $(1+\delta)r$. The interference comes from the transmitters outside this region. Divide the sensing region by concentric circles C_i, $i = 1,2,...$ centered at X_j. The radius of circle C_i is $r_i = (1+\delta)ri$. Denote A_i as the annulus formed by C_i and C_{i+1}. Next, we quantify the interference to X_j caused by the transmitters in each annulus.

Denote a_i as the number of simultaneous transmitters within a particular annulus A_i. Since the distance from X_j to any node in this annulus is larger than $(1+\delta)ri$, the interference from this annulus is:

$$P_f(A_i) = \sum_{k,X_k \in A_i} \frac{P_0}{|X_k - X_j|^\alpha} < \frac{a_i P_0}{((1+\delta)ri)^\alpha}. \tag{13.20}$$

$P_f(A_i)$ can be bounded once a_i is bounded. In doing so, we connect adjacent transmitters clockwise with line segments, as Figure 13.7a shows. Figure 13.7b gives an enlarged view of two adjacent transmitting nodes X_s and X_t. Connect the center of circle with the two nodes and extend the lines so that they intersect C_{i+1} at points T_1 and T_2. From T_2 draw a line parallel to $\overline{X_s X_t}$ and intersect $\overline{X_s T_1}$ at point T_3. Then we have:

$$|\overline{X_s X_t}| \leq |\overline{T_2 T_3}| < |\overset{\frown}{T_1 T_2}| + (1+\delta)r. \tag{13.21}$$

A feasible schedule under protocol model ensures that the distance of each line segment is at least $(2+\delta)r$. Summing up all the segments in annulus A_i and using the inequality in Eq. (13.21), we have:

$$\begin{aligned}(2+\delta)ra_i &\leq \sum_{s,t} |\overline{X_s X_t}| \\ &< 2\pi(1+\delta)r(i+1) + (1+\delta)ra_i \\ &\Rightarrow a_i < 2\pi(1+\delta)(i+1).\end{aligned} \tag{13.22}$$

Substitute Eq. (13.22) into Eq. (13.20), and sum up the interferences from all annuluses, we have:

$$\begin{aligned} P_f &= \sum_{i=1}^{\infty} P_f(A_i) < \sum_{i=1}^{\infty} \frac{2\pi P_0 (1+\delta)(i+1)}{((1+\delta)ri)^{\alpha}} \\ &= \frac{2\pi P_0}{r^{\alpha}(1+\delta)^{\alpha-1}} \sum_{i=1}^{\infty} \left(\frac{1}{i^{\alpha-1}} + \frac{1}{i^{\alpha}} \right) \\ &= \frac{2\pi P_0 \left(\zeta(\alpha-1) + \zeta(\alpha)\right)}{r^{\alpha}(1+\delta)^{\alpha-1}}, \end{aligned} \tag{13.23}$$

where $\zeta(\cdot)$ is the Riemann Zeta function. When $\alpha > 2$, $\zeta(\alpha) < \frac{\pi^2}{6}$, and $\zeta(\alpha-1)$ converges to a constant. Denote $c_2 = \zeta(\alpha) + \zeta(\alpha-1)$. Then, when $r = r_0$ and $\delta = \delta_0 > \sqrt[\alpha-1]{\frac{2\pi\beta c_2}{1-\beta r^{\alpha} N_0/P_0}} - 1$, Eq. (13.23) can be written into:

$$P_f < \frac{P_0}{r_0^{\alpha}\beta} - N_0. \tag{13.24}$$

Substitute Eq. (13.19 and Eq. (13.24) into Eq. (13.18), we obtain $SINR_j > \beta$. This proves that a feasible scheduling under protocol model with $r = r_0$ and $\delta = \delta_0$ is also feasible under physical model. Therefore, the network capacity achieved under protocol model when $r = r_0$ and $\delta = \delta_0$ can also be achieved under physical model.

13.4.1.3 Capacity Gain over Naive Transmission

Collorary 13.1 *In a wireless sensor network with N uniformly distributed nodes, CDG can achieve a capacity gain of N/M over baseline transmission under both interference models, given that sensor readings are K-sparse, and $M = c_1 K$.*

Denote λ_1 as the network capacity of baseline transmission. It is achieved when every node is allowed to transmit once every n_1 slots, and traffic is evenly distributed among n_2 one-hop neighbors of the sink. Then we have $\frac{W}{n_1} = \frac{N\lambda_1}{n_2}$. Denote λ_2 as the network capacity of CDG. If the same transmission schedule and routing structure are adopted, we have $\frac{W}{n_1} = \frac{M\lambda_2}{n_2}$. From these two equations, we can conclude that CDG can achieve a capacity gain of N/M over baseline transmission.

13.4.2 NS-2 Simulation

The network capacity analysis is based on scheduled medium access control (MAC). In practice, the computational and communication overhead of MAC scheduling is too high. Contention-based MAC is more often adopted in wireless sensor networks.

Table 13.1 Simulation parameters.

MAC protocol	802.11
Physical data rate	2 Mbps
Transmission range	15 meters
Interference range	25 meters
Payload size	20 bytes
RTS/CTS status	OFF
Retry limit	7
IFQ length	200
K/N (data sparsity)	0.05
$c_1 = M/K$	4

In order to understand how CDG performs in practical settings, we evaluate its performance through ns-2 [315] simulations and compare it with baseline transmission on two typical topologies: chain [316] and grid topologies [317]. Table 13.1 lists the main parameters used in the simulation. We adopt 802.11 instead of ZigBee because the implementation of 802.11 in ns-2 is well established.

The payload size of each packet is 20 bytes for both baseline transmission and CDG. Although both approaches can combine multiple messages in one packet and improve transmission efficiency, we keep only one message per packet because we are only interested in the comparison between them. Data sparsity is assumed to be 5%. For example, when $N = 1000$, $K = 50$, and we assume that the sink can recover the original data from $M = 200$ random measurements. In the best case, CDG should achieve a capacity gain of $N/M = 5$.

13.4.2.1 Chain Topology

The chain topology is composed of 1000 sensors and one sink locating at one extreme of the chain. The distance between any two adjacent nodes are 10 meters. Under the given transmission and interference range, nodes can only communicate with adjacent nodes, and may cause interfere to two-hop neighbors.

In the simulation, we vary the input interval and evaluate how output interval and packet loss ratio change accordingly. In general, as the input interval decreases, the output interval decreases and the packet loss ratio increases. However, if an input interval is not achievable, the output interval will cease to decrease and may slightly increase as a result of congestion collapse. We may infer the network capacity from the minimum achieved output interval.

Figure 13.8a shows that the minimum output interval of baseline transmission is 10.6 seconds per message, and it is achieved when the input interval is 10.2 seconds per message. There is a small gap between these two values because of network jitters and packet losses. Figure 13.8b shows the performance of CDG. The minimum output interval is 2.11 seconds per message achieved when the input interval is 1.92

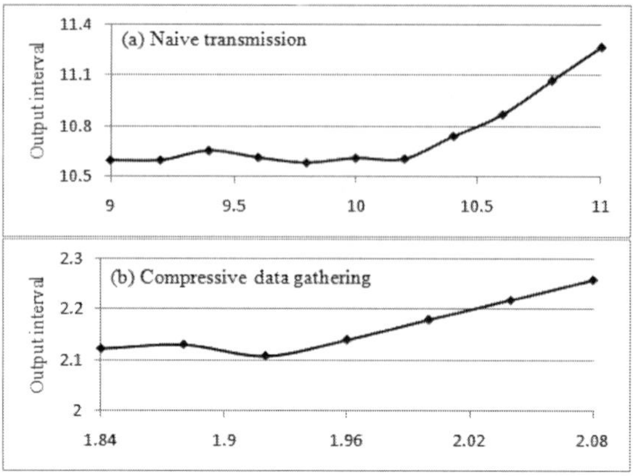

Figure 13.8 Output interval versus input interval in chain topology.

seconds per message. We can see that CDG can achieve a capacity gain of 5 over baseline transmission.

In addition, the packet loss ratio of CDG is zero when the input interval is 1.92 seconds per message and above. In contrast, even when the network is not overloaded, baseline transmission incurs a constant packet loss ratio between 3% to 4% as a result of traffic burst.

In this chain topology, CDG introduces an initial delay of 1.80 seconds. This is because transmission starts from the leaf node, which is 1000-hops away from the sink. This initial delay does not affect the network capacity because transmitting readings acquired at different time instance can be pipelined.

13.4.2.2 Grid Topology

The grid topology contains 1089 nodes in 33 rows by 33 columns. The distance between adjacent nodes in the same row or column is 14 meters. Therefore, any node not at the border of the network can communicate with four neighbors. Figure 13.9 shows a typical tree on the grid topology. The sink is in the middle of the network and four subtrees are represented by four different colors. The subtrees contain a similar number of sensor nodes, though not exactly the same. In the simulation, we assume that data from each subtree can be reconstructed from 55 random measurements.

Different from the chain topology where the routing path is deterministic, the grid topology produces changing routing trees in each test run. Therefore, we run three independent tests for each parameter setting and present the average results. In each test run, ten messages per node are collected at given intervals.

Figure 13.10a shows that baseline transmission achieves the minimum output interval of 5.93 seconds per message when the input interval is 4.7 seconds per message. Figure 13.10b shows that CDG achieves the minimum output interval of 2.54 seconds per message when the input interval is 2.2 seconds per message. The

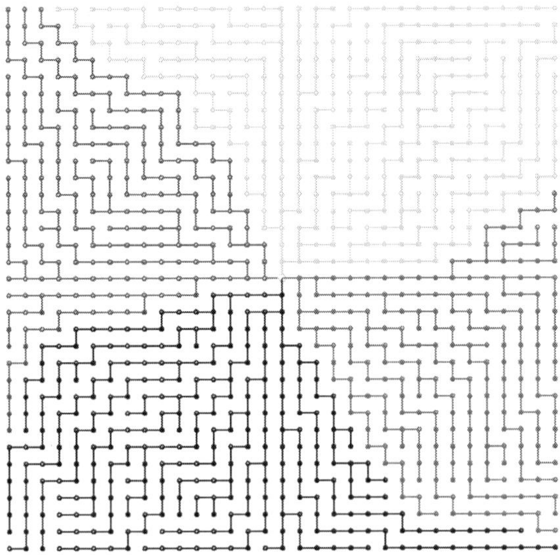

Figure 13.9 Grid topology and a typical routing tree.

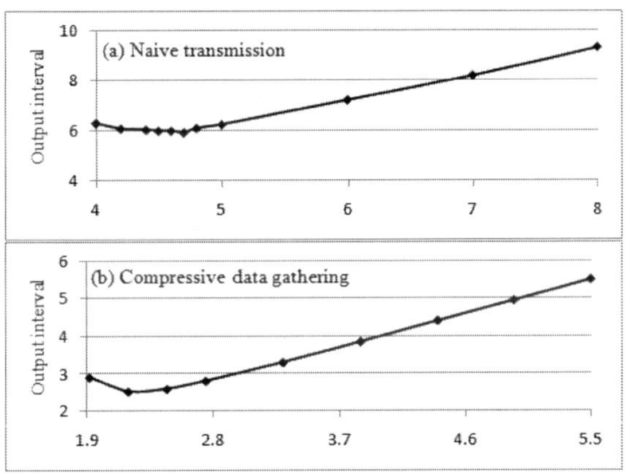

Figure 13.10 Output interval versus input interval in grid topology.

capacity gain is 2.3 instead of 5. The reason is that in contention based MAC, the transmission slots allocated to each node are not even. Nodes with heavier loads get more time slots to transmit. Therefore, baseline transmission transmits faster than what is assumed in scheduled MAC.

Figure 13.11 compares the packet loss ratio of the two approaches. Similar to the results in chain topology, CDG achieves near-zero loss ratio when the network is not overloaded. In contrast, the packet loss ratio in baseline transmission is much higher. Even when the input interval is 10 seconds per message, the packet loss ratio is still higher than 20%. In the grid topology, the initial delay of CDG is neglectable

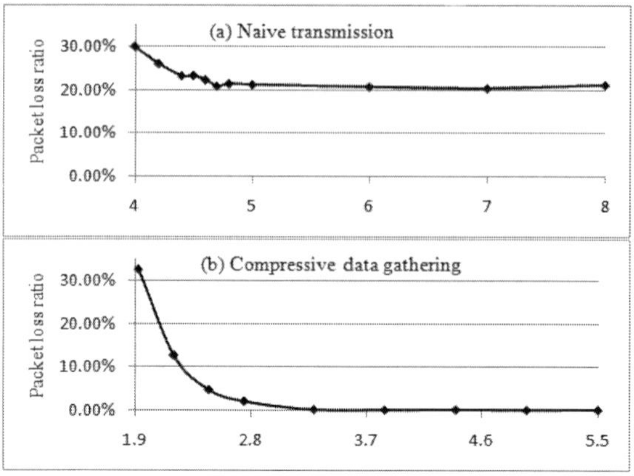

Figure 13.11 Packet loss ratio in grid topology.

because the tree depth is 32 hops. In our simulations, the average initial delay is less than 0.1 second.

13.5 Experiments on Real Data Sets

The previous section demonstrated the efficiency of CDG under the assumption that data are sparse and can be reconstructed from ideal random measurements. This section will show that sensor data are indeed sparse in reality. Further, data reconstruction is highly robust and efficient although real data are contaminated with noise.

13.5.1 CTD Data from the Ocean

The CTD (Conductivity, Temperature, and Depth) data come from National Oceanic and Atmospheric Administration's (NOAA) National Data Buoy Center (NDBC). CTD is a shipboard device consisting of many small probes. When collecting data, it is lowered down to the seafloor, and then measures data as it ascends. Although the CTD data are collected by one moving instrument, they demonstrate the same properties as if they were collected by a collection of sensors.

We look into the temperature data collected in the Pacific Sea at (7.0N, 180W) on March 29, 2008 [318]. The data set contains 1000 readings obtained at different sea depths. We plot the original data by a solid curve in Figure 13.12a. Since the readings are piece-wise smooth, they should be sparse in the wavelet domain. Figure 13.12b shows the 1000 coefficients after a 6-level 5/3 wavelet decorrelation. There are only 40 coefficients whose absolute value is larger than 0.2, accounting for only 4.0% of

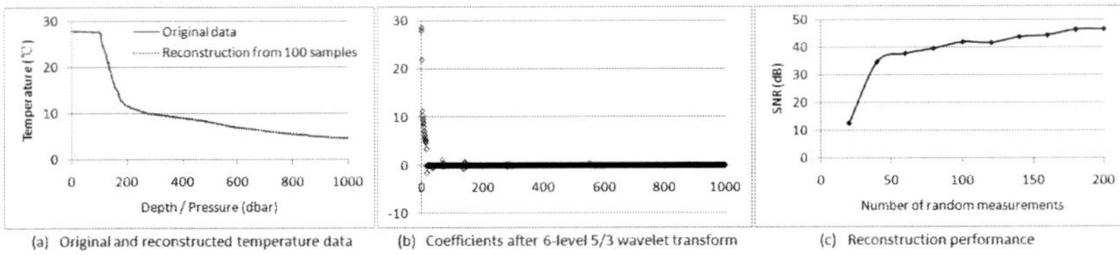

Figure 13.12 Results on temperature data from the Pacific Ocean.

the total coefficients. Although the rest of the coefficients are not strictly zero, we may set $K = 40$.

The compressive sensing theory suggests that data can be reconstructed with high probability from $M = 3K - 4K$ random measurements. Figure 13.12c shows the reconstruction performance with different numbers of random measurements. Each indicated data point is averaged over 10 test runs to avoid fluctuations. Apparently, the reconstruction precision increases as M increases. A steep rise is observed in both figures when M becomes greater than K. When $M = K \sim 40$, a reasonable reconstruction SNR of 35 dB can be achieved. This translates to a precision over 98%. When $M = 100$ and $M = 200$, the reconstruction precision is 99.2% (41.9 dB) and 99.5% (46.5 dB). The black dotted curve in Figure 13.12a shows the reconstructed data when $M = 100$.

13.5.2 Temperature in the Data Center

A contemporary practical application of wireless sensor networks is to monitor server temperatures in data centers. The temperature is an indication of server load and abnormal readings in temperature usually sound a note of warning. The sensor data used in this research are collected from a fraction of a data center as shown in Figure 13.13. Each rectangular shape represents a rack and the oval shape indicates a sensor placed at the top, middle, and bottom of the rack. As the figure shows, most of the racks are equipped with three sensors while some racks are not monitored and a few others have one or two malfunctioned sensors. There are 498 sensors in total. The data are measured every 30 seconds and transmitted to a sink through a baseline scheme. We analyze these data offline to see how much traffic would be reduced if CDG was used. In this network, each node only communicates with adjacent nodes. For simplicity, we assume that all 498 sensors form one subtree to the sink. The energy gain over a baseline scheme is similar if sensors form two or more subtrees.

An important observation on this set of data is that sensor readings exhibit little spatial correlations. Although the racks are physically close to each other, temperature readings are dominated by server loads instead of ambient temperature. Figure 13.14 plots a snapshot of the sensor readings. For clarity, we only show the sensor readings from the bottom of each rack (167 sensors in total) and put the data of each column side by side. Obviously, these data are not sparse in any intuitively known

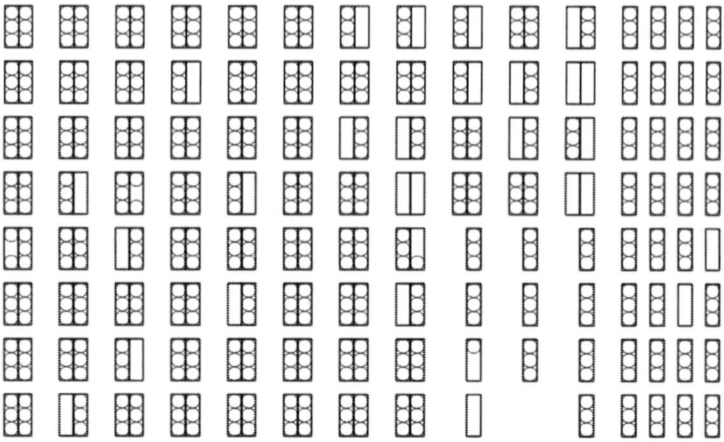

Figure 13.13 Rack and temperature sensor locations.

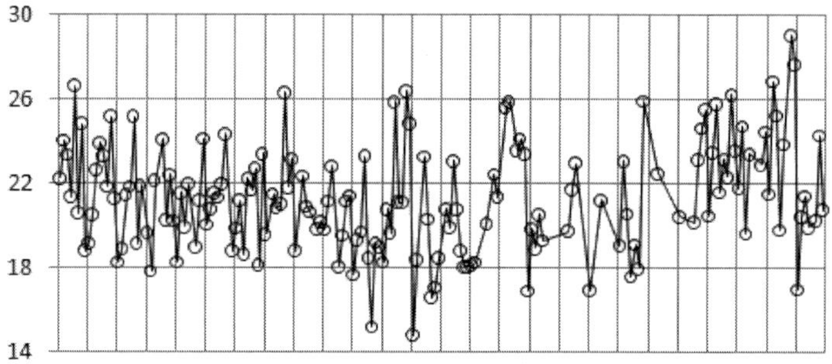

Figure 13.14 Temperature data of the lowest slot.

domain. We have also checked the entire data set containing sensor readings from all 498 sensors and they are not apparently sparse either. Therefore, conventional compression mechanisms will fail in this situation.

In fact, since the 498 sensors all take values between 10 to 30 degrees centigrade, we have elegantly reorganized d_i into an apparently sparse signal. In particular, we sort d_i in ascending order according to their sensing values at a particular moment t_0. The resulting $\mathbf{d}$ vector is piece-wise smooth and sparse in the wavelet domain. Furthermore, since server temperatures do not change violently, sensor readings collected within a relatively short time period can also be regarded as piece-wise smooth if organized in the same order. Figure 13.15a and Figure 13.16a show the ordered sensor readings 10 minutes and 30 minutes after t_0, respectively. They are generally in ascending order with only some small fluctuations. There are also a few significant spikes indicating abnormal temperature readings.

Based on proposed compressive data gathering scheme, we are able to reconstruct such noisy sparse signals with spikes from $M (M < N)$ random measurements.

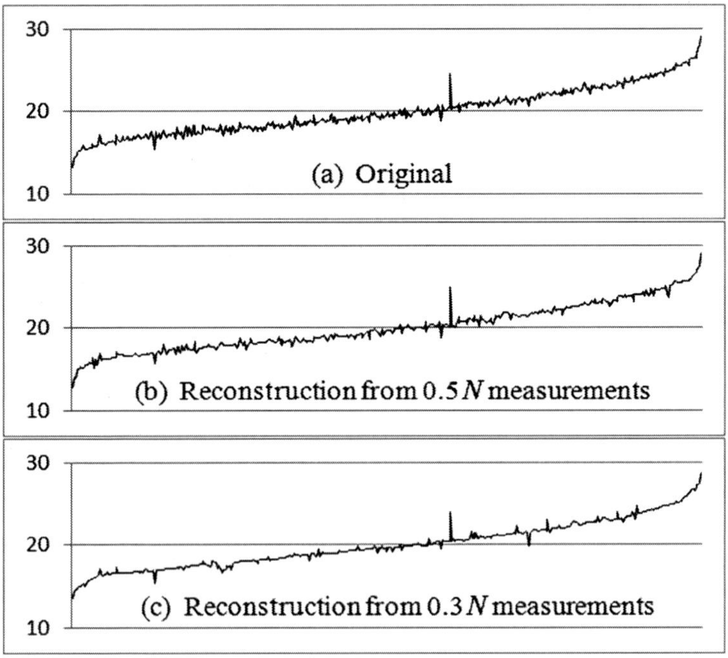

Figure 13.15 Original and reconstructed sensor readings at $t = t_0 + 10$.

Figure 13.15b,c and Figure 13.16b,c show the reconstruction results from $M = 0.5N$ and $M = 0.3N$ measurements at two time instances. The average reconstruction precision is over 98%. More importantly, the abnormal readings are accurately captured.

To cope with the situation that temporal correlation becomes weak when the time interval increases, we can refresh the ordering of d_i periodically. In particular, for every one or two hours, the sink requests $M(M = N)$ random measurements in one data gathering process. When $M = N$, the set of equations in Eq. (13.3) is solvable and the sink is able to obtain the exact values of d_i. Then, the sink can re-sort d_i and use this new ordering for data reconstruction in the subsequent hour or two.

We would like to point out that both conventional compression and distributed source coding are unable to exploit this type of sparsity, which is observed only at certain reshuffled ordering. In conventional compression, explicit data communication is required between correlated nodes. If correlated nodes are not physically close to each other, the communication between them may take multiple hops. This introduces high overheads and makes a compression procedure costly. In distributed source coding, nodes are classified into main nodes and side nodes. The sink allocates an appropriate number of bits to each node according to the correlation pattern. However, if the correlation pattern is based on changing sensor ordering, the sink needs to carry out these two tasks and communicate the results to every single node periodically. In contrast, the data gathering process in CDG is unaffected, even when

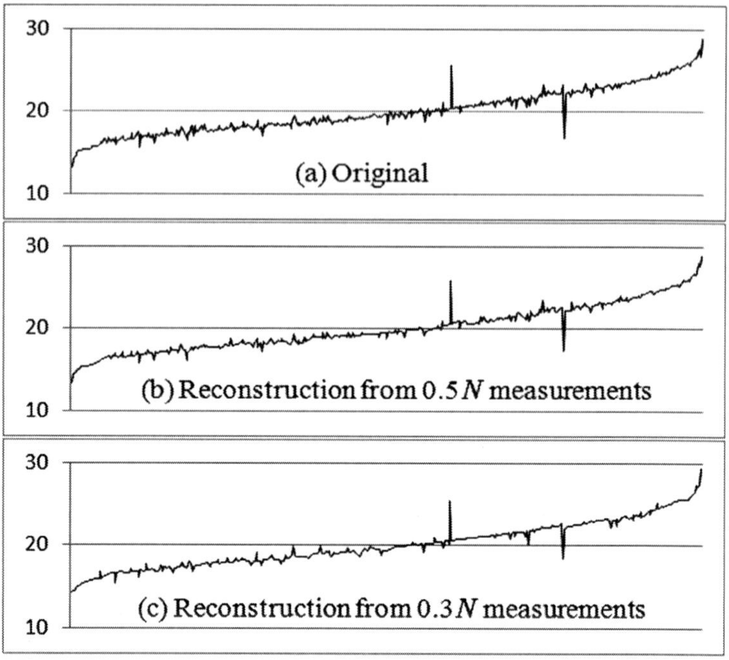

Figure 13.16 Original and reconstructed sensor readings at $t = t_0 + 30$.

the ordering of d_i changes. The knowledge of correlation is only used in the data reconstruction procedure.

Recall that CDG solves an l_1-minimization problem defined in Eq. (13.8) to reconstruct data. In previous sections, we have discussed how to select the Ψ matrix such that sensor readings $\mathbf{d}$ can be represented by a sparse vector $\mathbf{x}$ in Ψ domain. This section shows how $\mathbf{d}$ can be reorganized to be a sparse signal. This unprecedented flexibility of CDG demonstrates how CDG can achieve a compression ratio of two to three at bottleneck nodes when other conventional mechanisms fail.

13.6 Summary

We have described a novel scheme for energy efficient data gathering in large-scale wireless sensor networks based on the compressive sensing theory. We believe this is the first complete design to extend the original idea of the compressive sensing theory in combining data acquisition and efficient representation into one single process from single and a few sensors to massive sensor networks. Through such a combination, we have shown that the proposed scheme is not only able to achieve the needed sensor data compression; hence, lowering the global communication cost, but also to accomplish the desired load balancing; hence, extending the network lifetime, for large-scale sensor networks. In essence, the proposed scheme converts the traditional compress-then-transmit process into a compressive gathering (compress-with-transmission) process to address the two major technical challenges that today's large-scale sensor networks are facing. In addition, compressive data gathering provides an elegant solution to the tough problem of coping with abnormal reading within sensor data.

In the development of the proposed scheme, we have carried out the analysis of capacity for wireless sensor network when compressive data gathering is adopted. We have shown that CDG can achieve a capacity gain of N/M over baseline transmission. We have also designed ns-2 simulations to validate the proposed scheme when contention-based MAC is used. Furthermore, numerical studies based on real sensor data not only verified data sparsity in practical data acquisition, but also demonstrated the efficiency and robustness of the sensor data reconstruction with and without abnormal readings.

It should be noted that the proposed CDG is not suitable for small scale sensor networks when signal sparsity may not be prominent enough and the potential capacity gain may be too small. The proposed CDG is also more effective for networks with a stable routing structure. This is because frequent node failure or dynamic route change will lead to high control overhead that potentially cancel out the gain from data compression. We are currently investigating the extension of CDG to more challenging networking scenarios and the exploitation of fault tolerance of the compressive sensing principles to achieve more robust performance in sensor data gathering.

Chapter 14
Compressive Modulation

14.1 Introduction

In digital communications, as we have discussed in Chapter 3, modulation is the process of impressing a bitstream on a carrier wave by varying its amplitude, frequency, or phase relative to a certain point in time. Every state of modulated waveform lasts for a fixed period of time, which is referred to as a symbol. In order to convey K bits of data in one symbol, it needs to have 2^K different states. The symbol rate of a communications system is usually fixed. Thus, a sender realizes different transmission rates by adjusting K.

A symbol can also be conveniently represented as a point on a complex plane, called constellation diagram. So, there are 2^K possible points on the sender's constellation diagram. Due to channel distortion, the received symbol has a Euclidean shift from its transmitted position. Then, the demodulation process is to estimate the transmitted symbol through maximum likelihood detection, that is, select the closest among 2^K constellation points to the received symbol. When the noise is large, the received symbol point may be closer to other constellation points than the one being transmitted and thus an error occurs. The error resilience capability of a modulation scheme is usually determined by the minimum Euclidean distance between any two constellation points.

When the average symbol energy is fixed, increasing data rate will reduce the minimum Euclidean distance of a constellation and increase the error probability. Therefore, the rate adaptation problem arises as how to adjust the transmission rate by means of adjusting the number of constellation points such that the minimum Euclidean distance of the constellation is at least two times larger than the noise shift in most cases. Rate adaptation can be achieved through either *sender adaptation* or *receiver adaptation*.

Most existing rate adaption schemes adopt sender adaptation. In particular, the receiver estimates the channel condition from its received symbols and feeds it back to the sender. According to this information, the sender chooses the most appropriate constellation size. However, as we all know, a wireless channel is time varying and channel estimation cannot be both accurate and instant. Therefore, there is a dilemma for sender adaptation schemes about whether to make a prompt adaptation based on inaccurate channel parameters or to wait for accurate parameters and react slowly to the fast changing channel.

Receiver adaptation avoids this dilemma by letting the sender always transmit at a high rate using a fixed dense constellation. The receiver decides how many and which bits can be reliably decoded from the received symbol according to its perceived channel quality. Two issues immediately arise: (1) how the receiver decides which bits are correct; and (2) how to recover these corrupted bits. The first issue is usually addressed by using soft information to compute the confidence level of a bit when demodulating a received symbol. Only the bits with high confidence are selected by the demodulator. The second issue is more challenging. Since the positions of corrupted bits are randomly spread, it is difficult to use retransmissions to recover them without significant signaling overhead. Thus, a better way is to use a rateless code, also known as fountain code, to ensure reliable data transmission.

Rateless codes are binary codes that can generate unlimited number of encoding symbols, and have the property that the source bits can be recovered from any subset of encoding symbols, as long as the number of encoding symbols is equal to or slightly larger than the number of source bits. For receiver adaptation, the sender may continuously generate and transmit rateless symbols. The receiver waits to collect an adequate number of reliably received rateless symbols, and when decoding is successful, it signals the sender to stop transmission by sending an acknowledgement.

However, in receiver-based rate adaptation, the interaction between the binary rateless code and the soft-information decoding causes a severe performance degradation when the channel condition is poor and the soft information obtained by the demodulator cannot be used in rateless decoding. This problem is also referred to as the *mismatched decoding problem* by Martinez et al. [319].

This chapter presents a novel compressive modulation (CM) scheme that solves this problem by decoding directly from modulation symbols instead of from separate bits. CM achieves receiver side rate adaptation and smoothly adapts its rate to a wireless channel condition. The core component of CM is a random projection (RP) code for generating rateless multilevel symbols from binary streams. Based on the RP code, we develop a completely novel modulation scheme which is capable of achieving fine-grained receiver rate adaption over a large range of channel conditions. Since the rateless symbol generation process in RP code resembles the linear projection process in the emerging compressive sensing theory, we name our scheme as *compressive modulation* (CM).

14.2 Background

14.2.1 Rate Adaptation

Rate adaption addresses the problem of matching the transmission rate to the time-varying channel condition. It can be realized through both sender adaptation and receiver adaptation. Sender adaptation is the mainstream approach in which the sender selects the appropriate coding and modulation parameters according to the receiver's feedback on the channel condition. The core problem in this approach is which channel properties should be estimated and be used as the rate selection metric.

Channel signal-to-noise ratio (SNR) is the most intuitive metric, since the Shannon capacity is defined as a function of SNR. Receiver-based autorate (RBAR) [320] adapts a rate based on the per-packet SNR estimation during each ready to send/clear to send (RTS/CTS) exchange. Channel-aware rate adaptation (CHARM) [321] proposes to estimate channel SNR in a longer time frame by letting a potential receiver overhear data packets from its potential transmitter before the two parties actually communicate. However, it is well recognized that SNR estimation cannot be both accurate and instant.

Frame loss ratio is a more accurate but much slower metric that describes channel condition. SampleRate [322] uses a two-second time window to obtain the packet loss statistics. A robust rate adaptation algorithm (RRAA) [323] uses short-term observation of frame loss ratio. However, the frame loss ratio is a very coarse metric when the statistics are based on a small number of frames.

The recent work, SoftRate [324], uses per-bit confidences exported by the physical layer to estimate the interference-free bit error rate (BER) of received frames. Using this metric can react more quickly to channel variation than using SNR or frame loss ratio. However, in a fast fading channel where channel conditions may change even during the transmission of one packet, none of the above metrics is accurate enough for instant sender rate adaptation.

The alternative to sender adaptation is receiver adaptation, in which coding and modulation parameters remain unchanged, and only demodulation and decoding parameters are adapted to channel condition. Since both channel estimation and adaptation take place at the receiver, it completely avoids the feedback loop and overcomes the dilemma about estimation accuracy and speed.

The main challenge for receiver adaptation is how to design a "one-fits-all" modulation scheme such that the transmission rate can downgrade gracefully as the channel condition deteriorates. Simply using a dense constellation and transmitting at a high rate cannot achieve this goal because it is well-known that using more points in the constellation diagram than the channel condition permits will dramatically increase error probability.

Adaptive demodulation [325] is the most representative work of receiver adaptation. It employs a rateless code to protect source bits and combat high frequency of bit errors. In particular, source bits are first encoded by a Raptor code [326] to generate rateless bits. Then every 4 rateless bits form a modulation symbol, which is mapped to a square 16-QAM constellation with Gray code. Upon receiving a symbol, the receiver first decides how many bits, denoted by K, can be accurately retrieved from the modulation symbol at its perceived channel condition. Then it adjusts decision regions of possible bit combinations to retrieve the most reliable K bits. These rateless bits are then accumulated for decoding. Obviously, this approach has a saturation rate at 4 bits/Hz, because it used 16-QAM constellation at the sender.

Increasing the density of the sender constellation may relieve the rate saturation problem, but it will aggravate the mismatched decoding problem that degrades the decoding performance.

14.2.2 Mismatched Decoding Problem

The problem of mismatched decoding may exist in both sender adaptation and receiver adaption approaches as long as binary codes, whether or not they are channel codes or rateless codes, are used before modulation. This problem was most recently studied by Martinez et al. [319] in analyzing the performance of bit-interleaved coded modulation [327]. In this section, we will review this problem in the context of receiver rate adaptation with binary rateless codes.

Figure 14.1 shows the typical coding and modulation process in a receiver adaptation scheme. We denote the input bits of the system as a block $\mathbf{x} = (x_1, x_2, \ldots, x_N)$, with block length N. The sender generates the rateless bit $b_i (i = 1, 2, 3, \ldots)$ using rateless codes, such as the LT code [328] or Raptor code [326]. Each encoded bit b_i is the logical exclusive-OR (XOR) of a randomly selected subset of input bits. A bipartite graph is formed to describe the relationship between source bits and encoded bits. In order to achieve a good recovery performance, the design of graph connectivity needs to follow some guidelines. In addition, Raptor code includes a precoding stage which generates some redundant bits from the input bits. For simplicity, the precoding process is not shown in Figure 14.1 and the redundant bits are assumed as part of input bits.

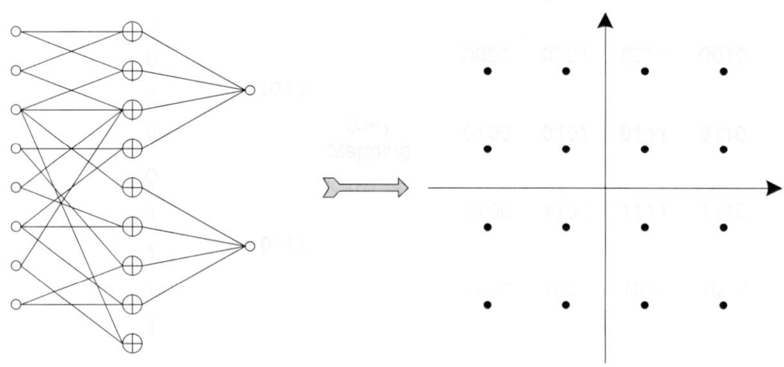

Figure 14.1 Rateless coding and symbol mapping for receiver adaptation.

The rateless bits are then concatenated in groups of K into a modulation symbol sequence $(s_1, s_2, s_3, \ldots)$. The size K is preselected and determines the maximum transmission rate. These symbols belong to a finite alphabet $S = \{S_1, S_2, \ldots, S_M\}$ with $M \triangleq |S| = 2^K$. Finally, the symbol set is mapped to square quadrature amplitude modulation (QAM) constellation using Gray code. Here, S_i denotes both the modulation symbol and its Euclidean coordinate to keep the notations simple. The noisy version of the transmitted symbol is obtained at the receiver due to channel noise. The received symbols can be represented by points in the constellation plane, and their Euclidean coordinates can be denoted by $r_i (i = 1, 2, 3\ldots)$.

The demodulation process starts from solving the maximum a posteriori probability problem based on the received symbols. In general, the transmitted symbol s_i ($s_i \in S$ and $i = 1, 2, 3, \ldots$) is assumed to be uniformly distributed. Then, s_i's posterior probability conditioned on its noisy version r_i can be expressed as

$$P(s_i = S_j | r_i) = \frac{f(\|S_j - r_i\|_2)}{\sum_{m=1}^{M} f(\|S_m - r_i\|_2)}, \tag{14.1}$$

where $f(\cdot)$ is the probability density function of channel symbol transition defined in the $2D$ constellation plane. If the channel is AWGN (additive white Gaussian noise), $f(\cdot)$ is an exponential function. $\|\cdot\|_2$ is the $2D$ Euclidean distance between two points in the constellation plane. Accordingly, the posterior probability of each bit $b_k (b_k \in \{0, 1\}$ and $k = 1, 2, \ldots, K)$ in s_i can be calculated as

$$P(b_k | r_i) = \sum_{j=1}^{M} P(s_i = S_j | r_i) P(b_k | s_i = S_j). \tag{14.2}$$

With $P(b_k | r_i)$, the rateless decoder is able to recover input bits in block $\mathbf{x}$ by a soft-decision iterative decoding. However, a rateless decoder requires that coded bits be independent from each other. This prerequisite does not hold if several coded bits are sent within one wireless symbol. This can be easily proved by contradiction. In particular, if the K bits $b_1 b_2 \ldots b_K$ in symbol s_i are independent, then the probability of $P(s_i = S_j | r_i)$ would be

$$P(s_i = S_j | r_i) = \prod_{k=1}^{K} P(b_k(S_j) | r_i). \tag{14.3}$$

$b_k(S_j)$ indicates that the k-th bit of s_i is equal to the corresponding digit value in S_j. The $P(s_i = S_j | r_i)$ calculated by Eq. (14.3) is obviously different from that obtained by Eq. (14.1).

The effect of information mismatch is minor if the noise shift is within the minimum Euclidean distance of a constellation. The problem of mismatched decoding is hardly noticeable in sender adaptation schemes with channel coding, because in these schemes the minimum Euclidean distance is adjusted to be larger than twice the noise shift with very high probability. However, in receiver adaptation, the problem becomes prominent because the sender always transmits at a high rate, that is, the minimum Euclidean distance of the constellation could be smaller than noise shift in a large portion of modulation symbols. For this reason, a receiver adaptation scheme will have a significantly degraded performance at low channel SNR if it aggressively uses a very dense constellation.

Mismatched decoding cannot be avoided if the receiver adaptation scheme uses existing rateless codes, such as LT codes and Raptor codes, because they can only decode from rateless bits. If there is a rateless encoder that can directly generate multilevel symbols, and a corresponding decoder that can directly decode from the multilevel symbols retrieved from each independent channel dimension, the problem of mismatched decoding vanishes and the soft information can be fully and accurately utilized. It is based on this understanding that we propose the completely new RP codes inspired by compressive sensing theory to fill in such seemingly unbridgeable gap.

14.3 Compressive Modulation

The proposed compressive modulation is considered as a fine-grained receiver rate adaptation scheme. As shown in Figure 14.2, the framework consists of RP coding, modulation, soft demodulation, and RP decoding. This section will describe these four modules in sequence, and then discuss the RP code design for wireless modulation.

14.3.1 Coding and Modulation

The key to realizing receiver rate adaptation is to design a multilevel rateless code and a gracefully degradable modulation scheme. RP code is designed as a rateless code that encodes binary digits into multilevel values. Let $\mathbf{x} = (x_1, x_2, \dots, x_N)$ be a block of input bits of length N, RP encodes them into a series of symbols $(s_1, s_2 \dots)$ by random projection:

$$s_i = \mathbf{c}_i \cdot \mathbf{x}^T, \tag{14.4}$$

where $\mathbf{c}_i$ is a low-density random vector. In particular, only L entries in $\mathbf{c}_i$ are nonzero, and they take values from a weight set $\{w_1, w_2, \dots, w_L\}$. The size of L and the values of weights are key design parameters of RP code. In Section 14.3.3, we will provide details about the design rules and our final choices for CM framework. The position

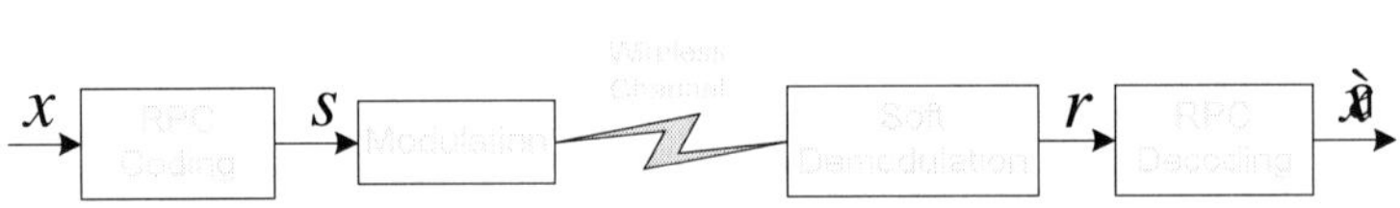

Figure 14.2 The framework of the proposed compressive modulation.

of the nonzero entry in $\mathbf{c}_i$ that takes weight w_l can be denoted by $i_l (l = 1, 2, \ldots, L)$. In input vector $\mathbf{x}$, only the entries at corresponding positions are sampled, so Eq. (14.4) can also be written into:

$$s_i = \sum_{l=1}^{L} w_l \cdot x_{i_l}. \tag{14.5}$$

This random projection process can be repeated for an arbitrary number of times and generate unlimited number of symbols. Each symbol carries virtually the same amount of information, and can be accumulated to perform decoding. Therefore, RP code is a rateless code.

Different from conventional rateless codes, which generate binary symbols by logical XOR, RP code generates multilevel values through arithmetic weighted sum, creating a finite alphabet $\mathscr{S}$:

$$\mathscr{S} = \{\sum_{l \in \Lambda} w_l | \Lambda \subseteq \{1, 2, \ldots, L\}\}. \tag{14.6}$$

In this alphabet, the minimum and the maximum values are $s_{min} = \sum_{w_l<0} w_l$ and $s_{max} = \sum_{w_l>0} w_l$.

We sequentially and evenly map symbol values to the modulated parameters of each channel dimension. In amplitude modulation, the mapping is from $[s_{min}, s_{max}]$ to $[-A, A]$, where A represents the maximum amplitude in use. Therefore, the actual amplitude used to transmit s_i is given in Eq. (14.7). In QAM modulation, two consequently generated symbols are represented by one constellation point in the complex plane.

$$a_i = -A + \frac{2A}{s_{max} - s_{min}} (s_i - s_{min}). \tag{14.7}$$

Figure 14.3 illustrates the coding and modulation process with simple parameter settings. In particular, the size of the weight set is $L = 3$, and the weights are $\{1, 1, 1\}$. Obviously, the range of symbol values is $[0, 3]$, so a 4×4-QAM is used for modulation.

For practical applications, RP code should have a larger L and more diverse weights than the parameters given above. As a result, the encoded symbols have a much larger alphabet size, and the constellation becomes much denser than any practical ones in current use. In addition, the symbol values generated by RP code are not uniformly distributed in $[s_{min}, s_{max}]$. They are concentrated in the middle of the range. This property has its pros and cons. On the one hand, when the mean energy of the constellation is given, the minimum Euclidean distance of the CM constellation is larger than that of an evenly distributed constellation with the same number of points. This partially cancels out the downside of using a dense constellation. On the other hand, our constellation demands a higher peak-to-noise ratio (PAR) of the wireless channel. Therefore, the RP code should be designed such that the corresponding PAR is kept within a reasonable range.

In the above example, when the input bits are purely random, that is, there is equal probability of a bit being 0 or 1, the percentage of encoded symbols 1 and 2 is three times of the symbols 0 and 3. Therefore, when the mean energy of the

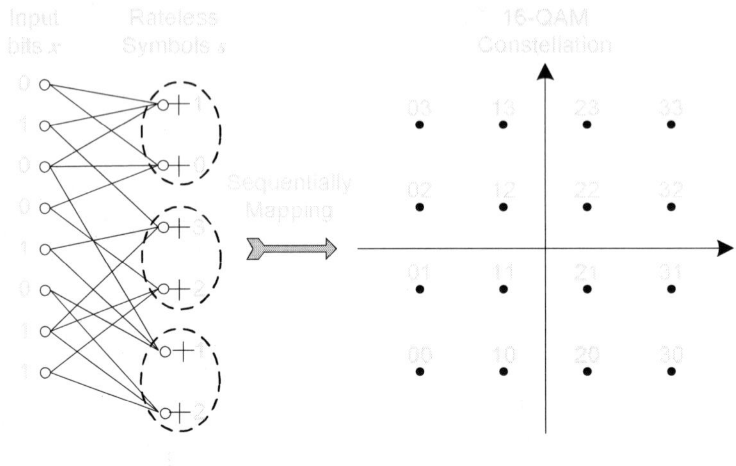

Figure 14.3 The proposed compressive modulation.

constellation is given, the minimum Euclidean distance of the CM constellation is $\sqrt{5/3}$ times longer than that in a normal 16-QAM constellation. The PAR of the CM constellation is also $\sqrt{5/3}$ times larger.

14.3.2 Soft Demodulation and Decoding

In the proposed CM, each rateless multilevel symbol s_i is modulated to one channel dimension. In amplitude modulation, the modulated amplitude a_i for s_i is given in (14.7). Due to channel noise, the amplitude of the received waveform, denoted as $\hat{a}_i$, is a noisy version of a_i. When a Gaussian channel is considered, we have:

$$\hat{a}_i = a_i + e_i^c, e_i^c \mathcal{N}(0, \sigma_c^2), \tag{14.8}$$

where e_i^c is the Gaussian noise with variance σ_c^2. Then the received rateless symbol r_i corresponding to s_i can be computed as:

$$r_i = s_{min} + \frac{s_{max} - s_{min}}{2A}(\hat{a}_i + A). \tag{14.9}$$

From Eq. (14.7), Eq. (14.8), and Eq. (14.9), we have:

$$r_i = s_i + \frac{s_{max} - s_{min}}{2A} e_i^c. \tag{14.10}$$

Denote $e_i = \frac{s_{max} - s_{min}}{2A} e_i^c$ as the symbol noise. Since s_{max}, s_{min}, and A are all constants for a given RP code design, e_i is also a Gaussian variable, and its variance is $\sigma^2 = \frac{s_{max} - s_{min}}{2A} \sigma_c^2$.

Through sequential mapping, CM successfully converts the Gaussian channel noise into Gaussian symbol noise, making it possible for the rateless decoder to recover input bits directly from multilevel symbols. It should be noted that CM utilizes and benefits from soft information. Although the received symbol r_i may not belong to the encoding alphabet $\mathscr{S}$, its value is in the range of $[s_{min}, s_{max}]$ in most cases.

Decoding from rateless symbols into binary digits is achieved through maximum a posteriori estimation. Assume that M symbols are accumulated for decoding, then the rateless decoding problem can be formulated as:

$$\begin{aligned} \hat{\mathbf{x}} = & \arg\max_{\mathbf{x}\in GF(2)^N} P(\mathbf{x}|r_1, r_2, \ldots, r_M), \\ \text{s.t.} \quad & \mathbf{r} = \mathbf{C}\mathbf{x}^T + \mathbf{e}, \end{aligned} \tag{14.11}$$

where $\mathbf{r} = (r_1, r_2, \ldots, r_M)^T$ is the vector of received symbols, and $e = (e_1, e_2, \ldots, e_M)^T$ is the corresponding noise vector. $\mathbf{C}$ is an $M \times N$ matrix, in which each row is a random projection vector.

The problem formulation of RP decoding resembles that of low-density parity-check (LDPC) decoding. It can be similarly solved by the belief propagation algorithm. However, since the symbol generation operator in RP code is arithmetic addition instead of logical XOR, the messages sent out by symbols are calculated in a significantly different fashion from that in a conventional BP algorithm.

The RP decoding algorithm is described as follows.

1. Initialization:
Input bits are assumed to be random. Initial probability of both $P(x_j = 0)$ and $P(x_j = 1)$ is 0.5. The message sent by every input node x_j to its connected symbol node s_i is initialized as

$$\begin{cases} u_{ji}^{(0)}(0) = P(x_j = 0) \\ u_{ji}^{(0)}(1) = P(x_j = 1) \end{cases} . \tag{14.12}$$

2. Iterations:
(i) *Messages sent out by symbol nodes:* According to Eq. (14.5) and Eq. (14.10), received rateless symbol r_i can be written into:

$$r_i = w_1 x_{i_1} + w_2 x_{i_2} + \ldots + w_L x_{i_L} + e_i. \tag{14.13}$$

Let $j = i_l$ and define $X_j = r_i - w_l x_j$. Then the message sent from the i-th symbol node to the j-th variable node in the round t is:

$$\begin{cases} v_{ij}^{(t)}(0) = P(x_j = 0|r_i) = P(X_j = r_i) \\ v_{ij}^{(t)}(1) = P(x_j = 1|r_i) = P(X_j = r_i - w_l) \end{cases} . \tag{14.14}$$

The distribution of X_j is the convolution product of the weighted distribution of every other variable node and e_i

$$P(X_j) = \{\bigotimes_{i_k, 0<k\leq L, k\neq l} P(w_k x_{i_k})\} \otimes P(e_i), \tag{14.15}$$

where the distribution of weighted variable is

$$\begin{cases} P(w_k x_{i_k} = 0) = u_{i_k i}^{(t-1)}(0) \\ P(w_k x_{i_k} = w_k) = u_{k_k i}^{(t-1)}(1) \end{cases}. \tag{14.16}$$

Inspired by Baron et al. [329], we implement the convolution through simple addition in the Fourier domain. We further simplify the decoding algorithm based on the prior knowledge that input digits are binary.
(ii) *Messages sent out by symbol nodes:* The message sent by the j-th variable node to the i-th symbol node in round t is:

$$\begin{cases} u_{ji}^{(t)}(0) = C_{ji} P(x_j = 0) \prod_{k \in T_j \setminus i} v_{kj}^{(t)}(0) \\ u_{ji}^{(t)}(1) = C_{ji} P(x_j = 1) \prod_{k \in T_j \setminus i} v_{kj}^{(t)}(1) \end{cases}, \tag{14.17}$$

where T_j is the index set of the symbol nodes connected with variable node j. If there is an edge between variable node j and symbol node i, then $i \in T_j$. C_{ji} is a normalization constant which ensures $u_{ji}^{(t)}(0) + u_{ji}^{(t)}(1) = 1$.

3. Stop and Output:
After T rounds of iteration, the belief propagation process stops and outputs the following posterior probabilities of the j-th variable node being 0 and 1:

$$\begin{cases} u_j^{(T)}(0) = C_j P(x_j = 0) \prod_{k \in T_j} v_{kj}^{(T)}(0) \\ u_j^{(T)}(1) = C_j P(x_j = 0) \prod_{k \in T_j} v_{kj}^{(T)}(1) \end{cases}. \tag{14.18}$$

Then the values of input binary digits are estimated by:

$$x_j = \begin{cases} 0 & u_j^{(T)}(0) \geq u_j^{(T)}(1) \\ 1 & \text{otherwise} \end{cases}. \tag{14.19}$$

RP decoding through belief propagation can fully utilize the statistics of channel noise. When the noise variance increases, successful demodulation and decoding can be achieved by accumulating more rateless symbols. Therefore, the proposed RP code has an inherent capability of rate adaptation. Suppose a block of N bits are successfully decoded from M rateless symbols that are transmitted within $\frac{M}{2}$ channel uses, the transmission rate can be calculated as $\frac{2N}{M}$. The granularity of rate adaptation is very fine, because the transmission can be adjusted on a per-symbol basis.

The belief propagation algorithm for RP decoding shares the same assumption with LDPC code about the connections between variable nodes and symbol nodes. The bipartite graph should have no loops longer than T. Because our symbols are generated randomly, the assumption cannot be guaranteed. Therefore, the damping method, Message Damping Belief Propagation (MDBP) [330] has been applied to the iterations for improving the recovery performance.

14.3.3 Design RP Codes

An RP code is a linear rateless code which encodes binary digits into multilevel symbols. It can be represented by a bipartite graph, where left nodes are binary input and right nodes are encoded multilevel symbols. Every right node has the same fixed node degree L and the weights of connecting edges are taken from the same set $\{w_1, w_2, \ldots, w_L\}$. The key design parameters for a RP code are the right node degree L and the set of edge weights.

As the pursuit for RP code arises from the modulation problem, the design objective naturally is to find a low-complexity code whose output symbols can be easily modulated by a digital modulator and achieves high transmission rate within typical channel SNR range.

We only consider symmetric weight sets which can be represented by $\{-w_{L/2}, \ldots, -w_1, w_1, \ldots, w_{L/2}\}$ and L is an even number. Symmetric weights generate zero-centric RP symbols which can be easily mapped to the constellation. In the rest of this section, we use only the positive weights to represent the symmetric weight set. Without loss of generality, we only select integral power of 2 as weights, that is, $\{w_1, \ldots, w_{L/2}\} \subseteq \{2^n | n \in \mathbb{N}\}$. As the input digits are binary, such weight selection is sufficient to generate any integer number.

In wireless communications, we can assume that the operational range of channel SNR is between 5 dB to 25 dB. Conventional rate adaption selects among binary phase shift keying (BPSK), quadrature phase shift keying (QPSK), 16-QAM, and 64-QAM modulations, which offer rates from 1 bits/Hz to 6 bits/Hz. As each modulation symbol contains two RP symbols, in order to achieve 6 bits/Hz, the average information carried in one RP symbol should be no less than 3 bits. Let the encoder generate M RP symbols, then the average symbol information can be calculated through the mutual information between RP symbols and the input bits.

$$\begin{aligned} I_{RP} &= \frac{1}{M} I(s_1, \ldots, s_M; x_1, \ldots, x_N) \\ &= \frac{1}{M} \{H(s_1, \ldots, s_M) - H(s_1, \ldots, s_M | x_1, \ldots, x_N)\}, \end{aligned} \tag{14.20}$$

where $H(x)$ is the entropy of variable x. However, RP code is considered as a rateless code. Symbols are generated on-the-fly with random projection vectors. Therefore, we can only obtain an upper bound of the average symbol information.

$$\begin{aligned} I_{RP} &\le I(s_i; x_{i,1}, \ldots, x_{i,L}) \\ &= H(s_i) - H(s_i | x_{i,1}, \ldots, x_{i,L}) \\ &= H(s_i) = -\sum_{k=1}^{T} p_k \log_2 p_k, \end{aligned} \tag{14.21}$$

where p_k is the distribution of RP alphabet $\mathscr{S}$. Equality holds if and only if RP symbols are independent. It can be shown that this upper bound only depends on the distribution of RP symbols.

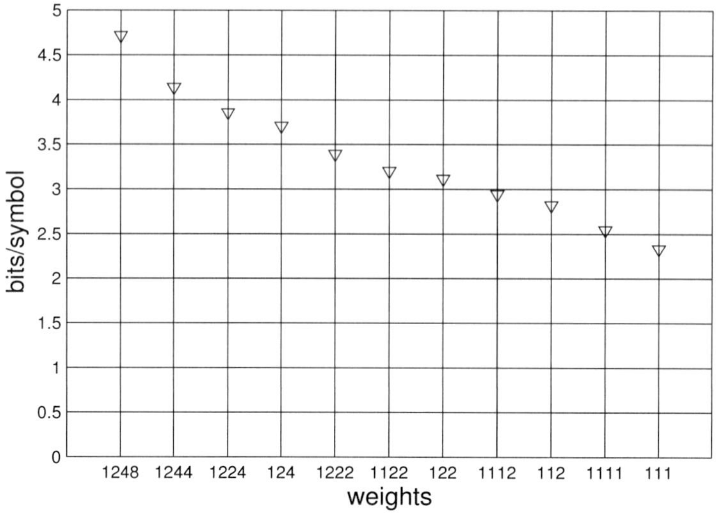

Figure 14.4 Upper bound of mutual information for considered weight sets.

In general, using large L can increase per symbol information. However, we prefer small L in RP code design for two reasons. First, in BP decoding, the number of message passing increases exponentially with the symbol degree L. A larger L also causes more loops in the bipartite graph that will impair BP decoding performance. Second, large L generally results in large alphabet size of RP symbols. When the modulation symbol energy is fixed, it decreases the minimum Euclidean distance of the constellation, and the transmission is susceptible to errors.

When $L \leq 4$, the upper bound of I_{RP} is less than 3 bits. Therefore, we consider the weight selection when $L \geq 6$. Figure 14.4 shows the upper bound of I_{RP} for all the possible weight sets when $L = 6$ and $L = 8$. We find that there are seven sets of weights whose symbol information is above 3 bits. Therefore, it may not be necessary to consider the selections when $L \geq 10$.

In the practical modulation, symbol information should be slightly more than the requested because mutual information is the upper bound. Therefore, we take $\{1,2,4,8\}$, $\{1,2,4,4\}$, $\{1,2,2,4\}$, and $\{1,2,4\}$ as candidates. In order to evaluate which set of weights is the most appropriate for modulation purpose, we set up simulation experiments at typical channel SNRs and measure their achievable rates. From Figure 14.5, we see that $\{1,2,4,4\}$ outperforms other selections in most cases. With this selection, the RP symbol range is $[-11,11]$. Therefore, we use a 23×23 QAM constellation in the proposed CM. The mapping is shifted by a small offset in both channel dimensions to avoid modulation symbols sitting on the origin.

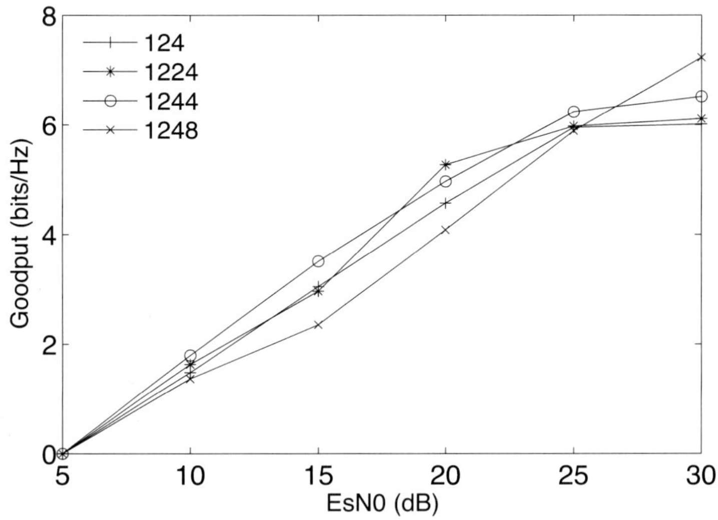

Figure 14.5 Rate versus SNR curves for considered weight sets.

14.4 Simulation Study

We first evaluate CM through simulations. Our simulation is set up on MATLAB® 2009a. We are particularly interested in finding out how CM adapts to different channel conditions. For wireless communications, both throughput and bit error rate (BER) are important performance metrics. We use the joint metric goodput as the evaluation metric. Let R_T be the throughput, and P_e be the bit error rate, then the goodput for a $1K$-byte block can be computed as: $R_G = (1-P_e)^N \cdot R_T$, where $N = 1024 \times 8$, is the number of bits in one block. Throughout the evaluation sections, the term rate or achievable rate is referred to as goodput if not otherwise indicated.

14.4.1 Rate Adaptation Performance

In this section, we study the granularity of CM rate adaptation. The block size of RP coding is 400 bits. It should be noted that the constellation point of CM is not uniformly distributed. The average symbol energy is normalized to 1 by its distribution, so the channel SNR can be controlled by adjusting noise power. We carry out experiments at integer SNR values from 5 dB to 30 dB. At each SNR value, 5000 randomly generated blocks are transmitted. Each transmission stops when the receiver has successfully decoded the input bits. The decoding interval is set to 10 RP symbols.

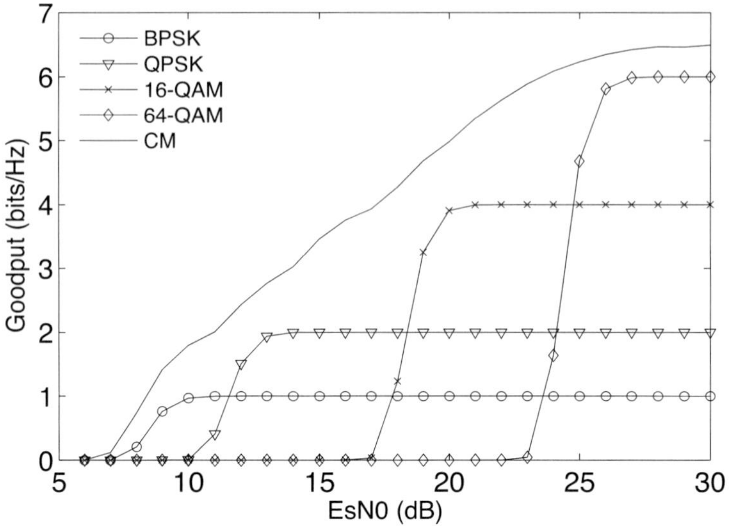

Figure 14.6 Goodput of CM without channel coding.

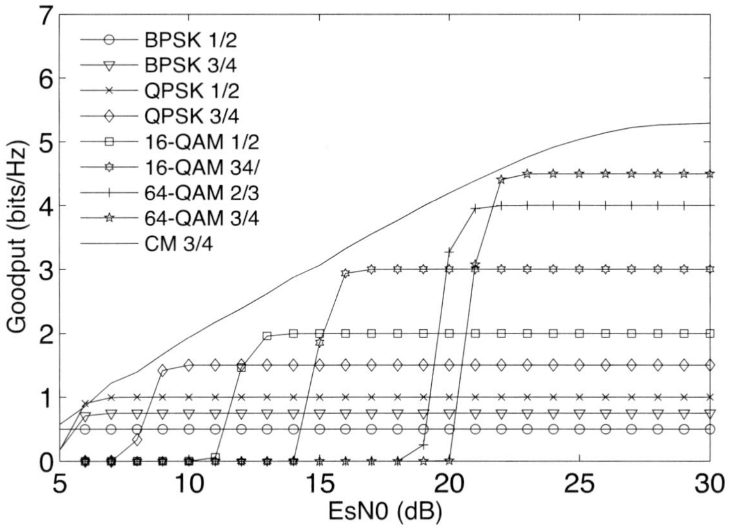

Figure 14.7 Goodput of CM with channel coding.

Figure 14.6 compares the goodput of CM with that of conventional modulation schemes, including BPSK, QPSK, 16-QAM, and 64-QAM. It shows the result when channel coding is not applied. Figure 14.7 shows the result with channel coding. In both cases, the proposed CM can achieve smooth rate adaptation, in contrast to the stair-shaped rates that can be achieved by switching between conventional modulations. The rate for CM is virtually continuous from 0 bits/Hz to 6.4 bits/Hz without channel coding and from 0.5 bits/Hz to 5.3 bits/Hz with 3/4 convolutional code as specified by IEEE [331].

For conventional modulation schemes adopted in IEEE 802.11a [331], the rate is almost zero if channel SNR is below a certain threshold and sharply rises to its saturate rate when channel SNR is above the threshold. The target of conventional rate adaptation is to make the rate as close to the envelope of all rate curves of candidate modulations as possible. However, the envelope can only achieve a discrete rate set due to its sharp rising characteristic. Even though the rate set can be extended by combining the modulation rate with channel coding rate, there is still a noticeable gap from that of CM. It is obvious that the proposed CM outperforms conventional rate adaptation. The gain over conventional rate adaptation is the accumulated area of the gap between the combined performance curves for the entire set of modulation schemes and that of CM over the range of channel SNRs.

14.4.2 Sensitivity to SNR Estimation

In RP decoding, the noise variance is an important input parameter. However, in practical system, channel SNR is difficult to estimate. Accurate estimation needs long term observation while short term estimation may suffer from unacceptable uncertainty. Therefore, we propose to evaluate how sensitive RP decoding is to imprecise noise variance.

In this simulation, transmissions are taking place at four typical SNR values: 10 dB, 15 dB, 20 dB, and 25 dB. For symbols received at a fixed SNR, we try to perform RP decoding at an imprecise SNR input. The difference between actual channel SNR and the SNR input is varied from -10 dB to 10 dB at a 2 dB step. For example, when data are transmitted over a 10 dB AWGN channel, we use $\{0, 2, 4, 6, 8, 10, 12, 14, 16, 18, 20\}$ dB as the SNR input to RP decoder. We run 500 tests for each parameter setting.

Figure 14.8 shows the mean and standard deviation of achieved goodput. It can be found that RP decoder is not sensitive to imprecise SNR input, especially when the input SNR underestimates the actual channel. The performance loss is not significant even when the input SNR is 10 dB lower than the actual value. However, if the input SNR is much higher than the actual SNR, a large variation in goodput is observed. In fact, an overestimated SNR unreasonably increases the confidence probability of RP symbols, causing oscillations in belief propagation that may finally result in decoding errors.

14.5 Testbed Evaluation

We evaluate CM performance using a SORA [332] platform. SORA is a programmable software radio platform built on commodity general-purpose PCs. Since

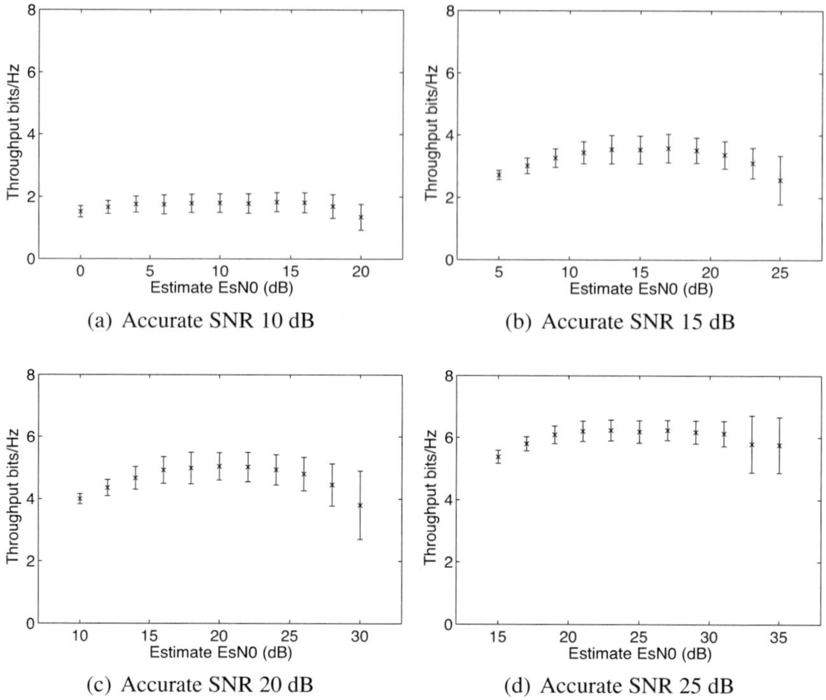

Figure 14.8 CM performance with inaccurate SNR estimation.

RP decoder requires a fair amount of computation for high-speed communication, we leave it as our future work to optimize it on software platform. In this chapter, we mainly evaluate the modulation and demodulation performance using an offline process. We assume instant ACK (acknowledge) after a block is successfully decoded at the receiver.

We evaluate CM under both a static scenario, where the channel changes slowly, and a mobile scenario, where the channel fades quickly due to multi-path effects. We compare CM to an "Oracle" rate adaptation scheme, which adjusts the modulation rate, for example, BPSK, QPSK, 16-QAM, or 64-QAM, based on posterior knowledge of channel SNR. Obviously, this "Oracle" rate adaptation has the upper bound performance compared to any practical rate adaptation schemes. We further compare CM to adaptive demodulation (ADM), the state-of-the-art receiver adaptation. In our comparisons, while we directly measure the performance of CM, we emulate "Oracle" and ADM based on the same channel traces as CM to ensure these schemes are compared under exactly the same channel conditions.

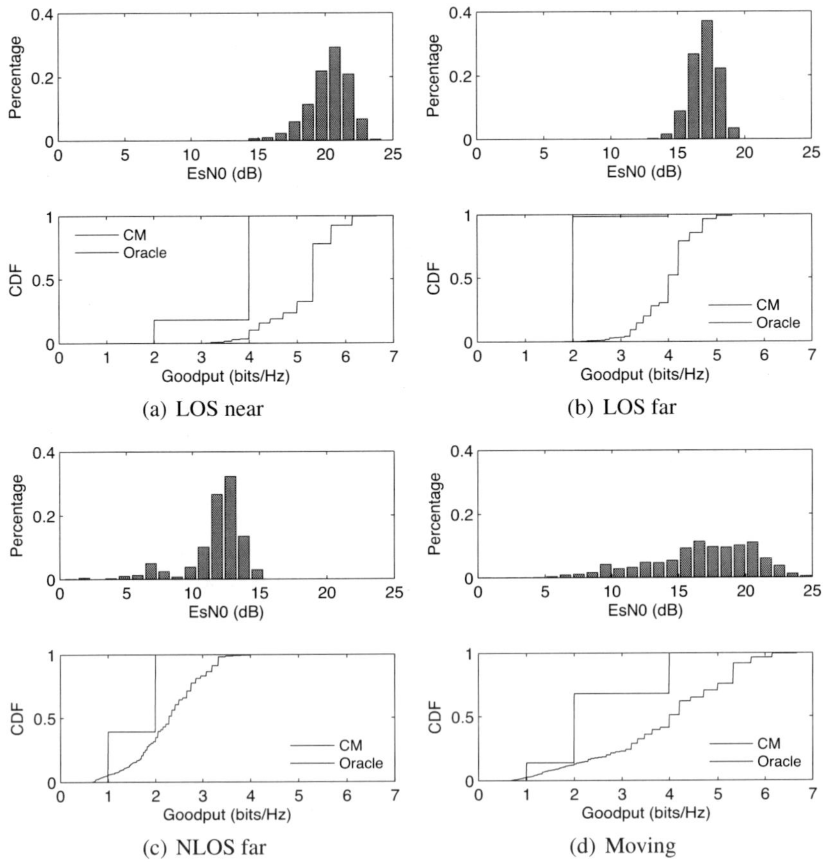

Figure 14.9 Performance comparison of CM to the "Oracle" sender-based rate adaption scheme. The "Oracle" scheme shows an upper bound performance of any practical rate adaptation scheme.

14.5.1 Comparison to Oracle

In a static scenario, we carry out the experiments with different sender-receiver distances, with and without line-of-sight (LOS) path. The upper-row figures in Figure 14.9a,b,c summarize the SNR statistics in three sets of experiments. The lower-row figures show the cumulative distribution function (CDF) of the achieved rate of both the CM and "Oracle" sender rate adaptation.

The results show that CM outperforms "Oracle" in all three settings. The average goodput CM achieved in three cases are 5.1 bits/Hz (LOS near), 4.0 bits/Hz (LOS near), and 2.3 bits/Hz (NLOS near). It yields a performance gain from 41.6% to 100% compared to "Oracle," which achieves an average goodput of 3.6 bits/Hz, 2.0 bits/Hz, and 1.6 bits/Hz under the same three conditions.

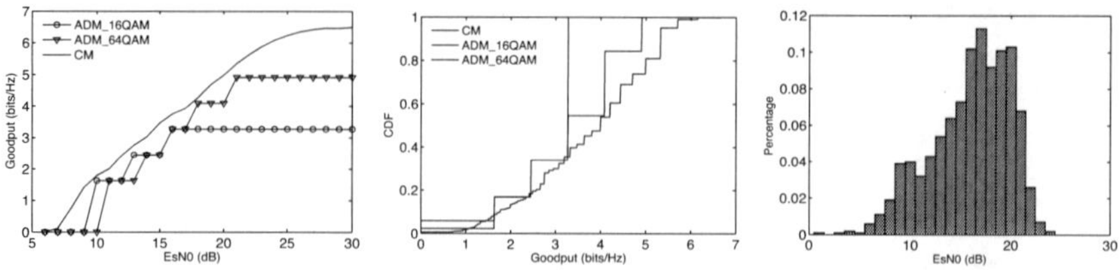

(a) SNR Statistics (b) Goodput Comparison (c) Simulation

Figure 14.10 Performance comparison of CM to the ADM receiver-based rate adaption scheme by using traced channel condition.

The mobile scenario shows the similar results, as shown in Figure 14.9d. CM achieves 56% gain over oracle adaptation in mobile scenario. The goodput of the two schemes are 3.9 bits/Hz and 2.5 bits/Hz, respectively. This demonstrates CM works well even when the channel condition changes very quickly.

The major performance gain of CM comes from its ability to seamlessly adjust its modulation rate, while conventional rate adaptation has only a limited set of modulation rates. Therefore, a low modulation rate has to be used even if the channel condition is much higher than the requirement, but still not enough to support a higher rate. Thus, the channel capacity is wasted. In contrast, CM achieves a modulation rate that best fits the instantaneous wireless channel and thus make full use of the channel capacity.

14.5.2 Comparison to ADM

We compare CM to the state-of-the-art receiver rate adaptation scheme, ADM. Both CM and ADM use the same transmission power at the sender. Figure 14.10a shows the channel SNR statistics of the experiment.

Our ADM implementation includes the original scheme based on fixed 16-QAM as described by Brown et al. [325]. This fixed 16-QAM modulation will become a bottleneck when the channel SNR is high. For example, when the SNR is higher than 21dB, 64-QAM can actually be used. Thus, we also extend their work to use fixed 64-QAM modulation. We compare CM with both variants of ADM.

Figure 14.10b shows the goodput achieved by CM, 16-QAM-, and 64-QAM-based ADM under the same channel trace. The average goodputs achieved by the three schemes are 3.75 bits/Hz, 2.82 bits/Hz, and 3.26 bits/Hz. The performance gains of CM over two ADM schemes are 33.0% and 15.0%, respectively.

To have a better understanding of why CM performs better, let's refer to Figure 14.10c. It shows simulated results of CM, 16-QAM-, and 64-QAM-based ADM with different channel SNRs. We can see that when we choose a low fixed modulation rate (i.e., 16-QAM) for ADM, it performs well with a low SNR range (close to CM when

SNR is less than 15 dB). But the rate is saturated when SNR becomes high and higher modulation rates should be used. However, if we fix a high modulation (i.e., 64-QAM) for ADM, its performance in low SNR range is bad, due to the aforementioned mismatched decoding problem. In a real wireless environment, the channel condition can change widely, and thus it is difficult for ADM to choose a single modulation rate for all SNR conditions. In contrast, CM can seamlessly adapt to a wide range of SNR and thus outperform ADM in general wireless conditions.

14.6 Related Work

This section discusses related work ranging from coded modulation to the recent development in compressive sensing.

14.6.1 Coded Modulation

Because the proposed CM introduces rateless multilevel codes into the modulation, it is natural to start our discussion from coded modulation even though most of them are not designed for rate adaptation. The trellis-coded modulation (TCM) proposed by Ungerbocke is the first scheme of coded modulation [333, 334]. It uses trellis codes to generate symbols from input bits with a certain redundancy and then multilevel symbols are mapped to discrete constellation points through set partitioning. At receiver, the Viterbi algorithm is adopted for optimal joint decoding of all bits in symbols. Inspired by this work, it is recognized that channel coding and modulation should be combined as an entity to improve performance.

Multilevel coded (MLC) modulation is another way to combine channel coding and modulation [335]. It first partitions input bits into several subsets and every subset is protected by a binary channel coding. Output bits at the same time instant from all channel encoders are concatenated as a symbol. At receiver, instead of optimally joint decoding of all bits in symbols, a sub-optimal multi-stage decoding (MSD) is proposed to achieve good performance with limited complexity [336]. The theoretical analyses by Wachsmann et al. [337] show that the multi-stage decoder can achieve channel capacity.

In order to be robust in a Rayleigh channel, Zehavi proposes bit-interleaved coded modulation (BICM) from TCM [338], where a bit interleave module separates trellis code from modulation. Trellis coded symbols are bit-wise interleaved with a depth exceeding the coherence time of the fading process. In order to limit information loss arising in this separation approach, soft information of coded bits are used in the decoder. Caire et al. present detailed analyses on BICM from the perspective of information theory and show that the loss in BICM is limited [327]. The fact that BICM is a mismatched decoding has been recently recognized [319].

The performance of coded modulation not only depends on how to combine them together but also depends on which channel codes are selected. The discovery of Turbo code [18] and the rediscovery of LDPC codes [16, 17] provide new research avenues for the further development on coded modulation. These modern codes can approach the capacity of binary-input channel with low complexity.

14.6.2 Compressive Sensing

The design of RP codes is inspired by the emerging compressive sensing theory [295, 311] and its recent development in using low-density matrix to generate measurements. Mathematically, let $\mathbf{x} = [x_1 x_2 \dots x_N]^T$ be a K-sparse signal, that is, there are only $K(K \ll N)$ nonzero elements in x_i's. Let us take M measurements of $\mathbf{x}$ through linear projection:

$$\mathbf{y} = \Phi \cdot \mathbf{x} \tag{14.22}$$

where Φ is an $M \times N(M < N)$ random matrix, then $\mathbf{x}$ can be perfectly reconstructed from $\mathbf{y}$ under certain conditions. The central problem in CS is how matrix Φ should be designed and what algorithm should be used to recover $\mathbf{x}$ from the underdetermined linear system defined in Eq. (14.22).

The random matrix Φ is often considered as a dense matrix in the early stage of CS theory development. Recently, several papers have reported the consideration of the CS reconstruction from a Bayesian inference if the statistical characterization of the signal is available [339–341]. This essentially bridges CS with low-density parity-check (LDPC) codes, although the operation in CS is arithmetic plus instead of logical XOR. Due to the complexity in belief propagation, CS reconstruction by Bayesian inference can be practically implemented only if Φ is a low-density matrix. Sarvotham et al. first propose a sparse Φ with elements drawn from $\{0,1\}$ and its decoding is similar to that of erasure code [342]. Wu et al. formulate the decoding from graph evolution [343]. Baron et al. further extend the sparse Φ with elements drawn from $\{0,1,-1\}$ and solve the reconstruction by belief propagation [329].

The proposed RP codes are not a simple extension to the low-density CS codes. In particular, the pursuit for multilevel rateless codes arises from the modulation problem in communications systems. Therefore, the code design needs to allow for fine-grained rate adaptation in a wide range of channel conditions. Our analysis on the design of RP codes illustrates that the sparse matrices Φ with elements either $\{0,1\}$ or $\{0,1,-1\}$ will create rate ceilings because the generated multilevel symbols carry little mutual information.

14.7 Summary

We have described in this chapter a completely novel compressive modulation scheme for receiver rate adaptation in wireless communication. It achieves fine-grained adaptation through a rateless multilevel code named RP code and a universal modulation constellation. The RP encoder generates multilevel symbols from input binary digits, and each encoded symbol is modulated to one channel dimension. The sequential mapping from RP symbol values to modulation parameters allows for Gaussian characterization of symbol errors, which can be readily used in the belief propagation-based RP decoding.

We carry out extensive simulations and obtain the achievable rates of proposed CM in the operational range of channel SNR. These results show that CM consistently outperforms the rate envelope of conventional modulation, whether or not the channel coding is applied. More importantly, we have implemented and evaluated CM on a software radio testbed. The performance gain of CM over oracle rate adaptation is above 40% in all four sets of experiments under both static and mobile settings. The trace based evaluation shows that CM achieves 33% gain over the state-of-the-art adaptive demodulation scheme. These promising results validate the practical use of the proposed CM in real wireless environments.

Chapter 15
Joint Source and Channel Coding

15.1 Introduction

The end-to-end performance of wireless communications depends on the effective processing in both the source and the channel. In practical communications systems, source coding is handled at the application layer, mostly for multimedia contents, and channel coding is most commonly performed at the physical layer (PHY). Guided by Shannon's separation principle, source coding and channel coding are usually accomplished independently. Hence, an implicit assumption at PHY is that the data to be transmitted are compressed source without redundancies. Contrary to this assumption, numerous applications inject uncompressed data into the network (e.g., e-mails, Web pages, and uncompressed files). The measurement on web pages [344] shows that there exist a substantial amount of compressible data in practical networking traffic. Undoubtedly, intelligently utilizing data redundancy in transmission will greatly boost communication efficiency.

A wireless channel is time-varying and subject to fading, additive noise, and interference. In order to increase spectrum efficiency, PHY has to adapt the transmission rate to varying channel conditions. In conventional adaptive modulation and coding (AMC), the sender selects the best combination of channel coding and modulation that achieves the highest transmission rate under the estimated channel condition. However, the success of AMC depends on the instant and accurate channel estimation, which cannot be obtained simultaneously. In addition, AMC only achieves stair-shaped rates due to the limited number of rate combinations. To overcome these two drawbacks, there are three recently published pieces of research on smooth and blind rate adaptation [345–347]. The adaptation is blind in the sense that channel information is not required at the sender, and the adaptation is smooth as it provides fine-grained rate adjustment. These smooth and blind rate adaptations are very valuable to the wireless communications, especially in fast fading channels. Unfortunately, none of them take into account possible redundancies in the source data.

As it is not practical to require all applications to compress their data before transmission, a natural question to ask will be whether it is possible to achieve high-efficiency compression at PHY. A straightforward answer would be no, because conventional data compression techniques, such as Huffman coding and arithmetic coding, need sufficient statistics for a specific data type. This requirement usually cannot be fulfilled because PHY contains small sets of hybrid data (the maximum IP packet size is 1.5 KB). In addition, application data have already been represented by streams of bits instead of logical data structures at PHY. It would require dramatic change to the entire network stack if the PHY needs high-level information of application data.

In this chapter, we will further study how our proposed CM can simultaneously achieve source compression, channel protection, and seamless rate adaptation over the existing network stack. As we have discussed in Chapter 14, CM operates on binary sources, and is widely applicable to time-varying wireless channels. The core of the proposed CM is a random projection (RP) code inspired by the compressive sensing (CS) theory [295,311]. The proposed RP code is significantly different from the random measurement matrices used in CS, because the source is binary and the generated symbols will pass the existing DAC (digital-to-analog converter) and be transmitted with the existing digital communication hardware. The RP code is also fundamentally different from the contemporary channel codes because it generates multilevel symbols instead of binary symbols through weighted sum operations. According to CS theory, the required number of symbols for successful decoding decreases as the source redundancy increases, and also decreases as the channel quality increases. Therefore, a source with higher redundancy can be decoded from less number of RP symbols and achieves a higher transmission rate. In contrast, more RP symbols need to be transmitted when channel condition worsens, and hence it leads to a lower transmission rate. Because the rate can be adjusted on a per symbol basis, the adaptation becomes seamless. It is worth noting that the proposed CM design can be easily integrated into the existing PHY layer. In particular, we have implemented CM in 802.11a PHY. It simply bypasses the forward error correction (FEC) and modulation in the existing design and directly generates the modulation symbols on the data subcarriers. Experiments over a software radio platform have proved the feasibility of our design.

Two technical problems in the proposed CM are studied in this chapter. First, we shall design code for binary sources with different levels of redundancy or sparsity. Three principles are established based on both source and channel considerations. In particular, we shall derive the upper bound of the achievable rate for any given RP code under source sparsity p and a given channel error distribution. Based on these principles, we design weight multisets of RP code for typical source sparsities. Although RP code has a saturation effect at high signal-to-noise ratios (SNRs), our evaluations show that the same weight multiset can be used over the typical SNR range (from 5 dB to 25 dB) for moderate source sparsity. Second, we shall design an algorithm to reduce the decoding complexity of RP code so that the decoding speed matches with the contemporary wireless communication rate. This is a linear time belief propagation (BP) algorithm for the decoding of RP code. Harnessing the binary-input property of the variable nodes, we propose to compute convolutions (at constraint nodes) in the time domain. We shall show that this is more efficient than the conventional computation in the frequency domain. Observing that the probability distribution functions (PDFs) of the constraint nodes have limited support, we devise a novel ZigZag deconvolution to further reduce the decoding complexity. We perform careful analysis to show that this decoding algorithm is nearly 20 times faster than the CS-BP algorithm for a typical code.

15.2 Related Work and Background

In this section, we will first discuss the related work on joint source-channel coding and rate adaptation, and then provide background information on the compressive sensing theory.

15.2.1 Joint Source-Channel Coding

Joint source-channel coding (JSCC), in its broad sense, covers all schemes that jointly consider source coding and channel coding under a unified resource constraint and/or toward a common optimization objective. Existing work on JSCC can be classified into three categories.

The first category usually applies to image/video transmission. Among them, some works have exploited the rate-distortion behavior of the source and study the problem of allocating the transmission rate between the source coder and the channel coder to minimize expected distortion [348–350] or optimize the power consumption [351]. Others have designed unequal error protection (UEP) for progressively coded sources. They are called JSCC because more important contents (source) are encoded with stronger channel codes [352]. Both variations of LDPC (low-density parity-check) codes, such as rate-compatible LDPC codes by Pan et al. [353], and rateless codes, such as Raptor codes over GF(2) by Aditya and Katti [354], and Raptor codes over GF(4) by Bursalioglu et al. [355], have been studied in the literature.

Works in the second category jointly consider the compression of correlated sources and channel protection of them. Liveris et al. [356] show that LDPC codes are able to compress correlated binary sources close to the Slepian-Wolf limit. Sartipi and Fekri [357] apply a similar idea to wireless sensor networks (WSNs) for data collection. Zhong and Garcia-Frias [358] and Xu et al. [359] use low-density generator matrix (LDGM) codes and Raptor codes for distributed JSCC.

More related to the proposed scheme are the works in the third category, which apply the same code for both compression and protection of a single source. Research has shown that conventional channel codes can be used for source compression. In particular, LDPC codes, Raptor codes, and Turbo codes [360–362] can be concatenated with a Burrows-Wheeler transform (BWT) to serve this purpose. Furthermore, Fresia et al. [363] use the concatenation of two independent LDPC codes in the transmitter, and implement a joint BP decoder at the receiver. Zhu and Alajaji [364] and

Cabarcas et al. [365] design Turbo code for transmitting nonuniform binary memoryless or identically distributed (i.i.d.) sources over noisy channels. However, all these codes are designed at a fixed rate. None of them is able to achieve smooth rate adaptation.

15.2.2 Coded Modulation

The proposed CM scheme is essentially an enhanced coded modulation with embedded compression, so we also briefly review related works on coded modulation. The trellis-coded modulation (TCM) proposed by Ungerboeck is the first scheme of coded modulation [333, 334]. By introducing trellis coding and set partitioning in the PHY layer, the minimum free Euclidean distance is maximized. Multilevel coded (MLC) modulation [335, 337] is another elegant way for coded modulation. For a M bits modulation, the channel coding is divided into M levels and each level is designed independently. The advantage of MLC is that there is no need to design the coding on Euclidean distance. Any binary code designed on Hamming distance can be used.

To address the issue of robustness under a Rayleigh channel, Zehavi proposes bit-interleaved coded modulation (BICM) [338] and Caire et al. [327] present a detail analysis on BICM from the information theoretic perspectives. On the one hand, bit interleaving increases the channel diversity, which equivalently increases the Hamming distance of coding. On the other hand, infinite depth interleaving tackles the mismatch decoding issue [319].

The discovery of Turbo code [18] and the rediscovery of LDPC codes [16, 17] open up new research avenues for advance development in contemporary coded modulation. With these capacity-achieving codes, most of the existing coded modulation schemes can approach the Shannon limit at the designed SNR. However, they lack rate adaptation capability, and have to rely on separate adaptive modulation to change the rate when channel varies dramatically. In addition, the capacity-achieving performance usually requires infinite block coding length and high complexity.

15.2.3 Rate Adaptation

Rate adaptation refers to the techniques which adjust the modulation, coding, power, and other protocol parameters based on varying channel conditions. The most well-known rate adaptation technique is adaptive modulation and coding (AMC), which has been extensively studied [323, 366–368] and practically applied to wireless systems. However, AMC has two major drawbacks. First, rate selection relies heavily on accurate and instant channel estimation, which cannot be simultaneously obtained in practice. Second, AMC is only able to achieve stair-shaped rates due to limited rate combinations.

Hybrid automatic repeat request (HARQ) [369] is a supplemental technique to AMC. The incremental redundancy scheme, also known as type II HARQ, can provide a smoother rate by employing either a rate compatible code, for example, punctured Turbo code [370] or rateless codes, for example, Raptor code [371]. However, both codes have a limited rate adaptation range.

A more recent thread of research is able to achieve smooth and blind rate adaptation in a wide range of channel conditions. Three independent works on this topic have been published in recent years. Gudipati and Katti [346] propose a scheme named Strider which appends a minimum distance transformer (MDT) after existing coding and modulation. The free distance of constellation can be transformed to fit the channel condition. Perry et al. [347] invent the spinal codes, which replace the state transition of convolution code in TCM by an infinite state hash function. The rateless property is achieved by refining the state transition. The previous work on rateless compressive modulation [345] proposes a novel bit to symbol mapping, and the averaged bit energy can be smoothly refined by accumulating symbols at the receiver. The rate adaptation is therefore seamless.

Though not explicitly stated, all these rate adaptation works assume that PHY data are incompressible. When this is not the case, that is, when data contain non-negligible redundancy, the end-to-end performance of these schemes becomes sub-optimal. Through extensive analyses, we discover that the RCM we developed earlier can be naturally extended to opportunistically exploit data compressibility. In other words, the bit-to-symbol mapping can be used for JSCC. When properly designed, RCM has the potential to significantly improve the system throughput.

15.2.4 Compressive Sensing

The bit-to-symbol mapping in RCM converts source bits into multilevel symbols through weighted sum operation. It is essentially a CS [295,311] scheme with binary input. Therefore, we first review this emerging theory with special emphasis on its connection with channel codes.

CS theory states that an n-dimensional signal having a sparse or compressible representation can be reconstructed from m linear measurements even if $m < n$ [295, 372]. Traditionally, the signals under consideration are sparse or compressible signals in $\mathbb{R}^N$, the projection matrix Φ is a dense matrix in $\mathbb{R}^m \times \mathbb{R}^n$, and CS reconstruction can be achieved by l_1-minimization [373], matching pursuit [374], iterative threshold methods [375], and subspace pursuit [376]. Although these algorithms can be extended to sparse Φ or the binary signal, the decoding complexity is high due to the nature of the iterative greedy algorithms.

It is observed that CS can be used as a joint source-channel coding scheme [377–381]. Bajwa et al. [378] and Chen et al. [381] examine the energy-performance design space of CS, and show that CS is robust to channel fading and noise. Duarte et al. [377] and Feizi et al. [379, 380] apply CS to compress correlated sources and protect the measurements under noisy channels. Their results show that a CS decoder can provide trade-offs between rate and decoding complexity. In addition, CS-based joint source-channel coding scheme has continuous rate-distortion performance in a noisy Gaussian channel. This research is done in the context of wireless sensor systems and the application of CS theory is rather "traditional," that is, the sources are highly sparse real-valued sensor readings and the CS decoding is based on computation-intensive convex optimization. These schemes are not readily extended to the transmission of generic binary data seen in the current physical layer of communications systems.

A less traditional application of CS theory is to use it purely as channel codes and to apply various restrictions to reduce its complexity. Sarvotham et al. [342] design the Sudocodes, which can be used as erasure codes for real-valued data. Both encoding and decoding complexity can be reduced by limiting Φ to sparse binary matrices. Wu et al. [343] and Liu et al. [382] consider binary source in CS, and derive closed-form formulation for its decoding. However, these works only consider noiseless case. CS-BP [329] provides a solution to decode noisy CS under certain constraints. It considers a sparse Φ whose nonzero entries are drawn from Rademacher distribution, and adopts BP algorithms for decoding.

The complexity of CS-BP decoding is much higher than LDPC decoding because the constraint nodes have to compute convolutions of several PDFs, since CS computes measurements via weighted sum operation instead of logical exclusive-OR (XOR). A standard solution will be to perform processing in frequency domain via fast Fourier transform (FFT) and inverse FFT (IFFT), in order to convert convolution to multiplication and deconvolution to division. However, such processing is not efficient for binary-input PDFs. A better solution to this complexity problem is critical for practical deployment of CM, so as to facilitate the broader application of CS.

15.3 Compressive Modulation (CM) for Sparse Binary Sources

Note that the physical layer is responsible for transmitting raw bits instead of logical data packets over a physical link. The redundancy of physical layer data can be easily modeled by unequal probabilities on the appearance of 0's and 1's [383]. In particular, we model the input bits as i.i.d. Bernoulli random variables with probability $P(b=1)=p$. Without loss of generality, we consider only the cases where $p \leq 0.5$, as a bitstream with $p > 0.5$ can be converted to a stream with $P(b=1)=1-p$ through simple bit flipping. Borrowing the terminology from the CS theory, we call p the source sparsity or we call the length-N bit vectors p-sparse.

For a sparse binary source, the bit-to-symbol mapping serves dual purposes in data compression and channel coding. We name this process random projection coding. According to the CS theory, the number of symbols (M) required to decode the source is proportional to the source sparsity p. Therefore, the modulation rate for sparse source, that is, $p < 0.5$, will be higher than that for non-sparse source ($p=0.5$). The compression gain is thus achieved.

In Chapter 14, we selected the weight multiset $\mathscr{W} = \{\pm 1, \pm 2, \pm 4, \pm 4\}$ for encoding non-sparse sources. However, for sparse sources, this weight selection may not be optimal. The reason is that the distribution of the symbol alphabet Ψ, denoted by $P(\Psi)$, plays a key role in achieving the communication efficiency. As $P(\Psi)$ is jointly decided by weight multiset and source sparsity, it is crucial to study how these two factors affect each other and how to select the proper weight multiset for different source sparsity.

15.3.1 Design Principles

This section outlines three design principles which should be enforced in RP code design. The first two principles concern individual RP symbols, and the third principle takes the dependency of RP symbols into consideration. Following these principles, which are necessary conditions to find an optimal code, we derive several guidelines to facilitate code design.

Principle 1 *The entropy of RP symbols should exceed half of the desired maximum transmission rate.*

One major objective of any physical layer design is to achieve high transmission rate. The achievable rate is bounded by the entropy of wireless symbols. As every two RP symbols constitute one wireless symbol, the entropy of RP symbols should exceed half of the desired maximum transmission rate.

We have defined $\Psi = \{\psi_1, \psi_2, \ldots \psi_K\}$ as the symbol alphabet and $P(\Psi)$ as the symbol distribution. In particular, we denote $P(\psi_k)$ as the probability that symbol s is equal to ψ_k. By definition, the entropy of RP symbols, denoted by $H(s)$, is:

$$H(s) = -\sum_{k=1}^{K} P(\psi_k) \cdot \log P(\psi_k). \tag{15.1}$$

It should be noted that both Ψ and $P(\Psi)$ are functions of weight multiset $\mathscr{W}$ and source sparsity p, such that $H(s)$ is a function of $\mathscr{W}$ and p too.

We consider two anchors on symbol entropy. First, the highest modulation rate used in Wireless Local Area Network (WLAN) is 6 bits/s/Hz achieved with 64-QAM. Therefore, each RP symbols should carry at least 3 bit information (if the source has been optimally compressed) to avoid early rate saturation. Second, the typical SNR range of WLAN is 5 dB to 25 dB, and the Shannon capacity for 25 dB additive white Gaussian noise (AWGN) channel is around 8.3 bits. Therefore, it is not necessary to consider weight sets whose RP symbol entropy is larger than 4.15 bits.

Principle 2 *The generated RP symbols should have a fixed mean regardless of source sparsity.*

We consider a communications system with an average (i.e., per wireless symbol) power constraint of $2E$, or equivalently a constraint of E per RP symbol. Let E_{RP} denote the average symbol energy before power scaling. We shall minimize E_{RP} so that each RP symbol can be scaled by a larger factor in transmission, and becomes more robust to channel noise.

Given a set of RP symbol Ψ and its distribution $P(\Psi)$, the mean and variance of the RP symbols can be calculated as:

$$\mathbb{E}[\psi_k] = \sum_{k=1}^{K} \psi_k \cdot P(\psi_k), \tag{15.2}$$

$$Var[\psi_k] = \sum_{k=1}^{K} |\psi_k - \mathbb{E}[\psi_k]|^2 P(\psi_k). \tag{15.3}$$

It is known that E_{RP} is minimized when each RP symbol is shifted by $\mathbb{E}[\psi_k]$ in the modulation constellation, and the minimum achieved is exactly the symbol variance. As it is not practical to shift symbols differently when source sparsity p varies, the RP symbols should have a fixed mean regardless of source sparsity.

Lemma 1 *The weight multiset has a zero mean is the sufficient and necessary condition for the generated RP symbols to have a fixed mean regardless of source sparsity* p.

Proof. The mean of RP symbols can be written into:

$$\mathbb{E}[\psi_k] = \mathbb{E}\left[\sum_{l=1}^{L} w_l b_l\right] = \sum_{l=1}^{L} w_l \cdot \mathbb{E}[b_l] = \sum_{l=1}^{L} w_l \cdot p, \tag{15.4}$$

which is a function of p. It does not change with p iff $\sum_{l=1}^{L} w_l = 0$, that is, the weights are zero-mean.

Principle 3 *Let* $\mathbf{G}_M$ *be the sub-matrix composed of the first M rows of the encoding matrix* $\mathbf{G}$. *Denote* $\mathbf{G}_M[i]$ *as the* i^{th} *column of* $\mathbf{G}_M$. *Given a weight multiset,* $\mathbf{G}$ *should be organized such that* $\min_i ||\mathbf{G}_M[i]||^2$ *is maximized for all possible M.*

CM can be viewed as a coded modulation scheme. Therefore, it is important to measure the free distance d_{free} of the code, which is defined as the minimum distance between any two codewords (vectors of RP symbols). A large free distance indicates that a sequence of wireless symbols are robust against channel noise.

Denote $\mathbf{s}(\mathbf{b})$ as the codeword generated from bit block $\mathbf{b}$ by a RP code, then the free distance of the code is:

$$d_{free}^2 = \min_{\mathbf{b}_1 \neq \mathbf{b}_2 \in \{0,1\}^N} ||\mathbf{s}_1(\mathbf{b}_1) - \mathbf{s}_2(\mathbf{b}_2)||^2. \tag{15.5}$$

Usually, the free distance is evaluated by taking an all-zero vector as reference. Since the RP encoding of an all-zero bit sequence will generate an all-zero codeword, the free distance can be deduced to a simpler form $d_{free}^2 = \min ||\mathbf{s}(\mathbf{b})||^2$. The minimum is usually taken when there is only one bit 1 in the source vector $\mathbf{b}$, and the minimum value is $||\mathbf{G}[i]||^2$.

However, the proposed CM is a rate adaptation scheme. The transmission could stop at any time before all the N symbols generated by $\mathbf{G}$ are transmitted. Suppose the decoding is successful after the first M symbols are transmitted, the actually encoding matrix used is $\mathbf{G}_M$. Therefore, we shall require that $\mathbf{G}_M$ creates a large free distance for all possible M.

15.3.2 Weight Selection

The achievable transmission rate of CM is a function of source sparsity, channel condition, and RP code design. Ideally, there is an optimal code for each source sparsity and channel condition. However, CM is designed for "blind" rate adaptation, that is, the channel condition is not known to the sender. Therefore, the objective of RP code design is to find a set of weights which achieve an overall high throughput for the primary SNR range of wireless channels.

We consider source sparsity between 0.1 and 0.5, and pick four representative values of source sparsity including 0.1, 0.15, 0.25, and 0.5 for the performance study. These values are selected such that their entropies are evenly spaced. For the sake of simplicity, we focus on the integer weights $\{1,2,4\ldots\}$, which are powers of 2. According to Principle 2, the weights should have zero-mean. A simple choice is to have positive-negative symmetric weights. In addition, the size of weight multiset does not need to exceed 20. According to the numerical results by Barron et al. [329], for weights $\{-1,1\}$ the optimal check node degree is $L_{opt} \approx 2/p$ beyond which the performance gains will become marginal. Therefore, $L_{max} = 20$ will be sufficient for source sparsity 0.1 and above.

Table 15.1 lists several candidate weight multisets that satisfy the above requirements and Principle 1. The entropies of the generated RP symbols are listed in the last column. Next, we could select among these weight multisets by two ways.

The first way is through MATLAB simulations. For each weight multiset and source sparsity combination, 10^6 bits are transmitted over AWGN channel with SNR ranging from 5 dB to 25 dB. These throughputs are shown in Figure 15.1. We

Table 15.1 Candidate weight multisets.

$\mathscr{W}_1$	$\{\pm1,\pm1,\pm1,\pm1,\pm1,\pm1,\pm1,\pm1,\pm1,\pm1\}$	6.42
$\mathscr{W}_2$	$\{\pm1,\pm1,\pm1,\pm1,\pm1,\pm1,\pm1,\pm1,\pm2,\pm2\}$	7.09
$\mathscr{W}_3$	$\{\pm1,\pm1,\pm1,\pm1,\pm1,\pm1,\pm2,\pm2\}$	6.90
$\mathscr{W}_4$	$\{\pm1,\pm1,\pm1,\pm2,\pm2,\pm2\}$	7.00
$\mathscr{W}_5$	$\{\pm1,\pm1,\pm1,\pm2,\pm4,\pm4\}$	8.37
$\mathscr{W}_6$	$\{\pm1,\pm2,\pm4,\pm4\}$	8.28
$\mathscr{W}_7$	$\{\pm1,\pm1,\pm4,\pm4\}$	7.91

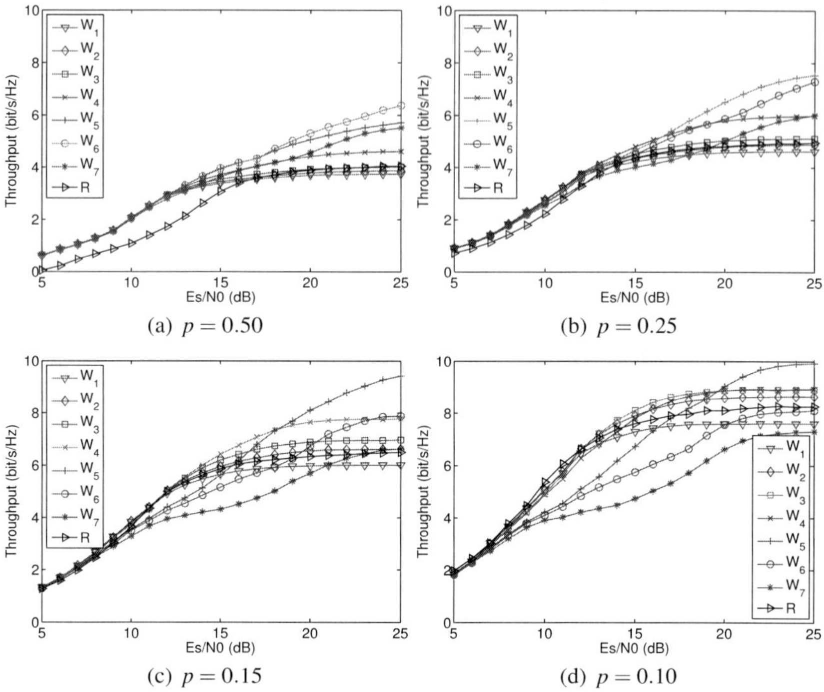

Figure 15.1 Weight selection according to throughput in a simulated AWGN channel.

observe that no single weight multiset can excel for all cases of source sparsity over the entire channel SNR range. However, if the primary SNR range is 5 dB to 20 dB, as in a typical WLAN, we could make the following choices: $\mathscr{W}_{0.5} = \mathscr{W}_6$, $\mathscr{W}_{0.25} = \mathscr{W}_5$, $\mathscr{W}_{0.15} = \mathscr{W}_4$, and $\mathscr{W}_{0.1} = \mathscr{W}_3$. For completeness, the black curve in Figure 15.1 shows the performance of using random weights drawn from the standard Gaussian distribution. We adopt the decoding algorithm proposed by Rangan [384]. Comparison shows that random weights do not perform well especially when the source is not sparse or has moderate sparsity.

Note that when the primary SNR range changes, different choices can be made. For example, if the channel varies between 18 dB to 25 dB, $\mathscr{W}_5$ (instead of $\mathscr{W}_4$) should be selected for source sparsity $p = 0.15$ according to Figure 15.1c. If the channel is always below 12 dB, and the source sparsity is 0.5, then according to Figure 15.1a, all the listed weight multisets have similar performance. In addition, the saturation effect which has been observed in linear encoding of real symbols also appears here. Different weight multisets have different saturation SNR and rates, and it is also affected by source sparsity.

The second way to select weights is through extrinsic information transfer (EXIT) chart analysis [385]. The EXIT chart has been widely used for the design and analysis of LDPC codes [386]. The RP codes resemble LDPC codes in that they can be represented by bipartite graphs too, where variable nodes are source bits and check nodes are RP symbols. We refer to Brink et al. [385, 386] for further details on extrinsic information processing.

As it is difficult to obtain a closed-form expression for the extrinsic information transfer in RP decoding, we measure the mutual information after each iteration of decoder processing. We use the notation by Brink [385] and write I_A for the average mutual information between the bits and the a priori probabilities. Similarly, we write I_E for the average mutual information between the bits and the extrinsic probabilities. However, since the source has sparsity, [Eq. (12) in 385] should be modified to:

$$\begin{aligned} I_A = (1-p)\int_{-\infty}^{+\infty} p_A(\xi|X=0)\log_2 \frac{p_A(\xi|X=0)}{p_A(\xi)}d\xi \\ +p\int_{-\infty}^{+\infty} p_A(\xi|X=1)\log_2 \frac{p_A(\xi|X=1)}{p_A(\xi)}d\xi, \end{aligned} \tag{15.6}$$

where $p_A(\xi) = p \cdot p_A(\xi|X=1) + (1-p)\cdot p_A(\xi|X=0)$. The extrinsic information I_E can be calculated similarly.

Due to the space limit, we only show the EXIT chart for source sparsity $p = 0.5$ and $p = 0.1$ and compare weight multiset $\mathscr{W}_3$ through $\mathscr{W}_6$ in Figures 15.2 and 15.3. For successful decoding, two curves in a chart should cross at the source entropy $(H(p))$. Figure 15.2 is obtained when SNR is 20 dB and 190 RP symbols are transmitted for a length-480 bit block (spectrum efficiency of 5.05 bit/s/Hz). It is clear from the figure that weight multiset $\mathscr{W}_3$ and $\mathscr{W}_4$ cannot decode successfully under this setting. $\mathscr{W}_5$ is very close to successful decoding, but the two curves do not reach 1 in Y-axis. $\mathscr{W}_6$ is the weight multiset we chose for $p = 0.5$, and it can be seen that it ensures successful decoding at such a high spectrum efficiency. In addition, the tunnel between the two curves is still quite wide, indicating that a higher spectrum efficiency could be achieved.

Figure 15.3 can be interpreted similarly. Since the source sparsity is 0.1, the two curves should cross at $H(0.1) = 0.469$. Under this test setting, $\mathscr{W}_5$ and $\mathscr{W}_6$ fail decoding, and $\mathscr{W}_3$ and $\mathscr{W}_4$ succeed. We further decrease the value of M to 120, and find that only $\mathscr{W}_3$ decodes successfully. This is also consistent with our previous selection.

15.3.3 Encoding Matrix Construction

According to Principle 3, the construction of the encoding matrix **G** affects CM performance, especially when the channel condition is good and the required number of RP symbols for decoding is small. Next, we present three steps to construct **G** that satisfies Principle 3. We will take $\mathscr{W}_6$ as an example.

First, we construct three elementary matrices A_1, A_2, and A_4. The structure of A_1 is shown as follows. Matrices A_2 and A_4 have the same structure, but the nonzero values are replaced with $+2/-2$ or $+4/-4$,

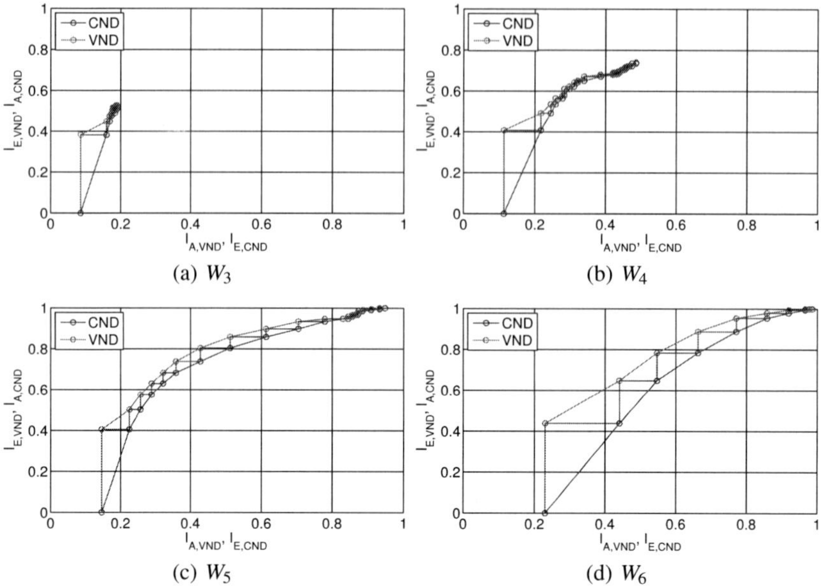

Figure 15.2 EXIT chart for different weight multisets when SNR = 20 dB, $p = 0.5$, $M = 190$ for $N = 480$.

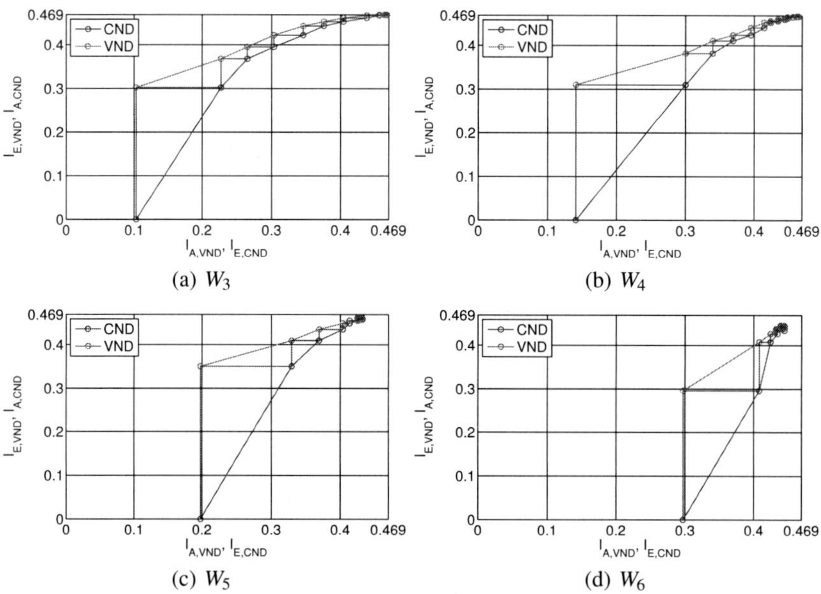

Figure 15.3 The EXIT chart for different weight multisets when SNR = 15 dB, $p = 0.1$, $M = 120$ for $N = 480$.

$$A_1 = \begin{bmatrix} +1 \; -1 & & & \\ & +1 \; -1 & & \\ & & \ddots & \\ & & & +1 \; -1 \end{bmatrix}.$$

Second, we form matrix $\mathbf{G}_0$ by stacking random permutations of A_1, A_2, and A_4 as follows:

$$\mathbf{G}_0 = \begin{bmatrix} \pi(A_4) & \pi(A_4) & \pi(A_2) & \pi(A_1) \\ \pi(A_2) & \pi(A_1) & \pi(A_4) & \pi(A_4) \\ \pi(A_4) & \pi(A_4) & \pi(A_1) & \pi(A_2) \\ \pi(A_1) & \pi(A_2) & \pi(A_4) & \pi(A_4) \end{bmatrix},$$

where $\pi(\cdot)$ denotes randomly permutated columns of a matrix. This matrix has a dimension of $N/2 \times N$, and ensures that $\min_i ||\mathbf{G}_{N/2}[i]||^2 = 1^2 + 2^2 + 4^2 + 4^2 = 37$. Further, $\min_i ||\mathbf{G}_{N/4}[i]||^2 = 1^2 + 4^2 = 17$. By using different permutation choices, we may construct a virtually unlimited number of matrix $\mathbf{G}_0$.

The third and final step in constructing $\mathbf{G}$ is to stack all the randomly generated $\mathbf{G}_0's$. In practice, we only stack two matrices to form an $N \times N$ matrix, and repeatedly use it when the channel condition is poor.

15.4 Belief Propagation Decoding

The RP code can be represented by a bipartite graph as LDPC code, where variable nodes are source bits and constraint nodes are RP symbols. Hence, the RP code can be decoded through belief propagation in a similar fashion. However, since RP symbols are generated by arithmetic addition rather than logical XOR, the belief computation at constraint nodes is much more complex than that in LDPC decoding. Reducing the computational complexity of the decoding algorithm has become the key to the success of the entire CM scheme.

We discover that, for binary variable nodes, the belief computation at constraint nodes can be more efficiently accomplished by direct convolution rather than the fast Fourier transform (FFT) as used in CS-BP. For each constraint node, we compute the convolution of the distributions for all its neighboring variable nodes, and then reuse the convolution for all outgoing messages by deconvolving the distribution of the node being processed. To save computational cost, the zero multiplication in the convolution is avoided. In addition, we propose ZigZag iteration to guarantee a unique solution to the deconvolution. Finally, we show through analysis that the proposed decoding algorithm is computationally more efficient than the CS-BP algorithm.

Let v and c denote the variable node and the constraint node, respectively. Throughout the algorithm description, we introduce some notations as listed in Table 15.2.

Table 15.2 Definitions for belief propagation.

$w(v,c)$	The weight on the edge between v and c.
$\mu_{v\to c}$	The message sent from v to c.
$\mu_{c\to v}$	The message sent from c to v.
p	The a priori probability of bit 1 for all variable nodes.
$p_v(\cdot)$	The Probability Density Function (PDF) of variable node v.
$p_c(\cdot)$	The PDF of constraint node c calculated based on the PDFs of all its neighboring variable nodes.
$p_{c\backslash v}(\cdot)$	The PDF of constraint node c calculated based on the PDFs of all its neighboring variable nodes except variable node v.

RP-BP Decoding Algorithm

1. *Initialization:* Initialize messages from variable nodes to constraint nodes with a priori probability
$$\mu_{v\to c} = p_v(1) = p.$$
2. *Computation at constraint nodes:* For each constraint node c, compute the probability distribution function $p_c(\cdot)$ via convolution Eq. (15.7). For each neighboring variable node $v \in n(c)$, compute $p_{c\backslash v}(\cdot)$ via deconvolution Eq. (15.8):
$$p_c = (*)_{v\in n(c)} \left(w(c,v)\cdot p_v\right), \tag{15.7}$$
$$p_{c\backslash v} = p_c \,\tilde{*}\, \left(w(c,v)\cdot p_v\right). \tag{15.8}$$
Then, compute $p_v(0)$ and $p_v(1)$ based on the noise PDF p_e and the received symbol s_c:
$$p_v(0) = \sum_i p_{c\backslash v}(i)\cdot p_e(s_c - i), \tag{15.9}$$
$$p_v(1) = \sum_i p_{c\backslash v}(i)\cdot p_e(s_c - i - w(c,v)). \tag{15.10}$$
Finally, compute the message $\mu_{c\to v}$ via normalization:
$$\mu_{c\to v} = \frac{p_v(1)}{p_v(0)+p_v(1)}. \tag{15.11}$$
3. *Computation at variable nodes:* For each variable node v, compute $p_v(0)$ and $p_v(1)$ via multiplication:
$$p_v(0) = (1-p)\prod_{u\in n(v)} (1-\mu_{u\to v}), \tag{15.12}$$
$$p_v(1) = p\prod_{u\in n(v)} \mu_{u\to v}. \tag{15.13}$$
Then for each neighboring constraint node $c \in n(v)$, compute $\mu_{v\to c}$ via division and normalization:

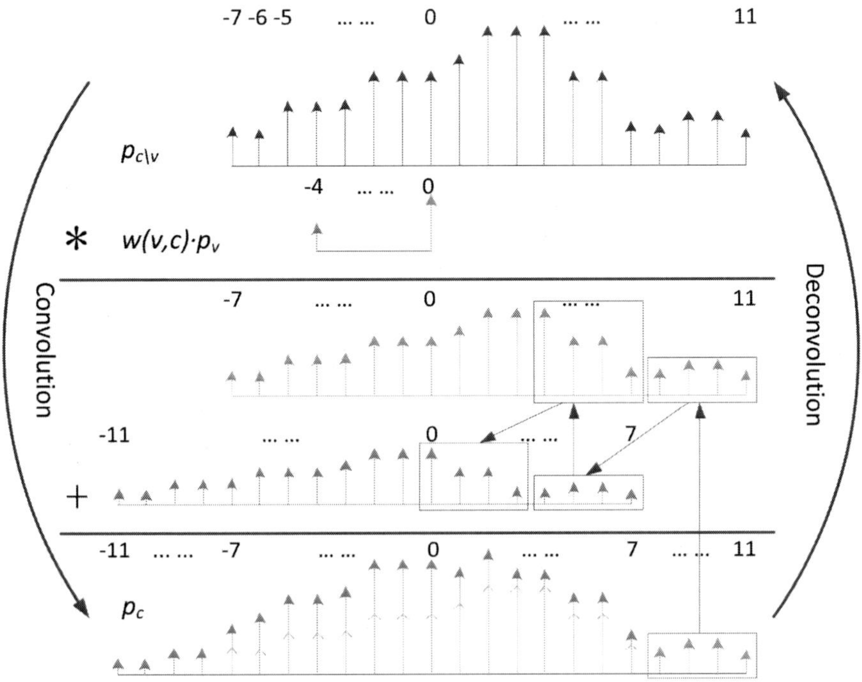

Figure 15.4 Convolution by shift addition (from top to bottom) and ZigZag deconvolution (from bottom to top).

$$\mu_{v\to c} = \frac{p_v(1)/\mu_{c\to v}}{p_v(0)/(1-\mu_{c\to v}) + p_v(1)/\mu_{c\to v}}. \tag{15.14}$$

Repeat steps 2 and 3 until the maximum iteration time is reached. We find that the maximum iteration time can be set to 15, beyond which any performance gain is marginal.

4. *Output:* For each variable node v, compute $p_v(0)$ and $p_v(1)$ according to Eq. (15.12), and output the estimated bit value via hard decision.

The main difference between this RP-BP decoding algorithm and the CS-BP decoding algorithm lies in the computation at the constraint nodes (step 2). First, we process the variable nodes and the noise node in separate steps, because the former is binary and the latter is continuously valued. Second, we propose a ZigZag deconvolution to compute $p_{c\backslash v}$ in Eq. (15.8). Figure 15.4 depicts the convolution by shift addition and ZigZag deconvolution.

The convolution flow is shown from top to bottom. In this example, the weight between variable node v and constraint node c is $w(v,c) = -4$. Hence, the PDF of $w(v,c)\cdot p_v$ only has two spikes at -4 and 0. The convolution of $w(v,c)\cdot p_v$ and any PDF $p_{c'}$ (including but not limited to $p_{c\backslash v}$) can be computed by

$$\left(p_{c'} * w(v,c)\cdot p_v\right)(n) = p_v(0)\cdot p_{c'}(n) + p_v(1)\cdot p_{c'}(n-w(v,c)). \tag{15.15}$$

The addition in the equation is shown in the middle of the figure.

Table 15.3 Complexity comparison between RP-BP and CS-BP.

$\mathcal{W}$	RP-BP		CS-BP	
	×	+	×	+
±(1244)	492	246	8192	9216
±(111244)	856	428	12288	13824
±(111222)	640	320	12288	13824
±(11111122)	954	477	16384	18432

The deconvolution flow is shown from bottom to top in Figure 15.4, in which the rectangles highlight the ZigZag process. In this figure we demonstrate the ZigZag deconvolution from right to left. In theory, it can be performed in both directions. However, due to the computational accuracy of the C program, the practical direction should be determined by the values of $p_v(0)$ and $p_v(1)$. When $p_v(0) > p_v(1)$, it should be computed by

$$p_{c\backslash v}(n) = \frac{p_c(n) - p_{c\backslash v}(n - w(v,c)) \times p_v(1)}{p_v(0)}. \tag{15.16}$$

When $p_v(0) \leq p_v(1)$, it should be computed by

$$p_{c\backslash v}(n) = \frac{p_{c\backslash v}(n + w(v,c)) \times p_v(0) - p_c(n + w(v,c))}{p_v(1)}. \tag{15.17}$$

Let $n_{\min}$ and $n_{\max}$ be the minimum and the maximum value of constraint node c. For $n \notin [n_{\min}, n_{\max}]$, $p_{c\backslash v}(n) = 0$. Thus, we can get a unique solution for the deconvolution by recursion.

Similar to CS-BP, the complexity of RP-BP algorithm is $O(N)$. However, the linear scaler in front of N in RP-BP is much smaller than CS-BP. Since most of the computation is taken by the constraint node message and other computation cost of both algorithms are the same, we compare the computation cost on constraint node in each iteration, as shown in Table 15.3. We can see that the computation cost of CS-BP is around 20 times of RP-BP.

15.5 Performance Evaluation

In this section, we will report the performance evaluation of the proposed CM through the simulations on MATLAB 2011b and the emulations based on traced channel state information. Four representative values of source sparsity including 0.1, 0.15, 0.25, and 0.5 are considered. The primary evaluation metric is throughput (in bit/s/Hz).

15.5.1 Implementation

We implement CM and three reference schemes as follows.

CM: For each block of information bits, the sender progressively transmits RP symbols with a given step size until an acknowledgment is received or the maximum number of transmission is reached. At the receiver end, the RP symbols are accumulated for RP-BP decoding. If the demodulated bits pass the cyclic redundancy check (CRC), an acknowledgment will be delivered to the sender. Otherwise, more RP symbols are needed to perform the next round of decoding.

The bit block length for the RP code is set to be $N = 480$, which is a multiple of all the possible weight multiset sizes. The progressive transmission step is set to be $\Delta M = 24$ and the maximum number of symbols for transmission is set to be $M_{max} = 1920$, which is equivalent to the rate for the Physical Layer Convergence Protocol (PLCP) header transmission in 802.11 PHY. In the PLCP header, we introduce 3 bits. One is the bit flipping indicator and the remaining two are the source sparsity indicator. These three bits can be placed in the reserved position of the PLCP header. Therefore, the overhead is negligible.

BICM with ideal source compression (denoted by BICM for brevity): We implement BICM [327], which is the state-of-the-art coded modulation scheme. In order to avoid the performance loss due to short block length, we implement BICM with 2304 block length and 23040 interleaver length, which is the longest block length setting defined in the WiMax standard [387]. The coding rates are $\frac{1}{2}$, $\frac{2}{3}$, $\frac{3}{4}$, and $\frac{5}{6}$, and the modulation schemes are QPSK, 16-QAM, 64-QAM, and 256-QAM. Altogether, they create 16 combinations and 14 identical rates.

As BICM does not have compression capability, we assume an ideal source compressor for BICM capable of representing a length-N source block with $NH(p)$ bits, where $H(p)$ is the source entropy. For simplicity, we just generate source blocks of length $NH(p)$ from non-biased Bernoulli trial, but count N bits for each block when computing the throughput.

HARQ with ideal source compression (denoted by HARQ for brevity): We implement type-II HARQ (also known as incremental redundancy) which is the state-of-the-art rate adaptation scheme. Rate-$1/3$ Turbo code is used as the mother code. As in Wideband Code-Division Multiple Access (WCDMA) and long-term evolution (LTE), the code length is 1024 bits, and the component encoder is based on recursive convolution code with polynomial $(13, 15)$ in octal. The puncture period is set to 8 for smooth rate adaptation, and the puncture pattern is the same as that employed by Rowitch and Milstein [370]. Rates corresponding to $\frac{8}{8+l}$ $(l = 0, 2, \ldots 16)$ are thus created. Due to the limited adaptation range of HARQ, three modulation schemes QPSK, 16-QAM, and 64-QAM are adopted. Different modulation and coding schemes create 27 combinations and 21 identical rates. The decoder we used at the receiver is the soft input Viterbi decoder with 8 iterations.

Similarly, we assume an ideal source compressor to be concatenated in front of HARQ. Note that the ideal rates are not practically achievable for either BICM or HARQ, because the ideal source compressor does not exist for short block length on the order of one thousand bits. However, we consider these two reference schemes as the upper bounds in the achievable rates by the modulation schemes for these two categories of rate adaptation.

JSCC: We aslo implement a practical joint source-channel coding scheme as described by Zhu and Alajaji [364]. It is essentially an HARQ scheme, so the same punctured Turbo code as in HARQ is used. The difference is that, instead of using a separate source encoder, JSCC utilizes the prior information at the decoder. Therefore, JSCC and HARQ become exactly the same when $p = 0.5$.

15.5.2 Simulations over an AWGN Channel

Simulations are carried out under static AWGN channels at integer channel SNRs (in dB) from 5 dB to 25 dB, which covers the main SNR range of most 802.11 systems. Figure 15.5 shows the results in terms of throughput versus the required channel SNR for BER less than 10^{-6}. For each scheme, we plot one curve for each sender setting (in terms of modulation and coding combination for BICM, modulation scheme for HARQ, and weight multiset for CM), and the curves beyond the effective range may be omitted for clarity. The Shannon limits, calculated by dividing the Shannon capacity by the source entropy, are also plotted.

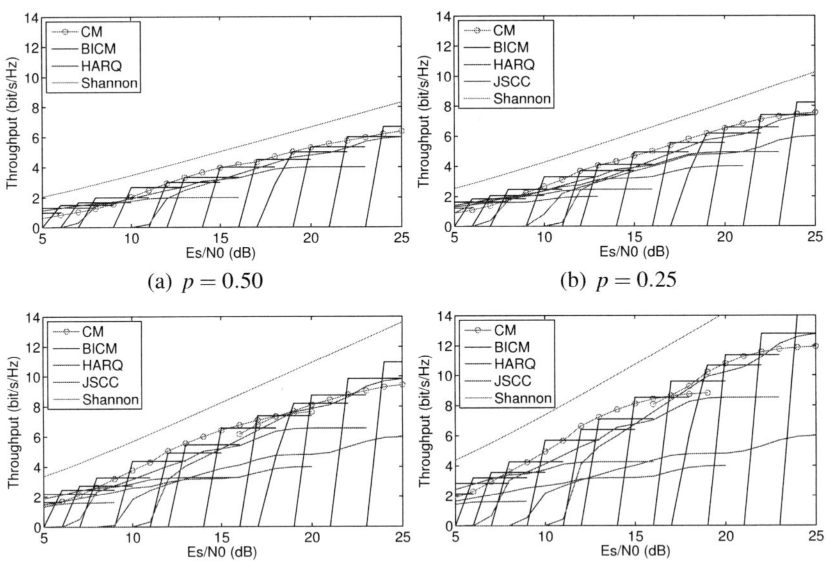

(a) $p = 0.50$ (b) $p = 0.25$ (c) $p = 0.15$ (d) $p = 0.10$

Figure 15.5 Comparing the throughput of CM and two reference schemes for different source sparsities.

Note that CM is not expected to achieve a higher throughput than the two reference schemes with ideal source coding. Through the simulations, we would like to show that CM has a much wider adaptation range than the two reference schemes while achieving a similar throughput as the ideal reference schemes. The performance of any practical implementations of separate source coding combined with BICM or HARQ would not achieve the throughput as the reference schemes, due to the inevitable loss in source coding.

Figure 15.5a,b show that, when the source is not sparse or has moderate sparsity, CM can use the same weight multiset for channels from 5 dB to 25 dB, with an adaptation range spanning over 20 dB. In contrast, for the same SNR range, BICM switches among 12 modulation and coding schemes, and HARQ switches among three modulation schemes. Such a wide adaptation range of CM brings significant benefits as the sender does not require channel state feedback, and avoids potential performance loss due to mismatch between the estimated and the actual channel states. The other reference scheme HARQ+JSCC has a similar adaptation range as the proposed CM, but its achievable rate is much lower.

Figure 15.5c,d show the results when the source sparsity is more prominent. We observe that the rate for CM saturates at lower SNR when p decreases. The saturation point is around 19 dB when $p = 0.15$ and is around 16 dB when $p = 0.1$. This suggests that, for very sparse data, CM may need to change the weight multiset when the channel becomes good. Figure 15.5c,d These two figures show that using two weight multisets is sufficient for CM to cover the main channel SNR range when $p = 0.15$ and $p = 0.1$.

15.5.3 Emulation in Real Channel Environment

We also implement the CM scheme on a software radio platform called SORA [332]. The evaluation scenarios include mobile Line-of-Sight (LOS), stationary Non-Line-of-Sight (NLOS) and mobile NLOS. For each scenario, the channel state is traced and a fair comparison between CM and two reference schemes HARQ and BICM are carried out over the traced CSI. We do not include JSCC in the emulation because the earlier simulations have shown that its achievable rate is much lower than that of the other schemes.

In this evaluation, CM selects a fixed weight multiset for the given source sparsity and does not adapt it to the channel condition (even when $p = 0.15$ and $p = 0.1$). Thus, CM completely avoid channel estimation and CSI feedback. The two reference schemes, as they need to switch between several modulation and coding settings, require channel feedback. We assume that the receiver could precisely measure the actual SNR from the received packet and provide immediate feedback, and then the sender uses the algorithm by Chen et al. [388] for rate adaptation.

Table 15.4 lists the average throughput of the three schemes under different source and channel settings. To provide more details of how each scheme behaves, we show the CDFs (cumulative distribution functions) of instant rates in Figures 15.6–15.8, respectively, for the three evaluated scenarios. Each point (d, T) on the curve can be

Table 15.4 Average rate in each scenario (bits/s/Hz).

Scenario	Scheme	$p = 0.5$	$p = 0.25$	$p = 0.15$	$p = 0.1$
	CM	4.74	5.83	7.23	8.50
LOS_M	HARQ	4.02	4.95	6.58	8.56
	BICM	3.64	4.49	5.97	7.76
	CM	3.74	4.79	6.18	7.42
NLOS_S	HARQ	2.93	3.62	4.81	6.25
	BICM	2.28	2.81	3.74	4.87
	CM	2.83	3.73	5.30	6.61
NLOS_M	HARQ	2.43	2.99	3.98	5.18
	BICM	2.16	2.66	3.54	4.60

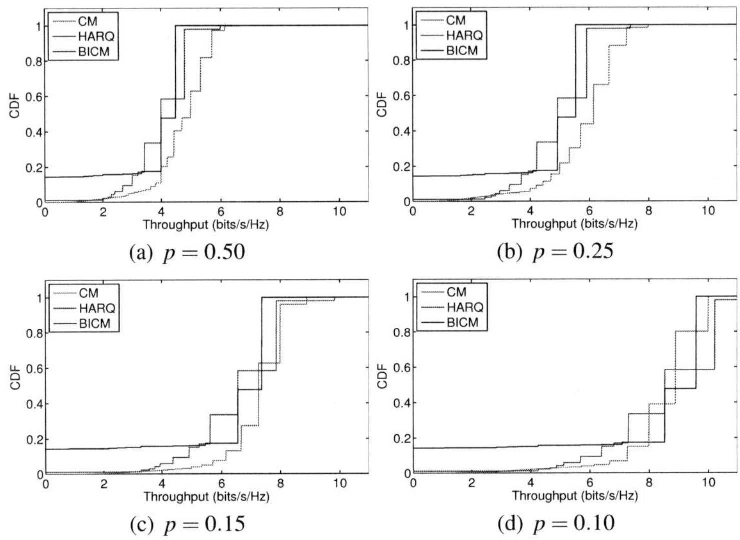

Figure 15.6 CDF of throughput in a mobile LOS case.

interpreted as "The rate in d portion of the time is below T." Therefore, the rightmost curve achieves the highest throughput.

Overall, CM performs the best and BICM the worst in the trace-driven emulation, despite the simulation results that suggest that BICM could achieve the highest rate in static channel settings. This clearly demonstrates the advantage of CM brought by using a fixed modulation. In the two NLOS scenarios, the performance gap between reference schemes and CM is even greater than that in LOS scenario, because the channel varies more dramatically and the two reference schemes make more improper decisions in rate selection. In NLOS static scenario, CM achieves 70% throughput gain over BICM when $p = 0.25$. In NLOS mobile scenario, CM achieves 33% throughput gain over HARQ when $p = 0.15$. These results amply confirm that CM has successfully achieved the design goal to embed data compression into seamless rate adaptation.

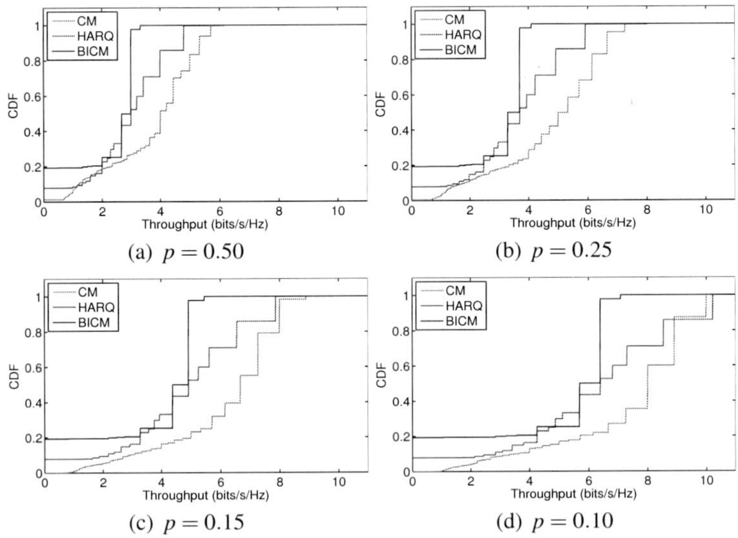

Figure 15.7 CDF of throughput in stationary NLOS case.

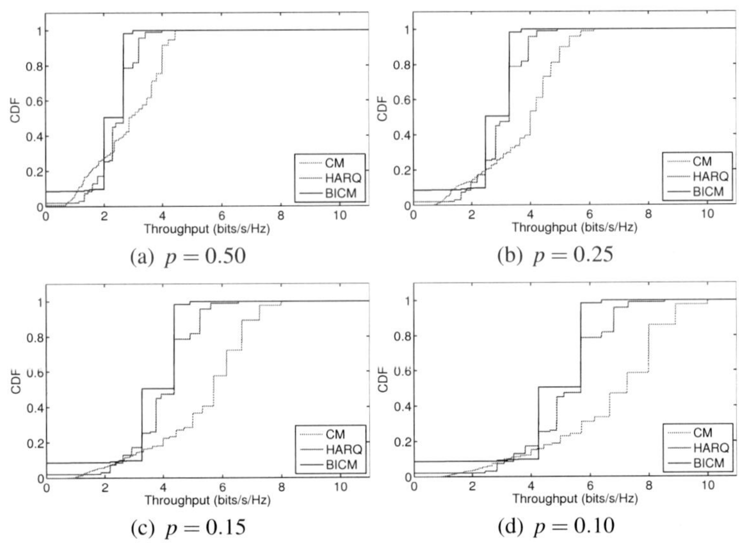

Figure 15.8 CDF of throughput in a mobile NLOS case.

15.6 Summary

We have described in this chapter a novel compressive coded modulation scheme capable of achieving simultaneously both source compression and seamless rate adaptation. In this novel scheme, data compression is elegantly embedded into coded modulation through a virtually rateless multilevel code named RP code. Through careful design, this RP encoder in turn generates multilevel symbols, which are sequentially and evenly mapped to a dense and universal constellation. Such innovative mapping maintains the statistics of channel noise to be used effectively during the process of RP-BP decoding.

We have also developed practical implementation of the proposed CM scheme. EXIT chart analysis is used for the weight selection of RP codes. Extensive simulations and emulations have been carried out to evaluate the performance of CM. The comparisons against conventional schemes show that CM has a much wider adaptation range than existing approaches, and consequently achieves a higher throughput under varying channel conditions. These promising results validate the practical feasibility of CM.

Part VI
Pseudo-Analog Transmission

In the current digital communications systems, image and video are compressed fully and thus the redundancy is completely removed in the compressed data. However, in the analog communications systems, image and video are not compressed at all. In this part, we are looking for new communications systems between digital and analog systems. A typical feature in such systems is that entropy coding is removed in source compression. Furthermore, when source is no longer entropy coded, conventional bit-based channel coding is unsuitable to source protection. Another typical feature in such systems is that channel coding is removed in data protection. Although the transmission is still implemented in a digital manner, the transmission behavior looks like that of analog systems. Therefore, we call this research pseudo-analog transmission.

Chapter 16 first presents a novel framework called DCast for distributed video coding and transmission over wireless networks, which is different from existing distributed schemes in three aspects. First, coset quantized DCT coefficients and motion data are directly delivered to the channel coding layer without syndrome or entropy coding. Second, transmission power is directly allocated to coset data and motion data according to their distributions and magnitudes without forward error correction (FEC). Third, these data are transformed by Hadamard and then directly mapped using a dense constellation (64K-QAM) for transmission without Gray coding. One of the most important properties in this framework is that the coding and transmission rate is fixed and distortion is minimized by allocating the transmission power. Thus we further propose a power distortion optimization algorithm to replace the traditional rate distortion optimization. This framework avoids the annoying cliff effect caused by the mismatch between transmission rate and channel condition. In multicast, each user can get almost the best quality matching its channel condition. Our experiment results show that the proposed DCast outperforms the typical solution using H.264 over 802.11 up to 8 dB in video peak signal-to-noise ratio (PSNR) in video broadcast. Even in video unicast, the proposed DCast is still comparable to the typical solution.

Chapter 17 challenges the conventional wisdom that video redundancy should be removed as much as possible for efficient communications. We discover that, by keeping spatial redundancy at the sender and properly utilizing it at the receiver, we can build a more robust and even more efficient wireless video communications system than existing ones. In the proposed framework, inter-frame (temporal) redundancy in video is removed at the encoder, but intra-frame (spatial) redundancy is retained. In doing so, pixel values after a transform-domain scaling are directly transmitted with amplitude modulation. At the receiver, spatial redundancy is utilized by image denoising. Note that denoising in our decoder is not a post-processing, but has to be immediately performed on channel output. We implement the video communications system, which we call *Cactus* on a SORA platform. The adopted image denoising algorithm is made real-time through graphic processing unit (GPU) implementation. Cactus is extensively evaluated in 802.11 a/g Wireless Local Area Network (WLAN) environment. On average, Cactus outperforms SoftCast by 4.7 dB in video PSNR and is robust to packet losses. In addition, Cactus is shown to be capable of transmitting high-definition videos in WLAN, and the performance is even better than an omniscient MPEG scheme.

Chapter 18 considers image delivery in multiple-input multiple-output (MIMO) broadcasting networks with diverse channel quality and varying numbers of antennas across receivers. In such systems, performance is normally constrained by the weakest users with either a low channel signal-to-noise ratio (SNR) or only a single receiver antenna. To address both dimensions of heterogeneity, we present a new analog image delivery system that adapts seamlessly along both dimensions simultaneously. The sender scales the DWT coefficients according to a power allocation strategy, and generates linear combinations of the coefficients using compressive sensing (CS), before transmitting them with amplitude modulation. On the receiving side, the received physical layer symbols are passed directly to the source decoder without conventional MIMO decoding, and the DWT coefficients are recovered using a CS decoder. There are two main contributions of our system. First, integrating CS into MIMO transmission ensures that the reconstructed image quality at the receivers is commensurate with both the channel SNR and the MIMO channel dimension. Second, a power allocation strategy is introduced to achieve a performance trade-off between receivers with different antenna numbers. Experimental results show that the presented system outperforms both the analog reference SoftCast and the conventional digital system known as HM-STBC. The average gain is 2.92 dB over SoftCast for single-antenna users and 1.53 dB over HM-STBC for two-antenna users.

Chapter 16
DCast: Distributed Video Multicast

16.1 Introduction

distributed video coding (DVC) [389–392] is an attractive scheme for video compression that has emerged over the past decade. Different from conventional video coding schemes, it utilizes cross-frame correlation only at decoder. This brings several unique advantages. First, DVC can shift intensive computation from encoder to decoder, which is appealing for low-complexity video encoding applications. Second, DVC framework is robust to transmission errors, which is desirable for wireless applications. Although it has been proven that the theoretical coding performance should be equivalent no matter what the source correlation is utilized at encoder or decoder for some typical sources [306, 393], the actual coding performance of DVC is still far inferior to that of the conventional H.264 standard [13].

In DVC, quantized transform coefficients are converted to bit planes and compressed to bits by syndrome or entropy coding [390, 392, 394]. The syndrome coding is implemented via channel codes (e.g., low-density parity-check codes). These channel codes are also typically applied for error protection in the physical (PHY) layer. Therefore, Xu *et al.* propose the first work on designing the joint source-channel coding for distributed video transmission [359]. Except for this, the transmission of distributed coded video still looks like that of conventional coded video. All data from the PHY layer have been corrected by channel coding and thus are error-free. But it does not fully take the advantage that distributed coded video is robust to transmission errors.

Recently, a joint video coding and transmission scheme, named SoftCast [395, 396], was proposed for wireless video multicasting. The key idea in SoftCast is that transform coefficients are not compressed by entropy coding. Instead, they are directly transmitted through a dense constellation after allocating a certain power, such that the received data can be decoded at any channel conditions. The decoded data is not error-free and its signal-to-noise ratio (SNR) is dependent on channel condition for a given transmission power. Although the video coding of SoftCast is simply through 2D or 3D transformation, the overall performance of SoftCast still outperforms the typical solution using H.264 over 802.11 in video multicast.

The current SoftCast only adopts 3D DCT to exploit cross-frame correlation. Many researches in scalable video coding have fully demonstrated that this is inefficient due to lack of motion alignment among frames [62, 79, 397]. However, motion compensation (MC) in H.264 is difficult to adopt in SoftCast because in SoftCast the reconstructed frames are determined by channel noise and the encoder cannot obtain the same reconstructed frames as the decoder. Thus this chapter presents a novel framework called DCast, which can not only utilize the cross-frame correlation by motion alignment but also keep the nice properties provided by SoftCast.

In the proposed DCast, transformed coefficients are first coset quantized and then are transmitted as SoftCast. Similar to other DVC frameworks, DCast utilizes the cross-frame correlation at the decoder. The proposed DCast has two different approaches to process motion vectors (MVs). Like most traditional DVC schemes, in the first approach motion vectors are estimated at the decoder. It does not need reference frames at the encoder and greatly reduces encoding complexity. But the side information may be inaccurate, thus leading to low coding efficiency. Several DVC schemes also propose to estimate motion vectors at the encoder and transmit them to the decoder for improving the quality of side information [398, 399]. In the second approach, motion vectors are estimated at the encoder and then transmitted to the decoder. The initial results of these two approaches have been reported in our conference papers [400, 401]. In this chapter we will focus our study on the second approach but both of them will be evaluated.

The key technical contribution is the proposed power distortion optimization. In the proposed DCast, each pair of quantized discrete cosine transform (DCT) coefficients or transformed motion vector is transmitted by a time slot and thus the transmission rate is fixed. The distortion is minimized by optimally allocating transmission power. This chapter evaluates the impact of channel noise on the distortion of motion vectors and then the impact of this distortion on the distortion of reconstructed video via the power spectrum approach [72]. Furthermore, a joint power optimization among coefficients and motion data is derived. Our experimental results show that the proposed DCast can outperform SoftCast up to 2 dB in video PSNR because it can better utilize the cross-frame correlation. Compared with the typical solution using H.264 over 802.11, the proposed DCast can gain up to 8 dB in video PSNR in multicast. Even in unicast, it is still comparable to the typical solution of H.264 over 802.11.

16.2 Related Works

16.2.1 Distributed Video Coding

To compress a source with its prediction only available at the decoder is a typical problem in distributed source coding (DSC). As shown in Figure 16.1, X is the

Figure 16.1 Compression of X when its side information S is available at the decoder.

source to be compressed (possibly representing the source video), and S is its side information (possibly representing the predicted frame). The theoretical foundations of DSC, the Slepian-Wolf theorem [306] and the Wyner-Ziv theorem [393], show that the source X can be efficiently compressed with its predictor S only available at the decoder. In practice, efficient DSC can be achieved by coset coding, Turbo coding, and LDPC coding [356, 402].

Accompanied by the advances of practical DSC solutions, DVC has emerged for a decade. Puri et al. propose a DVC framework called PRISM, which implements DVC by coset coding and supports motion estimation (ME) at the decoder [391,392]. The main attributes of PRISM include increased robustness to channel losses and more flexible sharing of computational complexity between encoder and decoder. Another DVC work is the low-complexity framework proposed by Aaron et al. [389, 390]. In this framework, the DVC is implemented by Turbo code, while the motion estimation at the decoder is based on motion compensated interpolation (MCI) and motion compensated extrapolation (MCE).

Although DVC has shown unique advantages in visual communication, its compression efficiency is still much lower than conventional framework such as H.264 [13]. In recent years, many researchers have focused on improving the performance of DVC. Enabling transform coding [403, 404] and intra/inter-mode selection [405–407] allow DVC to exploit not only inter but also intra-frame redundancy. Hash-based DVC lets the encoder send hash code to the decoder to improve the accuracy of ME and the side information quality [408]. Successive refinement schemes [409–412] perform ME and DVC decoding alternatively and recursively, such that the MVs and reconstruction frame are successively refined during decoding process. More accurate correlation estimation in DVC improves the utilization of the side information [413–416].

Compared with these DVC schemes, the proposed DCast does not use syndrome [390] or entropy coding, and directly delivers coefficients and motion vectors to the PHY layer. Furthermore, when coefficients and motion vectors are transmitted from encoder to decoder, they are allowed to be corrupted by channel noise. It is clear from our results that DVC is robust to noise added in the received data.

16.2.2 Distributed Video Transmission

The transmission of distributed coded video is usually similar to the transmission of conventional coded video in the PHY layer of wireless network. Coded binary data is first protected by channel coding and then is mapped to a constellation for transmission. When syndrome coding is adopted, DVC coding and channel coding can be jointly optimized. The first attempt to study DVC from a joint source-channel coding (JSCC) perspective is by Xu et al. [359]. It is a layered coding scheme, where the enhancement layer uses Raptor code for both video compression and data protection. In another framed-based JSCC scheme [398], the functionality of both DVC and channel coding is implemented universally by one error correction code.

In these JSCC schemes, distributed video transmission is actually processed as data transmission. The successfully decoded data in the PHY layer does not contain any transmission errors. Thus many bits are paid on channel coding to correct all transmission errors. Obviously, it does not fully take the advantage of distributed video coding. In the proposed DCast, quantized coefficients and transformed motion vectors are directly transmitted after allocating a certain power. Although the received data after decoding may still contain a certain amount of channel noise, it is more efficient on power consumption because some received noise can be tolerated by DVC.

16.2.3 SoftCast

SoftCast is a simple but comprehensive design for wireless video multicast, covering the functionality of video compression, data protection, and transmission in one scheme [395]. The SoftCast encoder consists of the following components: DCT transform, power allocation, Hadamard transform, and direct dense modulation. Transform removes spatial redundancy of a video frame. Power allocation minimizes the total distortion by optimally scaling transform coefficients. Hadamard transform is in some sense a precoding to make packets with equal power and equal importance. After that, the data is directly mapped into wireless symbols by a very dense quadrature amplitude modulation (QAM). The decoder uses the linear least square estimator (LLSE) algorithm to reconstruct the signal. All the components in SoftCast are linear operations and thus channel noise is directly transferred into reconstruction noise of the video. Therefore, SoftCast is asymptotically robust in the sense that each user can get the visual quality matching its channel condition.

However, SoftCast exploits intra-frame correlation only and thus is not efficient in the aspect of video compression. Recently, Aditya et al. [354] propose another video coding and transmission scheme, called Flexcast. It removes entropy coding from conventional video coding and adopts rateless channel coding for channel variation. Thus it has better coding efficiency. However, Flexcast is a unicast approach and can hardly multicast or broadcast video to the users with different SNRs simultaneously because of motion compensation. In a recent improved version of SoftCast, the utilization of 3D DCT partially enables inter-frame compression [396]. However, without motion alignment inter-frame correlation is still not fully exploited yet.

The proposed DCast not only fully utilizes the cross-frame correlation but also keeps the good properties of SoftCast. DCast enables inter-frame coding by DVC rather than conventional motion compensation. Instead of transmitting a video frame itself like SoftCast, DCast transmits coset codes of the video frame such that the frame can be reconstructed by utilizing the prediction frame as side information at decoder. This saves the transmission power (or equivalently increases the SNR) because coset data has typically much smaller magnitude than original data. Recently, we also noted that Kochman and Zamir [417] have studied the utilization of coset coding in the Wyner-Ziv Dirty-Paper problem and proved its optimality and asymptotical robustness in multicast. It can be considered in general as the theoretical foundation to support the proposed DCast.

16.3 Proposed DCast

DCast divides an input video sequence into groups of pictures (GOP). In each GOP, the first frame is intra coded, while the following frames are inter coded. The compression and transmission of intra frame in DCast is the same as in SoftCast, which consists of DCT, power allocation, and Hadamard transform. In the rest of this chapter, we will focus on the compression and transmission of inter frames. For simplicity, we mainly discuss the case with motion vectors estimated at encoder.

Figure 16.2 depicts the server side of DCast. DCast first transforms the current frame into DCT domain. Meanwhile, DCast performs ME and MC on the original video sequence to get predictions and MVs. Then DCast applies coset coding on the transform coefficients of the original frame to get, for each DCT coefficient, the coset data. The quantization step size of the coset coding is determined at encoder according to the estimated prediction noise of the decoder. The MVs of the current frame, in the form of a matrix, are also transformed by DCT. The coset data and the motion data are then scaled for power distortion optimization (PDO).

The scaling factors and other meta data are transmitted by using a conventional scheme consisting of variable length coding (VLC), forward error correction (FEC), and binary phase shift keying (BPSK) modulation. The scaled coefficients are transformed by Hadamard as precoding to make packets with equal power and equal importance. After that, the resulting coefficients are mapped to complex symbols directly by a dense constellation (64K-QAM): each coefficient is quantized into

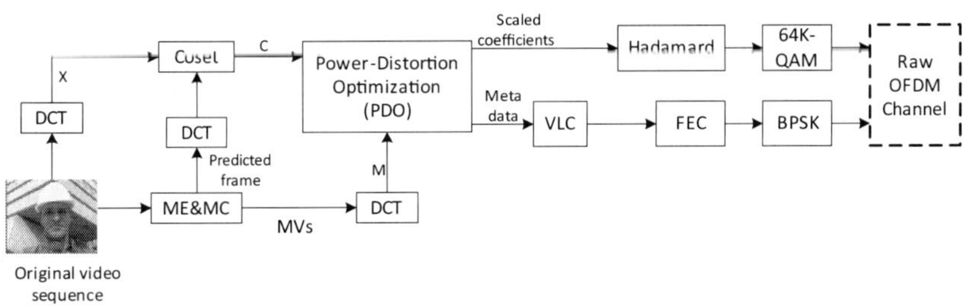

Figure 16.2 DCast server for inter frames.

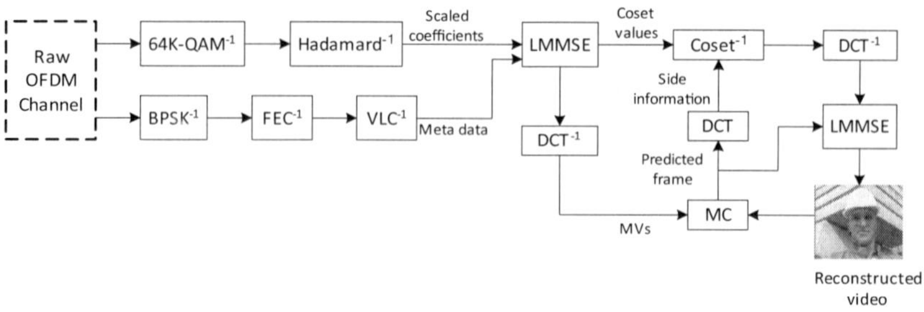

Figure 16.3 DCast receiver for inter frames.

8-bit integer number and every two integers compose one complex number of 64K possible values. At last, these complex numbers are passed into a raw orthogonal frequency division multiplexing (OFDM) module undergoing inverse fast Fourier transform (iFFT) and D/A conversion for transmission.

The receiver side of DCast is depicted in Figure 16.3. The raw OFDM module performs A/D conversion and FFT to reconstruct modulated data including both scaled coefficients and metadata. The metadata is demodulated and decoded first. Then the scaled coefficients are reconstructed by inverse 64K-QAM and inverse Hadamard transform. The inverse 64K-QAM here does nothing but splitting each complex value back into two real values. Each real value here is actually the 8-bit integer number plus channel noise.

After inverse Hadamard transform, the linear minimum mean square error (LMMSE) estimation of residual coefficients and MV coefficients is performed. Then the MVs are transformed back to a spatial domain by inverse DCT. After this, the MC module generates the predicted frame by the MVs and the reference frame. The predicted frame is transformed into a frequency domain by DCT. Then with the coset residues and the predictors, the coset decoding module recovers the DCT coefficients of the current frame. At last, the signals are transformed back to a spatial domain, and are linearly combined with the predicted signals by LMMSE to generate the final reconstruction.

16.3.1 Coset Coding

Coset coding is a typical technique used in DSC. It partitions the set of possible input source values into several cosets and transmits the coset index to the decoder. With the coset index and the predictor, the decoder can recover the source value by choosing the one in the coset closest to the predictor. Coset coding achieves compression because the coset index has typically lower entropy than the source value.

Let X be DCT coefficients of the original video frame. DCast encodes X to get coset values C. DCast divides the coefficients into 64 subbands according to the frequency. Let X_i be the ith subband of X, and C_i be the ith subband of C. For each

i, DCast quantizes the ith subband of X by a uniform scaler quantizer $\mathsf{Q_i}(\cdot)$ and gets the residual value [417] by

$$C_i = X_i - \mathsf{Q_i}(X_i) = X_i - \lfloor \frac{X_i}{q_i} + \frac{1}{2} \rfloor q_i. \tag{16.1}$$

This coset coding is actually throwing away the main part of X. In some sense C represents the detail of X.

At the client side, with the side information S (i.e., the predicted DCT coefficients) and the received coset value $\hat{C}$, the receiver reconstructs the DCT coefficients by coset decoding. Let S_i be the ith subband of S, and $\hat{C}_i$ be the ith subband of $\hat{C}$. Since S_i is close to X_i, $S_i - \hat{C}_i$ is around $X_i - C_i$. Thus $S_i - \hat{C}_i$ is around $\mathsf{Q_i}(X_i)$ from Eq. (16.1). The quantizers are carefully designed such that applying quantization $\mathsf{Q_i}(\cdot)$ on $S_i - \hat{C}_i$ we could get $\mathsf{Q_i}(X_i)$, that is,

$$\mathsf{Q_i}(X_i) = \mathsf{Q_i}(S_i - \hat{C}_i), \tag{16.2}$$

in high probability. Therefore, each subband of coefficients is decoded by

$$\hat{X}_i = \mathsf{Q_i}(S_i - \hat{C}_i) + \hat{C}_i, \tag{16.3}$$

where $\hat{X}$ is the reconstruction of X, and each $\hat{X}_i$ is the ith subband of $\hat{X}$. When the coset decoding is successful, that is, $\mathsf{Q_i}(X_i) = \mathsf{Q_i}(S_i - \hat{C}_i)$, the reconstruction noise is

$$\hat{X}_i - X_i = \hat{C}_i - C_i. \tag{16.4}$$

16.3.2 Coset Quantization

The value of each coset step q_i is crucial to the coding performance of DCast. If q_i is too small, the coset decoding may suffer failure. On the other side, if q_i is too large, the coset value C_i in Eq. (16.1) will be large and will consume a lot of transmission power to keep the distortion small. The value of each q_i is determined as follows. Injecting Eq. (16.1) into Eq. (16.2), we get

$$\begin{aligned} \mathsf{Q_i}(X_i) &= \mathsf{Q_i}(S_i - \hat{C}_i + C_i - X_i + \mathsf{Q_i}(X_i)) \\ &= \mathsf{Q_i}(X_i) + \mathsf{Q_i}(S_i - \hat{C}_i + C_i - X_i). \end{aligned} \tag{16.5}$$

To guarantee successful coset decoding, the last item should be 0. This means the quantization step q_i should satisfy

$$\frac{q_i}{2} \geq |S_i - X_i + C_i - \hat{C}_i|. \tag{16.6}$$

In this equation, $S_i - X_i$ is the prediction noise at decoder and $C_i - \hat{C}_i$ is the reconstruction noise of coset value C_i due to transmission. In this work, we assume they are independent Gaussian sources. We let each q_i be $2n$ times of the standard deviation of $S_i - X_i + C_i - \hat{C}_i$, that is,

$$q_i^2 = 4n^2 \sigma^2_{S_i - X_i + C_i - \hat{C}_i}, \tag{16.7}$$

and this guarantees the condition Eq. (16.6) is satisfied in probability

$$\mathsf{Pr} = \mathsf{erf}(n/\sqrt{2}). \tag{16.8}$$

Under the same assumption, the variance of $S_i - X_i + C_i - \hat{C}_i$ is the summation of the variance of $S_i - X_i$ and $C_i - \hat{C}_i$, that is,

$$\sigma^2_{S_i - X_i + C_i - \hat{C}_i} = \sigma^2_{S_i - X_i} + \sigma^2_{C_i - \hat{C}_i}, \tag{16.9}$$

and each q_i can be calculated by

$$q_i^2 = 4n^2(\sigma^2_{S_i - X_i} + \sigma^2_{C_i - \hat{C}_i}). \tag{16.10}$$

In our implementation, we let $n = 3$ such that the coset decoding is successful for more than 99.7% coefficients. In Eq. (16.10), $\sigma^2_{S_i - X_i}$ is the variance of the hypothetic residue between the source and the side information, and it is estimated by simulating at encoder a receiver with target channel SNR. $\sigma^2_{C_i - \hat{C}_i}$ is the distortion of coset value C_i due to transmission. It is also the distortion of the source X_i according to Eq. (16.5). $\sigma^2_{C_i - \hat{C}_i}$ is related to both the residue $\sigma^2_{S_i - X_i}$ and the channel SNR. The explicit expression of $\sigma^2_{C_i - \hat{C}_i}$ is given in Section 16.4.4.

16.3.3 Power Allocation

DCast transmits both coset values and motion information. Thus it has two levels of power allocation. The first allocation is between MV data and coset data. The second level means the allocation within MV coefficients or coset coefficients. The optimal power allocation between MV data and coset data is given in Section 16.4. The optimal power allocation within coset coefficients and MV coefficients are as follows.

Let P_{coset} be the total power for coset data, and g_{C_i} be the gain (scaling factor) of C_i. The problem is how to minimize the reconstruction distortion of X, by optimally allocating power among C_i. Under the assumption that the coset decoding is successful in high probability, the reconstruction distortion of X will be equal to the reconstruction distortion of C according to Eq. (16.10). This means the problem becomes how to minimize the reconstruction distortion of C, by optimally allocating power among C_i. Thus, the solution has similar form as the one in SoftCast [396], that is,

$$\tilde{C}_i = g_{C_i} C_i, \quad g_{C_i} = \left(\frac{P_{coset}}{\sigma_{C_i} \sum_j \sigma_{C_j}} \right)^{1/2}, \tag{16.11}$$

where $\tilde{C}$ is the coset value after power allocation, $\tilde{C}_i$ is the ith subband of $\tilde{C}$, and σ_{C_i} is the standard deviation of C_i. This power allocation tends to scale down large coefficients to get better performance under the constrained total power. The encoder calculates the variance $\sigma^2_{C_i}$ for each subband and transmits it to the decoder. With $\sigma^2_{C_i}$, both encoder and decoder calculate the gain g_{C_i} for each C_i by Eq. (16.11).

For MV data, DCast also performs power allocation. To apply power allocation, the encoder performs 2D DCT on MVs (the whole MV field) and gets transform coefficients M. Note that each MV contains horizontal and vertical components and the transform is actually applied to both components separately. Each coefficient M_i is then considered as a subband. The encoder applies a similar optimal power allocation over M, that is,

$$\tilde{M}_i = g_{M_i} M_i, \quad g_{M_i} = \left(\frac{P_{mv}}{\sigma_{M_i} \sum_j \sigma_{M_j}} \right)^{1/2}, \tag{16.12}$$

where $\tilde{M}$ is MV data after power allocation, $\tilde{M}_i$ is the ith subband of $\tilde{M}$, σ_{M_i} is the standard deviation of M_i, and P_{mv} is the total power for motion data. Since each subband of M only contains one coefficient, it is not efficient to transmit the variance of each subband. In light of this, DCast only transmits the average variance $\sigma^2_M = \frac{1}{n} \sum_i \sigma^2_{M_i}$ where n is the number of subbands. As shown in our previous work [401], the $\sigma^2_{M_i}$ and g_{M_i} are calculated by using σ^2_M. Under the assumption that the motion field is random Markov field, where the correlation coefficient between two neighboring MVs is ρ, each $\sigma^2_{M_i}$ can be calculated by

$$\sigma^2_{M_i} = \sigma^2_M V_{M_i}, \tag{16.13}$$

where V_{M_i} is the ith element of matrix V_M, and

$$V_M = \mathsf{diag}(\mathsf{2D\_DCT}(R^{(h)}))\mathsf{diag}(\mathsf{2D\_DCT}(R^{(w)}))^T \tag{16.14}$$

is a constant matrix for given ρ. Here, function $\mathsf{diag}(\cdot)$ produces the diagonal elements of the input matrix in the form of a column vector. $\mathsf{2D\_DCT}(\cdot)$ means 2D DCT transform. w and h are the width and height of the motion field, respectively, and

$$R^{(k)} = \begin{bmatrix} 1 & \rho & \cdots & \rho^{k-1} \\ \rho & 1 & \cdots & \rho^{k-2} \\ \vdots & \vdots & \ddots & \vdots \\ \rho^{k-1} & \rho^{k-2} & \cdots & 1 \end{bmatrix}. \tag{16.15}$$

The value of σ_M^2 is calculated at the encoder and is transmitted to decoder as mentioned in the previous section. Both the encoder and the decoder calculate the value of each $\sigma_{M_i}^2$ by Eqs. (16.13 to 16.15). In our experiments, we let $\rho = 0.7$ according to statistics over several different video sequences. With each $\sigma_{M_i}^2$, the optimal power allocation gain g_{M_i} for each subband is calculated at both encoder and decoder by Eq. (16.12). The decoder needs the value of g_{M_i} in Eq. (16.12) to reconstruct the signal.

16.3.4 Packaging and Transmission

Similar to SoftCast [396], DCast transmits not only a small amount of binary symbols but also mainly real value symbols. The organization of the symbol stream is as follows. The symbol stream consists of a header and a following data stream

$$symbol\_stream = \{header\_bitstream, data\_stream\}. \tag{16.16}$$

The header bitstream contains coset variances $\sigma_{C_i}^2$, quantization steps q_i, average MV variance σ_M^2 and other useful parameters,

$$\begin{aligned} header\_bitstream \leftarrow \; & \{coset\,variances, \\ & quantization\,steps, \\ & average\,MV\,variance, \\ & parameters\}. \end{aligned} \tag{16.17}$$

The header information is coded in a conventional way. The encoder applies 8-bits scalar quantization on σ_{C_i}, q_i, and σ_M, respectively. Then the quantization results are compressed by variable length coding (VLC). The VLC is the universal one used for coding motion vectors in H.264 [13]. The compressed header bitstream is transmitted by the standard 802.11 PHY layer at the lowest speed, that is, by using 1/2 convolutional code and BPSK modulation. This is to make sure that the header bits are decoded correctly when channel SNR is in typical working range (5–25 dB) of 802.11. Note that the size of the header is very small with respect to the whole data of one frame. According to our experiments, the proportion of the bandwidth required by the header is less than 3%.

The data stream contains coset data $\tilde{C}$ and MV data $\tilde{M}$. Similar to SoftCast [396], DCast applies Hadamard transform on the coset data $\tilde{C}$ and the MV data $\tilde{M}$ to create packets with equal energy. Coset data and MV data are mixed together and then every 64 numbers are grouped for Hadamard transform. This forms the data stream

$$data\_stream \xleftarrow{H} \{coset\,data,\, MV\,data\}. \tag{16.18}$$

Note that the data stream consists of real values rather than binary values. In the PHY layer, these real values are mapped to complex symbols directly by 64K-QAM constellation [396]. This constellation is a typical N-QAM constellation with N equal to 65,536 (256 by 256). Each input real value is quantized into an 8-bit integer number by uniform scalar quantizer. The dynamic range of the quantizer is formed by the minimal and maximal input values. It is calculated for each frame at encoder and sent to decoder as a parameter in Eq. (16.17). After this quantization, every two integers compose one complex number as the output of the 64K-QAM constellation. An inverse FFT is computed on each packet of symbols, giving a set of complex time-domain samples. These samples are then quadrature-mixed to passband in the standard way. The real and imaginary components are first converted to the analog domain using D/A converters; the analog signals are then used to modulate cosine and sine waves at the carrier frequency, respectively. These signals are then summed to generate the transmission signal.

In DCast, both MV data and coset data are transmitted by the aforementioned direct source channel mapping. This makes the system adaptive to the fluctuation of the channel SNR. Given the same transmitter, high SNR users would receive accurate MVs and coset values, and reconstruct high quality video. Meanwhile, low SNR users would receive noisy MVs and coset values, and derive noisy prediction frames based on the noisy MVs. However, the coset decoding in DCast has good tolerance to the noise of the prediction. Thus the low SNR users would still reconstruct the video.

16.3.5 LMMSE Decoding

The proposed approach contains two linear minimal mean square error (LMMSE) estimators, operating in transform domain and spatial domain, respectively.

The first LMMSE estimator is to reconstruct coset data C and MV data M in transform domain with minimum distortion. Let Y be the received signal after inverse Hadamard transform. Y contains the noisy version of coset data and MV data. Y can be written as:

$$Y = \begin{bmatrix} \dot{C} \\ \dot{M} \end{bmatrix}, \tag{16.19}$$

where $\dot{C}$ is the noisy version of coset data, $\dot{M}$ is the noisy version of MV data. Let $W^{(C)}$ and $W^{(M)}$ be the channel noise in $\dot{C}$ and $\dot{M}$, respectively. Let $\dot{C}_i$, $\dot{M}_i$, $W_i^{(C)}$, and $W_i^{(M)}$ be the ith subband of $\dot{C}$, $\dot{M}$, $W^{(C)}$, and $W^{(M)}$, respectively. We model each element in $W^{(C)}$ and $W^{(M)}$ as identically distributed (i.i.d.) Gaussian source with variance N_0. Each subband of $\dot{C}$ and $\dot{M}$ can be expressed as

$$\dot{C}_i = g_{C_i} C_i + W_i^{(C)}, \quad \dot{M}_i = g_{M_i} M_i + W_i^{(M)}. \tag{16.20}$$

Therefore, the LMMSE reconstruction of original signals is

$$\hat{C}_i = \frac{\sigma_{C_i}^2}{\sigma_{C_i}^2 g_{C_i}^2 + N_0} \dot{C}_i, \quad \hat{M}_i = \frac{\sigma_{M_i}^2}{\sigma_{M_i}^2 g_{M_i}^2 + N_0} \dot{M}_i. \tag{16.21}$$

And the reconstruction distortion of each subband is

$$\mathbb{E}\{(\hat{C}_i - C_i)^2\} = \frac{\sigma_{C_i}^2 N_0}{\sigma_{C_i}^2 g_{C_i}^2 + N_0}, \tag{16.22}$$

$$\mathbb{E}\{(\hat{M}_i - M_i)^2\} = \frac{\sigma_{M_i}^2 N_0}{\sigma_{M_i}^2 g_{M_i}^2 + N_0}. \tag{16.23}$$

The purpose of the second LMMSE estimator is to reconstruct each pixel x in spatial domain with minimum distortion. DCast decoder applies inverse DCT transform on coset reconstruction $\hat{X}$ and gets a pixel-domain preliminary reconstruction $\hat{x}$. $\hat{x}$ is considered as the first noisy version of x. DCast also has the predicted pixel s as the second noisy version of x. With $\hat{x}$ and s, the optimal LMMSE estimation x^* is given by:

$$x^* = \theta s + (1-\theta)\hat{x}, \tag{16.24}$$

where

$$\theta = \frac{\sigma_{\hat{x}-x}^2}{\sigma_{s-x}^2 + \sigma_{\hat{x}-x}^2}. \tag{16.25}$$

$\sigma_{\hat{x}-x}^2$ is the variance of $\hat{x}-x$, and σ_{s-x}^2 is the variance of $s-x$. In DCast, the prediction noise variance σ_{s-x}^2 is estimated at block level. Since $\hat{x}$ is close to x, σ_{s-x}^2 is estimated by calculating $\mathbb{E}\{(s-\hat{x})^2\}$. The variance $\sigma_{\hat{x}-x}^2$ is calculated as follows. According to Parseval's theorem and Eq. (16.5), we have

$$\sigma_{\hat{x}-x}^2 = \mathbb{E}\{(\hat{x}-x)^2\} = \mathbb{E}\{(\hat{X}-X)^2\} = \mathbb{E}\{(\hat{C}-C)^2\}, \tag{16.26}$$

where $\mathbb{E}\{(\hat{C}-C)^2\}$ is directly calculated by summation on Eq. (16.22).

16.4 Power-Distortion Optimization

In DCast, both MVs and coset values require power to transmit. Thus it is necessary to investigate the optimal power allocation between MVs and coset values. Let D be the reconstruction distortion, and P be the transmission power. Let P_{coset} and P_{mv} be the transmission power for coset values and MVs, respectively. The optimal power allocation is the one minimizing the reconstruction distortion D for a given power P, that is, the optimization problem is

$$\begin{aligned} \min \quad & D, \\ \text{s.t.} \quad & P_{mv} + P_{coset} \le P. \end{aligned} \tag{16.27}$$

16.4.1 Relationship between Variables

The distortion D is directly related to both the decoder prediction noise variance σ_{S-X}^2, and the coset transmission power P_{coset}. Intuitively, using larger transmission power P_{coset} decreases the variance of the coset error $\hat{C}-C$ at decoder. This means smaller D because the reconstruction error $\hat{X}-X$ equals to the coset error $\hat{C}-C$ according to Eq. (16.5). Meanwhile, larger σ_{S-X}^2 means lower quality of side information (SI), and lower quality SI leads to larger reconstruction distortion. Therefore, the distortion D should be a decreasing function of the coset power P_{coset} and an increasing function of the prediction noise variance σ_{S-X}^2.

Furthermore, the prediction noise variance σ_{S-X}^2 is related to the MV transmission power P_{mv}. We use two dimensional random vector $\Delta \sim \mathcal{N}(0, \sigma_\Delta^2 \mathbf{I}_{2\times 2})$ to model MV error, while $\sigma_\Delta^2 = \frac{1}{2}\mathbb{E}\{\Delta^T \Delta\}$ is the distortion of MV. Using larger transmission power P_{mv} decreases the MV distortion σ_Δ^2 and this means more accurate MVs. More accurate MVs produces higher quality of SI S at decoder, and hence smaller prediction noise variance σ_{S-X}^2. Thus the prediction noise variance σ_{S-X}^2 should decrease with the increase of the MV transmission power P_{mv}.

However, due to the power constraint, allocating more power to coset (i.e., larger P_{coset}) means less power to MV (i.e., smaller P_{mv}), and vice versa. This is why we need power distortion optimization. In the following part of this section, before solving Eq. (16.19), we will derive the relationship between

- MV transmission power P_{mv} and MV distortion σ_Δ^2
- MV distortion σ_Δ^2 and prediction noise variance σ_{S-X}^2
- Distortion D, coset power P_{coset}, and prediction noise variance σ_{S-X}^2

16.4.2 MV Transmission Power and Distortion

This section focuses on the relationship between motion vector (MV) transmission power P_{mv} and MV distortion σ_Δ^2. According to Parseval's theorem, the MV distortion σ_Δ^2 in spatial domain equals to the MV distortion in DCT domain, that is,

$$\sigma_\Delta^2 = \frac{1}{n_{mv}} \sum_i \mathbb{E}\{(\hat{M}_i - M_i)^2\}, \tag{16.28}$$

where n_{mv} is the number of MV coefficients. From Eq. (16.22), we get

$$\sigma_\Delta^2 = \frac{1}{n_{mv}} \sum_i \frac{\sigma_{M_i}^2 N_0}{\sigma_{M_i}^2 g_{M_i}^2 + N_0} \approx \frac{1}{n_{mv}} \sum_i \frac{N_0}{g_{M_i}^2}, \tag{16.29}$$

where the approximation is accurate when $P_{mv} \gg N_0$. Substituting Eq. (16.12) into Eq. (16.29), we get

$$\sigma_\Delta^2 \approx \frac{N_0(\sum_i \sigma_{M_i})^2}{n_{mv}P_{mv}}. \tag{16.30}$$

Then using Eq. (16.13), we get

$$\sigma_\Delta^2 \approx \frac{N_0\sigma_M^2(\sum_i V_{M_i}^{\frac{1}{2}})^2}{n_{mv}P_{mv}}. \tag{16.31}$$

By defining

$$\alpha_{mv} = (\frac{1}{n_{mv}}\sum_i V_{M_i}^{\frac{1}{2}})^2, \tag{16.32}$$

we can rewrite Eq. (16.31) as

$$\sigma_\Delta^2 \approx \frac{n_{mv}N_0\sigma_M^2\alpha_{mv}}{P_{mv}} = \alpha_{mv}\sigma_M^2\left(\frac{P_{mv}}{n_{mv}N_0}\right)^{-1}. \tag{16.33}$$

In this equation, σ_M^2 is the variance of the MV signal to transmit, $\frac{P_{mv}}{n_{mv}N_0}$ is the SNR for MV signal. Thus α_{mv} can be considered as the extra gain owning to the power allocation in Eq. (16.12). From this equation, the MV distortion σ_Δ^2 is proportional to the inverse of the MV transmission power P_{mv}.

16.4.3 MV Distortion and Prediction Noise Variance

This section focuses on the relationship between MV distortion σ_Δ^2 and prediction noise variance σ_{S-X}^2. Let $\dot{S}$ be the original decoder prediction when the MVs are perfectly received. The practical decoder prediction noise $S-X$ consists of two components: the original prediction noise $\dot{S}-X$, and the additional prediction noise $S-\dot{S}$ caused by erroneous MVs. In this chapter, we assume they are independent of each other, and therefore

$$\sigma_{S-X}^2 = \sigma_{\dot{S}-X}^2 + \sigma_{S-\dot{S}}^2. \tag{16.34}$$

Given that the $\dot{S}$ is a phase-shift version of S, $\sigma_{S-\dot{S}}^2$ can be analyzed by using power density. Similar to the derivation by Secker and Taubman [72], we have

$$\sigma_{S-\dot{S}}^2 = \frac{1}{4\pi^2}\int_{-\pi}^{\pi}\int_{-\pi}^{\pi} 2\Phi_{ss}(\omega)(1-\mathbb{E}\{cos(\omega^T\Delta)\})d\omega, \tag{16.35}$$

where $\Phi_{ss}(\cdot)$ is the power density function of side information, ω is two-dimensional frequency (in radians), and $\Delta \sim \mathscr{N}(0, \sigma_\Delta^2 \mathbf{I}_{2\times 2})$ is the MV error. For small σ_Δ^2, we have

$$1 - \mathbb{E}\{cos(\omega^T \Delta)\}) \approx \frac{1}{2}\mathbb{E}(\omega^T \Delta)^2 = \frac{1}{2}\sigma_\Delta^2 \omega^T \omega, \tag{16.36}$$

and thus

$$\sigma_{S-\dot{S}}^2 \approx \frac{1}{4\pi^2}\sigma_\Delta^2 \int_{-\pi}^{\pi}\int_{-\pi}^{\pi} \Phi_{ss}(\omega)\omega^T \omega d\omega. \tag{16.37}$$

We define

$$\gamma = \frac{1}{4\pi^2}\int_{-\pi}^{\pi}\int_{-\pi}^{\pi} \Phi_{ss}(\omega)\omega^T \omega d\omega, \tag{16.38}$$

and γ is a constant for a given video frame. Then we get

$$\sigma_{S-\dot{S}}^2 \approx \gamma \sigma_\Delta^2. \tag{16.39}$$

Substituting Eq. (16.39) into Eq. (16.34), we get

$$\sigma_{S-X}^2 = \sigma_{\dot{S}-X}^2 + \gamma \sigma_\Delta^2. \tag{16.40}$$

Therefore, the prediction noise variance σ_{S-X}^2 is linear to the MV distortion σ_Δ^2.

16.4.4 Distortion Formulation

The derivation of the distortion D is as follows. First, from Eq. (16.5) we have $\hat{X} - X = \hat{C} - C$ in high probability. Thus the distortion D approximately equals to the distortion of the coset value, that is,

$$D = \sigma_{\hat{X}-X}^2 \approx \sigma_{\hat{C}-C}^2. \tag{16.41}$$

Similar to Section 16.4.2, we can derive and express the coset distortion as

$$\sigma_{\hat{C}-C}^2 \approx \alpha_{coset} \sigma_C^2 \left(\frac{P_{coset}}{n_{coset} N_0}\right)^{-1}, \tag{16.42}$$

where α_{coset} is the coding gain of power allocation, σ_C^2 is the variance of C, and n_{coset} is the number of coset subbands.

In general, our DCast transmits the coset values of the source X over Gaussian channel, with the side information S at the receiver side. Therefore, for each subband, it forms a typical Wyner-Ziv Dirty-Paper problem, in which transmitting the coset values has been proven to be as efficient as transmitting the residue $S - X$ over the same channel (if assuming that $S - X$ is available to the encoder) [417]. Actually, according to the theorem by Kochman and Zamir [417] (the existence of good lattice), the coset value C of each subband has the same variance with the prediction residue $S - X$ of each subband, that is,

$$\sigma_{C_i}^2 = \mathbb{E}\{C_i^2\} = \mathbb{E}\{(S_i - X_i)^2\}. \tag{16.43}$$

Thus, the coset value and the prediction residue have the same variance in frame level, that is,

$$\sigma_C^2 = \mathbb{E}\{(S - X)^2\} = \sigma_{S-X}^2. \tag{16.44}$$

Therefore, Eq. (16.41), Eq. (16.42), and Eq. (16.44) implies

$$D = \sigma_{\hat{C}-C}^2 \approx \alpha_{coset} \sigma_{S-X}^2 \left(\frac{P_{coset}}{n_{coset} N_0} \right)^{-1}. \tag{16.45}$$

This means D is proportional to the prediction noise variance σ_{S-X}^2 and the inverse of coset power P_{coset}.

16.4.5 Solution

Substituting Eq. (16.33) and Eq. (16.40) into Eq. (16.45), we get

$$D = (\sigma_{\dot{S}-X}^2 + \gamma \alpha_{mv} \sigma_M^2 n_{mv} N_0 P_{mv}^{-1}) \alpha_{coset} n_{coset} N_0 P_{coset}^{-1}. \tag{16.46}$$

Then taking Eq. (16.46) into the problem Eq. (16.27), and solving the problem, we get

$$\begin{aligned} P_{mv} &= [(A^2 + A)^{1/2} - A]P, \\ A &= \frac{\gamma \alpha_{mv} \sigma_M^2 n_{mv} N_0 P^{-1}}{\sigma_{\dot{S}-X}^2}. \end{aligned} \tag{16.47}$$

Although it seems A contains so many variables, there is actually a quite straightforward way to estimate A. In A, σ_M^2 is the variance of the MV signal to transmit, $\frac{P}{n_{mv} N_0}$ is the SNR when all power is allocated to MV, and α_{mv} is the coding gain of the power allocation. This means that, if all power is allocated to MV, the MV distortion σ_Δ^2 will be $\alpha_{mv} \sigma_M^2 n_{mv} N_0 P^{-1}$ according to Eq. (16.33). Furthermore, Eq. (16.33) together with Eq. (16.39) implies that $\gamma \alpha_{mv} \sigma_M^2 n_{mv} N_0 P^{-1}$ is the variance of the additional prediction noise caused by erroneous MVs when all transmission power is allocated to MV. Therefore, the parameter A is estimated as follows. DCast simulates the transmission and decoding process to get a hypothetic side information S^*, which is the side information when all transmission power is allocated to MV data. DCast also calculates another hypothetical side information $\dot{S}$, which is the side information assuming the transmission of MVs are lossless. Since $S^* - \dot{S}$ is the additional prediction noise caused by erroneous MVs, we have

$$\sigma^2_{S^*-\dot{S}} = \gamma \alpha_{mv} \sigma^2_M n_{mv} N_0 P^{-1}. \qquad (16.48)$$

With Eq. (16.48), the solution Eq. (16.47) is rewritten as

$$P_{mv} = [(A^2 + A)^{1/2} - A]P, \qquad (16.49)$$
$$A = \frac{\sigma^2_{S^*-\dot{S}}}{\sigma^2_{\dot{S}-X}}.$$

Therefore, for optimal power distortion optimization, the encoder first estimates $\sigma^2_{S^*-\dot{S}}$ and $\sigma^2_{\dot{S}-X}$, and then calculates optimal MV transmission power P_{mv} by Eq. (16.49).

16.5 Experiments

In our experiments, we evaluate the performance of the proposed DCast in video streaming applications including both unicast and multicast. We compare DCast with SoftCast [395, 396] and conventional frameworks. We have implemented two versions of SoftCast based on 2D DCT and 3D DCT respectively, that is, SoftCast2D [395] and SoftCast3D [396].

We have also implemented two conventional frameworks. One uses H.264 as video encoder and the other uses a DVC codec named Witsenhausen-Wyner Video Codec (WWVC) [399]. Both of the two frameworks use the standard 802.11 PHY layer with FEC and QAM modulations. We use JM14.2 software as H.264 codec. For error resilience, the intra MB refresh rate is set to be 10%. Each video slice is packed into one RTP packet. We set the maximal slice size to be 1192 bytes such that the length of RTP packet is no greater than 1200 bytes. The WWVC coded bitstream is also packed into RTP packet of maximal length 1200 bytes. We append to each RTP packet a 32-bit CRC, and then encode each packet separately. Similar to the experiments by Jakubczak and Katabi [396], for error protection we apply on each packet an outer Reed-Solomon code with the same parameters (188/204) used for digital TV. Each packet is individually interleaved between the outer Reed-Solomon code and the inner FEC in accordance with the same recommendation. For inner FEC, we generate the 1/2 convolutional code with polynomials $\{133, 171\}$ and puncture it to get 2/3 and 3/4 convolutional codes. The FEC coded bits are mapped to the complex symbols by BPSK, QPSK, 16-QAM, or 64-QAM. The complex symbols are then transmitted over OFDM. We assume the channel noise is Gaussian and the channel bandwidth is 1.15 MHz. The FEC decoding is by soft Viterbi algorithm. After the FEC decoding and RS decoding, the decoder performs CRC check for each RTP packet and forward those error-free packets to video decoders. The WWVC decoder performs Wyner-Ziv decoding and is able to reconstruct the video frames when

Table 16.1 Summary of the four frameworks.

Frameworks	SoftCast2D	SoftCast3D	DCast	H.264/WWVC
GOP	IIII...	-	IPPP...	IPPP...
Reference frames	0	-	1	1
ME	N	N	Y	Y
ME block size	-	-	fixed	variable
ME search range	-	-	32×32	32×32
MV precision	-	-	$1/4$	$1/4$
DCT	2D	3D	2D	2D
Coding delay	1 frame	4 frames	1 frame	1 frame
Modulation	OFDM	OFDM	OFDM	OFDM
Constellation	64K-QAM, BPSK	64K-QAM, BPSK	64K-QAM, BPSK	BPSK, QPSK, 16- or 64-QAM
FEC rate	1/2 (BPSK only)	1/2 (BPSK only)	1/2 (BPSK only)	1/2, 2/3, 3/4
RS rate	-	-	-	188/204

the reference frames have some error. The H.264 decoder can also tolerate a small percentage of RTP packet loss, by utilizing error concealment. In our test, we have configured the H.264 decoder to use the most complex error concealment method in JM14.2, the motion copy one, to get the best reconstruction quality.

The test video sequences are standard CIF sequences (352×288, 30Hz), including "akiyo," "bus," "coastguard," "crew," "flower," "football," "foreman," "harbour," "husky," "ice," "news," "soccer," "stefan," "tempete," "tennis," and "waterfall." To evaluate the average performance of each framework, we also create a monochrome 512-frame test video sequence, called "all_seq," by combining the first 32 frames of the above 16 test sequences.

For DCast, H.264, and WWVC, the GOP structure is "IPPP" and the GOP length is 32. In the following tests, all the PSNR results are for all the frames including both intra and inter frames. The number of reference frames for inter frame is 1. In DCast, the intra-frame coding is exactly the same as SoftCast with 2D DCT and the inter-frame coding is by the proposed scheme. The transmission power allocated to an intra frame is set to be 4 times the power of an inter frame. According to our experiments, this approximately makes intra and inter frames have similar video PSNR. The search range of ME is 32×32 and the MV precision is $1/4$ pixel. In ME, DCast uses only 8×8 block size, while H.264 and WWVC use all the 7 block size from 4×4 to 16×16. Table 16.1 gives a summary of the techniques and configurations of these frameworks.

16.5.1 PDO Model Verification

This test is to verify the models of power distortion optimization (PDO) in Section 16.4. We use "all_seq" as test sequence. In the first test, we fix the coset transmission

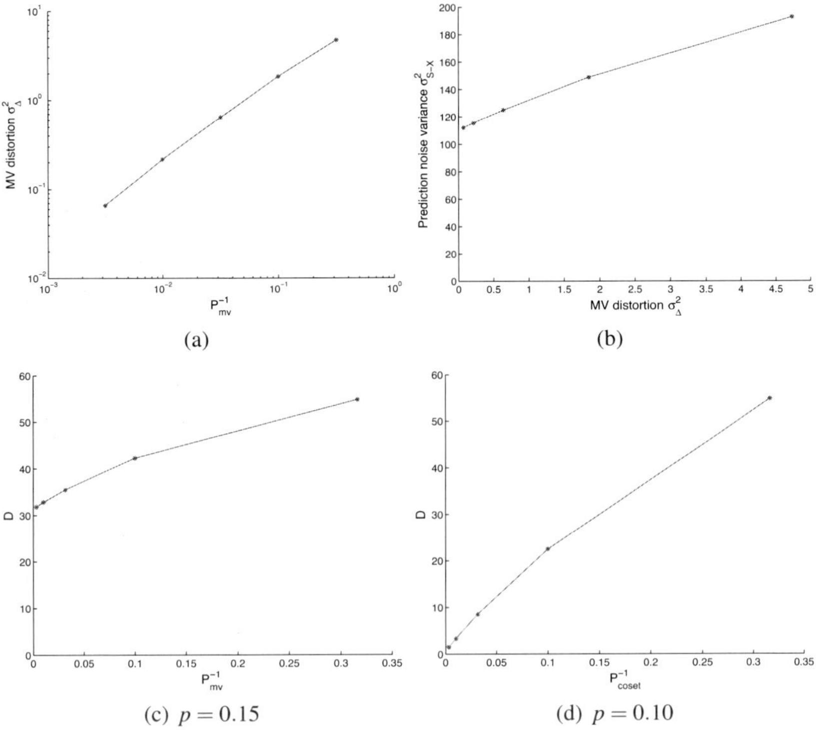

Figure 16.4 Verification of the models of power distortion optimization in Part IV. P_{coset} and P_{MV} are the transmission power of coset data and MV data respectively. D is reconstruction distortion.

power P_{coset} and let the MV transmission power P_{MV} change. The channel noise power N_0 is set to 1. The results are given in Figure 16.4. Figure 16.4a shows the relation between the MV transmission power P_{MV} and the MV distortion σ_Δ^2. According to the result, the inverse of P_{MV} is proportional to the MV distortion. This confirms Eq. (16.33). Figure 16.4b shows the linear relation between the MV distortion σ_Δ^2 and the prediction noise variance σ_{S-X}^2. This verifies the model of Eq. (16.40). Figure 16.4c shows the relation between the MV transmission power P_{MV} and the reconstruction distortion D. They are approximately in linear relation as shown in Eq. (16.46).

In the second test, we fix the MV transmission power P_{MV} and let the coset transmission power P_{coset} change. The channel noise power N_0 is set to 1. The result is given in Figure 16.4d. The reconstruction distortion D is proportional to the inverse of the coset transmission power P_{coset}. This verifies the model in Eq. (16.45) and Eq. (16.46).

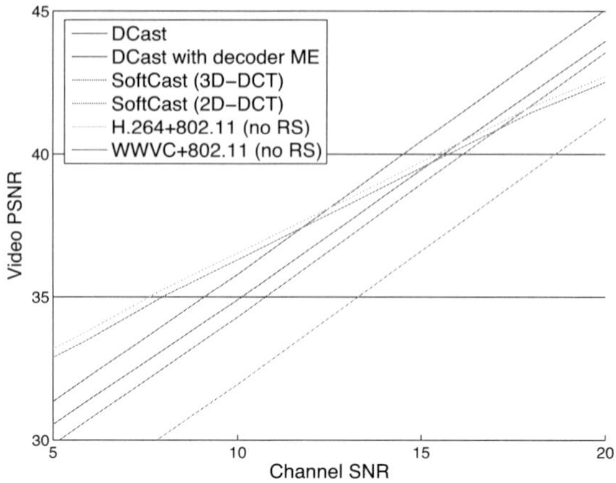

Figure 16.5 Unicast performance comparison. Both the encoder and decoder are assumed to know the channel SNR.

16.5.2 Unicast Performance

This test is to compare unicast performance among all the above frameworks. In this test the input video is "all_seq" and the channel SNR is 5 dB−20 dB. Both the encoder and decoder are assumed to know the channel SNR. For each channel SNR, the parameters of DCast are optimally tuned. The total transmission power is optimally allocated to the coset data and motion data as explained in Section 16.4. The conventional framework is assumed to be able to choose the best combinations of the FEC and the QAM methods recommended by 802.11 according to the channel SNR, to get the maximal bit rate for the source coding layer. The source coding layer, that is, the H.264 codec, performs rate control to utilize the bit rate as much as possible.

The experimental result is given in Figure 16.5. This figure compares the reconstruction quality of all the five frameworks at the different channel SNRs. The reconstruction quality is measured by PSNR. DCast is uniformly 4 dB better in video PSNR than SoftCast2D at all channel SNRs, mainly due to enabling inter-frame prediction. DCast gains about 1.5 dB in video PSNR over SoftCast3D, which mainly comes from motion alignment. Compared with an H.264-based framework, DCast is about 0.8 dB worse in video PSNR at the low channel SNR but is about 2.9 dB better in video PSNR at the high channel SNR. A WWVC-based framework performs slightly lower than the H.264-based framework. In this test, we also implement another version of DCast in which the ME is performed at the decoder by motion compensated extrapolation [390]. Like most other DVC frameworks, the DCast with ME at the decoder has a low encoding complexity but a high decoding complexity. Compared with the conventional framework, the DCast with ME at the decoder is about 1.6 dB worse in video PSNR at the low channel SNR but is 1.7 dB better in video PSNR at the high channel SNR.

Note that the result in Figure 16.5 does not mean that DCast can outperform H.264 in compression efficiency. H.264 is a video coding standard, while DCast is a wireless video transmission framework. H.264 has high compression efficiency but the coded stream is not robust to error. This is why the H.264 coded stream needs additional FEC bits to protect. DCast may not be as efficient as H.264 in video compression, but it is robust to channel noise. Thus, it can skip FEC and can use a dense 64K-QAM modulation, and can achieve high system efficiency.

16.5.3 Evaluation of Each Module

DCast has several modules such as coset coding, motion estimation (ME), and power distortion optimization (PDO). In the following test, we incrementally turn off these modules in DCast to evaluate their contribution. In this test the input video is "all_seq" and the channel SNR is 5 dB to 15 dB. The test results are given in Figure 16.6. In this figure, "PDO off" means that there is no PDO. The encoder utilizes a straightforward power allocation, where the total transmission power is equally allocated between the motion data and coset data, that is, $\frac{P_{mv}}{N_{mv}} = \frac{P_{coset}}{N_{coset}}$. "ME off" means that there is no ME and the decoder uses a previous reconstructed frame directly as side information. Note that there are dependencies between the three modules (coset, ME, and PDO). When the ME is disabled, the PDO must be off because there is no MV to transmit. When coset coding is disabled, the ME should be disabled too because the decoder no longer needs side information. Furthermore, when all three modules (coset, ME, and PDO) are off, the DCast becomes the same as SoftCast2D. According to the results in Figure 16.6, the contributions of coset coding, ME, and PDO are about 2.7 dB, 0.8 dB, and 0.5 dB in video PSNR, respectively.

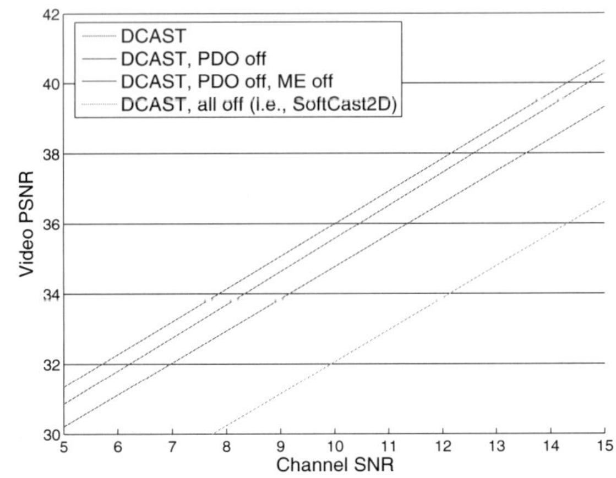

Figure 16.6 Evaluation of each module. The contributions of coset coding, ME, and PDO are about 2.7 dB, 0.8 dB, and 0.5 dB in video PSNR, respectively.

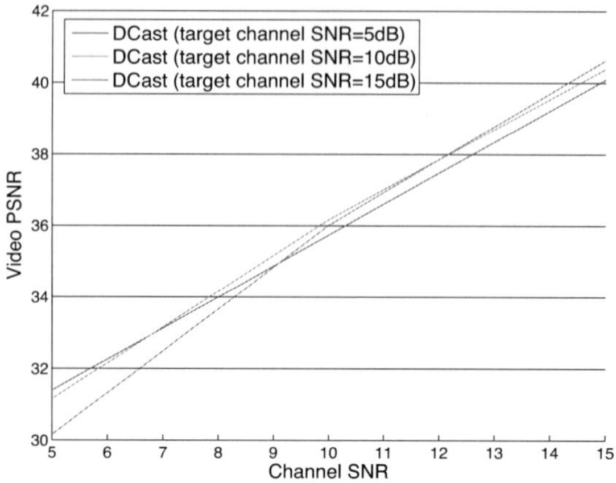

Figure 16.7 Robustness test. DCast is configured to be optimized for the target channel SNR of 5 dB, 10 dB, and 15 dB, respectively, and then tested under the different channel SNR.

16.5.4 Robustness Test

In practical wireless applications, the channel SNR may not be perfectly known to the encoder. In this test, we will evaluate the performance of DCast in this situation. The input video is "all_seq" and the channel SNR is 5 dB to 15 dB. We let DCast optimize for the target channel SNR of 5 dB, 10 dB, and 15 dB, respectively. The video PSNR are compared in Figure16.7. According to the results, each of the three encoders performs best when the practical channel SNR matches its optimization target, but performs worse than the best one when the practical channel SNR does not match the target. The one optimized for the 15 dB channel performs 1 dB lower in video PSNR than the other two when the practical channel SNR is 5 dB, mainly due to unsuccessful coset decoding.

We then compare DCast with the conventional frameworks based on H.264 and WWVC. We still assume that only the decoder knows the channel SNR. DCast is optimized for a target channel SNR of 5 dB in this test. For conventional framework, we implement all the eight recommended combinations of channel coding and modulation of 802.11. We calculate the corresponding bit rates, respectively, according to the bandwidth, and set the bit-rate constraint to the H.264 encoder and WWVC encoder for rate control. Both the video bit rates and the channel bit rates (the bit rates after RS coding and FEC) under the eight transmission approaches are given in Table 16.2 (note that WWVC and H.264 have same bit-rate constraints). For DCast, there is no bit rate but only a channel symbol rate. Note that all the frameworks consume the same bandwidth and transmission power.

The video PSNR of each framework under the different channel SNR is given in Figure 16.8. In Figure 16.8a, all eight conventional transmission approaches suffer a serious cliff effect. For example, the approach "H.264,1/2FEC,16-QAM" performs well when channel SNR is between 13 dB to 14 dB, but is not good when

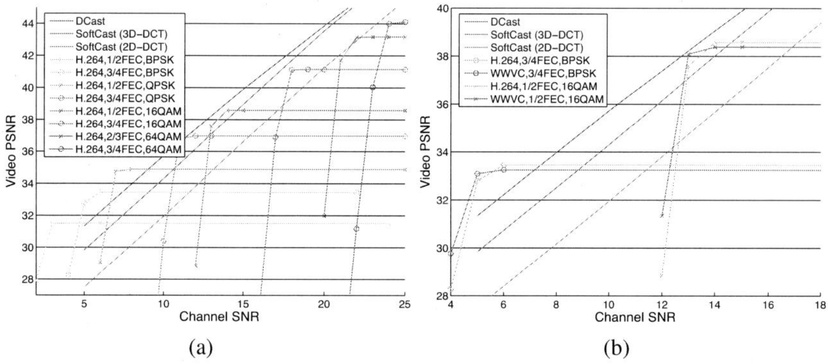

Figure 16.8 Robustness comparison between DCast and (a) H.264 and (b) another DVC framework: WWVC. Channel SNR is unknown to all the encoders. The DCast encoder is optimized for channel SNR of 5 dB.

channel SNR is out of this range. When the channel SNR becomes more than 14 dB, the reconstruction quality does not increase. When the channel SNR becomes 12 dB, the reconstruction quality drops quickly. When the channel SNR becomes even lower, the video decoder cannot work since almost all received RTP packets have bit errors. Note that the cliff effect can be partially mitigated in a layered approach [418], combining the scalable video extension of H.264 and a hierarchical modulation PHY layer. However, as shown by Jakubczak and Katabi [396], the layered approach needs a higher channel SNR than the single layer approach to achieve the same PSNR. Figure 16.8b shows the performance of WWVC based framework. Although WWVC can benefit from Wyner-Ziv decoding and achieves some gain over H.264 in erroneous situations, it still suffers a serious cliff effect.

In contrast, the three all-in-one frameworks do not suffer the cliff effect. When the channel SNR increases, the reconstruction PSNR increases accordingly, and vice versa. DCast is still the best one among the three all-in-one frameworks. At low channel SNR, DCast is 1.5 dB and 4 dB better in video PSNR than SoftCast3D and SoftCast2D, respectively. However, when the channel SNR increases, the gain of DCast decreases. When channel SNR is 25 dB, DCast performs similar to SoftCast3D and gains only about 2.5 dB in video PSNR over SoftCast2D. Compared with the unicast result in Figure 16.5, the performance of DCast becomes 1.5 dB worse in video PSNR at high channel SNR. This is mainly due to the fact that the optimization of DCast (including both the PDO and the coset quantization step) is for 5 dB channel SNR in this test. Figure16.9 gives the performance comparison on different video sequences.

16.5.5 Multicast Performance

We then let all the frameworks serve a group of three receivers with diverse channel SNRs. The channel SNR for each receiver is 6 dB, 12 dB, and 18 dB, respectively.

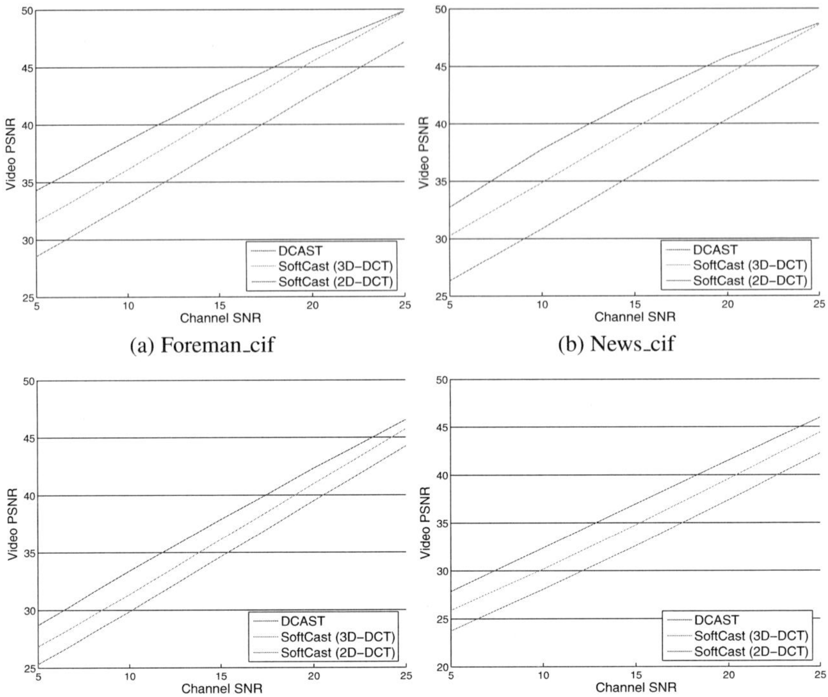

(a) Foreman_cif (b) News_cif (c) Bus_cif (d) Flower_cif

Figure 16.9 Multicast performance on different video sequences.

The test result is shown in Figure 16.10. In conventional frameworks based on H.264 and WWVC, the server transmits the video stream by using 3/4 FEC and BPSK. It cannot use a higher transmission rate because otherwise the 6 dB user will not be able to decode the video. Due to this, although the other two receivers have better channel conditions, they will also only receive a low speed 802.11 signal and reconstruct low quality video. In SoftCast and DCast, the server can accommodate all the receivers simultaneously. Using DCast, the 6 dB user can get a slightly lower reconstruction quality than using H.264- or WWVC-based conventional frameworks. However, the 12 dB and 18 dB users get 4 dB and 8 dB better reconstruction quality, respectively, by using DCast than conventional frameworks.

Figure 16.11 compares the multicast performance of four frameworks, with respect to the range of receiver SNR. The range of receiver SNR is defined as the difference of the maximal and minimal channel SNR of the users in the group. The average channel SNR of the users in the group is 14 dB. When the channel SNR range is 0 dB, that is, the channel SNR of all the users are equally 14 dB, DCast, SoftCast3D, and H.264 frameworks perform similarly. However, when the users' channel SNRs become diverse, the performance of the H.264 framework drops quickly.

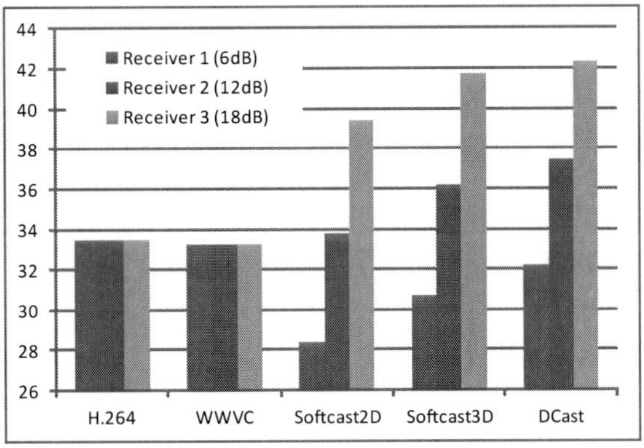

Figure 16.10 Multicast to three receivers.

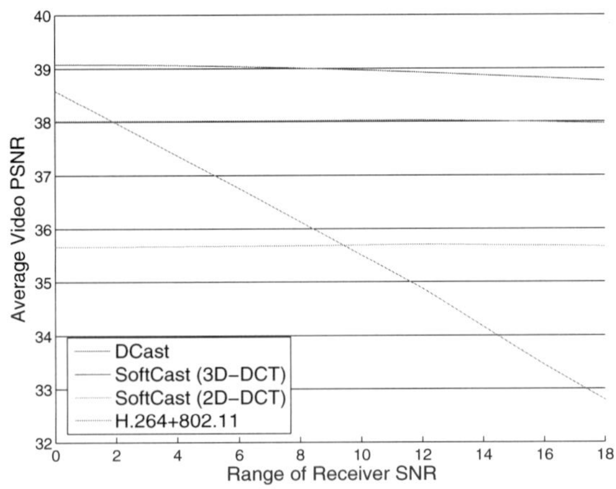

Figure 16.11 Serving a group of receivers with diverse channel SNRs. The average channel SNR of each group is 14 dB.

The visual quality comparison is shown in Figure 16.12. The channel SNR is set as 5 dB. DCast has a clearly better visual quality than both SoftCast2D and SoftCast3D. In all the tests including unicast and multicast, DCast performs better than both SoftCast2D and SoftCast3D. Moreover, DCast does not introduce frame delays as SoftCast3D and is applicable for real-time video multicast like SoftCast2D.

16.5.6 Complexity and Bit Rate

The proposed DCast allows ME to be performed at the encoder. Therefore the encoder would be in high complexity but the decoder would be in low complexity. Table 16.2 shows the average encoding time and decoding time per frame in millisecond. The test machine has a Pentium (R) Dual-Core CPU E5300 @ 2.60

(a) Original (b) SoftCast2D (c) SoftCast3D (d) DCast

Figure 16.12 Visual quality comparison with channel SNR as 5 dB.

GHz, 2G internal memory and Microsoft Windows XP Professional 5.1.2600, with Service Pack 3. The input video is "all_seq" with CIF size at 30 frames per second. DCast has less encoding time than H.264 codec (JM14.2) probably because DCast has no mode decision and no entropy coding. As to the decoding time, DCast is comparable to the H.264 codec.

Table 16.2 also shows the video bit rate and channel bit rate of H.264 solutions. For example, when the modulation is BPSK, the channel bit rate is equal to the

Table 16.2 Comparison of complexity and bit rate.

Schemes	Encode Time	Decode Time	Video Rate	Channel Rate	Symbol Rate
H.264+1/2FEC+BPSK	387 ms	7 ms	530 Kb/s	1.15 Mb/s	1.15 M/s
H.264+3/4FEC+BPSK	387 ms	8 ms	795 Kb/s		
H.264+1/2FEC+QPSK	406 ms	9 ms	1060 Kb/s	2.3 Mb/s	
H.264+3/4FEC+QPSK	389 ms	10 ms	1590 Kb/s		
H.264+1/2FEC+16-QAM	381 ms	11 ms	2120 Kb/s	4.6 Mb/s	
H.264+3/4FEC+16-QAM	385 ms	14 ms	3180 Kb/s		
H.264+2/3FEC+64-QAM	371 ms	15 ms	4240 Kb/s	6.9 Mb/s	
H.264+3/4FEC+64-QAM	427 ms	16 ms	4770 Kb/s		
DCast	304 ms	10 ms	-	-	

channel symbol rate, that is, 1.15 M/s. If the FEC is 1/2 convolutional code and the RS code is 188/204, then the video bit rate is $1.15M \times \frac{1}{2} \times \frac{188}{204} = 530$ Kb/s. When the modulation is QPSK and the FEC is 3/4 convolutional code, then the channel bit rate is 2.3Mb/s and the video bit rate is 1590 Kb/s. The decoding time of H.264 codec depends on the video bit rate. Basically, the decoding time becomes longer when the bit rate increases. The DCast framework has no bit rate but a universal channel symbol rate. Its decoding time is fixed and is similar to the decoding time of the H.264 decoder at a bit rate of 1590 Kb/s.

16.6 Summary

In this chapter, we present a novel framework called DCast for distributed video coding and transmission over wireless networks. DCast first presents a new design on how to efficiently transmit distributed coded video data over the Gaussian channel. Furthermore, we also propose a new power distortion optimization for the proposed DCast.

DCast avoids the annoying cliff effect of conventional frameworks caused by the mismatch between the transmission rate and channel condition. A single DCast server can accommodate multiple users with diverse channel SNRs simultaneously in multicast without sacrificing any user's coding performance. As shown in the experiments, DCast performs competitively with the H.264 framework in unicast but gains up to 8 dB in video PSNR in multicast.

DCast, as a unique DVC framework, does not utilize some sophisticated video coding tools such as variable block ME, intra mode, or mode decision. How to enable these tools to further improve the performance of DCast is one possible future work. Furthermore, the DCast in this chapter is mainly designed and optimized for the Gaussian channel. Another opportunity for future work is to extend the proposed DCast to the fading channel, which may require more complicated channel estimation and power distortion optimization.

Chapter 17
Denoising in Communications

17.1 Introduction

In 2011, mobile video traffic exceeded 50% of mobile traffic for the first time, and it is predicted to increase 25-fold in the next five years, according to Cisco Visual Networking Index (VNI) [3]. Wireless video communications are facing a dilemma in achieving efficiency and robustness. On the one hand, videos in their raw format are huge in size, and they need to be efficiently compressed for transmission. On the other hand, compressed video sequences have too little redundancy left, and therefore are susceptible to channel errors.

Direct application of Shannon's separation theorem [4] suggests that source redundancy should be completely removed and channel coding is responsible for adding redundancy against noise. However, joint source-channel coding (JSCC) suggests to keep a certain amount of source redundancy and has been shown to achieve better performance at limited complexity and delay. This inspires us to consider the following questions: How much source redundancy should be retained in wireless video communications in order to achieve both efficiency and robustness? Is it possible to skip channel coding and completely rely on source redundancy for channel protection?

Interestingly, the answer to the second question is a resounding YES, and the answer to the first question becomes clear after we carefully examine the two types of redundancy in video and their respective characteristics. Our research finds that: (1) Inter-frame (temporal) redundancy should be removed as much as possible at the encoder for high efficiency, while intra-frame (spatial) redundancy should be retained to protect videos against channel noise. (2) Residual frames should be transmitted in spatial domain (e.g., scaled pixel values) instead of transform domain (i.e., coefficients) through analog transmission to combat losses and noises. (3) The key to fully utilize the source redundancy is to perform image denoising at the decoder based on both source and channel characteristics.

Based on these findings, we propose a hybrid digital-analog video communications system called *Cactus*. At the encoder, temporal redundancy is removed by motion compensated temporal filtering [62]. The motion information is entropy coded and protected by the strongest channel codes during transmission. Pixel values in residual frames are transmitted using amplitude modulation. In order to minimize the mean squared error (MSE) under the average power constraint, a transform-domain scaling is performed. However, we emphasize that the sender should transmit scaled pixel values instead of transform-domain coefficients. This allows the receiver to fully utilize the source redundancy through applying image denoising techniques. In particular, Cactus employs a median filter [419] to deal with packet losses and block matching with 3D transform (BM3D) [420] to deal with additive noises.

We have implemented Cactus on a SORA [332] platform and have evaluated it in 802.11a/g-based wireless LAN environments. It confirms that our design achieves high received video quality and is robust to channel variations. In addition, Cactus allows for graceful degradation in a wide range of receiver signal-to-noise ratios (SNRs), and therefore can be readily used for multicasting. Trace-driven experiments show that Cactus outperforms a recent analog mobile video system SoftCast by 4.7 dB in average video peak signal-to-noise ratio (PSNR). In addition, Cactus is shown to be capable of transmitting high-definition videos (720 p) in WLAN, and the performance is even better than an omniscient Moving Picture Experts Group (MPEG) scheme.

17.2 Background

Visual content, including both still images and motion pictures (videos), contain a huge amount of redundancy. In the spatial domain, near pixels are likely to have similar values, and every small patch is likely to have many similar patches in the same image. In the temporal domain, successive frames usually have small differences, especially when the frame rate is high. In image/video processing, both types of redundancy are well understood, precisely modeled, and fully exploited in various applications. This section will introduce visual redundancy through the review of two related topics, namely image denoising and video coding. We do not intend to provide a comprehensive review of the two topics here. There is more information for interested readers [13, 14, 421, 422].

17.2.1 Image Denoising

The basic principle behind any image denoising technique is that natural images contain certain structural redundancy while noise does not. The structural redundancy could be *local* or *nonlocal*. The assumption for *local* structural redundancy is that nearby pixels (belonging to the same object) usually have similar gray level values. The assumption for *nonlocal* structure redundancy is that every small window in a natural image has many similar windows in the same image. Interestingly, this regularity and periodicity assumption turns out to be very general and accurate.

The basic denoising techniques can be classified into spatial-domain smoothing and transform-domain thresholding. The median filter [419] is a well-known spatial-domain denoising technique which harnesses the local redundancy. It replaces the noisy pixel with the median of neighboring pixel values. In the presence of impulsive noise (or salt-and-pepper noise), the median filter has been proved to be very effective. Gaussian smoothing is another spatial-domain denoising technique. The representative methods include SUSAN noise filter (SNF) [423] and the bilateral filter [424]. Both of them use the Gaussian mean of the neighborhood as the denoised value. The main drawback of local smoothing techniques is that they are not able to preserve fine structures, details, and texture [421]. Based on this observation, the nonlocal means (NLM) [421] filter considers both local and nonlocal redundancies. The denoised value of a pixel is a mean of the values of all pixels whose Gaussian neighborhood looks like the neighborhood of this pixel.

Transform-domain thresholding techniques [425, 426] typically assume that the signal has a sparse representation in a transform domain, such as the discrete cosine transform (DCT), wavelet, or curvelet. Hence, one could preserve the few coefficients whose magnitudes are larger than a threshold and discard the rest, which is more likely introduced by random noise. These techniques are basically based on the local redundancy assumption, but have the ability to preserve certain fine structures. Actually, the nonlocal redundancy can be incorporated into transform-domain thresholding techniques by prediction, or block matching. The BM3D (block matching with 3D transform) denoising algorithm [420] exploits nonlocal redundancy by selecting sets of blocks similar to a given reference block and grouping them into 3D data arrays. There is also a video version of the BM3D algorithm, which finds matching blocks in both current and adjacent frames. By such, the temporal redundancy is also exploited.

It should be noted that, although the topic of image denoising has been extensively studied, and there exist many mature algorithms, they can hardly be applied to digitally encoded images or videos. This is due to the mismatch between noise assumptions in denoising algorithms and the error patterns in digital transmissions. However, if the visual contents are transmitted with analog modulation and the noise is additive, most of the existing image denoising techniques can directly apply.

17.2.2 Video Compression

Videos usually have a very large data volume in their raw formats. Video compression significantly reduces the data size so that they can be practically stored and/or transmitted. This is achieved by removing spatial and temporal redundancy. Modern video compression standards all rely on two basic techniques: block-based motion compensation (MC) and transform domain based compression [13, 14].

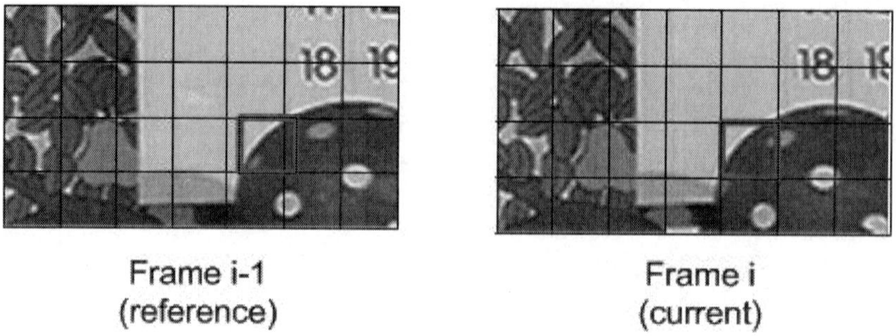

Figure 17.1 Demonstration of temporal redundancy by part of successive frames in standard video test sequence *mobile*.

Figure 17.1 shows part of two successive frames in *mobile* sequence. Obviously, there is significant temporal redundancy. This redundancy is reduced by block-based MC. The assumption behind it is that each patch of the current picture can be modeled as a translation of the picture at some previous time. In particular, when encoding frame i (current frame), it uses the reconstruction of frame $i-1$ as a reference. For each block in the current frame (e.g., the block marked by a bold box), it searches for similar blocks in the reference frame. Once the best match (the bold-boxed block in frame $i-1$) is found under the SAD (sum of absolute differences) criterion, the encoder will call it a prediction. The difference between the original block and the prediction is called residual. Both residual and motion vectors (i.e., amplitude and direction of the displacement between current block and its best match) need to be encoded in the following steps.

Spatial redundancy is utilized by encoding both I-frames (i.e., frames coded without reference to any other frame) and the residual of P-frames (i.e., frames coded using MC prediction from one reference frame) and B-frames (i.e., frames coded using bidirectional MC). Encoding of I-frames has an intra-prediction mode, which reduces the local redundancy through prediction. Figure 17.2 shows part of an I-

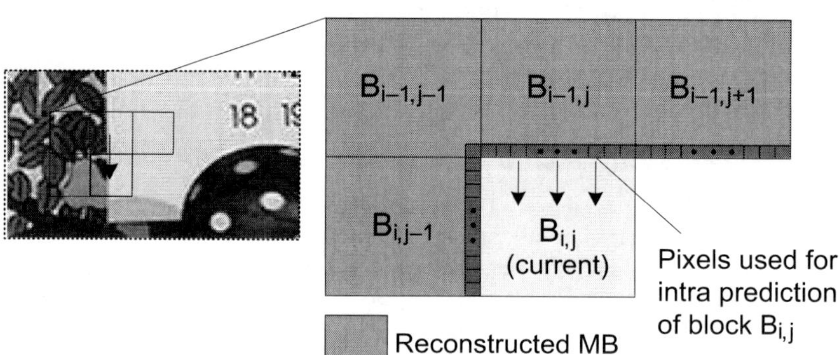

Figure 17.2 Utilizing spatial redundancy by intra prediction.

frame. Now the encoder is processing block $B_{i,j}$, and all the gray blocks shown in the figure have been encoded and reconstructed. The pixel in the bottom-right corner of $B_{i-1,j-1}$, the pixels in the bottom row of $B_{i-1,j}$, and pixels in the right column of $B_{i,j-1}$ are used for the prediction of $B_{i,j}$. In this example, the prediction direction is vertical. Then each column in $B_{i,j}$ will be subtracted with the corresponding predicted value, and only the residual needs to be encoded.

It is observed that both the original frames and residuals have spatial redundancy. Based on the sparse representation assumption, the encoder could perform transform-based compression. The discrete cosine transform (DCT) is known as the best estimate of the optimal Karhunen-Loève transform (KLT), and therefore is widely used in video coding standards. Note that the transform only decorrelates the pixels, the reduction of redundancy is actually performed by the following quantization and entropy coding steps. The motion information generated from both intra prediction and inter prediction is entropy coded too.

Note that the MC in present video coding standards [13] is based on a closed-loop prediction, that is, the prediction is based on the reconstructed frame at the decoder, not the original frames. In the conventional digital transmission paradigm, the transmission is assumed to be lossless if channel coding provides enough protection. Thus, the encoder could implement a decoder, and create the reconstruction from the bitstream it generated. However, in a hybrid digital-analog transmission scheme, the encoder is not able to know the exact reconstructed frame at the receiver, not to mention that in a multicast session different receivers will receive different reconstructions. In this case, the closed-loop prediction will bring drifting errors.

In the video coding literature, there is an open-loop alternative for inter-frame prediction, which is called motion-compensated temporal filtering (MCTF) [62]. The prediction in MCTF is based on original frames, so drifting errors are avoided. This allows us to evaluate the design choice to remove temporal redundancy at the encoder. Unfortunately, intra-frame prediction does not have an open-loop alternative. Direct application of the closed-loop prediction is not possible too. This is because the long prediction path (from top-left corner to the right-bottom corner) will create dramatic drifting errors if blocks are transmitted through analog modulation.

17.3 System Design

17.3.1 System Overview

We are seeking for a joint source channel coding design for wireless video communications. The intuition behind it is that source redundancy may be used for channel protection, but the trade-off between robustness and efficiency needs to be balanced. Figure 17.3 provides an overview of the designed hybrid digital-analog communications system named Cactus.

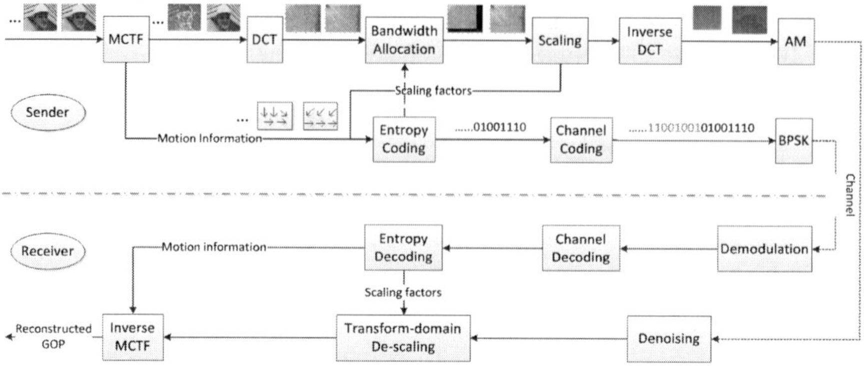

Figure 17.3 Cactus overview.

At the sender, a video sequence is first divided into group of pictures (GOP). Commonly used GOP sizes vary from 4, 8, 16 to 32 depending on the application requirements. We select GOP size equaling to 8 in our system. Each GOP is first decorrelated in the temporal axis via motion-compensated temporal filtering (MCTF). The motion information, including mode and motion vector, needs to be faithfully received by every receiver, so they are entropy coded and transmitted using a robust digital scheme. We adopt the combination of 1/2-rate channel coding and binary phase shift keying (BPSK) modulation.

The temporally filtered frames are then transformed into THE frequency domain by DCT. According to the remaining channel bandwidth budget, a certain portion of the coefficients need to be discarded. This resource allocation is performed on a GOP basis. The remaining coefficients in each frame are then divided into 10 L-shaped chunks, and are scaled accordingly. The scaling parameters are transmitted through digital methods too.

Finally, inverse DCT is performed on each frame. This is a key step in our design in order to fully utilize the spatial redundancy, because the loss of pixels is more friendly to image denoising algorithms than the loss of frequency coefficients. The scaled pixel values are interleaved and transmitted with amplitude modulation. In particular, every two pixel values are transmitted as the I and Q components of a complex symbol. It should be noted that the amplitude modulation we used is actually *pseudo-analog*, because we use a discrete modulation constellation, except that it is much denser than the commonly used 16-QAM or 64-QAM. This pseudo-analog implementation allows our design to be easily integrated into an existing network stack.

At the receiver, the digitally transmitted symbols are processed with a sequence of inverse operations including demodulation, channel decoding, and entropy decoding. Correct motion information and metadata can be obtained. Meanwhile, the receiver directly reads the scaled pixel values from the I/Q components of wireless symbols, and pieces together all the frames. Denoising is immediately applied on the scaled frames. Then transform-domain descaling is performed for each individual frame.

Finally, frames from the same GOP are processed with inverse MCTF to output the reconstructed video sequence.

17.3.2 Sender Design

17.3.2.1 Reduction of Temporal Redundancy

For natural video sequences, motion compensation (MC) is an essential step to remove temporal redundancy. However, we have discussed in Section 17.2.2 that the closed-loop prediction used in the current video coding standards is not suitable for analog transmission, due to the fact that the encoder is unable to obtain the exact reconstruction at the receiver. SoftCast simply adopts 3D-DCT to get around this problem, but transform without motion alignment cannot fully exploit the temporal correlation.

In our system, we adopt an alternative approach called MCTF [62] to reduce temporal redundancy. MCTF is essentially motion-aligned temporal transform. It is attractive to our system because it is based on an open-loop prediction model, that is, the prediction is based on original pixel values, not the reconstructed ones. It has been shown that the drifting errors are much smaller than its closed-loop counterpart.

Figure 17.4 demonstrates the lifting structure of a 2-layer 5/3 temporal filter for the ith GOP when the GOP size is 4. The even frames (frame $4i+2$ and $4i+4$) are set as high-pass frames. For each block in a high-pass frame, two similar blocks are identified in the previous and following frames. The average of these two blocks creates a prediction of the current block, so that the high-pass component is computed by subtracting the prediction from the current block. After the first-layer high-pass frames are generated, the first-layer low-pass frames can be computed by adding one fourth of the high-pass components from the two adjacent frames to the current frame. It can be seen that each high-pass frame is generated from 3 original frames and each low pass frame is generated from 5 original frames, so this process is called 5/3 filter. Similar processing steps are applied to the two low-pass frames to perform the second layer temporal filtering. We implement the Barbell-lifting MCTF proposed by Xiong et al. [397], and perform 3-layer filtering for each 8-frame GOP.

17.3.2.2 Bandwidth Allocation and Reduction

We define bandwidth ratio, denoted by ρ, as the ratio of channel bandwidth to source bandwidth. In our system, the digital transmission of motion information will occupy a certain portion of bandwidth. The exact amount can be computed from the result of entropy coding. When (BPSK, $1/2$) is used, each entropy coded bit takes two complex symbols to transmit. The remaining bandwidth, denoted by the ratio ρ_c, is used to transmit pixels. When $\rho_c < 1$, not all pixel values can be transmitted, and

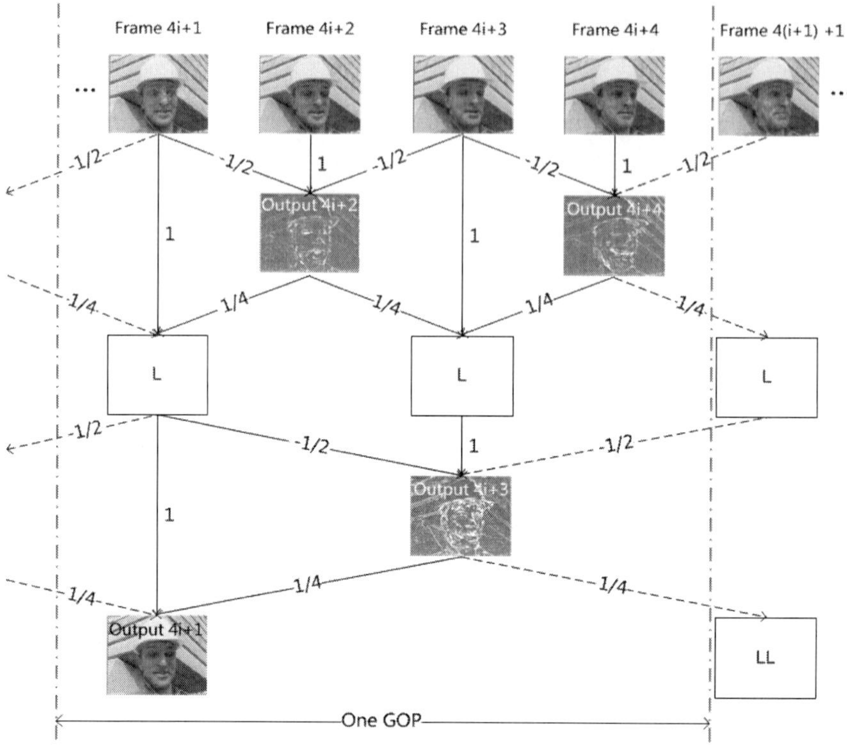

Figure 17.4 Lifting structure of a 2-layer 5/3 temporal filter for GOP size 4.

the sender needs to decide how to reduce the bandwidth usage and how to allocate bandwidth among frames.

It is well understood from digital image/video coding that the truncation of data should be based on *energy*. Therefore, we perform DCT for each individual frame. As the low-pass and high-pass frames in a GOP differ drastically in energy, the bandwidth allocation should be per GOP basis. A straightforward solution, which divides the transform coefficients into equal-sized blocks and discards the least-energy blocks, cannot be applied in our design. This is because we transmit scaled pixel values instead of DCT coefficients. Even though a right portion of DCT coefficients are discarded and padded with zeros, the number of pixels after inverse DCT does not change.

We solve this problem by transmitting a down-sampled frame. It is based on an interesting property of DCT. Let I be an image with resolution $W \times H$, and C be its DCT coefficients. If we truncate C into a $W' \times H'$ matrix C' where $C'(w,h) = C(w,h)$ for all $1 \leq w \leq W'$ and $1 \leq h \leq H'$, then the inverse DCT transform of C' using a $W' \times H'$ transform matrix will create $I'_{W' \times H'}$, which is a down-sampled image of I. Therefore, transmitting I' instead of I achieves bandwidth reduction.

17.3.2.3 L-Shaped Chunk Division and Scaling

To optimally transmit the pixels under MSE criterion in a power-constrained system, one should first decorrelate the pixel values through transform, then each transform coefficient should be scaled by a factor which is inversely proportional to the fourth root of its variance [427]. As it is not practical to scale each coefficient individually, Jakubczak and Katabi [395] propose to group nearby coefficients into chunks and model the values in each chunk as random variables (RVs) from the same distribution. Then the coefficients in the same chunk will be scaled by the same factor. The scaling factors (which is also called metadata) need to be reliably transmitted to the receiver for decoding.

We propose a new adaptive L-shaped chunk division method. The motivations are twofold. First, in the previous step of our system, bandwidth deduction will discard L-shaped coefficients from the peripheral of the frame. Second, we observe that transform coefficients decay rapidly from low-frequency to high-frequency, and those belonging to a similar frequency band are more likely to have similar values.

The problem can be mathematically described as follows. Let P be the total power budget. Divide the transform coefficients into M chunks, and let λ_i and g_i denote the variance and scaling factor of the ith chunk. It is known that:

$$g_i = \lambda_i^{-\frac{1}{4}} \sqrt{\frac{P}{\sum_i \sqrt{\lambda_i}}} \tag{17.1}$$

An optimal chunk division should minimize $\sum_i \sqrt{\lambda_i}$. For L-shaped chunk division, the adjustable parameters are r_j $(j = 1, 2, ...M-1)$, which are the positions of chunk boundaries.

We adopt an iterative approach to search for the optimal set of $\{r_j\}$. The initial values of $r_j^{\prime}s$ are evenly spaced. Then the algorithm iteratively updates the parameters one by one. In updating r_j, the values of r_{j-1} and r_{j+1} are fixed. Figure 17.5 shows

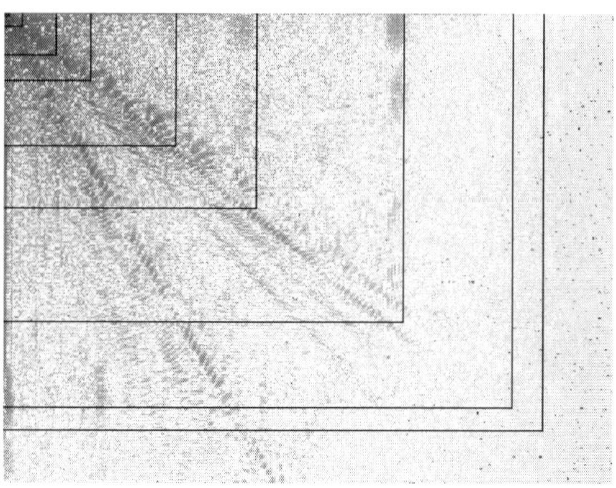

Figure 17.5 L-shaped chunk division for the first frame of Foreman.

(a) Channel input

(b) Channel output

(c) After denoising by median filter

(d) After denoising by BM3D

Figure 17.6 Transmitting a low-pass frame of Foreman over a 5 dB AWGN channel and 1% loss rate.

our chunk division for the first frame of *Foreman* when $M = 10$. In this case, only 20 metadata (10 scaling factors and 10 chunk boundaries) need to be transmitted.

Actually, both bandwidth reduction and power scaling are performed on transform domain. Therefore, the sender should perform an inverse discrete cosine transform (IDCT) after these two steps. Transmitting scaled pixel values does not change the overall power because IDCT is an orthonormal transform. Figure 17.6a shows the channel input for the first frame of *Foreman*. The original frame is 8-bit grayscale (pixel values are from 0 to 255). After transform-domain scaling, the pixel values range from -8.82 to 10.46 for this particular frame. We amplify each value by 10 times (then plus the shift 128) just for the viewing purpose.

17.3.3 Receiver Design

One key finding in our research is that source redundancy can provide channel protection under the premise that it is fully utilized at the receiver. We propose to use image denoising techniques at the receiver, and emphasize that denoising should be immediately applied to channel output.

The denoising processes for low-pass and high-pass frames are identical. We use different denoising techniques to deal with packet losses and random-valued noises. In particular, we adopt the classic median filter [419] to handle losses. Under ideal interleaving, packet loss creates randomly dispersed pixel "holes" in the frame. These holes are filled with the median of surrounding eight pixel values. We have tried more advanced median filters such as the directional weighted median filter [428], but the performance improvement is marginal at moderate packet loss ratios.

Then BM3D [420] is adopted to reduce the random noise for two reasons. First, BM3D is the state-of-the-art denoising algorithm. Second, there is a video version of BM3D which utilizes temporal redundancy to denoise. This provides an alternative to our MCTF design and could help us to evaluate whether and in which cases temporal redundancy should be removed at the encoder.

The complete BM3D algorithm has two estimate steps: basic estimate and final estimate. Each estimate is again composed of two steps: block-wise estimate and aggregation. In a block-wise estimate, each block finds similar blocks in a large neighborhood and stacks them in a 3D array. Then, 3D transformation, hard thresholding (Weiner filtering in final estimate), and inverse 3D transformation are consecutively performed to generate estimates for all the involved pixels. After all the blocks are processed, overlapping estimates are aggregated through weighted sum operation.

Figure 17.6 uses an example to illustrate the denoising process in our decoder. We assume an additive white Gaussian noise (AWGN) channel with 5 dB receiver SNR and additional 1% loss rate. Figure 17.6b shows the channel output where white dots indicate the lost pixels. The entire image is contaminated with noise, but interestingly, most image features are still recognizable. This phenomenon supports our argument that spatial redundancy can provide channel protection, and image denoising is the necessary step to utilize such redundancy. Figure 17.6c,d shows the result after the median filter and BM3D, respectively. The resulting image is very similar to channel input.

After denoising, transform-domain de-scaling is performed on each frame. This is accomplished by DCT transform, scaling, and inverse DCT transform. If the frame size is smaller than the regular size, indicating a portion of the coefficients have been dropped, the decoder will pad zeros to form a frame of regular size, then perform inverse DCT. The de-scaled frames and decoded motion information will then be used to reconstruct the GOP by inverse MCTF.

17.4 Implementation

17.4.1 Cactus Implementation

The Cactus system is composed of application-layer CODEC (coder and decoder) and physical-layer modules. The Cactus encoder only needs the available bandwidth information from the channel, which can be predetermined. In the Cactus encoder, we use a reference C code for MCTF, and implement all the other modules, including transform, bandwidth allocation, and reduction, L-shaped chunk division and scaling, entropy coding, and channel coding, in MATLAB. All the modules except MCTF could process CIF (352×288) videos in real time. However, we believe that MCTF can be run in real time, because it has very similar processing steps and complexity as the hierarchical-B coding structure in H.264 [397]. The latter already has a real-time implementation x.264 [429], which can encode four or more 1080 p streams in real time on a single consumer-level computer. In particular, the two schemes have similar computational complexity in motion estimation step (both find motions in previous and following frames), which is known to be the most time-consuming module in a video encoder.

We implement Cactus on orthogonal frequency division multiplexing (OFDM) physical layer (PHY) defined in IEEE 802.11 a/g. Specifically, the channel is divided into 64 subcarriers and 48 of them are used to transmit modulation symbols. To reduce the overhead of a Physical Layer Convergence Protocol (PLCP) header, we use 100 OFDM symbols in each PLCP frame for data transmission. Therefore, the total number of modulation symbols in each transmission is 4800. Metadata is transmitted at the lowest rate 0.5 bits/s/Hz (BPSK with 1/2 coding) in 802.11 a/g.

To resist packet loss, the adjacent symbols from a picture are pseudo-randomly shuffled across different PLCP frames. We limit the shuffling in a GOP of video frames to reduce the decoding delay. We generate the shuffle mapping through sorting a set of random numbers between 0 and 1. The sorted index is used for shuffle mapping. The random numbers can be produced by a predefined random seed and it does not introduce additional overhead. The shuffled symbols are sequentially placed on each OFDM symbol. Therefore, when a PLCP frame is lost, it creates randomly dispersed "holes" in the video frame, which can be easily processed by the median filter.

In the Cactus decoder, we implement channel decoding, entropy decoding, and transform-domain de-scaling in MATLAB. The inverse MCTF has a much lower computational complexity than MCTF encoder. Therefore, the decoder also can be implemented in real time. We use the median function in MATLAB to perform the median filter denoising, and use the MATLAB code published by Matteo Maggioni, Alessandro, and Foi [430] to perform BM3D denoising for all the evaluations. The processing time for one CIF video frame is around 1.4 seconds using Intel Core Quad CPU (Q9550) 2.83 GHz.

17.4.2 GPU Implementation of BM3D

We note that the current implementation of BM3D has a high computational complexity. Fortunately, it is very suitable for parallel computing (e.g., a specially designed chip, a field programmable gate array [FPGA], or a GPU). In order to validate that our system can run in real time, we implement BM3D through GPU NVIDIA

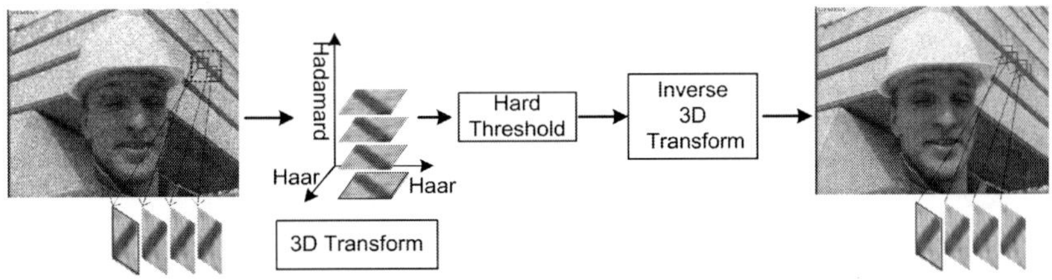

Figure 17.7 GPU implementation of the basic estimate step of the BM3D algorithm.

Table 17.1 Memory and core usage in one SM by each BM3D processing step.

Module	Data	Size (KB)	Cores
Block Matching	1 block	1.5	192
Haar	24 blocks	6.1	192
Hadamard	48 blocks	12.3	192
Inverse Hadamard	48 blocks	12.3	192
Inverse Haar	16 blocks	4	128
Blending	384 blocks	9.3	192

GTX680. It has 8 streaming multiprocessors (SMs). Each SM has 192 cores and 64 KB shared memory in which 48 KB can be used to store data. We implement BM3D in GPU following two optimization rules. The first rule is to fully utilize all 192 cores. Second, because accessing the display memory (size up to 2 GB) is slow, the data processed by the 192 cores should not exceed the SM's memory size, which is 48 KB.

We implement the basic estimate step of BM3D as shown in Figure 17.7. Every 8 × 8 block looks for matching blocks in a given rectangle region. The original block and matched blocks are organized in a 3D array. Then it is transformed by 2D Haar and 1D Hadamard transform. The noise is removed by hard thresholding. Finally, inverse transforms are performed, and pixel values corresponding to the same position are aggregated. Table 17.1 shows the memory usage, core usage, and involved data size for each SM. It can be seen that we make full use of 192 cores in almost all the processing steps. All the eight SMs perform identical operations.

We evaluate our GPU implementation of BM3D over 16 CIF test video sequences under 5 dB AWGN channel. The denoising results are listed in Table 17.2. The anchor results in the second column is achieved by the official MATLAB code [430]. On average, our implementation has 0.19 dB loss in video PSNR. This is due to two simplifications. First, we do not implement the final estimation step of the complete BM3D algorithm. Second, the original 2D biorthogonal transform is replaced by the Haar wavelet transform. The last column in the table shows the processing speed in fps. On average, GPU can process CIF videos at the speed of 35 fps, which is almost 50× of the CPU speed. It verifies the feasibility to use BM3D as part of a real-time video communications system.

Table 17.2 Reconstructed PSNR and speed of GPU implementaion for CIF videos under 5 dB AWGN channel.

Sequence	Anchor (dB)	GPU (dB)	Speed (fps)
Akiyo	42.92	42.33	33.60
Bus	32.32	32.20	35.60
Coastguard	35.12	34.89	33.97
Crew	39.63	39.53	35.00
Ower	32.32	32.23	35.07
Football	31.97	31.89	39.46
Foreman	37.48	36.97	34.83
Harbour	32.95	32.81	34.46
Husky	26.79	26.81	34.76
Ice	38.54	38.20	35.66
News	38.80	38.53	34.50
Soccer	36.03	35.88	35.49
Stefan	33.53	33.32	35.75
Tempete	33.20	33.31	34.43
Tennis	35.62	35.52	34.27
Waterfall	38.04	37.79	33.64
Average	35.33	35.14	35.03

17.5 Evaluation

17.5.1 Settings

Wireless environment: Evaluations are carried out with SORA (equipped with WARP radio board) over 802.11 a/g-based WLAN. The carrier frequency is 2.4 GHz. The channel bandwidth is 12 MHz and the data bandwidth is around 11.4 MHz. We define bandwidth ratio ρ as the ratio of channel bandwidth to source bandwidth.

We perform 52 test runs. In each test run, we transmit data generated by the Cactus encoder. The receiver records the received wireless symbols. These symbols are not only used for Cactus decoding, but are compared with the exact channel inputs to generate the traced channel noise. The traced data are labeled from 1 to 52 (according to the time they are obtained). Our comparisons with reference schemes are trace-driven to ensure fairness.

The effect of packet loss is evaluated by assuming an interferer who sends packets at constant intervals.

Video source: We create two monochrome video sequences of different resolutions for our evaluation. The common intermediate format (CIF) sequence has a resolution of 352×288, and the frame rate is 30 fps (frame per second). Hence, the source bandwidth is 1.52 MHz (in complex symbols). This sequence is created by extracting the first 32 frames from the following 16 standard video test sequences including *akiyo, bus, coastguard, crew, ower, football, foreman, harbour, husky, ice, news, soccer, stefan, tempete, tennis, waterfall*. Hence, it has 512 frames in total. It is similar to the test sequence used in SoftCast [395], with the only difference that the resolution used in SoftCast is 352×240.

The other HD (720 p) sequence has a resolution of 1280×720, and the frame rate is 30 fps too. Hence, the source bandwidth is 13.8 MHz. In order to transmit it in a 11.4 MHz channel, bandwidth compaction is needed and the ratio is around 0.826. This sequence contains the first 32 frames from 10 standard video test sequences, including *Intotree, Shields, Stockholm, City, Jets, Panslow, Parkrun, Sheriff, ShuttleStart, Spincalendar*. Therefore, the total length is 320 frames.

Reference schemes: Two reference schemes are considered. The first one is SoftCast. Our implementation has only one difference from that described by Jakubczak and Katabi [395]. In order for a fair comparison with Cactus, which uses GOP size 8, the GOP size of SoftCast is set to 8 too. We actually evaluated both schemes when the GOP size is increased to 16, and found that both schemes will have a 0.3–0.5 dB performance gain in PSNR. SoftCast needs to transmit metadata with digital method too. There are 64 variances per each video frame. We do not actually transmit these metadata for SoftCast.

The other reference is based on H.264 or MPEG4 AVC digital video coding standard [13]. We adopt a publicly available encoder called x264 [429] to encode test sequences at different rates and obtain a R-D (rate-distortion) curve. Similarly, GOP size is set to 8 for fairness. In the case of multicast, we simply call it MPEG. In the case of multicast, we name it Omni-MPEG because we assume the sender in this scheme can immediately obtain the SNR of the previous packet, and use this SNR to guide the rate selection of the next packet. The possible rates are those defined in 802.11 a/g. We then calculate the goodput rate for an entire test run, and find the corresponding distortion from the R-D curve, as if the encoder had known the channel conditions in advance. The performance of Omni-MPEG provides an upper bound for the conventional digital schemes in Unicast.

We do not compare with the scalable video coding (SVC) extension of H.264/AVC because it has been shown in SoftCast that the performance is inferior to SoftCast in all cases.

Performance metric: We evaluate the video delivery quality with the standard peak signal-to-noise ratio (PSNR) in dB. We compute the PSNR for each video frame by $PSNR = 10log_{10}\frac{255^2}{MSE}$, where MSE is the mean squared error of all pixels. Then the PSNR is averaged across frames.

17.5.2 Micro-Benchmarks

Micro-benchmarks verify our design choices. The results are obtained with the CIF video sequence, and the bandwidth ratio ρ is 1, if not otherwise stated.

Use of temporal redundancy: Cactus removes temporal redundancy by MCTF, and encodes the motion information into digital streams. The digital stream will share bandwidth with the analog transmission of scaled pixel values. We first examine the bandwidth percentage of motion information assuming a very robust transmission

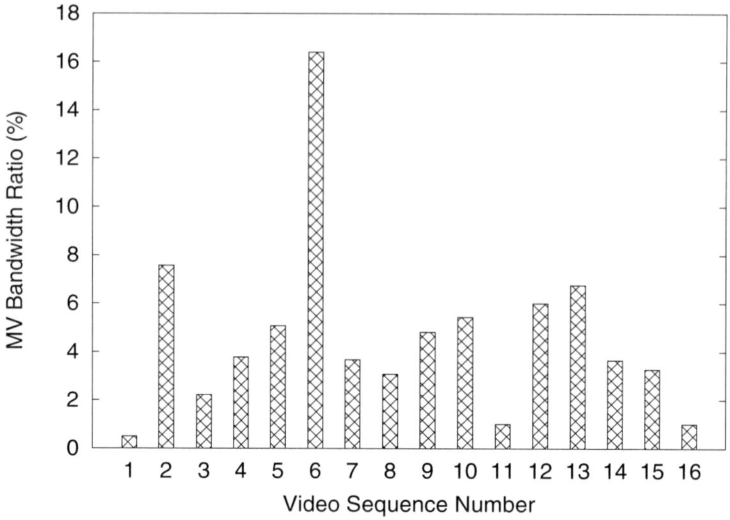

Figure 17.8 The bandwidth percentage used by transmitting motion information in (BPSK, 1/2).

scheme (BPSK, 1/2), and check whether the use of bandwidth is worthy. Figure 17.8 shows the bandwidth ratio and we can see that the amount of motion information differs greatly among sequences. Sequence #6 (*football*) has very complex motions, while Sequence #1 (*akiyo*) and #11 (*news*) have simple or small motions.

Then we verify our claim that temporal redundancy should be removed at the encoder, and examine in what cases this claim does not hold. We compare the final design of Cactus (MCTF at the encoder and BM3D at the decoder) against two alternatives. Method (DCT3d, BM3D) uses 3D-DCT at the encoder. Hence the temporal redundancy is not fully exploited. Method (DCT3d, VBM3D) exploits temporal redundancy at the decoder by using video BM3D. We have mentioned earlier that video BM3D searches for matching blocks not only in the current frame but in adjacent frames as well. Both alternatives have more bandwidth than Cactus to transmit coefficients, because the encoder does not generate motion information.

We run this test over Trace 6 (average receiver SNR = 9.97 dB) for all 16 CIF sequences. On average, Cactus has 1.67 dB gain over (DCT3d, VBM3D), and the latter has 2.39 dB gain over (DCT3d, BM3D). Figure 17.9 presents four representative sequences. Most of the skipped sequences share the same trend as *news*. For this sequence, Cactus performs the best, and using video BM3D brings certain gain over using BM3D. Both *bus* and *ice* show an interesting result that video BM3D does not bring any gain over BM3D. This means that most patches in the video have enough similar patches in the same frame to smooth out the noise. This is exactly the case in *bus* where the trees, fences, and bricks have similar patterns, and in *ice* too where the ice patches resemble each other.

The *football* is the only sequence that Cactus does not perform the best. This is because the motion information occupies too much bandwidth (around 16%), and

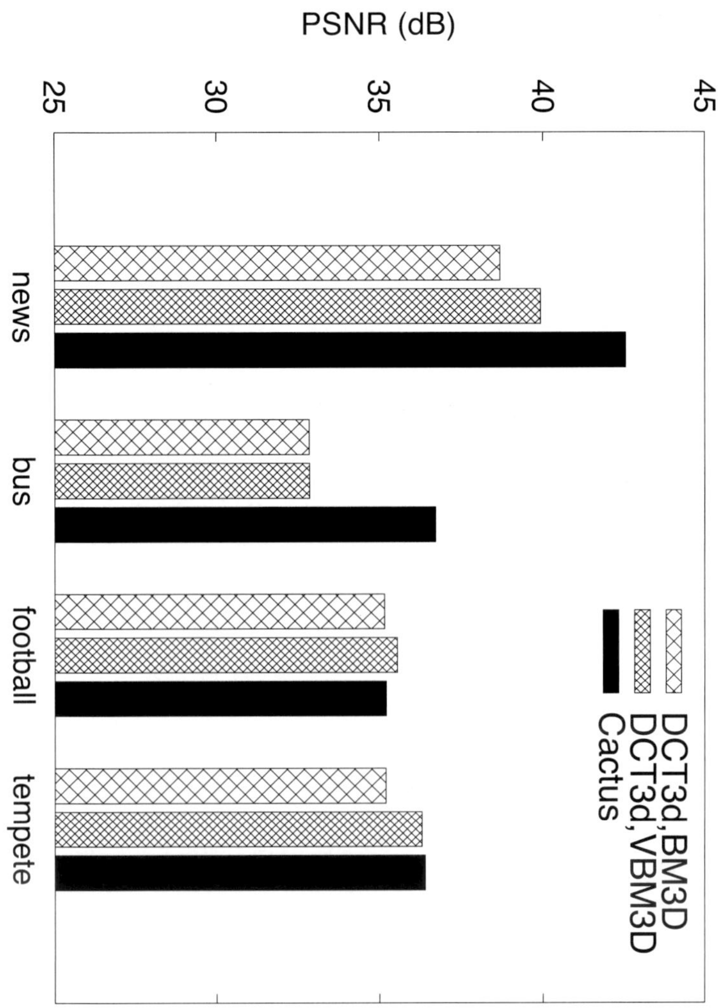

Figure 17.9 Three schemes, which use temporal redundancy differently, are compared on four representative sequences.

too many coefficients are dropped, which introduces loss. This suggests that there exists a trade-off point beyond which the temporal redundancy is better utilized at the receiver.

Image denoising: Image denoising is a key module in the Cactus decoder. This experiment evaluates the performance gain brought by this module, and examines the impacting factors. Figure 17.10 shows the average denoising gain for all 16 CIF sequences under different receiver SNRs. The gain shows a clear decreasing trend. This suggests that for a Cactus receiver, if the measured channel condition is

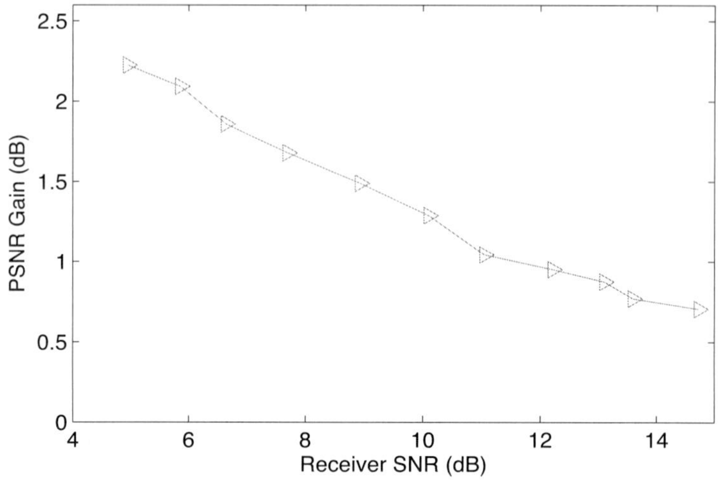

Figure 17.10 Denoising gain as a function of receiver SNR.

better than a certain SNR, it may turn off the denoising module without much loss in performance.

The denoising gain also depends on the video characteristics. Figure 17.11 shows the PSNR gain on each of the sequences, and the value presented is averaged over all 32 frames and all traces. Sequence #10 (*ice*) benefits the most from denoising, with an average gain over 2.5 dB. This is because the frames in this sequence are all very smooth. Sequence #16 (*waterfall*) gains the least from denoising, because the frames contain too much texture and details.

Transmission in spatial domain versus in transform domain: Actually, Cactus could choose between transmitting in spatial domain (scaled pixel values) or in transform domain (scaled DCT coefficients). The two choices cost the same transmission power as DCT transform is orthonormal. We have embraced the spatial-domain transmission in the final design of Cactus, and this experiment is going to verify this choice.

We compare our spatial-domain transmission with two transform-domain alternatives, one with Hadamard transform (as used in SoftCast) and the other without any transform. In this evaluation, we let $\rho_c = 1$, that is, there is just enough channel bandwidth to transmit all the pixels or coefficients. We do this simplification because the dimension of Hadamard matrix has to be a power of 2. We run the experiments for video Sequences #10 and #16 on Trace 8 (receiver SNR = 7.24 dB). These two sequences are chosen because they benefit the most and the least from image denoising as shown by the previous experiment.

Figures 17.12 and 17.13 show the comparison results for Sequences #10 and #16, respectively. It is not surprising that transmitting *ice* in spatial domain significantly outperforms the other two choices, since *ice* is very friendly to image denoising. The gain over the transform-domain transmission with Hadamard is as high as 7.44 dB

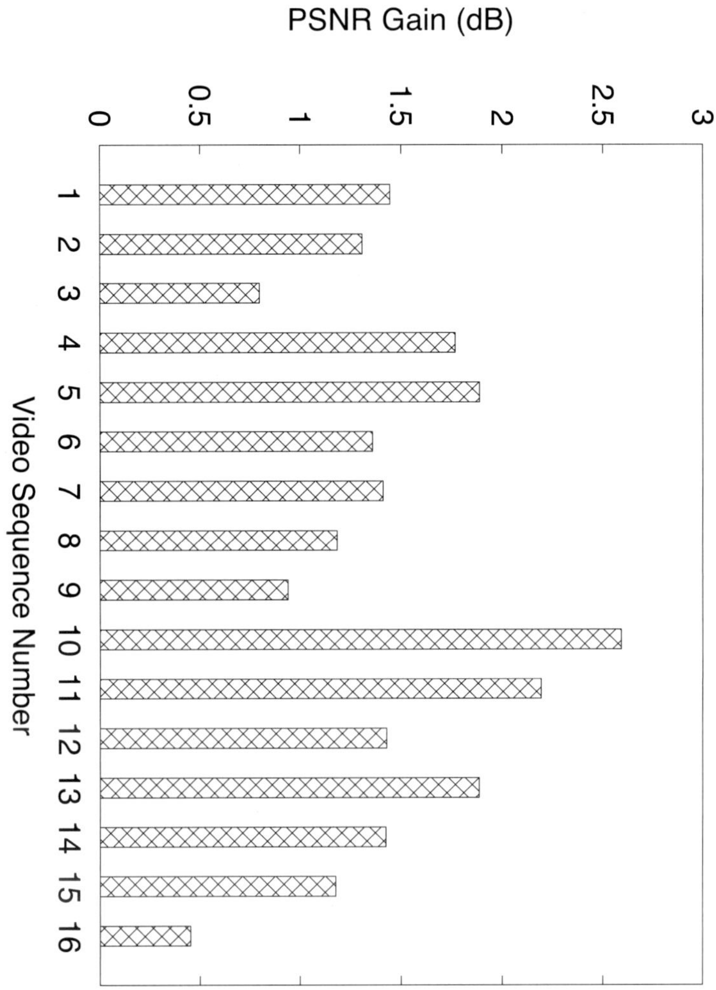

Figure 17.11 Denoising gain on different sequences.

when the packet loss ratio is 0.1. The experiment on *waterfall* obtains slightly different results. When the packet loss ratio is small, transmitting the video in spatial domain does not bring any gain. However, this is the sequence which benefits the least from image denoising. Even for this sequence, Cactus outperforms the other choices in most cases. This experiment validates our choice to transmit video in spatial domain.

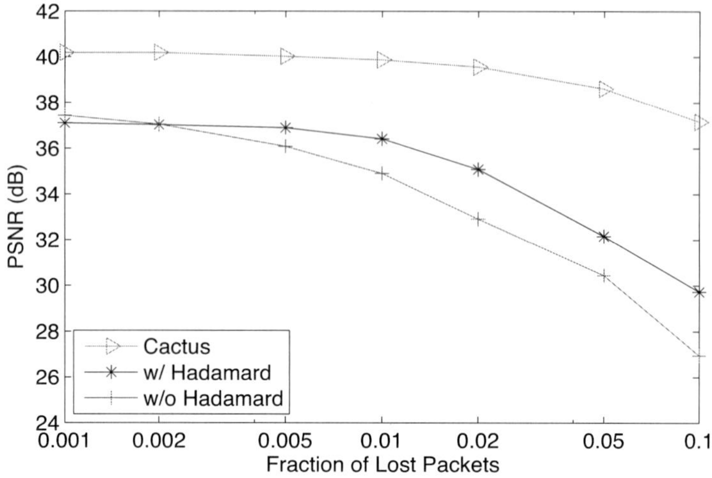

Figure 17.12 Comparing transmitting Sequence #10 (*ice*) in spatial domain or transform domain under different loss rates.

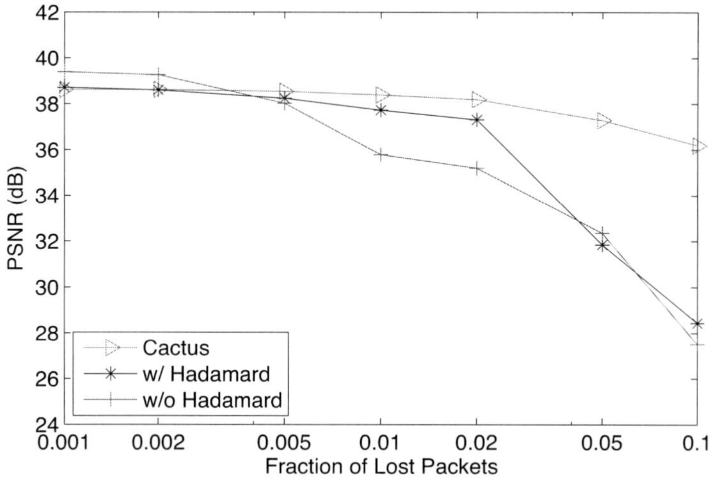

Figure 17.13 Comparing transmitting Sequence #16 (*waterfall*) in spatial domain or transform domain under different loss rates.

17.5.3 Comparison against Reference Systems

Figure 17.14 compares the performance of Cactus against two reference schemes, namely SoftCast and Omni-MPEG under multicast scenario. We run 52 traces over the CIF sequence to emulate a 52-receiver multicast session. For Cactus and SoftCast, in each test run, we compute the average receiver SNR across PHY packets

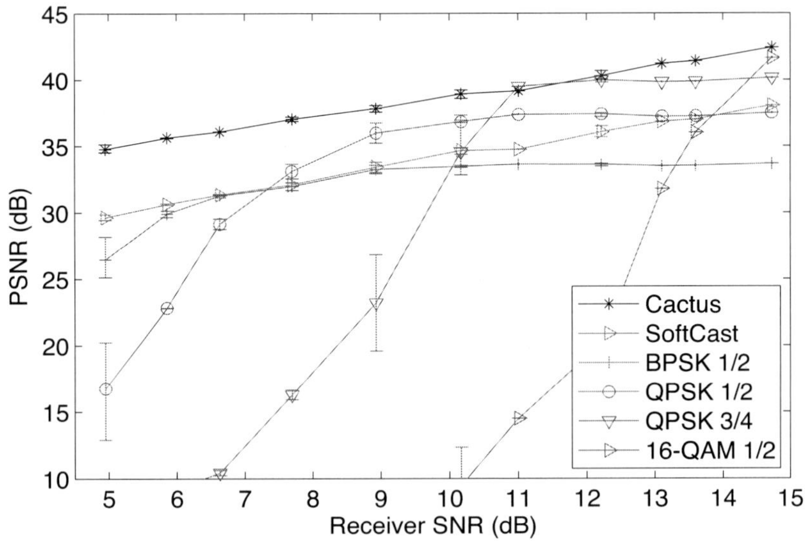

Figure 17.14 Compare Cactus against two reference schemes for a CIF sequence in a multicast session when bandwidth ratio $\rho = 1$.

and average video PSNR across sequences. To plot the video PSNR as a function of receiver SNR, we divide the receiver SNR range into 1 dB bins, and average all the (receiver SNR, PSNR) pairs whose receiver SNR fall into the same bin. Results show that although both Cactus and SoftCast achieve graceful rate adaptation, Cactus consistently outperforms SoftCast in video PSNR, and the average gain is 4.7 dB.

For MPEG, the sender has to fix a transmission rate in each test run. We run four tests, using the lower half of the PHY rates defined in 802.11a/g. Once the transmission rate is fixed, the video PSNR can be found in the R-D curve. We run each trace for each PHY rate, if the instantaneous receiver SNR is higher than expected, transmission is successful. Otherwise, the receiver can get nothing. We average the PSNR along each trace, and also plot (receiver SNR, PSNR) in bins. Figure 17.14 clearly shows that Cactus outperform MPEG because the latter suffers from the threshold effect.

Note that although Cactus transmits motion information through the digital method, they are always protected with the lowest rate (1/2) channel coding and transmitted using BPSK modulation. Therefore, it is insensitive to channel variations.

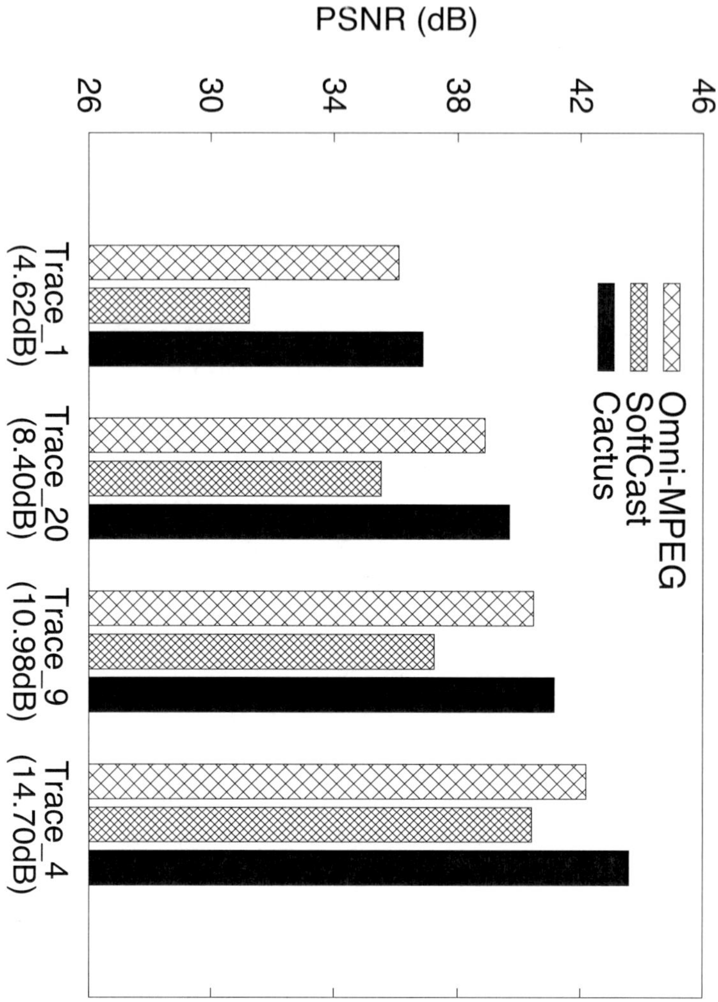

Figure 17.15 Compare three schemes for a 720 p sequence under bandwidth ratio 0.82.

17.5.4 Transmitting High-Definition Videos

High-definition videos have huge size in its raw form. It is important to evaluate whether a nearly uncompressed video communications system could transmit such a huge amount of data under the current wireless channel condition. The video source we use for this experiment is 720 p (1280×720), and the source bandwidth is 13.8 MHz. As the available channel bandwidth is 11.4 MHz, bandwidth compaction is needed, and $\rho = 0.82$.

We pick four traces whose average receiver SNR ranges from 4.62 dB to 14.7 dB to carry out this experiment. Figure 17.15 shows the average video PSNR achieved

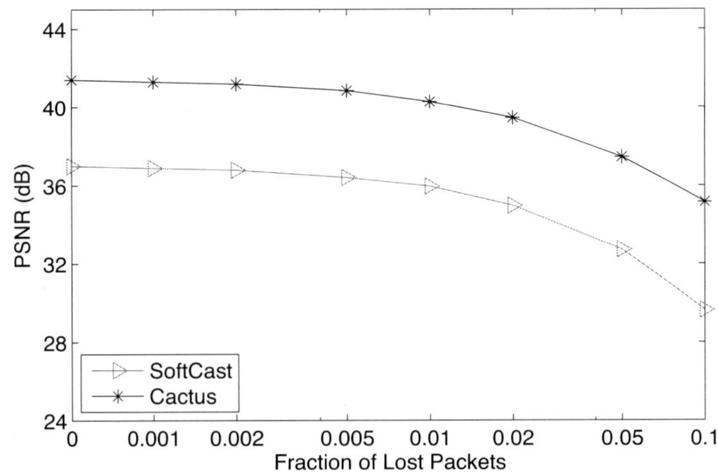

Figure 17.16 Comparing the error resilience capability of Cactus and SoftCast under trace #16 (receiver SNR = 13.59 dB).

by Cactus and the two reference schemes. It is very encouraging that Cactus achieves better performance than Omni-MPEG even in this Unicast transmission scenario. In the case of multicast, the gain of Cactus will be more significant, since both Cactus and SoftCast transmit the same content in these four traces, and they would allow different receivers to get a video quality that is commensurate with their channel conditions.

Cactus consistently outperforms SoftCast too. When the receiver SNR is 4.62 dB, Cactus achieves 5.6 dB gain over SoftCast. We also observe an interesting trend that, as the receiver SNR increases, the performance difference between Cactus and SoftCast decreases. When the receiver SNR is 14.70 dB, the performance gain decreases to 3.18 dB. This is due to the fact that image denoising is more useful in poor channels. The denoising effect is more prominent for HD videos than CIF videos because there is more spatial redundancy in HD videos.

We would like to mention that although bandwidth expansion is also a possible case in theory, it is less likely to happen in practice, especially for HD videos. Therefore, we do not evaluate such cases in this work.

17.5.5 Robustness to Packet Loss

This experiment evaluates whether Cactus is robustness to packet loss. The traces used in this experiment is #16, which has a receiver SNR of 13.59 dB, and #22, which has a receiver SNR of 5.82 dB. The packet loss is simulated by assuming an interferer who transmits packets at constant intervals. We evaluate both Cactus and

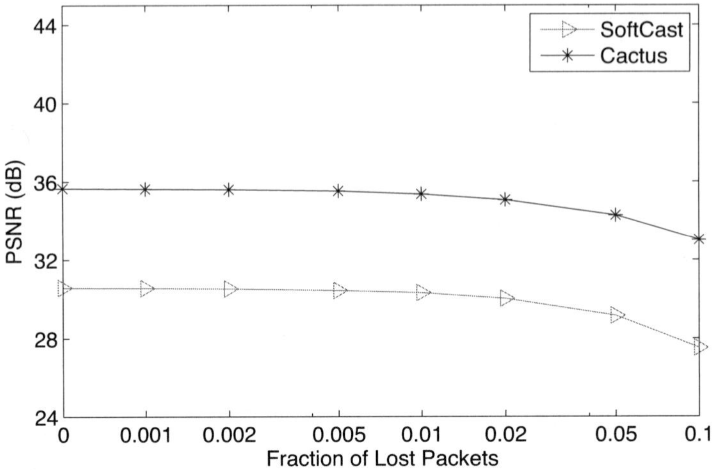

Figure 17.17 Comparing the error resilience capability of Cactus and SoftCast under trace #22 (receiver SNR = 5.82 dB).

SoftCast when the packet loss ratios are 0.001, 0.002, 0.005, 0.01, 0.02, 0.05, and 0.1.

Figures 17.16 and 17.17 show the average video PSNR achieved by Cactus and SoftCast as a function of loss rate under channel trace #16 and #22, respectively. In both figures, the x-axis is in logarithm. When the channel condition is poor (trace #22), both Cactus and SoftCast are not sensitive to packet losses. The PSNR loss of Cactus is less than 1 dB when the loss ratio is 0.02. However, when the channel condition is good (trace #16), the transmissions are sensitive to losses.

Though not very obvious in the figures, when the packet loss ratio is 0.1, the PSNR loss of Cactus with respect to no loss case is 6.3 dB for trace #16 and 2.6 dB for trace #22. Under the same packet loss ratio, the PSNR loss of SoftCast with respect to its no loss case is 7.4 dB and 3.1 dB, respectively. This shows that Cactus has even higher robustness to packet loss than SoftCast.

17.6 Related Work

The design of Cactus essentially belongs to joint source-channel coding (JSCC). JSCC is an extensively studied topic both from an information theoretical perspective and for the specific application on video communications.

In the category of JSCC for digital video communications, Cheung and Zakhor [431] proposed to distribute the available source and channel coding bits among the subbands to minimize the expected distortion. This is analogous to the transform-domain scaling (power allocation) in Cactus design. He et al. [350] proposed a JSCC scheme which determines the optimal β in source coding, which is the percentage of blocks coded without prediction, based on the channel parameters such as bandwidth and bit error rate (BER). A higher β implies retaining more redundancy in the source. Flexcast [354] replaces the entropy coding module in H.264 with a rateless code, thus to achieve graceful quality degradation in video Unicast.

Research on analog JSCC mostly considers general source and channel. For instance, it is well-known that a Gaussian source achieves the capacity of an AWGN channel [432]. Gastpar et al. [433] observed that channel coding is not necessary in some cases for optimal communication, but the source and the channel have to be matched in a probabilistic sense. Compression techniques, like vector quantization (VQ) [434] and Wyner-Ziv coding [417, 435], have been adopted in hybrid digital-analog transmissions to match source and channel. Recently, Kochman and Zamir [436] showed that, by combining prediction and modulo-lattice arithmetic, one can match any stationary Gaussian source to any colored-noise Gaussian channel, hence achieve Shannon's capacity limit. They also pointed out that the analog transmission scheme is more robust than its digital counterpart and is not sensitive to exact channel knowledge at the sender.

Recently, an analog mobile video system named SoftCast has been proposed [395]. In SoftCast, 3D-DCT is performed on a group of pictures, and the transform coefficients are transmitted as the I and Q components of a complex symbol after power scaling. As quantization and entropy coding are skipped, SoftCast keeps most of the source redundancy. Such a scheme is robust to channel variations and is capable of achieving graceful degradation in a wide range of channel conditions. However, an important fact that has been neglected in SoftCast is that the retained source redundancy should be actively utilized at the receiver. In addition, SoftCast does not explore the possibility to remove partial redundancy at the encoder.

17.7 Summary

We have described in this chapter an efficient and robust wireless video communications system — Cactus. Our design validates that it is possible in video communications to skip channel coding and rely solely on source redundancy for channel protection. We have also successfully used Cactus for the transmission of HD videos. Surprisingly, Cactus achieves better performance than Omni-MPEG even in the rate adaptation enabled unicast scenario, which demonstrates the great potential of such hybrid digital-analog scheme.

Chapter 18
MIMO Broadcasting with Receiver Antenna Heterogeneity

18.1 Introduction

Multimedia broadcasting has recently attracted a great deal of research interest [437, 438] along with growing wireless communication capabilities [439, 440] and the demand for audio and video streaming [2]. However, broadcasting systems traditionally suffer from heterogeneous channel conditions across receivers. A transmission scheme should normally ensure that all receivers can receive broadcasted information, resulting in the system performance often being constrained by the worst channel. This is likely to be exacerbated in future systems featuring multiple-input multiple-output (MIMO) technologies, which have increasingly become building blocks for high-capacity wireless link technologies such as 802.11n [439], WiMAX [440], and 3GPP LTE [441]. Devices in such networks may be equipped with varying numbers of antennas. We refer to this as *receiver antenna heterogeneity*.

Figure 18.1 shows an exemplified scenario. With two antennas, the base station can send two concurrent blocks of information that are either distinct, to achieve a higher rate, or duplicate, for better reliability. The former is only possible if the receiver also has two antennas. If we design a system based on the higher rate capability of the two-antenna receiver B1, the single-antenna receivers, A1 and A2, are likely to get no information, since they would have to solve an under-determined linear system. Conversely, some of B1's channel capacity would be wasted. Ideally, a transmission strategy is needed that allows both types of users to receive information whose quality is commensurate with the antenna setting and channel SNR simultaneously.

Some previous work has considered channel SNR heterogeneity or antenna heterogeneity. Hierarchical modulation (HM) schemes [418, 442] and SoftCast [396, 443] consider SNR heterogeneity for single antenna systems. Although the latter can achieve a smooth rate increase with the channel SNR (CSNR), it does not readily extend to addressing antenna heterogeneity. Diversity embedded space time code [444], designed for a broadcast system with four transmitting antennas and varying receiver

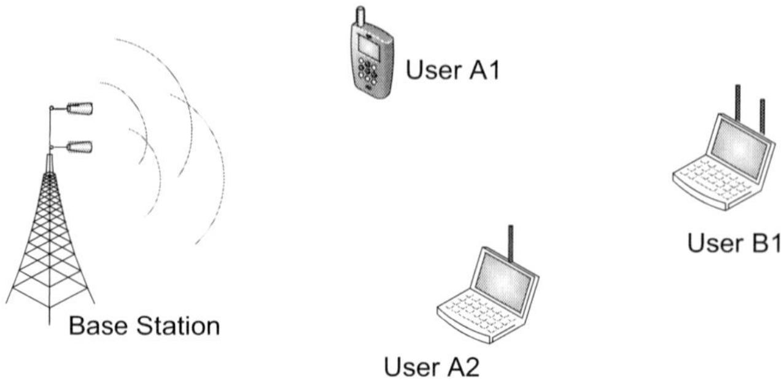

Figure 18.1 A MIMO broadcast system with receiver antenna heterogeneity.

antenna numbers, does address the antenna heterogeneity issue. The scheme separates the source into layers of different priorities, and superimposes different layers by embedding a high diversity code within a high rate code. Each receiver derives an amount of information matching its antenna number. Various superimposed MIMO codes have also been designed for cooperative channels [445], 2×1 MIMO [446], and 2×2 MIMO systems [447]. In these schemes, the strong users are expected to decode both layers of data with the help of relay nodes or good channel conditions, with weak users only able to derive the base layer. As with HM, however, only a few discrete levels of recovery quality can be achieved in all these schemes, and the bandwidth of the weak receivers is reduced by the transmission of low-priority data.

In this chapter we present a new analog image broadcast system in such MIMO systems with receiver antenna heterogeneity. As in conventional systems, raw image is first decomposed into compressible wavelet coefficients to remove spatial correlation. Unlike conventional systems, which proceed with quantization and entropy coding, our system employs a power allocation strategy to redistribute the transmit power among coefficients of different wavelet bands. This step scales the coefficients with respect to their contribution to the total distortion and balances the performance trade-off between the single-antenna and two-antenna users. The scaled coefficients are then sampled by a Hadamard matrix to generate linear combinations of wavelet coefficients, or *measurements*, before proceeding with amplitude modulation (AM) and pseudo-analog transmission. By *pseudo-analog*, we mean that the transmission is actually implemented on digital communication hardware but the constellation is much denser and the performance of AM is analogous to that in analog transmission. At the receiver, the received physical layer symbols are passed directly to the image decoder without conventional MIMO receiver processing, and the wavelet coefficients are decoded using a weighted l_1-minimization based decoding algorithm [448]. Finally, the image is recovered after an inverse wavelet transform.

Our system design makes two major contributions. First, our system addresses the two dimensions of heterogeneity simultaneously by integrating compressive sensing (CS) into MIMO transmission. By applying CS, receivers with an insufficient number of antennas can solve the under-determined problem with an approximate solution. The decoding performance scales with the quality and the number of received measurements, which is determined by the CSNR and the antenna setting respectively. While the generic framework has proved feasible in our previous work using an abstract statistical source model [449], the work in this chapter considers performance optimization for a practical end-to-end system for image transmissions. Second, we introduce a power allocation strategy to achieve a performance trade-off between single-antenna and two-antenna users in a way that is more flexible than in conventional digital systems. Our system is evaluated with extensive simulations and compared to an extension of SoftCast to MIMO systems and two conventional layered source-channel schemes.

18.2 Background and Related Work

18.2.1 Multi-Antenna Systems

Due to their high spectral efficiency or improved reliability, MIMO technologies have been increasingly incorporated as an important building block for next generation wireless networks [439–441]. By exploiting the spatial dimension across multiple antennas at the sender and the receiver, MIMO could potentially improve a wireless system's capacity, range, and reliability [450–452]. In such a setup, each of the n_t transmit antennas could transmit an independent packet fragment, called a *spatial stream*, concurrent with other antennas. Each of the n_r receiver antennas would measure a linear combination of the transmitted signals. The original streams could be decoded as long as enough linear combinations are received, that is, $n_r \geq n_t$, through a process analogous to solving simultaneous linear equations. This is referred to as *spatial multiplexing*. The maximum number of streams that can be supported is $M = min(n_t, n_r)$. This $n_t \times n_r$ system could potentially achieve M times the 1×1 rate.

Even with a single antenna at the receiver end, but multiple antennas on the transmitter, the system could benefit from improved reliability, or *diversity gain*, by having antennas transmit redundant information. Even better, more diversity could be achieved through space time block codes (STBC) [452]. The Alamouti scheme provides an example for the 2×1 case [453], extending over two time slots, where the two antennas transmit the symbols x_1 and x_2 respectively in the first time slot, and $-x_2^*$ and x_1^* in the second slot.

A fundamental challenge of point-to-point MIMO transmission lies in how to derive the optimal diversity-multiplexing trade-off [454]. For broadcasting, there is also a performance trade-off among different users. First, if all users have the same number of antennas, the trade-off is mainly across different CSNRs. Assuming a rich scattering environment that favors MIMO, the situation is then analogous to traditional single-antenna broadcasting systems, where sufficient reliability must be provided for the lowest-SNR users. Second, if receivers have varying numbers of antennas, that is, there is *receiver antenna heterogeneity*, the transmitter is normally constrained to a single spatial stream to ensure decoding at the single-antenna receivers. Therefore, the system can normally only benefit from diversity gain from the extra transmit antennas, and the multiplexing capacity of multi-antenna users would be wasted. This exacerbates the heterogeneity challenge in a broadcasting system. We need a scheme that will strike a balance between achieving diversity gain for low-SNR or single-antenna users and multiplexing gain for multi-antenna users when channel condition permits.

18.2.2 Layered Source-Channel Schemes

In conventional broadcasting systems, the overall performance tends to tailor to the receiver with the worst channel condition, since every user in the system needs the broadcasting content. In a multimedia broadcasting system, *multi-resolution coding* is often used to successively refine the source rate-distortion trade-off, such as the scalable video coding extension of H.264 (SVC) [455] and JPEG 2000 [456, 457]. The entropy coded bits are separated into layers with varying levels of importance. The base layer data are always expected to achieve acceptable quality, while higher quality is derived from the enhancement layers for users with sufficient channel conditions.

At the channel, superposition coding is recognized as an efficient approach [418, 442] to deal with diverse channel conditions in a broadcasting system. The PHY implementation of superposition coding is also known as hierarchical modulation (HM), where part of the symbol is decoded first and its effects are removed before the remaining symbol is decoded. Multi-resolution coding at the source naturally combines with superposition coding at the channel, such that layered source data with different priorities are mapped to different parts of the HM symbol. The weak receivers only aim to decode high-priority but low-quality content, while strong receivers are also expected to decode the enhancement layers.

There have been several efforts to combine superposition coding with multi-antenna technologies. For multiple-input single-output (MISO) systems, it is natural to transmit HM modulated signals with STBC to harness diversity gain [458, 459]. For MIMO systems, the main issue is whether users with good channel conditions can derive multiplexing gain. Given that a pure spatial multiplexing scheme is too aggressive for low-SNR users, a number of schemes [444,445,447] have been proposed to use multiplexing and STBC for different parts of an HM symbol, corresponding to different source layers. The more important data are sent via STBC to ensure reliable transmission, while the less important data, coded at a higher rate, are sent via spatial multiplexing.

However, these layered source-channel schemes only alleviate the cliff effects of digital systems and introduce new design challenges. First, multi-resolution coding imposes a particular decoding order and dependency between layers, and the bandwidth of the weak receivers is reduced by the transmission of the enhancement layer data. Second, due to the power difference requirement between layers for successful decoding, only two or three levels can be embedded in an HM symbol, which constrains the granularity of code rates. Third, STBC codes for MIMO systems with more than two transmit antennas cannot achieve full diversity. Therefore, such systems are still far from fully accommodating users with diverse antenna settings and channel conditions.

18.2.3 Compressive Sensing

Compressive sensing (CS) [295, 372], also known as *compressive sampling*, is an emerging theory that deals with the acquisition and recovery of sparse or compressible signals. According to CS theory, even when the number of sampled measurements of a source vector is smaller than the source dimension, which gives rise to an ill-posed problem, source decoding from the measurements is still feasible. It has been used for multimedia coding and transmission in wireless communication [460–463]. The CS measurements of raw signals are transmitted and the receiver is expected to achieve fine-grained quality scalability with the number and noise energy of the received measurements.

We first introduce the basic process of CS based image transmission. Let $\mathrm{X}_{W\times H}$ denote a $W \times H$ image that we vectorize into a column vector $\mathbf{x}$ of size $N \times 1$ ($N = W \times H$). Transforms like discrete cosine transform (DCT) and discrete wavelet transform (DWT) [464] can convert the image signal from the spatial domain to the frequency domain, in which images are normally regarded as compressible for CS sampling and reconstruction [465, 466]. Hence, the signal $\mathbf{x}$ can be represented by a set of coefficients $\theta_1, \theta_2, \cdots, \theta_N$ in the frequency domain. Let us sort the coefficients in descending order according to their absolute values, and stack them into an N-dimensional signal $\theta = (\theta_1, \theta_2, \cdots, \theta_N)^T$. Then θ is a compressible signal if $|\theta_i|$ decays with $i^{-1/p}$, which means that it obeys:

$$|\theta_i| \lesssim C_0 \cdot i^{-1/p} \tag{18.1}$$

where $\lesssim$ means "less than or approximately equal to" and C_0 is a constant. Previous research has shown that, for natural images, p is approximately 1.67 [467]. It should be noted that the parameter p dominates the degree to which a signal is compressible. A smaller p indicates a more compressible signal, as the magnitudes decay faster.

In compressive sensing-based image transmission, the measurements of the signal $\mathbf{x}$ are transmitted instead of the raw signal. In particular, we obtain these measurements by $\mathbf{m} = (m_1, m_2, \ldots, m_k)^T = \Phi\mathbf{x}$, where Φ is a $k \times n$ matrix, which is referred to as a *sensing matrix* or a *measurement matrix*. The received measurements at the users are always contaminated by channel noise. When matrix Φ satisfies certain conditions, the reconstruction of signal $\mathbf{x}$ can be achieved by solving an l_1-minimization problem, which is a relaxed form of the l_0 norm-based minimization problem of a combinatorial nature. Candes et al. [448] constructed a weighted l_1-minimization problem, and showed that using weights that are inversely proportional to the true signal magnitudes can improve CS reconstruction performance.

Multimedia coding via compressive sensing has three desirable properties that motivate its emerging application in wireless transmission. First, the generated measurements are democratic, which ensures that the transmitted measurement symbols have equal importance [468]. Second, for a compressible source that is not strictly sparse, CS recovery performance scales with the number of available measurements when they have similar noise levels [460,469]. Third, the lower the noise level of the received measurements, the better the CS decoding performance [470]. These properties illustrate that multimedia transmission based on compressive sensing could achieve fine-grained quality scalability matching the channel condition and avoiding the cliff effect in conventional multimedia broadcasting. However, to the best of our knowledge, no existing work applies compressive multimedia broadcasting to MIMO systems with receiver antenna heterogeneity.

18.2.4 SoftCast

Unlike most digital systems, SoftCast [396,443] uses a linear codec that combines video and channel coding to achieve graceful video recovery performance in line with the CSNR. It avoids a digital scheme of prespecified bit rates altogether. It starts with a DCT transform, and then allocates power weights based on the energy distribution of these DCT coefficients. After power allocation, a Hadamard transform-based whitening is applied across packets to ensure that transmitted packets have equal power and importance. In addition to the linear encoder, the transmitted signal strength is designed to be proportional to the source magnitude, and hence graceful decoding performance can be achieved for users with diverse channel quality.

However, SoftCast's decoder always constructs an approximately full-rank matrix to feed into a linear least square estimator (LLSE) [427]. The decoding performance degrades significantly when the number of received symbols is far smaller than the source dimension, such as in our scenario of a single-antenna setup. Therefore, SoftCast cannot be readily extended to our setting while maintaining the same performance as in a homogeneous case. Since we only consider image transmission in this work, we will mainly refer to the 2D version of SoftCast [443] hereafter.

18.3 Compressive Image Broadcasting System

We propose an analog image delivery system over MIMO links with receiver antenna heterogeneity as shown in Figure 18.1. The base station is equipped with two antennas, that is, $n_t = 2$, and single-antenna and two-antenna receivers coexist, that is, $n_r = 1$ or $n_r = 2$. For simplicity, we will refer to such a setup as a $2 \times \{1,2\}$ MIMO system throughout the paper. We assume the number of available time slots is $T = N/4$ to transmit an image of size $N = W \times H$ in this system.

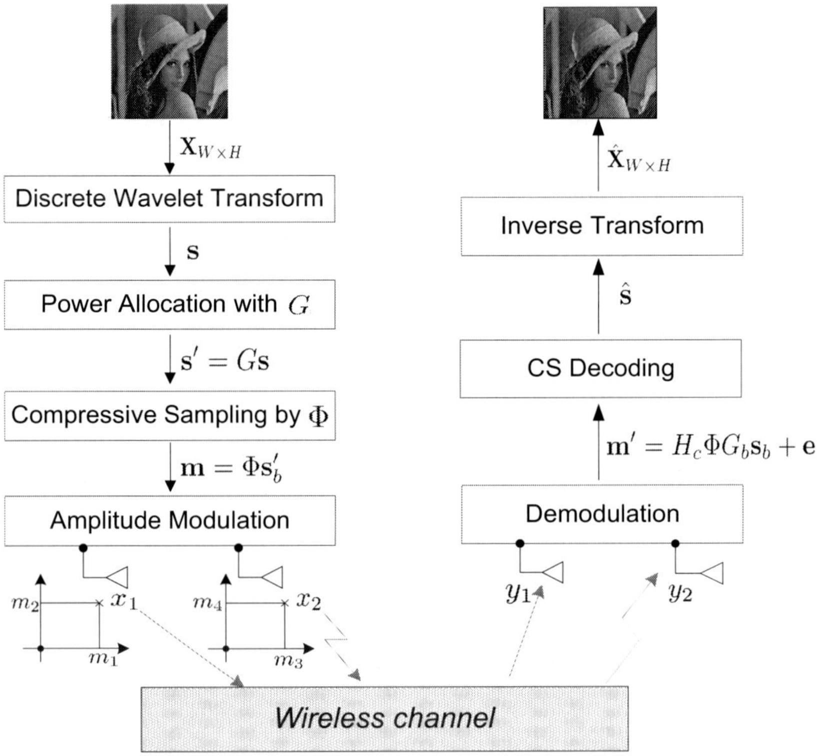

Figure 18.2 Block diagram of the compressive image broadcasting system.

Figure 18.2 depicts a block diagram of our proposed compressive image broadcasting system. There are four steps on the encoder side, DWT transform, power allocation, compressive sampling to generate measurements, and amplitude modulation and transmission of the measurements. The main decoding steps are CS decoding and inverse transform.

18.3.1 The Encoder and Decoder

The encoder first transforms the image to the frequency domain to remove the spatial correlation of the raw image using DWT, as in the JPEG 2000 standard [457]. We use orthogonal Daubechies wavelets [471] as the DWT basis, with a filter length of 8 and decomposition level $L = 5$.

Next, the encoder redistributes power across the DWT coefficients of different frequency bands. The wavelet coefficients are grouped into chunks, and we calculate the variance of the coefficients within each chunk. The total transmit power is split among these chunks according to their variance information, with consideration given to the antenna heterogeneity across receivers.

The scaled coefficients are then divided into a number of base vectors for the compressive sampling stage, from which measurements are generated by using a Hadamard sensing matrix. Note that this sampling process preserves the power distribution across coefficients after the power allocation step.

Finally, the measurements are directly transmitted as real (I) or imaginary (Q) components of complex wireless signals instead of being quantized and digitized. A distinct complex symbol is sent from each transmit antenna via spatial multiplexing. The per-chunk variance information is also broadcasting to all receivers as metadata, but through a reliable digital modulation and coding scheme.

At the receiver, the real and imaginary parts of the complex symbol on each antenna are separated and collected into the overall measurement vector. Regular MIMO decoding is skipped, and the original coefficients per vector are recovered via a weighted l_1-minimization CS decoder with the help of the variance metadata. The raw image is then reconstructed from the coefficients through an inverse DWT.

18.3.2 Addressing Heterogeneity

The power allocation strategy and the CS module combine to address CSNR heterogeneity and receiver antenna heterogeneity simultaneously. Our power allocation strategy adjusts the source data so that the subsequent CS module is more effective, specifically tuning the performance trade-off between receivers with varying antenna numbers.

CS theory makes it feasible to decode under-determined systems at single-antenna users, as shown when we previously proved that the MIMO channel would not invalidate CS recovery [472]. On the other hand, multi-antenna receivers are able to benefit from additional antennas via higher rates, because the CS recovery quality scales with the number of received measurements. Furthermore, the lower the noise level in each measurement, the higher the CS decoding performance. With the antenna setting determining the number of received measurements, and the CSNR governing the quality of each measurement, the overall recovery quality will then be simultaneously in line with the antenna number and the CSNR.

18.4 Power Allocation

Since we transmit the DWT coefficient via amplitude modulation, the energy carried in a coefficient can serve as protection at the channel to guard against noise. Therefore, transmit power allocation directly affects the reconstructed image's quality. In particular, we need to adjust the relative power differences between the coefficients of different frequency bands. This is essential for two reasons. First, the energy of DWT coefficients is concentrated in low-frequency bands. If the coefficients are transmitted verbatim without any power redistribution, low-frequency coefficients would consume most of the transmit power, so much so that high-frequency coefficients would be left with very little power budget. Second, receivers with different capabilities would favor different power allocation strategies. Since we target a heterogeneous broadcast system with diverse antenna settings, we need to balance the reconstruction performance of all types of users with a single scheme. This is a distinctive feature of our system.

18.4.1 Power Scaling Factors

It has been proven that, for a SISO (single-input single-output) AWGN channel and a linear decoder, linear scaling proportional to $\lambda^{-1/4}$ achieves optimal performance under the mean square error (MSE) criterion [396,427]. Here, λ denotes the variance of the random variable representing the transform coefficients. Although this works well for SoftCast, it is suboptimal for our system for two reasons. First, the optimality of a power allocation scheme depends on the decoding algorithm. Unlike SoftCast, our system adopts CS decoding instead of LLSE. Second, since single-antenna and two-antenna users coexist in our target broadcasting system, the power allocation strategy should balance the achievable multiplexing capability of two-antenna users and the diversity benefit for single-antenna users.

Let us first provide some intuition to this problem. The purpose of power allocation is to adjust the compressibility of the DWT coefficients $\mathbf{s}$. As mentioned earlier in Section 18.2, the parameter p dominates the degree to which the signal $\mathbf{s}$ is compressible. Therefore, we limit our discussion to scaling operations in which the power scaling factors are exponential functions of the coefficients, or mathematically:

$$g_i = C_1 \cdot s_i^{2\alpha} \tag{18.2}$$

where g_i is the scaling factor for coefficient s_i, and C_1 is a constant to ensure that the allocated power satisfies the total transmit power constraint. The constant 2 in the exponent is included so that α can be compared directly to its counterpart in SoftCast.

Since we have assumed that the compressibility parameter of the wavelet coefficients $\mathbf{s}$ is p, the scaled coefficient s_i' after power adjustment is:

$$|s_i'| = g_i \cdot |s_i| = C_1 \cdot |s_i|^{1+2\alpha} \lesssim C_1 \cdot C_0^{1+2\alpha} \cdot i^{-\frac{1}{p/(1+2\alpha)}} \tag{18.3}$$

The compressibility parameter now becomes $p/(1+2\alpha)$. Since a smaller p indicates higher compressibility, a positive α improves compressibility and a negative α does the opposite. From a channel protection perspective, on the other hand, a larger α tends to provide more protection to low-frequency bands. Conversely, the smaller the α value, the better protected the high-frequency bands.

Table 18.1 The effect of different numbers of chunks.

Number of chunks	1	16	25	70	259
PSNR Gain (dB)	0	2.77	2.83	2.96	3.19
Overhead (%)	0.14	2.40	3.07	6.69	21.11

From Eq. (18.3), we can see that setting $\alpha = -1/2$ means flattening out all coefficients, and setting $\alpha = 0$ means transmitting the coefficients as they are. In SoftCast, α is selected to be $-1/4$, which sets larger power scaling factors for high-frequency coefficients to avoid overprotecting the low-frequency bands. In our system, however, two types of receivers have different preferences. In particular, two-antenna users receive as many CS measurements as the number of unknown coefficients. They prefer a power scaling factor that properly amplifies high-frequency coefficients as in SoftCast, while single-antenna users only receive half of the CS measurements and would prefer a larger α, which suppresses the high-frequency coefficients, because those coefficients cannot be decoded anyway. Therefore, the best α that optimizes the entire multicast performance would depend on the proportion of the two types of receivers. In our design, we simply assume an equal number of single-antenna and two-antenna receivers, and choose α based on extensive simulations. We finally select $\alpha = -1/8$, which strikes a good balance for system performance. More detailed simulation results will be presented in the next section.

18.4.2 Aggregating Coefficients

A practical problem arises as the power scaling parameters need to be reliably transmitted to the receivers. If we select unique scaling factors for each of the coefficients and transmit them all, it immediately makes our system a trivial design at a huge overhead, because all s_i can be directly computed from the scaling factors. Fortunately, it has been shown [396] that adjacent coefficients in the same frequency band can be viewed as samples of the same zero-mean Gaussian distributed random variable. Therefore, we can group the coefficients into chunks and perform power scaling on a per-chunk basis. All coefficients in the same chunk share the same scaling factor g_i, in the computing of which the s_i^2 in Eq. (18.2) will be replaced by the variance of chunk i, denoted by λ_i. Then, the variance information used to calculate scaling factors are broadcasting to all receivers through a reliable modulation and coding scheme.

Obviously, smaller chunks allow for finer-grained adjustment, but they will incur higher overhead when transmitting metadata. Table 18.1 shows the bandwidth overhead and the corresponding PSNR gain for the image *Lena* divided into different numbers of chunks. Here, the overhead is evaluated as the ratio of time slots for transmitting metadata to that of transmitting DWT coefficients. We find that using 70 chunks achieves a good trade-off between overhead and performance gain.

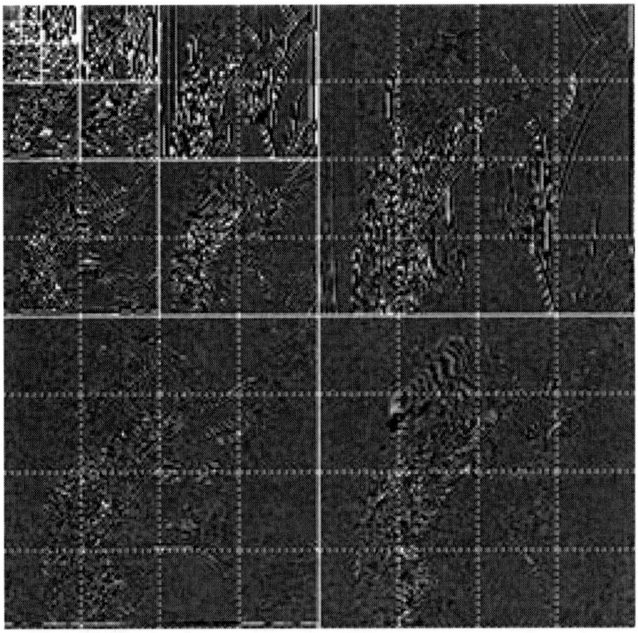

Figure 18.3 Pyramidal structure of 5-level wavelet decomposition and chunk division for *Lena*.

The variance metadata are quantized into integers and then encoded by Exponential-Golomb codes. After this step, they are transmitted using $1/2$ rate channel coding, QPSK, and STBC. The chunk division details are shown in Figure 18.3. The 10 upper left wavelet bands correspond to 10 chunks, while each of the remaining 6 bands is split into several chunks of equal size. The chunks in the upper-left corner are smaller than the others because there are fewer coefficients in the lowest-frequency band, and coefficients from different frequency bands should not be mixed into the same chunk.

In summary, in the power allocation module, the DWT coefficients are divided into 70 chunks, and the transmit power is allocated among these chunks according to their variance. The matrix representation of the power allocation process is:

$$\mathbf{s}' = \mathbf{G}\mathbf{s} \tag{18.4}$$

where G is a diagonal power weight matrix with diagonal elements g_i Eq. (18.2), while $\mathbf{s}$ and $\mathbf{s}'$ are the image coefficient vectors before and after power scaling.

18.5 Compressive Sampling

After power allocation, the CS sampling module generates measurements through linear projections. However, the number of DWT coefficients to represent an image is usually in the hundreds of thousands. Performing CS sampling and decoding on such a large vector is impractical. Since coefficients in the same chunk have the same statistical characteristics, we may form short vectors with the same compressibility characteristic by taking one element from each chunk.

However, due to the wavelet transform, not all chunks are the same size. We tackle this problem with an approximation. In particular, we merge the 7 upper left bands into one sampling chunk, while still computing and transmitting the variance for each band separately, and retain the other 63 chunks. As such, we form 64 new chunks of equal size. Since CS decoding favors a larger source dimension for better recovery performance [466], we generate compressible source vectors of length 256 by taking 4 coefficients from each chunk at a time. $n = 256$ is chosen to strike a balance between good recovery performance and decoding complexity.

Let $\mathbf{s}'_b$ be one such source vector of length n and $\mathbf{s}_b$ be the corresponding vector before power allocation. Therefore $\mathbf{s}'_b = \mathrm{G}_b\mathbf{s}_b$ where G_b is the corresponding submatrix of G. Then we can obtain the linear projections m_i of $\mathbf{s}'_b$ by:

$$\mathbf{m} = (m_1, m_2, \ldots, m_n)^T = \boldsymbol{\Phi}\mathbf{s}'_b \tag{18.5}$$

where $\boldsymbol{\Phi}$ is an $n \times n$ sensing matrix.

To ensure information capture capability and CS decoding performance, $\boldsymbol{\Phi}$ should satisfy some requirements, such as the restricted isometry property (RIP) and mutual coherence [297, 473]. The most commonly used sensing matrix ensembles, such as random Gaussian matrices and Hadamard matrices, have normalized or approximately normalized columns. This means that the transmit power distribution among different coefficients will not be changed in the sampling process. We use a Hadamard sampling matrix in the proposed system. It has been shown [474,475] that a Hadamard sampling matrix has the same desirable properties and comparable recovery performance as the optimal random Gaussian matrix ensemble, but allows us to generate measurements at a lower complexity. It is known that using the Hadamard sampling matrix may impose some constraints on the size of the source vector. We circumvent this problem by dividing the image coefficients into length-256 vectors according to the method described above.

18.6 Amplitude Modulation and Transmission

Since the measurements generated after the power allocation and linear sampling steps are compact and resilient representations of the original coefficients, further channel coding, such as that defined in the conventional 802.11 PHY layer, is not necessary. Instead, pairs of these linear projections are directly mapped to the I and Q components of the complex signal to be transmitted. This is the same pseudo-analog modulation as proposed in SoftCast. Each transmit antenna at the base station sends out a distinct complex symbol.

Mathematically, the sender transmits x_1^t and x_2^t on the two antennas in the tth time slot, where

$$\begin{aligned} x_1^t &= m_{4t-3} + j\, m_{4t-2} \\ x_2^t &= m_{4t-1} + j\, m_{4t}. \end{aligned} \tag{18.6}$$

For each source vector with length n, the sender generates n measurements and takes $T_0 = n/4$ time slots to transmit them.

18.7 The CS Decoder

For each receiver, the received signal quality is determined by the matrix channel defined by the n_t transmit antennas and n_r receiver antennas. Within the broadcasting system, the source transmission strategy is independent of the exact channel details between the sender and any receiver, and we assume such knowledge is not available to the sender. However, the receivers can normally obtain such knowledge through the frame preamble. Let H^t be the $n_r \times n_t$ channel matrix in the tth time slot, and each matrix element $h_{i,j}^t$ denotes the path gain from transmit antenna j to receiver antenna i.

For a single-antenna receiver, or a 2×1 case, the path gain from the two transmit antennas to the receiver antenna are $h_{1,1}^t$ and $h_{1,2}^t$, respectively, in the tth time slot. Then the received signal y^t in the tth time slot is:

$$y^t = h_{1,1}^t x_1^t + h_{1,2}^t x_2^t + e^t \tag{18.7}$$

where e^t is the Gaussian noise. From y^t, the receiver obtains two new measurements:

$$\begin{pmatrix} m_{2t-1}^t \\ m_{2t}^t \end{pmatrix} = \mathrm{H}_{2\times 1}^t \begin{pmatrix} m_{4t-3} \\ m_{4t-2} \\ m_{4t-1} \\ m_{4t} \end{pmatrix} + \begin{pmatrix} \Re(e^t) \\ \Im(e^t) \end{pmatrix}. \tag{18.8}$$

Matrix $\mathrm{H}_{2\times 1}^t$ can be written as:

$$\mathrm{H}_{2\times 1}^t = \begin{pmatrix} \Re(h_{1,1}^t) & -\Im(h_{1,1}^t) & \Re(h_{1,2}^t) & -\Im(h_{1,2}^t) \\ \Im(h_{1,1}^t) & \Re(h_{1,1}^t) & \Im(h_{1,2}^t) & \Re(h_{1,2}^t) \end{pmatrix} \tag{18.9}$$

where $\Re(\cdot)$ and $\Im(\cdot)$ are the real and imaginary parts of a complex number.

For two-antenna receivers, or 2×2 cases, each receiver derives four new measurements per time slot, 2 on each antenna:

$$\begin{pmatrix} m'_{4t-3} \\ m'_{4t-2} \\ m'_{4t-1} \\ m'_{4t} \end{pmatrix} = \mathrm{H}^t_{2\times2} \begin{pmatrix} m_{4t-3} \\ m_{4t-2} \\ m_{4t-1} \\ m_{4t} \end{pmatrix} + \begin{pmatrix} \Re(e^t_1) \\ \Im(e^t_1) \\ \Re(e^t_2) \\ \Im(e^t_2) \end{pmatrix} \tag{18.10}$$

where matrix $\mathrm{H}^t_{2\times2}$ can be written as:

$$\mathrm{H}^t_{2\times2} = \begin{pmatrix} \Re(h^t_{1,1}) & -\Im(h^t_{1,1}) & \Re(h^t_{1,2}) & -\Im(h^t_{1,2}) \\ \Im(h^t_{1,1}) & \Re(h^t_{1,1}) & \Im(h^t_{1,2}) & \Re(h^t_{1,2}) \\ \Re(h^t_{2,1}) & -\Im(h^t_{2,1}) & \Re(h^t_{2,2}) & -\Im(h^t_{2,2}) \\ \Im(h^t_{2,1}) & \Re(h^t_{2,1}) & \Im(h^t_{2,2}) & \Re(h^t_{2,2}) \end{pmatrix} \tag{18.11}$$

After T_0 time slots, the single-antenna receiver gets $2T_0$ new measurements. Stacking these new measurements to form a vector $\mathbf{m}_1$:

$$\begin{aligned} \mathbf{m}_1 &= \begin{pmatrix} \mathrm{H}^1_{2\times1} & \mathbf{0} & \cdots & \mathbf{0} \\ \mathbf{0} & \mathrm{H}^2_{2\times1} & \cdots & \mathbf{0} \\ \vdots & \vdots & \ddots & \vdots \\ \mathbf{0} & \mathbf{0} & \cdots & \mathrm{H}^{T_0}_{2\times1} \end{pmatrix} \begin{pmatrix} m_1 \\ m_2 \\ \vdots \\ m_{4T_0-1} \\ m_{4T_0} \end{pmatrix} + \mathbf{e} \\ &= \mathrm{H}_c \Phi \mathbf{s}'_b + \mathbf{e} \\ &= \mathrm{H}_c \Phi \mathrm{G}_b \mathbf{s}_b + \mathbf{e} \end{aligned} \tag{18.12}$$

The two-antenna receiver gets $4T_0$ new measurements, and the measurement vector $\mathbf{m}_2$ can be represented similarly.

For both types of users, the conventional MIMO decoding at the channel is skipped, and the receiver collects the raw signal samples for the measurements across all of its antennas over T_0 time slots. Decoding over many slots permits CS operations over a much larger matrix than a typical MIMO channel matrix, whose dimension is far too small to satisfy CS requirements.

Since the variance information can be sent to the decoder as metadata, we adopt the weighted l_1-minimization decoder [448], instead of the standard l_1-minimization for better recovery performance. Define $w_i = 1/\sqrt{\lambda_i}$, where λ_i is the known variance of the ith source element at the decoder. Then the minimization problem for decoding the source vector $\mathbf{s}_b$ is:

$$\hat{\mathbf{s}}_b = \min_{\mathbf{s}_b \in \mathbb{R}^n} \|\mathrm{W}\mathbf{s}_b\|_{l_1} \quad s.t. \|\mathbf{m}_1 - \mathrm{H}_c \Phi \mathrm{G}_b \mathbf{s}_b\| < \varepsilon \tag{18.13}$$

where W is the diagonal matrix with w_i on the diagonal and ε represents the noise power at the receiver.

The weighted l_1-minimization decoding problem for two-antenna users can be presented similarly. As we stated before, $T_0 = n/4$ time slots are needed to transmit each source vector. Therefore, the problem to be solved for single-antenna users is under-determined, and the CS decoding algorithm based on Eq. (18.13) is quite efficient. Although the linear system for two-antenna users has full rank, we find that the performance of the weighted l_1-minimization based CS decoder matches that of the optimal linear decoder LLSE.

18.8 Simulation Evaluation

This section serves two purposes. First, we evaluate the performance of the proposed compressive image broadcasting system, and validate some key design parameters. Second, we compare our system with alternatives, and demonstrate how well we achieve the design goal, that is, every receiver obtains image recovery quality that is commensurate with its antenna setting and channel quality.

In line with our previous description, we mainly focus on the $2 \times \{1,2\}$ MIMO setting and will discuss cases with more antennas later. Block Rayleigh fading channel with perfect receiver CSI is assumed if not otherwise stated. For clarity and without loss of generality, we assume that the two receiver antennas of two-antenna users have the same average per receiver antenna SNR. The recovery performance is evaluated using the standard objective measure peak signal-to-noise ratio (PSNR).

18.8.1 Micro-Benchmarks for Our System

We first evaluate the two main components of our system separately, the transmit power allocation strategy and the compressive sensing decoding algorithm. In particular, we evaluate the effect of different power scaling factors. In addition, the LLSE decoder and the l_1-minimization based CS decoder with or without reweighting are also evaluated. We use three 256×256 gray images, *Cameraman, Lena*, and *Peppers*, which are frequently used as test images in related literature.

18.8.1.1 Performance of Power Allocation

Recall that the sender scales the DWT coefficients by λ^{α}, where λ is the variance of the coefficients of a chunk and α is a parameter to be determined, which adjusts the coefficients' compressibility. When the α value is too large (> 0), too much power is distributed to the significant coefficients. Similarly, when α is too small ($< -1/4$), too much power is distributed to the insignificant coefficients. In either case, the recovery performance is bad for both single-antenna and two-antenna users. Therefore, we mainly consider α in the range $[-1/4, 0]$. Further simulation results show that a minor change of α has little influence on system performance. Therefore, we test five values of α ($-1/4$, $-3/16$, $-1/8$, $-1/16$, and 0) and compare their performance. The exact choice also depends on the decoder and the CSNR.

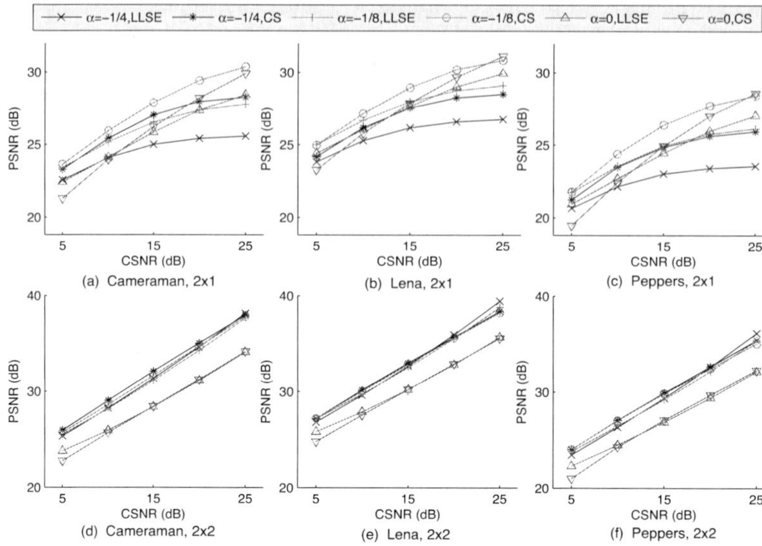

Figure 18.4 The recovery PSNR (dB) for images *Cameraman*, *Lena*, and *Peppers* under various power scaling factors and decoding methods. The subfigures in the two rows indicate the performance for single-antenna (labeled 2×1) and two-antenna users (labeled 2×2), respectively.

Figure 18.4 shows the received image PSNR for both single-antenna users (marked with 2×1) and two-antenna users (marked with 2×2) under different parameter settings. Here we only show the performance when α equals $-1/4$, $-1/8$, and 0, respectively, for the sake of clear presentation. From Figure 18.4d,e,f, we can see that $\alpha = 0$ performs the worst for two-antenna users, introducing a 2.4–3.8 dB loss in the received image PSNR. This confirms our assumption that transmitting wavelet coefficients verbatim tends to overprotect low-frequency bands and is suboptimal.

From Figure 18.4a,b,c, we have two findings. First, unlike the 2×2 cases where the LLSE decoder and the CS decoder have similar performances, the CS decoder shows its superiority in 2×1 cases. This is because single-antenna users receive only half as many measurements as there are coefficients. The CS decoder is designed precisely for such an underdetermined system and therefore outperforms the LLSE decoder. Second, among the five α settings, $\alpha = -1/8$ achieves the best performance in most channel conditions. As discussed earlier, the parameter used in SoftCast $\alpha = -1/4$ is unfavorable to single-antenna users, and it incurs a PSNR loss of as much as 2.4 dB when the CSNR is 25 dB.

Given all the combinations of the α settings and the decoding algorithms, the option of $\alpha = -1/8$ combined with CS decoding achieves the best performance for both single-antenna users and two-antenna users.

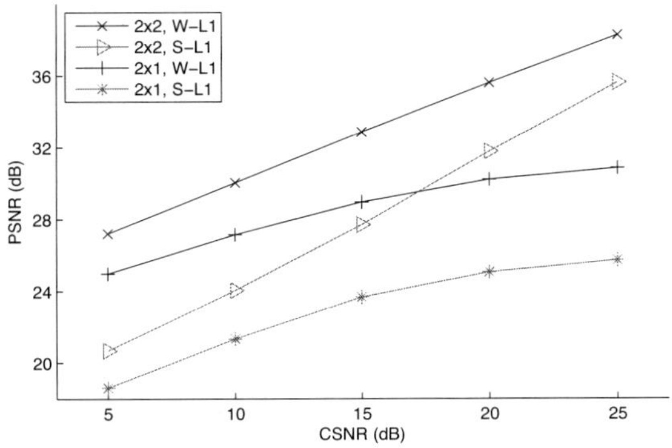

Figure 18.5 Recovery PSNR of weighted l_1 and standard l_1 decoders.

18.8.1.2 Performance of the Decoder

Next, we compare the performance of using *weighted* l_1-minimization (referred to as W-L1) against using the *standard* l_1-minimization CS decoder (referred to as S-L1) for decoding. The results for transmitting image *Lena* for different channel conditions are presented in Figure 18.5. We can observe that the weighted l_1-minimization using the coefficient variance metadata wins by more than 3 dB in PSNR.

18.8.1.3 Impact of Channel Estimation Errors

In practical systems, the receiver may not obtain perfect channel state information. We evaluate the impact of misestimated channel matrices in our system. Without loss of generalization, we assume the channel estimation error h_e is zero-mean complex Gaussian distributed. Hence, the estimated path gain from transmit antenna j to receiver antenna i at some time slot can be written as:

$$h'_{i,j} = h_{i,j} + h_e. \tag{18.14}$$

The energy ratio of estimation error to actual channel parameter, which could be denoted by ρ_h, determines the magnitude of channel estimation error. We evaluate the impact of misestimated channel matrices by comparing the system performance between $\rho_h = 0$ and $\rho_h = 0.01$, where $\rho_h = 0$ means that there is no channel estimation error. The experimental results for transmitting image *Lena* are shown in Figure 18.6. They show that the channel estimation error has a larger impact on high-SNR receivers than low-SNR receivers. When the channel estimation error parameter ρ_h is 0.01, the reconstruction PSNR is decreased only by 0.1 dB to 0.6 dB for single-antenna users and 0.1 dB to 0.9 dB for two-antenna users, when the SNR is between 5 dB to 20 dB. The loss is larger (around 2.6 dB) for two-antenna users when channel SNR equals to 25 dB.

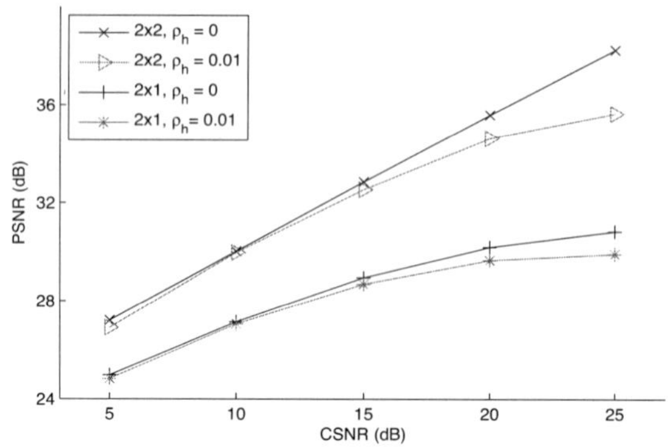

Figure 18.6 The effect of channel estimation error on image recovery PSNR values.

18.8.2 Performance Comparison with Other Broadcast Systems

In this section, we compare our proposed system with representatives of two classes of broadcast systems. One reference system is SoftCast, which adopts a similar analog framework and was originally designed for SISO broadcast. We extend it to MIMO settings and adopt a spatial multiplexing based MIMO transmission to transmit the Hadamard whitened linear projections. The other class encompasses conventional digital systems that adopt layered source-channel schemes. We also use more test images in addition to the three previously tested.

18.8.2.1 Comparison with SoftCast

SoftCast implements power allocation with $\alpha = -1/4$ and adopts LLSE decoding for all receivers. Figure 18.7 shows the comparison between our system and Soft-

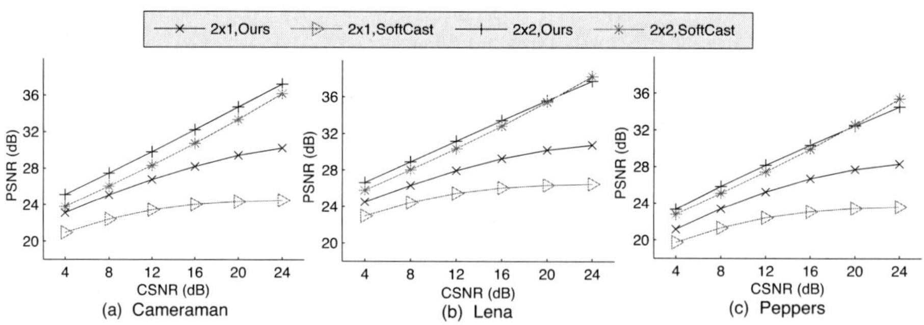

Figure 18.7 Our system versus SoftCast using images *Cameraman*, *Lena*, and *Peppers* under different CSNRs.

Cast when transmitting the three test images. Not surprisingly, our system achieves significant gains of up to 4.96 dB over SoftCast for single-antenna users. Meanwhile, our system does not incur any performance degradation for two-antenna users. For test image *Cameraman*, our system even achieves a 1 dB gain in 2×2 cases because DWT compacts energy much better than frame-based DCT for this image. This comparison clearly demonstrates the importance of power allocation and weighted CS decoding in a MIMO broadcast system with antenna heterogeneity.

18.8.2.2 Comparison with Conventional Digital Systems

We compare our system with two layered source-channel schemes. They adopt the same layered source coding but different MIMO transmission strategies.

In these two schemes, the state-of-the-art image codec JPEG 2000 [456, 457] is used to generate a layered source. A (204,188) short Reed-Solomon code is employed as suggested by the DVB-H standard. The encoded bitstream is then divided into the base layer and the enhancement layer. The encoded packets of both layers are subjected to convolutional coding and interleaving before modulation. The convolutional coding rate for the base and the enhancement layers is set to $1/2$ and $2/3$ respectively in our simulation. Correspondingly, the receiver performs de-interleaving, de-convolution, and RS decoding to obtain the JPEG 2000 stream.

Regarding MIMO transmission, our first reference system, referred to as HM-STBC, relies on STBC to fully exploit transmit diversity from multiple antennas on the base station [458, 459]. In this system, the channel coded bitstream is modulated into HM symbols. We implemented two typical HM variants, namely QPSK-in-16-QAM (i.e., hierarchical 4/16-QAM, QPSK for both layers) and QPSK-in-64-QAM (i.e., hierarchical 4/64-QAM, QPSK, and 16-QAM for the two layers respectively) [418]. The parameter that governs the ratio of protection of the base over the enhancement layer is set to 2. The HM symbols are sent using the Alamouti scheme.

Our second reference system, referred to as SP-MIMO, employs superposition MIMO coding [445, 447]. In this system, the base layer data are coded with the Alamouti scheme for higher reliability, and spatial multiplexing is used to code the enhancement layer to leverage the higher capacity of strong receivers. The transmitter combines them into HM symbols with a parameter ρ [445], which is the ratio of the amplitude of the second layer to that of the first layer. The value of ρ determines the extent of protection of the two layers in signal superposition, and based on experiments we set $\rho = 0.3$ to balance protecting both layers. As an example, if x_1 and x_2 are two successive symbols for the base layer data, and x_i, for $i = 3, 4, 5, 6$ represents successive enhancement layer data symbols, then SP-MIMO transmits $(x_1 + x_3)$ and $(x_2 + x_4)$ on the two antennas in the first time slot, and $(-x_2^* + x_5)$ and $(x_1^* + x_6)$ in the second slot, where '+' means superposition. Weak receivers with low CSNRs or a single antenna are expected to decode x_1 and x_2, while two-antenna receivers with good channel conditions can also decode x_3 to x_6. To ensure that SP-MIMO sends the enhancement layer at the same rate as in HM-STBC, we use both BPSK and QPSK to match the two HM settings in HM-STBC.

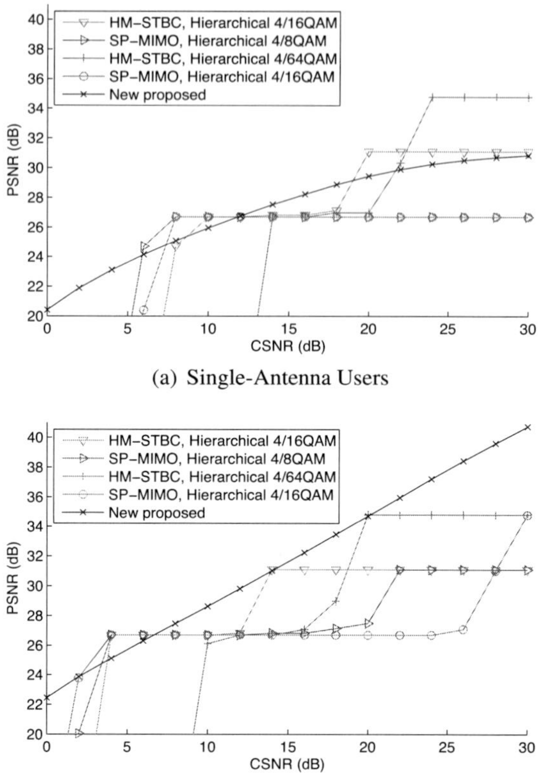

(a) Single-Antenna Users

(b) Two-Antenna Users

Figure 18.8 Our system versus two conventional reference systems, HM-STBC and SP-MIMO, using the image *Cameraman*.

Figure 18.8 (image *Cameraman*) and Figure 18.9 (image *Lena*) show the performance of our system as well as the two reference digital systems. For the same enhancement layer data rate, we can make three major observations between the two reference digital systems. First, for both single-antenna and two-antenna users, the CSNR required to decode the base layer in SP-MIMO is always lower than for HM-STBC. Second, single-antenna users cannot decode the enhancement layer at all in SP-MIMO, but they can in HM-STBC when the channel condition permits it. Third, in order to decode the enhancement layer, two-antenna users in SP-MIMO need much higher CSNRs than in HM-STBC.

When comparing our system with HM-STBC and SP-MIMO, an immediate observation is that the two digital systems only obtain stair-shaped incremental increases in quality while our system has a very smooth quality scaling behavior. Furthermore, in our system, two-antenna users always derive higher PSNRs over the single-antenna users when the per-antenna CSNR is the same. In HM-STBC and

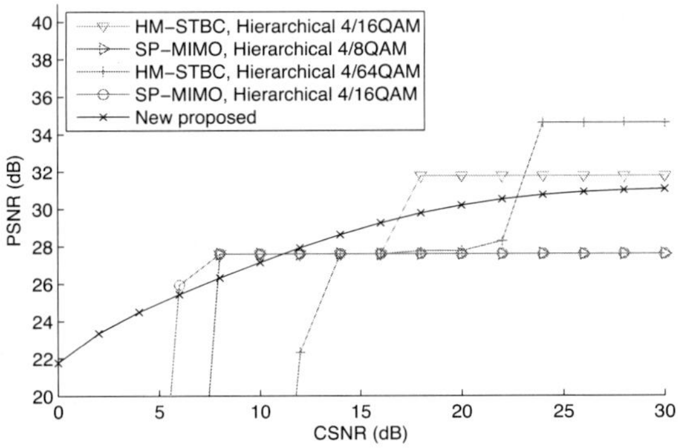

(a) Single-Antenna Users

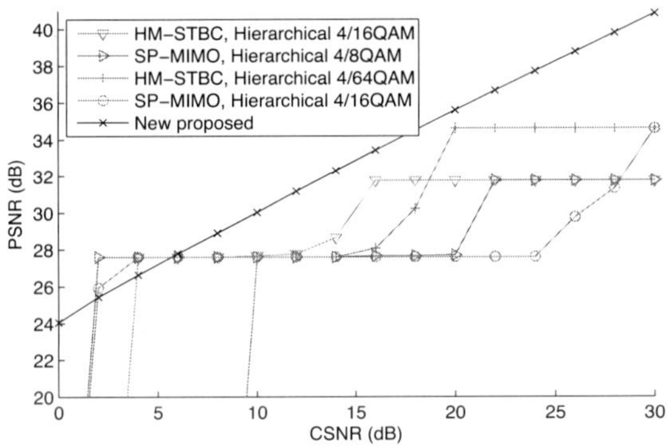

(b) Two-Antenna Users

Figure 18.9 Our system versus two conventional reference systems, HM-STBC and SP-MIMO, using the image *Lena*.

SP-MIMO, however, the additional antenna is underutilized in most cases, because the system performance stops improving when the CSNR exceeds a threshold (known as the *level-off effect* of digital systems).

Between our system and SP-MIMO, we find that both single-antenna and two-antenna users in our system can derive higher PSNRs at most CSNRs. The exceptions are when the CSNR is just high enough to allow the decoding of the base layer data, for example, at 6 dB for single-antenna users and 4 dB for two-antenna users.

Between our proposed system and HM-STBC, our system always performs better for two-antenna users at CSNRs of 8 dB or higher. At a CSNR of 26 dB, the PSNR from our system is more than 3 dB higher than that for the QPSK-in-16-QAM variant of HM-STBC, and about 7 dB higher than the QPSK-in-64-QAM variant of HM-STBC. For single-antenna receivers, our system is superior over HM-STBC at low CSNRs. However, when the channel is good enough to allow single-antenna users to correctly receive enhancement layer bits, our system incurs some performance loss. This is the cost our system has to pay in order to trade-off the performance

(a) Ours, 2x1, 26.8dB (b) SoftCast, 2x1, 23.4dB (c) J2K, 2x1, 26.7dB

(d) Ours, 2x2, 29.8dB (e) SoftCast, 2x2, 28.3dB (f) J2K, 2x2, 26.7dB

Figure 18.10 Visual quality comparison of decoded *Cameraman* at CSNR = 12 dB. The images in the top and bottom rows show the perceived quality by the single-antenna (labeled 2×1) and two-antenna users (labeled 2×2).

between single-antenna and two-antenna users. However, although HM-STBC with hierarchical 4/64-QAM can achieve high PSNR at high CSNR, the single-antenna users can barely get anything at a CSNR below 15 dB. In contrast, our system has a significantly larger operational range.

Figures 18.10 and 18.11 show the visual quality of *Cameraman* and *Lena*, respectively. Our scheme consistently produces better visual quality than SoftCast, especially for single-antenna users. Compared to HM-STBC, although our system delivers images that appear less smooth in some places, the image details are better preserved.

18.8.2.3 Overall Performance in a Broadcasting Session

We compare the overall performance of our system with two reference systems, namely SoftCast and HM-STBC, in a broadcasting session. We assume 100 single-antenna users and 100 two-antenna users in the session, with average per-antenna SNR 15 dB, and variance 4 dB. Table 18.2 lists the average received PSNR for eight test images. In the table, J2K denotes the digital system, which encodes images with the JPEG 2000 standard and transmits them using HM-STBC scheme. We do not show the results for SP-MIMO, because previous experiments have shown that it significantly underperforms our system.

From the last row of the table, we can see that our system achieves the highest PSNR among the three systems for both types of receivers. On average, our system

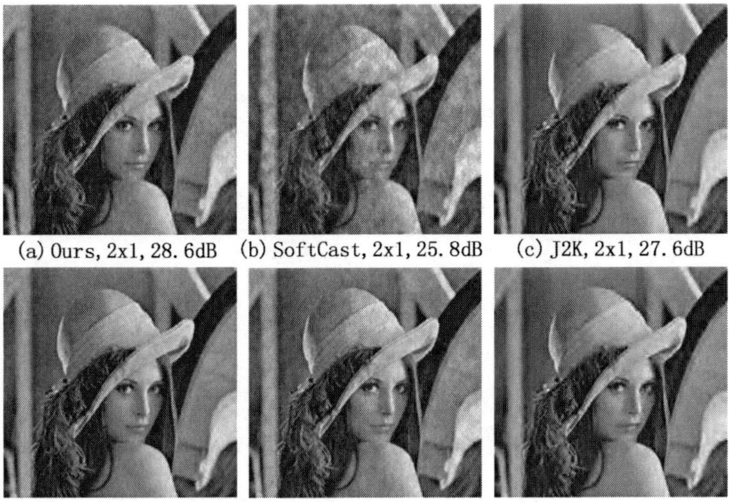

(a) Ours, 2x1, 28. 6dB (b) SoftCast, 2x1, 25. 8dB (c) J2K, 2x1, 27. 6dB

(d) Ours, 2x2, 32. 3dB (e) SoftCast, 2x2, 31. 6dB (f) J2K, 2x2, 27. 6dB

Figure 18.11 Visual quality comparison of decoded *Lena* at CSNR = 14 dB.

Table 18.2 Comparison between our system, SoftCast, and J2K.

	Single-Antenna Users			Two-Antenna Users		
	Ours	SoftCast	J2K	Ours	SoftCast	J2K
Cameraman	27.73	23.76	27.14	31.58	30.13	29.56
Lena	28.76	25.77	28.55	32.77	32.18	30.08
Peppers	26.16	22.80	25.94	29.73	29.30	28.49
Boat	29.42	28.02	28.39	33.32	34.13	31.26
Couple	30.26	26.25	30.86	33.32	32.22	33.23
Girl	30.45	27.17	31.35	33.62	33.39	33.16
House	32.41	29.86	33.24	35.91	36.32	34.45
Tree	26.48	24.72	25.76	30.23	31.07	28.01
Average	**28.96**	**26.04**	**28.90**	**32.56**	**32.34**	**31.03**

obtains 2.92 dB gain over SoftCast for single-antenna users, and 1.53 dB gain over HM-STBC for two-antenna users. As more advanced techniques on CS reconstruction for images are proposed, the compressive image broadcast system is expected to achieve even better performance. For example, it is reported in a recent body of work [476] that the reconstruction PSNR for *Lena* (256×256) could be as high as 29.16 dB and 33.58 dB when the number of available CS measurements is only 15% and 30% of the original pixel numbers. This suggests great potential for the proposed system.

18.9 Summary

In this chapter we have presented a compressive image broadcast system to simultaneously address heterogeneity across receiving antenna numbers and diverse channel SNRs. By integrating compressive sensing into the MIMO transmission of multimedia data, our framework ensures decoding for single-antenna users while two-antenna users can achieve multiplexing gain. By designing a suitable power allocation strategy, we can reach a performance trade-off between single-antenna and two-antenna users in a flexible way. Simulation results demonstrate the advantages of our system when compared with an analog system extended from SoftCast and two other conventional digital systems.

Our system is likely more beneficial for systems with more antennas. A digital approach would be even more constrained in such a scenario, since we would need to divide the source into more layers. Hybrid multiplexing and STBC schemes would also be limited by the inherent difficulty of designing efficient STBC codes for large channel dimensions. The LLSE decoder in SoftCast would suffer further if the number of received measurements becomes even smaller, for example, for a 3×1 or 4×1 scenario. In comparison, the main components in our system naturally cope with higher channel dimensions, merely using slightly different parameter choices for two possible reasons. First, the most suitable power scaling factor may be different. Second, it is likely that the system should send more measurements than are needed by the user with the most antennas. This would trade-off more multiplexing gain for diversity gain and help the single-antenna users.

Finally, although this work focuses on image transmission, we can treat images as intra-coded video frames without considering temporal correlation among successive frames. Our system can be extended to video broadcasting by using a 3D wavelet transform, which requires consideration of the trade-off between system performance and the decoding complexity.

Part VII
Future Work

This book starts from information theory. I would like to end by discussing the developments of information theory in future. Although most of our research presented in this book is technical and solution-oriented, the fundamental problems behind them are closely related to information theory. Shannon's information theory started from the end-to-end telegraph and has been developed for more than 60 years. With the advanced compression and communication technologies presented in this book, it is a right time to discuss the possible theoretical breakthrough in information theory. The contents presented in this part are our thoughts that we are strenuously working on or plan to do soon. I hope that we can present some solid progress on theoretical research in the near future.

Chapter 19
Computational Information Theory

19.1 Introduction

Since Shannon published his seminal paper "A Mathematical Theory of Communication" in 1948 [4], information theory has developed for more than 60 years. In Shannon's time, the communications system was a simple end-to-end telegraph. There were two fundamental problems to be solved. The first one was entropy-approached source coding. This problem was solved early on with arithmetic coding. The second problem was to study how to complete transmissions approaching the capacity of a channel, especially an end-to-end channel. When Turbo code was invented in 1993 [18, 19] and LDPC code was re-invented in 1996 [17], the problem of capacity-approached channel codes was also solved.

With the later emergence of networks, a natural extension of Shannon's information theory was toward more complicated channel models, with multiple senders and receivers in a network. This is called network information theory [477]. However, once channel models are extended beyond the end-to-end structure, it is hard to derive the close-form channel capacity. Theoretical results can be derived only in simple network channel models, for example, broadcast channels, multiple access channels, and 2-to-2 channels. For generic network channel models, only a max-flow min-cut theorem is available. Recently, there has been little progress in network information theory.

Our research into compression and communications has motivated us to consider another extension of Shannon's information theory. In Chapter 11, we study the SIFT-based image compression. The central question of this work is how to use the correlation of external images included in a large-scale database. Our scheme is totally different from existing schemes for image and video compression. In Chapter 17, we study the use of internal correlations of a video for channel denoising in the physical layer of wireless networks. One extension of this research is to use the correlation of external images in a large-scale database for channel denoising. Both of these areas of research point to new opportunities for information theory when a large amount of data is available.

Therefore, the extension we are currently considering is new source models. We call them *Cloud sources*. When you compress and transmit a source, we assume that the channel and receiver have a large-scale database available. Shannon's information theory already studied correlated sources. For a pair of correlated sources, joint source coding can improve compression efficiency. When a correlated source is available at the receiver, distributed source coding (DSC) can use this correlation at the receiver as efficiently as it is used at the sender. Conditional entropy is used to study these cases. Cloud sources are significantly different from previous correlated sources and thus will require corresponding source coding and channel coding, as well as joint source and channel coding.

The fundamental difference is that in Cloud sources you do not know which ones are correlated to the input source and how they are correlated. An important assumption in Cloud sources is that, for any input source, you can find a few correlated sources from a large-scale database with a probability close to one. At the same time, most sources in the large-scale database are not correlated to the input source. For example, Facebook has 220 billion photos [1]. When you upload one photo to Facebook, most likely you can find some correlated ones from among the 220 billion photos. But the percentage of photos correlated to the uploaded photo in the database is small and almost close to zero. In other words, most photos are not correlated to the uploaded photo. Before using the correlation of Cloud sources for compression and transmission, you have to identify the correlated ones and estimate their correlation.

According to Shannon's separate source and channel coding, an input source is usually compressed. However, once the source is fully compressed, compressed data looks like random bits and cannot be used to identify which ones are correlated in the Cloud sources. In order to find the correlated sources, the input source cannot be fully compressed and some redundancy has to be kept or introduced. If the existing redundancy can be further utilized during transmission, the system performance is still optimal. Therefore, for using the correlation of Cloud sources, a distinguishing feature is that the input source cannot be fully compressed and joint source and channel coding is needed. The existing redundancy can bring the additional benefit of finding the correlated sources in the Cloud sources. If the correlated sources can be used in source and channel coding, the system performance will most likely be much better without external correlation. Since the source model needs a computation to find the correlated sources, we call this extension *computational information theory*.

19.2 Cloud Sources

In the computational information theory, an important concept is Cloud sources. Let us define them as C, which contains a large number of sources $C = \{c(1), c(2), ..., c(N)\}$. N is a huge number. As we have mentioned, Cloud sources have two properties.

Definition 11 (Correlation Property) *For any input source s, the probability of finding a few correlated sources in the Cloud sources is close to one, namely,* $P\{C_s = \phi\} = \varepsilon$. C_s *is a set of sources correlated to the input source s.* ϕ *is an empty set and* ε *is a small number close to zero.*

How is the correlated set C_s selected? It depends on the correlation definition. We can still use the conditional entropy to characterize the correlation. For example, if $c_s(i)$ is a correlated source in C_s, the conditional entropy is $H(s|c_s(i)) \leq kH(s)$. k is a factor less than 1. The first property indicates that there is always a correlation between any input source s and Cloud sources C. Obviously, Cloud sources are not traditional sources studied in information theory. They are the hottest concept in Big Data, which contains almost all the sources that you will probably need.

Definition 12 (Diversity Property) *For any input source s,* C_s *is the set of sources correlated to s in Cloud sources C.* $|C_s|/|C|$ *is close to zero.*

$|\cdot|$ is the number of elements in a set. This property actually poses a large diversity constraint in Cloud sources C. If all sources in the Cloud sources are similar or belong to one category, the problem may become easy because you can arbitrarily take some of them for source and channel coding. However, because only a very small set in C is correlated to input source s, it is important to keep or introduce some redundancy in s so as to accurately find the correlated sources C_s in C.

Although Facebook has been mentioned in this book as an example of Cloud sources, strictly speaking, only the collection of large-scale data is hard to say that they are Cloud sources. The collected data has to be analyzed and reorganized as structured data before it can serve as Cloud sources. First, the collection of data should be classified into different categories according to their statistics. As shown in Figure 19.1, images are classified according to the layout and histogram of pixel values. But such a simple classification is not enough for efficiently exploiting the correlation in Cloud sources.

From our experiences, there are at least four types of metadata needed from low level to high level to make use of Cloud sources efficiently. According to our research presented in Chapters 11 and 12, local descriptors, such as SIFT, SURF, and CHoG are the first type of important metadata for an individual source in Cloud sources. An example of local SIFT descriptors is shown in Figure 19.2a. This type of metadata is used to identify the correlated sources and estimate the geometric transformations among the correlated sources. They are much better than the global transformation between two correlated sources because the corresponding SIFT descriptors actually indicate the matched regions of correlated sources. It allows multiple transformations between two correlated sources. Furthermore, they are scale-invariant and rotation-invariant, and less sensitive to illumination.

More importantly, the second type of metadata is the correlation of sources in Cloud sources. Since the number of sources is huge, it is an intensive computation to analyze their correlations. They have to be computed off-line and stored as part of the Cloud sources. According to our research presented in Chapter 12, the analysis of the correlation of the sources in Cloud sources can build many relationship graphs,

Figure 19.1 Part of the images in Cloud sources.

as shown Figure 19.2b. Every graph can be further simplified as a tree and, according to the tree, individual sources can be reorganized as a source sequence. Such models are still too simple to characterize local matching regions among sources. Furthermore, since Cloud sources are dynamic, it is necessary to study how to dynamically modify the tree models to achieve a better trade-off between computational cost and modeling accuracy.

Further extending our research presented in Chapter 12, the more accurate description of the correlation among sources should be scene information. It may be difficult for generic content to reconstruct scenes. But with modern technical developments, it has become feasible for some constrained scenarios. For some popular locations (e.g., tourist sites), as shown in Figure 19.3a, a large number of collected images and videos are available on the Internet and the technology of structure from motion can be used to reconstruct the 3D scenes from the collected data [478, 479], as shown in Figure 19.3b. For surveillance video, it is also possible to reconstruct 3D scenes using this technology because the scenes are constrained. Indoors, Microsoft Kinect data, including both depth and RGB data, can be used to reconstruct 3D scenes [480]. With the scene description, it is possible to code every object in the scene with the correct geometric transformation. Every object can find correlated external objects instead of correlated external sources.

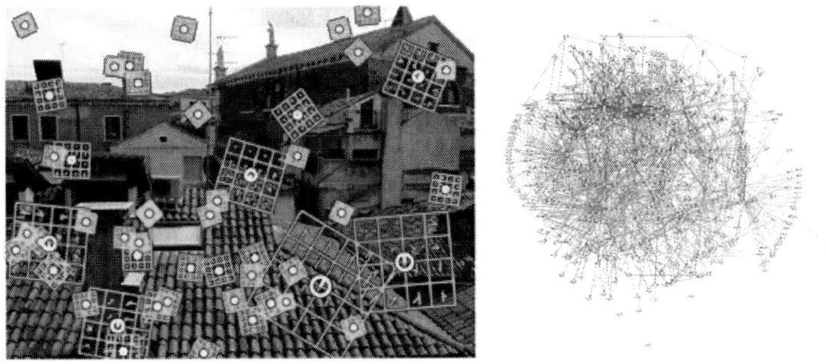

(a) (b)

Figure 19.2 Metadata in Cloud sources.

(a) (b)

Figure 19.3 Structure from motion.

In addition to the above three types of metadata, the highest level of metadata in Cloud sources is still semantic. Although there is a gap when getting the semantics from media content, some technologies have been developed to propagate human annotations on individual sources to a mass of correlated sources. From this viewpoint, it is easier to solve the semantic problem in Cloud sources than a single source. With the annotated Cloud sources, the semantic-based coding of input sources is becoming possible.

19.3 Source Coding

In this section, we first discuss how to efficiently compress the metadata. After that, we discuss the coding of Cloud image sources and Cloud video sources, respectively. One more challenging problem is how to compress an input source by using Cloud sources at the receiver. This is distributed cloud-based coding.

19.3.1 Coding of Metadata

In Shannon's information theory, the coding of metadata is studied relatively less than other areas, probably because metadata is easy to compress. However, in computational information theory, there is rich metadata that has to be generated offline and stored with Cloud sources, including local feature descriptors, source correlation, scene descriptors, and so on. Therefore, it is important to study how to efficiently compress various types of metadata in computational information theory.

The compression of local feature descriptors has been surveyed in Chapter 11, Section 11.2.2. It has been a popular research topic in the past decade. There are several different types of technologies that have been developed, including the dimension reduction of feature vectors, the quantization of feature vectors, the design of low-dimension feature vectors, and the compression of feature vectors via image compression. These technologies are often evaluated according to the accuracy of data retrieval, namely, the curve of the recall rate versus the compressed size of feature vectors. The recall rate is a reasonable measurement for data retrieval but not for source coding. One local feature vector represents a block of source data. The dimension reduction and quantization can be viewed as a kind of vector quantization to feature vectors. Instead of evaluating the distortion of local feature vectors, it would be better to evaluate the distortion of sources caused by the compression of feature vectors. The source distortion may be caused by source mismatch but most likely by the decrease of matched accuracy.

Therefore, a key research problem in the coding of local descriptors is how to measure the distortion of source data through the distortion of feature vectors. It depends on the purpose of using the feature vectors. When local descriptors are used to analyze the correlation of sources in Cloud sources, the distance of sources within one cluster and the distance of sources between clusters is a kind of distortion measurement. When local descriptors are used to retrieve external images for reconstructing one source, as we have done in Chapter 11, the reconstruction quality is another kind of distortion measurement. Once the distortion is defined, the corresponding rate of distortion optimization should be studied via the approaches for analyzing vector quantization.

As we have discussed in Chapter 12, the correlation of sources in Cloud sources is described by graphs and trees. The graphs and trees should be compressed losslessly. Our current work only studies the correlation of entire sources. This is not precise enough to characterize the correlation in Cloud sources. As a matter of fact, matched local descriptors represent corresponding local regions in two sources. In addition to the correlation of sources by weighting all matched regions, it is expected to develop the technology representing the correlation of local regions in Cloud sources. If so, one source is included in multiple graphs and trees of correlated regions. Obviously, the correlation will become considerably complicated. In addition, since Cloud sources are dynamic, many sources are continuously added and some sources may be deleted. The graphs and trees of correlated regions should be easy to adjust with a low computational cost.

The scene descriptors are depth images if they are reconstructed from Microsoft Kinect data. The compression of depth images has been developed based on the technologies of image and video coding [481,482]. The scene descriptors are a 3D point cloud if they are reconstructed by the structure from motion approach. In general, a 3D point cloud consists of some discrete feature points in the scene. These points may not be precisely measured in the scene. Several approaches have been proposed to compress the point cloud of 3D objects [483,484]. But the scene descriptors have a much larger scale than 3D objects. More importantly, in contrast to the compression of 3D objects, the compression of scene descriptors is not used for reconstructing scenes. Instead, it provides us 3D information at the region level. The regularization should be allowed before compression.

The above methods for the compression of scenes are still at the signal level. The more compact description of scenes should be object-based and even semantic. If so, the compression problem becomes simple. But it is still difficult to accurately extract objects, especially with semantics. It is outside the scope of information theory.

19.3.2 Coding of Cloud Image Sources

In Shannon's information theory, source coding is mainly studied for communication. Storage is seldom considered as an independent research problem. However, in the Cloud era, compression for large-scale storage is becoming more and more important. For example, Facebook has stored 220 billion images in its data center and this amount is continuously increasing at several hundred million every day. Currently, every image is stored as a JPEG file independently. Theoretical total size of independently compressing Cloud sources C is equal to

$$R_C^I = \sum_{i=1}^{N} H(c(i)). \tag{19.1}$$

The superscript I means independent compression. At least 1 million 1T hard disks would be necessary for the images in Facebook. This obviously creates a huge cost in power supply and air conditioning. Every hard disk has to be replaced every three years in order to make the storage safe.

As with the properties discussed in Section 19.2, every source in the Cloud sources has a few correlated ones, but overall the sources display a huge diversity. According to the correlation property of Cloud sources, the most efficient way to compress them is jointly. The theoretical size of joint compression is

$$R_C^J = H(c(1), c(2), \ldots, c(N)), \quad \{c(1), c(2), \ldots, c(N)\} \in C. \tag{19.2}$$

The superscript J means joint compression. However, N is huge and it is impractical to jointly compress all the sources in Cloud sources because of random access requirements and computational costs. According to the diversity property, it is not necessary because there is no gain for jointly compressing non-correlated sources. A feasible method would be to jointly compress correlated sources only.

$$R_C^C = R_{C_1}^J + R_{C_2}^J + \cdots + R_{C_K}^J, \quad C = C_1 + C_2 + \cdots + C_K. \tag{19.3}$$

The superscript C means correlated joint compression. C_1, C_2, ..., C_K are the set of correlated sources. Note that $R_C^C \longrightarrow R_C^J$.

In Chapter 12, we proposed analyzing the correlation of sources in the large-scale storage and organizing the sources as many trees. The correlated sources are described in the same tree, where the relationship among parents and children characterizes their correlation. Instead of compressing them one by one, we proposed compressing sources, described in the same tree, as a pseudo sequence. Our experimental results show that the compressed size of a pseudo sequence is only one-tenth of the size of independent compression. It indicates that using the special source coding designed for Cloud sources can save 90% of storage in the data center. An additional cost is that, when one source is accessed, multiple sources may need to be decoded. If the depth of a tree is constrained, such a computational cost should be able to be offered by the data center.

Our research presented in Chapter 12 demonstrates the technical feasibility and the advantages of coding Cloud sources jointly. However, the trees are built by analyzing the correlation of sources using local descriptors. It should not be optimal from the viewpoint of information theory. To simplify the computation of building graphs, it is fine to use correlation analysis to establish initial graphs. It is better to introduce the rate distortion optimization in the subsequent processing.

1. *Building graphs* — Some sources are included in multiple graphs. RDO should be used to evaluate in which graph the sources should be and whether the graphs should be merged together.
2. *Conversion from a graph to a tree* — When a graph is converted to a tree for the generation of a pseudo sequence, the rate distortion cost should be evaluated too. It may greatly benefit the trees with many sources and strong correlation.

For a given model of Cloud sources, the rate distortion theory should be studied to guide how to efficiently form different clusters of sources and how to efficiently compress them.

If correlated local regions are considered in joint compression instead of correlated sources, one source may be included in different correlated sets. For example, one source contains two object regions. It can be categorized into the sets of two different objects. Obviously, it will cause repeated compression of the source. In particular, when the number of sources in Cloud sources is huge, it may become impossible to find the optimal classification by using the correlated local regions. The graph building is a much tougher problem than what we considered in Chapter 12.

An interesting but fundamental theoretical problem for the coding of Cloud sources is whether the compression size R_C^J will converge when the number of sources in Cloud sources is huge. If this is the case, when more sources are added to the Cloud sources, the compressed size does not increase clearly because all the information carried by the added sources has been included in the Cloud sources. We only need to store the descriptions of how to use the Cloud sources to reconstruct the added sources. After that, the added sources can be thrown away. Although it is difficult to prove mathematically, it is not difficult to understand that the Cloud sources already contain all the information in the world. It should be reachable for the specified content. Google street images should help with this if the company captures the images of all streets in the world. Based on this conjecture, the cost of Cloud storage is a constant no matter how much data is uploaded by users.

19.3.3 Coding of Cloud Video Sources

The coding of Cloud video sources can be distinguished from the coding of Cloud image sources in that the frames in a video already present strong correlation. Is it also necessary to use the correlation of external videos in the Cloud? It seems reasonable to answer this question with a no. In particular, with the developments of the last two decades, the technologies to use the correlations between frames are mature and have a very good performance. In the HEVC standard, after motion compensation from a correlated frame, the energy of residual signals is small with random distribution. It requires few bits to compress. However, we also observe that the intra coded data is a big part of HEVC. It is mainly caused by scene changes and covered/uncovered regions. Obviously, externally correlated videos can help greatly reduce this part of the data.

However, it is not easy to use externally correlated videos to compress a video in the existing hybrid coding framework. Here we present our idea for Cloud video storage and video surveillance. In state-of-the-art video coding standards, at most four references are allowed. It is limited by memory cost and computational cost for motion compensation. However, in Cloud video storage and video surveillance, we propose increasing the number of references to millions or even more. Then, for any video, almost all of the frames can be coded by prediction from the references because most of the content in the video exists in the references. The references are actually Cloud sources discussed in this chapter. Although this idea sounds interesting and promising, there are several tough technical problems to solve.

First, where are the large number of references stored? Obviously, it is impossible to store them in the memory. Instead, they should be stored in hard disks as compressed data. In other words, every reference is compressed as intra-coded data and stored in the storage. Once it is selected as the reference for compressing a frame, it will be loaded into the memory and decoded immediately. In one data center, there are many CPU cores available and the computation for decoding an intra-coded reference is not high even for HEVC. It should not be a problem to simultaneously decode hundreds of references in a short time. In addition, the references should be compressed with a small quantization step and thus will be of high quality. Although frames in a video present strong correlation, they are compressed as inter frames and the quantization step is not small. In this case, the external references still have a better chance of being selected as a reference than as coded frames in the video.

Second, how are the external references selected for compressing a video? Obviously, it is impossible to select one external reference from a large number of candidates by rate-distortion optimization. Similar to our work on compressing Cloud image sources, we extract local feature descriptors from every reference and store them as metadata. For a frame to code, we also extract its local feature descriptors. Therefore, similar to content-based image retrieval, several external references are first selected by matching local feature descriptors. The matching process can be done in a very short time even for millions of references. The selected external references and their internal references are further selected by rate-distortion optimization to compress every block.

Third, which frames should be selected as external references? Obviously, they should be high-resolution and high-quality frames. The most challenging problem in this process is how to select them according to its content and the content of the selected references. One approach is similar to the correlation analysis by using local feature descriptors in Chapter 12. When a frame is not correlated to any stored references, it should be compressed as an intra frame with a small quantization size and marked as a new external reference. Once the frame is selected as an external reference, it cannot be deleted even when the video is deleted; otherwise some videos using the frame as a reference cannot be decoded correctly. In the beginning, the number of references will increase rapidly. We hope that the increase is not linear and will become slower and slower with the increasing number of references.

19.3.4 Distributed Coding Using Cloud Sources

If an input source is compressed by conventional source coding, it cannot utilize the correlation of Cloud sources at all because all the information has been included in the compressed data. The theoretical coding rate should be $H(s)$. The input source can only be reconstructed by decoding the compressed data. Therefore, this approach cannot benefit from the correlation of Cloud sources. One solution for using Cloud sources is the scheme we presented in Chapter 11. We describe images by local feature descriptors and thumbnails. The input images can be reconstructed from Cloud sources via the compressed descriptions. The significance of this work is to demonstrate that high-quality images can be reconstructed in this way. However, from the

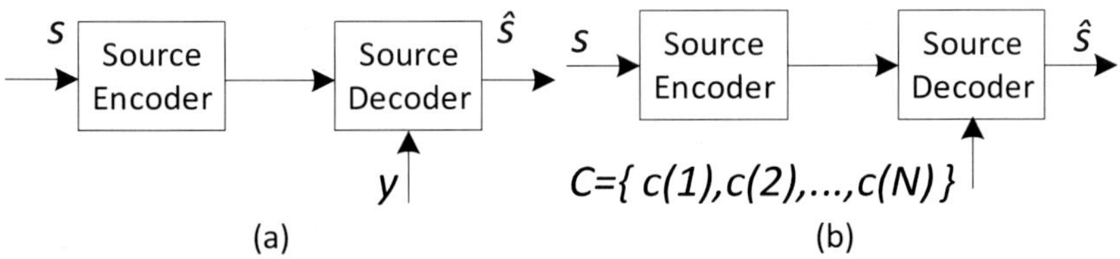

Figure 19.4 Distributed source coding without/with Cloud sources.

viewpoint of Shannon's information theory, it does not look like source coding. Furthermore, it cannot achieve a good reconstruction under the MSE criteria.

Because Cloud sources are assumed to be available at the receiver, one more reasonable solution should be distributed source coding (DSC). In conventional DSC, as shown in Figure 19.4a, a correlated source y, called side information, has been identified at the receiver. Input source s is first coset quantized and then is compressed by channel codes (e.g., Turbo code and LDPC code). The theoretical coding rate should be $H(s \mid y)$, which is less than $H(s)$. The compressed data is used to correct the correlated source y toward the input source s at receiver. When Cloud sources are available at the receiver, as shown in Figure 19.4b, the side information is not yet identified. Furthermore, since the number of sources is huge and most of them are not correlated to input source s, the distributed coding rate $H(s|C)$, directly using C, should be close to $H(s)$ because the compressed data by channel codes cannot be used to identify which of the ones in the Cloud sources are correlated to s.

To identify the correlated sources, similar to the scheme presented in Chapter 11, local feature descriptors of the input source s have to be sent to the receiver. Let us assume that the correlated sources C_s can be found from C with the received local feature descriptors. In general, any one of the correlated sources should not be the best side information that we can get. With our approach presented in Chapter 11, the best side information can be generated for combining the most matched parts of all the correlated sources. The distributed coding rate in this case should be $H(s \mid C_s)$. y, used in the conventional DSC, can be viewed as one correlated source in C_s. Therefore, $H(s \mid C_s)$ is less than $H(s \mid y)$. This is the advantage of Cloud sources. A key technical problem here is how to generate accurate side information. Since the number of combinations of correlated patches is big, the generation should be guided or checked by received parity data.

Theoretically, the above scheme should not be optimal because local feature descriptors are only used for data retrieval. From the viewpoint of compression, they are overhead. As we have discussed in Chapter 11, SIFT feature vectors are extracted in the feature regions of input source s. Every SIFT vector is a 128-dimension histogram of the gradient in different directions. Can we slightly change the generation of feature vectors and make them the linear combinations of some samples in the feature regions? If so, the feature vectors are generated by arithmetical calculations on samples. The channel coded data for reconstruction is the result of logical calculations on samples. Similar to our work in compressive modulation presented in

Chapter 14, they can be represented as the mixed logical and arithmetical nodes on the left side of a bipartite graph for joint reconstruction of the source nodes on the right side. A key technical problem here is how to design the modified feature vectors that are still accurate at identifying correlated sources.

Another alternative solution is to remove the syndrome coding from DSC. At the same time, channel coding in the traditional digital transmission is also removed. Quantized data is directly transmitted by the pseudo-analog approach presented in Chapter 16. If so, the quantized data remains part of the source redundancy and the transmission does not remove the redundancy. Therefore, with the remaining redundancy, the received quantized data can be used to retrieve the correlated sources. But the challenge here is whether the quantized data can be used for correlated source retrieval as efficiently as local feature descriptors because it has been quantized.

19.4 Channel Coding

As we have discussed, in order to use the correlation of Cloud sources, input sources cannot be fully compressed. For uncompressed or partially compressed sources, current transmission approaches are inefficient because they cannot use the remaining source redundancy. Furthermore, channel coding introduces new redundancy. It will further degrade the transmission performance. The pseudo-analog transmission presented in Chapters 16 and 17 should be one promising solution. In addition to the power allocation and channel denoising that we have discussed, currently we do not have a channel protection approach in the pseudo-analog transmission, such as channel coding in the current digital transmission.

Before we start to discuss detailed technologies to reduce channel errors, we would like to define the pseudo-analog transmission more strictly. As previously mentioned, pseudo-analog transmission is still implemented in a digital manner and transmitted data is also digital. But compared with current digital transmission, it removes quantization, entropy coding, and channel coding. All of these are non-linear processes in the current digital transmission, while the other processes are linear. The strict definition for the pseudo-analog transmission should be linear digital transmission. These two definitions indicate the same transmission approach but from different viewpoints. They can be used interchangeably hereafter.

19.4.1 Power Allocation and Bandwidth Matching

Power allocation and bandwidth matching are two basic components in linear digital transmission. As discussed in Chapter 16, transmission power is directly allocated to elements of the source data. For a given total transmission power, the power allocation should take source properties, channel models and conditions, and even receiving antennae into account. In Chapter 16, an algorithm has been presented to allocate the power to every element of source data to minimize the mean square error of received data. This is a way to take source redundancy into account in the transmission because the power is actually allocated according to source distribution. Next, the channel model and conditions should be considered in power allocation. In particular, a Gaussian channel and fading channel definitely need different strategies for power allocation. Furthermore, since more and more devices are equipped with multiple antennae, this factor should also be considered in power allocation.

In the linear digital transmission, one element of the source data uses one channel transmission. If the number of elements needed to transmit in a unit of time (source bandwidth) is equal to the number of channel transmissions in a unit of time (channel bandwidth), we call this matched source and channel bandwidth in the linear digital transmission. But in real applications, source bandwidth seldom matches channel bandwidth. When source bandwidth is larger than channel bandwidth, source elements have to be reduced and this is called bandwidth compression. If source bandwidth is less than channel bandwidth, source elements have to be increased and this is called bandwidth expansion. Currently, bandwidth compression is achieved by dropping some of the less important elements, while bandwidth expansion is achieved by repeating some of the important elements.

Obviously, modifying the number of elements for source and channel bandwidth matching is not optimal. For bandwidth compression, the elements of source data should have more than two choices (transmitted or not transmitted). In the quantization adopted in nonlinear digital transmission, less important elements are reduced in size before transmission. It means that some of the less important elements are still transmitted and not dropped totally. Similarly, for bandwidth expansion, most of the likely important elements may need to be transmitted by non-integer channel usages. One idea for increasing and decreasing the channel usage of every element at fractional times is to introduce network coding. For example, let us assume two elements $s(1)$ and $s(2)$. If we transmit $s(1)$, $s(2)$, and $s(1)+s(2)$, every element has 1.5 channel usages. If we transmit $s(1)$ and $s(1)+s(2)$, the element $s(1)$ has 1.5 channel usages and $s(2)$ has 0.5 channel usages. It is desirable to develop optimal network coding for source and channel bandwidth matching.

In the nonlinear digital transmission, the joint optimization of power allocation and bandwidth matching is achieved by separate source coding and channel coding. The source data is first compressed toward the given channel bandwidth and the resulting data is compressed bits. In general, complicated elements require more bits to describe and easy elements use few bits. It is equal to allocate bits to different elements. After the bit allocation, all the resulting bits are equally important in transmission and thus channel coding protects them by uniformly introducing parity bits. Finally, transmission power is averagely allocated to transmitted bits. However, entropy coding and channel coding have been removed in the linear digital transmission. We have to find a new way to achieve joint optimization in the linear digital transmission.

Power and channel bandwidth are two different resources for transmission. We should study the joint allocation of transmission power and channel bandwidth to the source elements. For an element $s(1)$, we can select to transmit it twice or increase its transmission power with one channel usage. The fundamental theoretical problem in the joint optimization is determining the relationship between transmission power and channel usage. With a formulated relationship, we can alternatively choose increasing transmission power or channel usages for important elements or decreasing transmission power or channel usages for less important elements. This is a new research topic that has appeared in linear digital transmission.

Next, a wild thought presented here is that source and channel bandwidth matching can use Cloud sources as well. From the viewpoint of signal processing, bandwidth compression and expansion are similar to down-sampling and up-sampling one source, respectively. But the down-sampling is not a reversible process. In bandwidth compression, once the source is down-sampled in order to match channel bandwidth, the receiver cannot get the source at the original resolution anymore because the down-sampling process drops a lot of the source detail. Our recent research has shown that if Cloud sources are available and correlated sources can be found, the up-sampled results can be significantly improved [485]. In other words, we should further study bandwidth compression and even bandwidth expansion with the assumption of Cloud sources being available.

The number of Cloud sources is huge. But in the network nodes (e.g., router and base station), it is difficult to have enough storage for the Cloud sources. Therefore, bandwidth compression and expansion using Cloud sources are impossible in network nodes. One possible solution is to reduce the size of Cloud sources by learning. For example, the hallucination technology in computer vision can be applied here [486, 487]. We can first train a data library for one category of sources in Cloud sources. The library includes many pairs of low-resolution patches and matched high-resolution patches. Thus, the size is no longer too large and can be stored on the network nodes. At a network node, one down-sampled patch can find a corresponding high-resolution patch for up-sampling the received source.

19.4.2 Multiple Level Channel Coding

Conventional channel coding such as LDPC codes and Turbo codes used in the nonlinear digital transmission mainly targets binary data. In linear digital transmission, source data is transmitted element by element. The protection should be operated on multilevel elements. Although conventional channel coding can be extended to multilevel Galois fields $GF(q)$ [488], they cannot fully benefit from the redundancy existing in the source data except for using the probability distribution. The elements of the source data implicate complicated and rich redundant information. It would be desirable to develop a new channel coding for linear digital transmission. It supports taking multilevel data as input and can fully employ source redundancy as prior information in channel decoding.

Random projection codes presented in Chapters 14 and 15 have shown that they can be used for error protection and source compression in the physical layer of wireless networks. This is a promising solution for channel coding in linear digital transmission. Although they currently only take binary data as input, they can be readily extended to multilevel data and generate multilevel parity data with the weighted sum of randomly selected source elements. If source elements are a part of the output, the channel coding in the linear digital transmission is systematic; otherwise it is non-systematic. The topic of how to design multilevel random projection codes with optimal degree and weighting parameters is interesting. In addition, it may be better not to generate parity data by randomly selecting source elements.

The decoding algorithm cannot be based on the belief propagation presented in Chapters 14 and 15 because input data is not binary. But we can use the algorithms for the reconstruction in compressive sensing. One advantage of these algorithms is that we can readily adopt the redundancy existing in source data as prior information. Some regulations such as sparsity and low rank can be added in the optimization to improve reconstructed quality. Furthermore, with more prior information, required parity data can be reduced for error correction. Some results in compressive sensing show that images can be reconstructed at high quality with only 20% of the measurements. The challenge is the decoding complexity. The complexity of such algorithms is very high, especially for both input data and weighted sums as multilevel values. The source redundancy can be used to reduce the decoding complexity because the solving space is reduced with prior information.

19.4.3 Channel Denoising

As shown in Figure 19.5, Cloud sources C are available in the channel decoder to suppress and even remove channel noise. In channel decoding, input signal $\hat{y}$ is equal to y plus noise z. Both source encoding from s to x and channel encoding from x to y make s become random bits. As we mentioned earlier, if input source s is fully compressed and compressed data x cannot be used to identify the correlated sources in the Cloud sources, it is difficult for channel decoding to benefit from the correlation of Cloud sources. In the linear digital transmission, entropy coding and channel coding have been removed. The transmission is similar to the pseudo-analog scheme presented in Chapter 17.

If source coding and channel coding are similar to that presented in Chapter 17, input source s is not compressed at all or is compressed partially. Therefore, y with additive noise z has the potential to identify the correlated sources in Cloud sources. We can directly extract local features from received data $\hat{y}$. Our initial results show that when channel SNR is higher than a threshold (e.g., 5 dB), extracted local features can still be used for correlated source retrieval. But it is important to study local feature descriptors robust to channel noise and to further reduce the channel SNR threshold. One solution could be to consider feature extraction with power allocation in communication together. The power allocation provides an active way to protect against channel noise.

Here, channel coding is neither traditional Turbo codes and LDPC codes nor our random projection codes. Instead, it becomes a denoising process similar to the

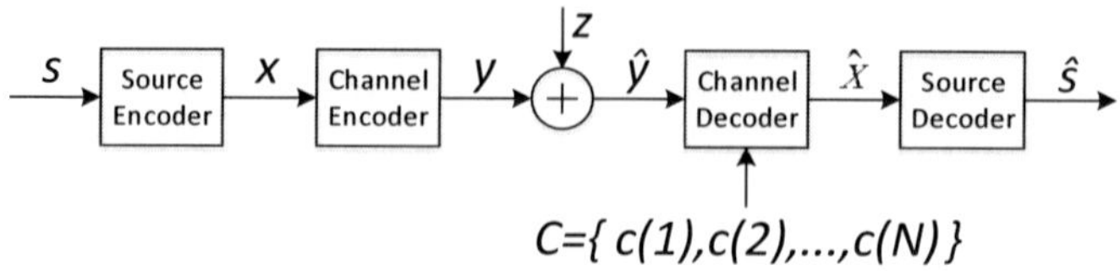

Figure 19.5 Channel denoising using Cloud sources.

approach presented in Chapter 17. Because channel noise is random, we can use the existing correlation among pixels or coefficients to suppress additive channel noise. Our experimental results show that the received quality is even better than that of the state-of-the-art digital systems. However, this scheme has revealed limitations because it only uses the correlation within input source. If channel SNR is too low (i.e., channel noise is too large), all received data contains large channel noise and it will be difficult to get good results by denoising.

As shown in Figure 19.5, when Cloud sources are available in the channel decoding, using correlated sources without channel noise can be a good way to suppress large additive channel noise. Once the correlated sources are identified, they can be added in the algorithm of denoising (e.g., BM3D) and generate 3D correlated data with the received noise image. Since the correlated sources do not contain channel noise, the results of denoising should be much better than when using the received noise data only. One technical problem is that only the received data contains noise. After a 3D transform, the noise energy is not concentrated at a high frequency. More intelligent noise removing approaches need to be studied.

19.5 Joint Source and Channel Coding

From the pseudo-analog schemes presented in Chapters 16 and 17, since the input source is not fully compressed and still has some redundancy, the transmission has to exploit this redundancy in order to make the system performance optimal. From the viewpoint of Shannon's rate-distortion theory, digital systems provide an extreme solution, where the source is fully compressed, while analog systems provide the other extreme, where the source is uncompressed. In computational information theory, we try to achieve the optimal rate-distortion performance at the middle points of the curve. Therefore, it is important to jointly design source and channel coding. The solution presented in Chapters 16 and 17 allocates the transmission energy according to the distribution of the transmitted source.

As shown in Figure 19.6, when Cloud sources are available in both the channel and receiver, it is important to study how to efficiently use the correlation of Cloud sources with joint source and channel coding. If the source coding is distributed as discussed in Section 19.3.3 and the channel coding is denoised as discussed in

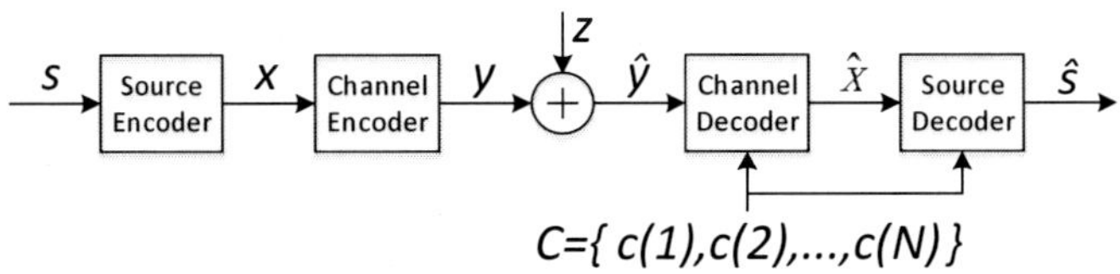

Figure 19.6 Joint source and channel coding with Cloud sources.

Section 19.4, we have to study how to efficiently use the correlation of Cloud sources in source coding and channel coding.

19.6 Summary

This chapter presents our thoughts about the future developments of visual data compression and communication. The core idea is to consider compression and transmission with the available Cloud sources. Based on our current research, some new schemes are discussed but more effort is needed to prove that they can be effective. No matter what the performance of these new schemes is at the moment, we believe that the computational information theory is a move in the right direction toward developing Shannon's information theory.

Appendix A
Our Published Journal and Conference Papers Related to This Book

This appendix lists our conference and journal publications closely related to the research content presented in this book. They are organized according to five parts.

A.1 Scalable Video Coding

1. Feng Wu, Shipeng Li, Ya-Qin Zhang, "A framework for efficient progressive fine granular scalable video coding," *IEEE Transactions on Circuits and Systems for Video Technology*, special issue on streaming video, vol. 11, no. 3, pp. 332–344, 2001.
2. Lin Luo, Feng Wu, Shipeng Li, Zixiang Xiong, Zhenquan Zhuang, "Advanced motion threading for 3D wavelet video coding," *Signal Processing: Image Communication*, special issue on subband/wavelet video coding, vol. 19, no. 7, pp. 601–616, 2004.
3. Ruiqin Xiong, Jizheng Xu, Feng Wu, Shipeng Li, "Barbell-lifting based 3-D wavelet coding scheme," *IEEE Transactions on Circuits and Systems for Video Technology*, vol. 17, no. 9, pp. 1256–1269, 2007.
4. Feng Wu, Honghui Sun, Guobin Shen, Shipeng Li, Ya-Qin Zhang, Bruce Lin, Ming-Chieh Li, "SMART: An efficient, scalable and robust streaming video system," *EURASIP Journal on Applied Signal Processing*, special issue on Multimedia over IP and Wireless Networks, vol. 2004, no. 2, pp. 192–206, 2004.

A.2 Directional Transforms

1. Wenpeng Ding, Feng Wu, Xiaolin Wu, Shipeng Li, Houqiang Li, "Adaptive directional lifting-based wavelet transform for image coding," *IEEE Transactions on Image Processing*, vol. 16, no. 2, pp. 416–427, 2007.

2. Hao Xu, Jizheng Xu, Feng Wu, "Lifting-based directional DCT-like transform for image coding," *IEEE Transactions on Circuits and Systems for Video Technology*, vol. 17, no. 10, pp. 1325–1335, 2007.
3. Jizheng Xu, Feng Wu, Jie Liang, Wenjun Zhang, "Directional lapped transforms for image coding," *IEEE Transactions on Image Processing*, vol. 19, no. 1, pp. 85–97, 2010.
4. Xiulian Peng, Jizheng Xu, Feng Wu, "Directional filtering transform for image/intra-frame compression," *IEEE Transactions on Image Processing*, vol. 19, no. 11, pp. 2935–2946, 2010.
5. Jizheng Xu, Feng Wu, Wenjun Zhang, "Intra-predictive transforms for block-based image coding," *IEEE Transactions on Signal Processing*, vol. 57, no. 8, pp. 3030–3040, 2009.

A.3 Vision-Based Compression

1. Dong Liu, Xiaoyan Sun, Feng Wu, Shipeng Li, Ya-qin Zhang, "Image compression with edge-based inpainting," *IEEE Transactions on Circuits and Systems for Video Technology*, vol. 17, no. 10, pp. 1273–1287, 2007.
2. Huanjing Yue, Xiaoyan Sun, Jingyu Yang, Feng Wu, "Cloud-based image coding for mobile devices — Toward thousands to one compression," *IEEE Transaction on Multimedia*, vol. 15, no. 4, pp. 845–857, 2013.
3. Lican Dai, Huanjing Yue, Xiaoyan Sun, Feng Wu, "IMShare: Instantly sharing your mobile landmark images by search-based reconstruction," *ACM Multimedia*, pp. 579–588, 2012.
4. Zhongbo Shi, Xiaoyan Sun, Feng Wu, "Photo album compression by using local features for cloud storage," *IEEE Journal of Emerging and Selected Topics in Circuits and Systems*, 2014.

A.4 Compressive Communication

1. Chong Luo, Feng Wu, Jun Sun, Changwen Chen, "Compressive data gathering for large-scale wireless sensor networks," *International Conference on Mobile Computing and Networking (MobiCom)*, pp. 145–156, 2009.
2. Chong Luo, Feng Wu, Jun Sun, Chang Wen Chen, "Efficient measurement generation and pervasive sparsity for compressive data gathering," *IEEE Transactions on Wireless Communication*, vol. 9, no. 12, pp. 3728–3738, 2010.
3. Hao Cui, Chong Luo, Kun Tan, Feng Wu, Changwen Chen, "Seamless rate adaptation for wireless networking," *ACM International Conference on Modeling, Analysis and Simulation of Wireless and Mobile Systems (MSWiM)*, pp. 437–446, 2011.

4. Hao Cui, Chong Luo, Jun Wu, Chang Wen Chen, Feng Wu, "Compressive coded modulation for seamless rate adaptation," *IEEE Transactions on Wireless Communication*, vol. 12, no. 10, pp. 4892–4904, 2013.
5. Min Wang, Jun Wu, Saifeng Shi, Chong Luo, Feng Wu, "Fast decoding and hardware design for binary-input compressive sensing," *IEEE Journal of Emerging and Selected Topic in Circuits and Systems*, vol. 2, no. 3, pp. 591–603, 2012.

A.5 Pseudo-Analog Transmission

1. Xiaopeng Fan, Feng Wu, Debin Zhao, Oscar C. Au, "Distributed wireless visual communication with power distortion optimization," *IEEE Transactions on Circuits and Systems for Video Technology*, vol. 23, no. 6, pp. 1040–1053, 2013.
2. Hao Cui, Zhihai Song, Chong Luo, Ruiqin Xiong, Feng Wu, "Cactus: Keeping redundancy for efficient and robust wireless video communications," *ACM International Conference on Modeling, Analysis and Simulation of Wireless and Mobile Systems (MSWiM)*, pp. 273–278, 2013.
3. Xiaolin Liu, Wenjun Hu, Chong Luo, Feng Wu, "Compressive Image Broadcasting in MIMO Systems with Receiver Antenna Heterogeneity," *Signal Processing: Image Communication*, 2014.
4. Feng Wu, Xiulian Peng, Ji-zheng Xu, "Linecast: Line-based distributed coding and transmission for broadcasting satellite images," *IEEE Trasactions on Image Processing*, vol. 23, no. 3, pp. 1015–1027, 2014.

References

1. Gigaom. Facebook has 220 billion of your photos to put on ice, 2012. http://gigaom.com.
2. Cisco. Cisco visual networking index: Forecast and methodology (2011–2016), 2012. http://www.cisco.com.
3. Cisco. Cisco visual networking index: Global mobile data traffic forecast (2012–2017), 2012. http://www.cisco.com.
4. C. E. Shannon. A mathematical theory of communication. *The Bell System Technical Journal*, 27:379–423, 1948.
5. T. M. Cover and J. A. Thomas. *Elements of Information Theory*. John Wiley & Sons, 2006.
6. D. A. Huffman. A method for the construction of minimum redundancy codes. *Proc. IRE*, 40:1098–1101, 1952.
7. T. K. Moon. *Error Correction Coding*. John Wiley & Sons, 2005.
8. ITU-T. H.261: Video codec for audiovisual services at p $\times$ 384 kbit/s — Recommendation h.261, 1988. http://www.itu.int.
9. ISO. Iso/iec 11172-2: 1993 — Information technology — Coding of moving pictures and associated audio for digital storage media at up to about 1.5 mbit/s — Part 2: Video, 1993. http://www.iso.org.
10. ISO. Iso/iec 13818-2: 1996 — Information technology — Generic coding of moving pictures and associated audio information: Video, 1996. http://www.iso.org.
11. ISO. Iso/iec 14496-2: 1998 — Information technology — Coding of audio-visual objects — Part 2: Visual, 1998. http://www.iso.org.
12. ISO. ISO/IEC 14496-10: 2003 — Information technology — Coding of audio-visual objects — Part 10: Advanced video coding, 2003. http://www.iso.org.
13. T. Wiegand, G. J. Sullivan, G. Bjontegaard, and A. Luthra. Overview of the H.264/AVC video coding standard. *Circuits and Systems for Video Technology, IEEE Transactions on*, 13(7):560–576, 2003.
14. G. J. Sullivan, J. Ohm, Woo-Jin Han, and T. Wiegand. Overview of the high-efficiency video coding (HEVC) standard. *Circuits and Systems for Video Technology, IEEE Transactions on*, 22(12):1649–1668, 2012.
15. A. P. Godse and U. A. Bakshi. *Analog Communication*. Technical Publications Pune, 2009.
16. R. G. Gallager. *Low-Density Parity-Check Codes*. M.I.T. Press, 1963.
17. D. J. C. MacKay and R. M. Neal. Near Shannon limit performance of low density parity check codes. *Electronics Letters*, 32(18):1645–1946, 1996.
18. C. Berrou, A. Glavieux, and P. Thitimajshima. Near Shannon limit error-correcting coding and decoding: Turbo-codes. In *Communications, IEEE International Conference on*, volume 2, pages 1064–1070, 1993.
19. C. Berrou and A. Glavieux. Near optimum error correcting coding and decoding: Turbo-codes. *Communications, IEEE Transactions on*, 44(10):1261–1271, 1996.
20. L. Bahl, J. Cocke, F. Jelinek, and J. Raviv. Optimal decoding of linear codes for minimizing symbol error rate. *Information Theory, IEEE Transactions on*, 20(2):284–287, 1974.
21. J. Lu. Signal processing for internet video streaming: A review. In *Image and Video Communication and Processing, SPIE International Conference on*, volume 3974, pages 246–258, 2000.
22. G. J. Conklin, G. S. Greenbaum, K. O. Lillevold, A. F. Lippman, and Y. A. Reznik. Video coding for streaming media delivery on the Internet. *Circuits and Systems for Video Technology, IEEE Transactions on*, 11(3):269–281, 2001.
23. D. P. Wu, Y. Hou, W. W. Zhu, Y. Q. Zhang, and J. M. Peha. Streaming video over the Internet: Approaches and directions. *Circuits and Systems for Video Technology, IEEE Transactions on*, 11(3):282–300, 2001.
24. J. F. Arnold, M. R. Fracter, and Y. Q. Wang. Efficient drift-free signal-to-noise ratio scalability. *Circuits and Systems for Video Technology, IEEE Transactions on*, 10(1):70–82, 2000.
25. L. P. Kondi, F. Ishtiaq, and A. K. Katsaggelos. On video SNR scalability. In *Image Processing, International Conference on*, volume 3, pages 934–938, 1998.

26. B. R. Lee, K. K. Park, and J. J. Hwang. H.263-based SNR scalable video codec. *Consumer Electronics, IEEE Transactions on*, 43(3):614–622, 1997.
27. A. Puri, L. Yan, and B.G. Haskell. Temporal resolution scalable video coding. In *Image Processing, IEEE International Conference*, volume 2, pages 947–951, 1994.
28. G.J. Conklin and S.S. Hemami. A comparison of temporal scalability techniques. *Circuits and Systems for Video Technology, IEEE Transactions on*, 9(6):909–919, 1999.
29. H. Katata, N. Ito, and H. Kusao. Temporal-scalable coding based on image content. *Circuits and Systems for Video Technology, IEEE Transactions on*, 7(1):52–59, 1997.
30. Stephan Wenger. Temporal scalability using p-pictures for low-latency applications. In *Multimedia Signal Processing, IEEE Second Workshop on*, pages 559–564, 1998.
31. U. Benzler. Spatial scalable video coding using a combined subband-DCT approach. *Circuits and Systems for Video Technology, IEEE Transactions on*, 10(7):1080–1087, 2000.
32. Q. Hu and S. Panchanathan. A comparative evaluation of spatial scalability techniques in the compressed domain. In *Electrical and Computer Engineering, 1996. Canadian Conference on*, volume 1, pages 474–477, 1996.
33. K. Illgner and F. Muller. Multiresolution video compression: Motion estimation and vector field coding. In *Global Telecommunications Conference, 1996. GLOBECOM '96. 'Communications: The Key to Global Prosperity*, volume 3, pages 1478–1482, 1996.
34. Tihao Chiang, Huifang Sun, and J.W. Zdepski. Spatial scalable HDTV coding. In *Image Processing, 1995. Proceedings., International Conference on*, volume 2, pages 571–574, 1995.
35. MPEG requirement group. MPEG-4 requirements version 8, 2001. ISO/IEC JTC1/SC29/WG11, w2194, Dublin.
36. R. Neff and A. Zakhor. Matching pursuit video coding at very low bit rates. In *Data Compression Conference, 1995. DCC '95. Proceedings*, pages 411–420, 1995.
37. R. Neff and A. Zakhor. Matching pursuit video coding. I. Dictionary approximation. *Circuits and Systems for Video Technology, IEEE Transactions on*, 12(1):13–26, 2002.
38. R. Neff and A. Zakhor. Matching-pursuit video coding. II. Operational models for rate and distortion. *Circuits and Systems for Video Technology, IEEE Transactions on*, 12(1):27–39, 2002.
39. W. Li. Status report on bit plane coding, 1998. ISO/IEC JTC1/SC29/WG11, m3792, Dublin.
40. Y. Chen, C. Dufour, H. Radha, R. Cohem, and M. Buteau. Request for fine granular video scalability for media streaming applications, 1998. ISO/IEC JTC1/SC29/WG11, m3792, Dublin.
41. J. M. Shapiro. Embedded image coding using zero trees of wavelet coefficients. *Signal Processing, IEEE Transactions on*, 41(12):3445–3462, 1993.
42. A. Said and W. A. Pearlman. A new, fast, and efficient image codec based on set partitioning in hierarchical trees. *Circuits and Systems for Video Technology, IEEE Transactions on*, 6(3):243–250, 1996.
43. J. Ohm. Description of core experiments in MPEG-4 video, 1998. ISO/IEC JTC1/SC29/WG11, N2474, Atlantic City.
44. Y. Chen, H. Radha, and R. Cohem. Results of experiment on fine granular scalability with wavelet encoding of residuals, 1998. ISO/IEC JTC1/SC29/WG11, m3988, Atlantic City.
45. B. Schuster. Fine granular scalability with wavelets coding, 1998. ISO/IEC JTC1/SC29/WG11, m4021, Atlantic City.
46. J. Liang, J. Yu, Y. Wang, M. Srinath, and M. Zhou. Fine granularity scalable video coding using combination of MPEG-4 video objects and still texture objects, 1998. ISO/IEC JTC1/SC29/WG11, m4025, Atlantic City.
47. E. Miloslavsky, S. Cheung, and A. Zakhor. SNR scalability using matching pursuits, 1998. ISO/IEC JTC1/SC29/WG11, m3833, Atlantic City.
48. M. Benetiere and C. Dufour. Matching pursuits residual coding for video fine granular scalability, 1998. ISO/IEC JTC1/SC29/WG11, m3833, Atlantic City.
49. W. Li. Bit-plane coding of DCT coefficients for fine granularity scalability, 1998. ISO/IEC JTC1/SC29/WG11, m3989, Atlantic City.
50. J. Macnicol, M. Frater, and J. Arnold. Results on fine granularity scalability, 1999. ISO/IEC JTC1/SC29/WG11, m5122.

51. S. Li, F. Wu, and Y. Q. Zhang. Study of a new approach to improve FGS video coding efficiency, 1999. ISO/IEC JTC1/SC29/WG11, M5583.
52. F. Ling, W. Li, and H.-Q. Sun. Bitplane coding of DCT coefficients for image and video compression. In *Image and Video Communication and Processing, SPIE International Conference on*, pages 500–508, 1999.
53. T. K. Tan, K. K. Pang, and K. N. Ngan. A frequency scalable coding scheme employing pyramid and subband techniques. *Circuits and Systems for Video Technology, IEEE Transactions on*, 4(2):203–207, 1994.
54. Yao Wang and Qin-Fan Zhu. Error control and concealment for video communication: A review. *Proceedings of the IEEE*, 86(5):974–997, 1998.
55. R. Talluri. Error-resilient video coding in the ISO MPEG-4 standard. *Communications Magazine, IEEE*, 36(6):112–119, 1998.
56. J. R. Yee and Jr. Weldon, E. J. Evaluation of the performance of error-correcting codes on a Gilbert channel. *Communications, IEEE Transactions on*, 43(8):2316–2323, 1995.
57. M. Vetterli and J. Kovacevic. *Wavelets and Subband Coding.* Englewood Cliffs, NJ: Prentice-Hall, 1995.
58. D. Taubman and A. Zakhor. Multirate 3-D subband coding of video. *Image Processing, IEEE Transactions on*, 3(5):572–588, 1994.
59. A. Wang, Z. Xiong, P. A. Chou, and S. Mehrotra. Three-dimensional wavelet coding of video with global motion compensation. In *Data Compression, International Conference on*, pages 404–413, 1999.
60. J.-R. Ohm. Three-dimensional subband coding with motion compensation. *Image Processing, IEEE Transactions on*, 3(5):559–571, 1994.
61. Jo Yew Tham, Surendra Ranganath, and A. A. Kassim. Highly scalable wavelet-based video codec for very low bit-rate environment. *Selected Areas in Communications, IEEE Journal on*, 16(1):12–27, 1998.
62. Seung-Jong Choi and J. W. Woods. Motion-compensated 3-D subband coding of video. *Image Processing, IEEE Transactions on*, 8(2):155–167, 1999.
63. Beong-Jo Kim, Zixiang Xiong, and W. A. Pearlman. Low bit-rate scalable video coding with 3-D set partitioning in hierarchical trees (3-D SPIHT). *Circuits and Systems for Video Technology, IEEE Transactions on*, 10(8):1374–1387, 2000.
64. B. Pesquet-Popescu and V. Bottreau. Three-dimensional lifting schemes for motion compensated video compression. In *Acoustics, Speech, and Signal Processing, 2001. Proceedings. (ICASSP '01). 2001 IEEE International Conference on*, volume 3, pages 1793–1796, 2001.
65. Lin Luo, Jin Li, Shipeng Li, Zhenquan Zhuang, and Ya-Qin Zhang. Motion compensated lifting wavelet and its application in video coding. In *Multimedia and Expo, 2001. ICME 2001. IEEE International Conference on*, pages 365 368, 2001.
66. A. Secker and D. Taubman. Motion-compensated highly scalable video compression using an adaptive 3D wavelet transform based on lifting. In *Image Processing, 2001. Proceedings. 2001 International Conference on*, volume 2, pages 1029–1032, 2001.
67. Peisong Chen and J. W. Woods. Bidirectional MC-EZBC with lifting implementation. *Circuits and Systems for Video Technology, IEEE Transactions on*, 14(10):1183–1194, 2004.
68. M. Flierl and B. Girod. Video coding with motion-compensated lifted wavelet transforms. *Signal Process.: Image Commun.*, 19(7):561–575, 2004.
69. P. Chen and J. W. Woods. Improved MC-EZBC with quarter-pixel motion vectors, 2002. ISO/IEC JTC1/SC29/WG11, M8366, Fairfax, VA.
70. J.-R. Ohm. Motion-compensated wavelet lifting filters with flexible adaptation. In *Digital Communication, International Workshop on*, pages 113–120, 2002.
71. D. S. Turaga, M. Van der Schaar, and B. Pesquet-Popescu. Complexity scalable motion compensated wavelet video encoding. *Circuits and Systems for Video Technology, IEEE Transactions on*, 15(8):982–993, 2005.
72. A. Secker and D. Taubman. Highly scalable video compression with scalable motion coding. *Image Processing, IEEE Transactions on*, 13(8):1029–1041, 2004.

73. D. Turaga, M. van der Schaar, Y. Andreopoulos, A. Munteanu, and P. Schelkens. Unconstrained motion compensated temporal filtering (UMCTF) for efficient and flexible interframe wavelet video coding. *Signal Process.: Image Commun.*, 20(1):1–19, 2005.
74. J. Xu, Z. Xiong, S. Li, and Y.-Q. Zhang. Three-dimensional embedded subband coding with optimal truncation (3-D ESCOT). *Appl. Comput. Harmonic Anal., International Journal*, 10:290–315, 2001.
75. Jizheng Xu, Zixiang Xiong, Shipeng Li, and Ya-Qin Zhang. Memory-constrained 3D wavelet transform for video coding without boundary effects. *Circuits and Systems for Video Technology, IEEE Transactions on*, 12(9):812–818, 2002.
76. Lin Luo, Feng Wu, Shipeng Li, and Zhenquan Zhuang. Advanced lifting-based motion-threading techniques for 3-D wavelet video coding. In *Visual Communication and Image Processing, SPIE International Conference on*, pages 707–718, 2003.
77. Lin Luo, Feng Wu, Shipeng Li, and Zhenquan Zhuang. Layer-correlated motion estimation and motion vector coding for the 3D-wavelet video coding. In *Image Processing, 2003. ICIP 2003. Proceedings. 2003 International Conference on*, volume 2, pages 791–794, 2003.
78. A. Secker and D. Taubman. Highly scalable video compression using a lifting-based 3D wavelet transform with deformable mesh motion compensation. In *Image Processing. 2002. Proceedings. 2002 International Conference on*, volume 3, pages 749–752, 2002.
79. A. Secker and D. Taubman. Lifting-based invertible motion adaptive transform (LIMAT) framework for highly scalable video compression. *Image Processing, IEEE Transactions on*, 12(12):1530–1542, 2003.
80. I. Daubechies and W. Sweldens. Factoring wavelet transforms into lifting steps. *Journal of Fourier Analysis and Applications*, 4:247–269, 1998.
81. W. Sweldens. The lifting scheme: A new philosophy in biorthogonal wavelet construction. In *Wavelet Applications in Signal and Image Processing, SPIE International Conference on*, 1995.
82. B. Girod. Motion-compensating prediction with fractional-pel accuracy. *Communications, IEEE Transactions on*, 41(4):604–612, 1993.
83. JPEG. Verification model ad-hoc, JPEG 2000 VM 8.5, 2000. ISO/IEC JTC1/SC29/WG1, N1878.
84. T. Wiegand and B. Girod. Lagrange multiplier selection in hybrid video coder control. In *Image Processing, 2001. Proceedings. 2001 International Conference on*, volume 3, pages 542–545, 2001.
85. P. Chen and J. W. Woods. Contributions to interframe wavelet and scalable video coding, 2002. ISO/IEC JTC1/SC29/WG11, M9034, Shanghai, China.
86. J. Xu, S. Li, and Y.-Q. Zhang. Three-dimensional shape-adaptive discrete wavelet transforms for efficient object-based video coding. In *Visual Communication and Image Processing, SPIE International Conference on*, volume 4067, pages 336–344, 2000.
87. Lin Luo, Feng Wu, Shipeng Li, Zixiang Xiong, and Zhenquan Zhuang. Advanced motion threading for 3-D wavelet video coding. *Signal Process.: Image Commun.*, 19:601–616, 2004.
88. R. Xiong, F. Wu, S. Li, Z. Xiong, and Y.-Q. Zhang. Exploiting temporal correlation with adaptive block-size motion alignment for 3-D wavelet coding. In *Visual Communication and Image Processing, SPIE International Conference on*, volume 5308, pages 144–155, 2004.
89. Bo Feng, Jizheng Xu, Feng Wu, Shiqiang Yang, and Shipeng Li. Energy distributed update steps (EDU) in lifting based motion compensated video coding. In *Image Processing, 2004. ICIP '04. 2004 International Conference on*, volume 4, pages 2267–2270, 2004.
90. R. Xiong, F. Wu, J. Xu, S. Li, and Y.-Q. Zhang. Barbell-lifting wavelet transform for highly scalable video coding. In *Picture Coding Symposium*, pages 237–242, 2004.
91. Ruiqin Xiong, Jizheng Xu, Feng Wu, Shipeng Li, and Ya-Qin Zhang. Layered motion estimation and coding for fully scalable 3D wavelet video coding. In *Image Processing, 2004. ICIP '04. 2004 International Conference on*, volume 4, pages 2271–2274, 2004.
92. X. Ji, J. Xu, D. Zhao, and F. Wu. Architectures of incorporating MPEG-4 AVC into three-dimensional wavelet video coding. In *Picture Coding Symposium*, 2004.
93. MPEG. Call for proposals on scalable video coding technology, 2003. ISO/IEC JTC1/SC29/WG11, N6193.

94. MPEG. Registered responses to the call for proposals on scalable video coding, 2004. ISO/IEC JTC1/SC29/WG11, M10569.
95. MPEG. Subjective test results for the CFP on scalable video coding technology, 2004. ISO/IEC JTC1/SC29/WG11, N6383.
96. MPEG. Exploration experiments on tools evaluation in wavelet video coding, 2005. ISO/IEC JTC1/SC29/WG11, N6914.
97. N. Mehrseresht and D. Taubman. A flexible structure for fully scalable motion-compensated 3-D DWT with emphasis on the impact of spatial scalability. *Image Processing, IEEE Transactions on*, 15(3):740–753, 2006.
98. D.S. Turaga and M. Van der Schaar. Content-adaptive filtering in the UMCTF framework. In *Acoustics, Speech, and Signal Processing, 2003. Proceedings. (ICASSP '03). 2003 IEEE International Conference on*, volume 3, pages 621–624, 2003.
99. L. Song, J. Xu, H. Xiong, and F. Wu. Content adaptive update for lifting-based motion-compensated temporal filtering. *Electronics Letters*, 41(1):14–16, 2005.
100. B. Girod and Sangeun Han. Optimum update for motion-compensated lifting. *Signal Processing Letters, IEEE*, 12(2):150–153, 2005.
101. Yihua Chen, Jizheng Xu, Feng Wu, and Hongkai Xiong. An improved update operator for H.264 scalable extension. In *Multimedia Signal Processing, 2005 IEEE 7th Workshop on*, pages 1–4, 2005.
102. D. Taubman and A. Secker. Highly scalable video compression with scalable motion coding. In *Image Processing, 2003. ICIP 2003. Proceedings. 2003 International Conference on*, volume 3, pages 273–276, 2003.
103. D. Taubman. High performance scalable image compression with EBCOT. *Image Processing, IEEE Transactions on*, 9(7):1158–1170, 2000.
104. D. Taubman, E. Ordentlich, M. Weinberger, and G. Seroussi. Embedded block coding in JPEG 2000. *Signal Process.: Image Commun.*, 17(1):49–72, 2002.
105. Shih-Ta Hsiang and J.W. Woods. Embedded image coding using zeroblocks of subband/wavelet coefficients and context modeling. In *Circuits and Systems, 2000. Proceedings. ISCAS 2000 Geneva. The 2000 IEEE International Symposium on*, volume 3, pages 662–665, 2000.
106. F. Lazzaroni, A. Signoroni, and R. Leonardi. Embedded morphological dilation coding for 2-D and 3-D images. In *Visual Communication and Image Processing, SPIE International Conference on*, volume 4671, pages 923–934, 2002.
107. H. Schwarz, T. Hinz, H. Kirchhoffer, D. Marpe, and T. Wiegand. Technical description of the HHI proposal for SVC CE1, 2004. ISO/IEC JTC1/SC29/WG11, M11244.
108. R. Schafer, H. Schwarz, D. Marpe, T. Schierl, and T. Wiegand. MCTF and scalability extension of H.264/AVC and its application to video transmission. In *Visual Communication and Image Processing, SPIE International Conference on*, volume 5960, pages 343–354, 2005.
109. M. Flierl and B. Girod. Generalized B pictures and the draft H.264/AVC video-compression standard. *Circuits and Systems for Video Technology, IEEE Transactions on*, 13(7):587–597, 2003.
110. H. Schwarz, D. Marpe, and T. Wiegand. Analysis of hierarchical B pictures and MCTF. In *Multimedia and Expo, 2006 IEEE International Conference on*, pages 1929–1932, 2006.
111. Ruiqin Xiong, Jizheng Xu, Feng Wu, Shipeng Li, and Ya-Qin Zhang. Spatial scalability in 3-D wavelet coding with spatial domain MCTF encoder. In *Picture Coding Symposium*, pages 583–588, 2004.
112. N. Mehrseresht and D. Taubman. Spatial scalability and compression efficiency within a flexible motion compensated 3D-DWT. In *Image Processing, 2004. ICIP '04. 2004 International Conference on*, volume 2, pages 1325–1328, 2004.
113. R. Xiong, Jizheng Xu, Feng Wu, Shipeng Li, and Ya-Qin Zhang. Optimal subband rate allocation for spatial scalability in 3-D wavelet video coding with motion aligned temporal filtering. In *Visual Communication and Image Processing, SPIE International Conference on*, volume 5960, pages 381–392, 2005.

114. Ruiqin Xiong, Jizheng Xu, Feng Wu, Shipeng Li, and Ya-Qin Zhang. Subband coupling aware rate allocation for spatial scalability in 3-D wavelet video coding. *Circuits and Systems for Video Technology, IEEE Transactions on*, 17(10):1311–1324, 2007.
115. Ruiqin Xiong, Jizheng Xu, Feng Wu, and Shipeng Li. Studies on spatial scalable frameworks for motion aligned 3-D wavelet video coding. In *Visual Communication and Image Processing, SPIE International Conference on*, volume 5960, pages 189–200, 2005.
116. Ruiqin Xiong, Jizheng Xu, Feng Wu, and Shipeng Li. In-scale motion aligned temporal filtering. In *Circuits and Systems, 2006. ISCAS 2006. Proceedings. 2006 IEEE International Symposium on*, pages 3017–3020, 2006.
117. R. Xiong, J. Xu, and F. Wu. A new method for inter-layer prediction in spatial scalable video coding, 2006. Joint Video Team of ITU-T VCEG and ISO/IEC MPEG, JVT-T081, Klagenfurt, Austria.
118. R. Xiong, Jizheng Xu, Feng Wu, and Shipeng Li. Generalized in-scale motion compensation framework for spatial scalable video coding. In *Visual Communication and Image Processing, SPIE International Conference on*, volume 6508, 2006.
119. Ruiqin Xiong, Jizheng Xu, Feng Wu, and Shipeng Li. Macroblock-based adaptive in-scale prediction for scalable video coding. In *Circuits and Systems, 2007. ISCAS 2007. IEEE International Symposium on*, pages 1763–1766, 2007.
120. Ruiqin Xiong, Jizheng Xu, and Feng Wu. In-scale motion compensation for spatially scalable video coding. *Circuits and Systems for Video Technology, IEEE Transactions on*, 18(2):145–158, 2008.
121. Ruiqin Xiong, Jizheng Xu, Feng Wu, and Shipeng Li. Adaptive MCTF based on correlation noise model for SNR scalable video coding. In *Multimedia and Expo, 2006 IEEE International Conference on*, pages 1865–1868, 2006.
122. P. Chen and J. W. Woods. Contributions to interframe wavelet and scalable video coding, 2002. ISO/IEC JTC1/SC29/WG11, m9034, Shanghai, China.
123. Y. Wu. Fully scalable subband/wavelet video coding system, 2005. Ph.D. dissertation, Rensselaer Polytechnic Inst., Troy, NY.
124. M. Wien and H. Schwarz. Testing conditions for SVC coding efficiency and JSVM performance evaluation, 2005. Joint Video Team of ISO/IEC MPEG and ITU-T VCEG, JVT-Q205, Poznan, Poland.
125. M. Wien and H. Schwarz. AHG on coding EFF and JSVM coding efficiency testing conditions, 2007. Joint Video Team of ISO/IEC MPEG and ITU-T VCEG, JVT-T008, Klagenfurt, Austria.
126. M. Wien and H. Schwarz. JSVM 6 software, 2006. Joint Video Team of ISO/IEC MPEG and ITU-T VCEG, JVT-S203, Geneva, Switzerland.
127. E. Feig, H. Peterson, and V. Ratnakar. Image compression using spatial prediction. In *Acoustics, Speech, and Signal Processing, 1995. ICASSP-95., 1995 International Conference on*, volume 4, pages 2339–2342, 1995.
128. H. Kondo and Y. Oishi. Digital image compression using directional sub-block DCT. In *Communication Technology Proceedings, 2000. WCC — ICCT 2000. International Conference on*, volume 1, pages 985–992, 2000.
129. W. B. Pennebaker and J. L. Mitchell. *JPEG Still Image Data Compression Standard*. New York: Van Nostrand, 1993.
130. D. Taubman and M. Marcellin. *JPEG 2000: Image Compression Fundamentals, Standards, and Practice*. Norwell, MA: Kluwer, 2001.
131. A. Ikonomopoulos and M. Kunt. High compression image coding via directional filtering. *Signal Processing*, 8(2):179–203, 1985.
132. D. Taubman and A. Zakhor. Orientation adaptive subband coding of images. *Image Processing, IEEE Transactions on*, 3(4):421–437, 1994.
133. Haibo Li and Zhenya He. Directional subband coding of images. In *Acoustics, Speech, and Signal Processing, 1989. ICASSP-89, 1989 International Conference on*, pages 1823–1826, 1989.
134. R.H. Bamberger and M. J. T. Smith. A filter bank for the directional decomposition of images: Theory and design. *Signal Processing, IEEE Transactions on*, 40(4):882–893, 1992.

135. T. T. Nguyen and Soontorn Oraintara. A directional decomposition: Theory, design, and implementation. In *Circuits and Systems, 2004. ISCAS '04. Proceedings of the 2004 International Symposium on*, volume 3, pages 281–284, 2004.
136. Yue Lu and M. N. Do. The finer directional wavelet transform [image processing applications]. In *Acoustics, Speech, and Signal Processing, 2005. Proceedings. (ICASSP '05). IEEE International Conference on*, volume 4, pages 573–576, 2005.
137. E. J. Candes. Monoscale ridgelets for the representation of images with edges, 1999. Dept. Statistic, Stanford Univ., Tech. Reporter.
138. M. N. Do and M. Vetterli. The finite ridgelet transform for image representation. *Image Processing, IEEE Transactions on*, 12(1):16–28, 2003.
139. E. J. Candes and D. L. Donoho. Curvelets, multiresolution representation, and scaling laws. In *SPIE Wavelet Applications in Signal and Image Processing VIII*, volume 4119, 2000.
140. V. Velisavljevic, P.-L. Dragotti, and M. Vetterli. Directional wavelet transforms and frames. In *Image Processing. 2002. Proceedings. 2002 International Conference on*, volume 3, pages 589–592, 2002.
141. M. N. Do and M. Vetterli. The contourlet transform: An efficient directional multiresolution image representation. *Image Processing, IEEE Transactions on*, 14(12):2091–2106, 2005.
142. F. C. A. Fernandes, R. L. C. van Spaendonck, and C. S. Burrus. A new framework for complex wavelet transforms. *Signal Processing, IEEE Transactions on*, 51(7):1825–1837, 2003.
143. F. C. A. Fernandes, M. B. Wakin, and R. G. Baraniuk. Non-redundant, linear-phase, semi-orthogonal, directional complex wavelets. In *Acoustics, Speech, and Signal Processing, 2004. Proceedings. (ICASSP '04). IEEE International Conference on*, volume 2, pages 953–956, 2004.
144. F. G. Meyer and R. R. Coifman. Brushlets: A tool for directional image analysis and image compression. *Journal of Application and Computer Harmonic Analysis*, 5:147–187, 1997.
145. D. Wang, L. Zhang, and A. Vincent. Curved wavelet transform for scalable video coding, 2004. ISO/IEC JTC1/SC29/WG11, M10535, Munich.
146. Chang-N Zhang and X. Wu. A hybrid approach of wavelet packet and directional decomposition for image compression. In *Electrical and Computer Engineering, 1999 IEEE Canadian Conference on*, volume 2, pages 755–760, 1999.
147. P. Carre, E. Andres, and C. Fernandez-Maloigne. Discrete rotation for directional orthogonal wavelet packets. In *Image Processing, 2001. Proceedings. 2001 International Conference on*, volume 2, pages 257–260, 2001.
148. R. Vargic. An approach to directional wavelet construction and their use for image compression. In *Video/Image Processing and Multimedia Communications 4th EURASIP-IEEE Region 8 International Symposium on VIPromCom*, pages 201–204, 2002.
149. Chin-Hwa Kuo, Tzu-Chuan Chou, and Tay-Shen Wang. An efficient spatial prediction-based image compression scheme. *Circuits and Systems for Video Technology, IEEE Transactions on*, 12(10):850–856, 2002.
150. W. Sweldens. The lifting scheme: A custom-design construction of biorthogonal wavelets, 1994. Technical Report 1994:7, Industrial Mathematics Initiative, Department of Mathematics, University of South Carolina.
151. D. Taubman. Adaptive, non-separable lifting transforms for image compression. In *Image Processing, 1999. ICIP 99. Proceedings. 1999 International Conference on*, volume 3, pages 772–776, 1999.
152. N. V. Boulgouris and M. G. Strintzis. Orientation-sensitive interpolative pyramids for lossless and progressive image coding. *Image Processing, IEEE Transactions on*, 9(4):710–715, 2000.
153. N. V. Boulgouris, D. Tzovaras, and M. G. Strintzis. Lossless image compression based on optimal prediction, adaptive lifting, and conditional arithmetic coding. *Image Processing, IEEE Transactions on*, 10(1):1–14, 2001.
154. R. L. Claypoole, G. M. Davis, W. Sweldens, and R. G. Baraniuk. Nonlinear wavelet transforms for image coding via lifting. *Image Processing, IEEE Transactions on*, 12(12):1449–1459, 2003.

155. Hongliang Li, Guizhong Liu, and Zhongwei Zhang. Optimization of integer wavelet transforms based on difference correlation structures. *Image Processing, IEEE Transactions on*, 14(11):1831–1847, 2005.
156. O. N. Gerek and A. E. Cetin. A 2-D orientation-adaptive prediction filter in lifting structures for image coding. *Image Processing, IEEE Transactions on*, 15(1):106–111, 2006.
157. W. Ding, F. Wu, and S. Li. Lifting-based wavelet transform with directionally spatial prediction. In *Picture Coding Symposium*, 2004.
158. Chuo-Ling Chang, A. Maleki, and B. Girod. Adaptive wavelet transform for image compression via directional quincunx lifting. In *Multimedia Signal Processing, 2005 IEEE 7th Workshop on*, pages 537–540, 2005.
159. L. Breiman, J. H. Friedman, R. A. Olshen, and C. J. Stone. Classification and regression trees, 1984. The Wadsworth Statistics/Probability Series, Belmont, CA.
160. L. Yaroslavsky. Fast signal sinc-interpolation and its applications in signal and image processing. In *SPIE 14th Annual Symposium Electronic Image 2003*, volume 4667, 2002.
161. Wenpeng Ding, Feng Wu, Xiaolin Wu, Shipeng Li, and Li Houqiang. Adaptive directional lifting-based wavelet transform for image coding. *Image Processing, IEEE Transactions on*, 16(2):416–427, 2007.
162. Bing Zeng and Jingjing Fu. Directional discrete cosine transforms for image coding. In *Multimedia and Expo, 2006 IEEE International Conference on*, pages 721–724, 2006.
163. P. Kauff and K. Schuur. Shape-adaptive DCT with block-based DC separation and delta;DC correction. *Circuits and Systems for Video Technology, IEEE Transactions on*, 8(3):237–242, 1998.
164. Wen-Hsiung Chen, C. Smith, and S. Fralick. A fast computational algorithm for the discrete cosine transform. *Communications, IEEE Transactions on*, 25(9):1004–1009, 1977.
165. Wen-Hsiung Chen and C. Smith. Adaptive coding of monochrome and color images. *Communications, IEEE Transactions on*, 25(11):1285–1292, 1977.
166. C. Loeffler, A. Ligtenberg, and George S. Moschytz. Practical fast 1-D DCT algorithms with 11 multiplications. In *Acoustics, Speech, and Signal Processing, 1989. ICASSP-89., 1989 International Conference on*, pages 988–991, 1989.
167. F. A. M. L. Bruekers and A. W. M. van den Enden. New networks for perfect inversion and perfect reconstruction. *Selected Areas in Communications, IEEE Journal on*, 10(1):129–137, 1992.
168. T. D. Tran. Fast multiplierless approximation of the DCT. In *Proc. 33rd Annu. Conf. Inform. Science System*, pages 933–938, 1999.
169. Jie Liang and T. D. Tran. Fast multiplierless approximations of the DCT with the lifting scheme. *Signal Processing, IEEE Transactions on*, 49(12):3032–3044, 2001.
170. P. P. Vaidyanathan and P.-Q. Hoang. Lattice structures for optimal design and robust implementation of two-channel perfect-reconstruction QMF banks. *Acoustics, Speech and Signal Processing, IEEE Transactions on*, 36(1):81–94, 1988.
171. T. H. Cormen, R. L. Rivest, C. E. Leiserson, and C. Stein. *Introduction to Algorithms*. Cambridge, MA: MIT Press, 2001.
172. J. Kovacevic and M. Vetterli. Nonseparable multidimensional perfect reconstruction filter banks and wavelet bases for RN. *Information Theory, IEEE Transactions on*, 38(2):533–555, 1992.
173. E. J. Candes and D. L. Donoho. Curvelets — A surprisingly effective non-adaptive representation for objects with edges, 1999. in Curve and Surface Fitting, A. Cohen, C. Rabut, and L. L. Schumaker, Eds. Saint-Malo: Vanderbilt University Press.
174. J. K. Romberg, M. Wakin, and R. Baraniuk. Multiscale wedgelet image analysis: Fast decompositions and modeling. In *Image Processing. 2002. Proceedings. 2002 International Conference on*, volume 3, pages 585–588, 2002.
175. T. T. Nguyen and Soontorn Oraintara. A multiresolution directional filter bank for image applications. In *Acoustics, Speech, and Signal Processing, 2004. Proceedings. (ICASSP '04). IEEE International Conference on*, volume 3, pages 37–40, 2004.

176. Yilong Liu, T. T. Nguyen, and Soontorn Oraintara. Low bit-rate image coding based on pyramidal directional filter banks. In *Acoustics, Speech and Signal Processing, 2006. ICASSP 2006 Proceedings. 2006 IEEE International Conference on*, volume 2, pages 437–440, 2006.
177. S. Esakkirajan, T. Veerakumar, V. S. Murugan, and R. Sudhakar. Image compression using contourlet transform and multistage vector quantization. *GVIP Journal*, 6(1):19–28, 2006.
178. R. Ansari, A. E. Cetin, and S. H. Lee. Subband coding of images using nonrectangular filter banks. In *Applications of Digital Image Processing XI, SPIE*, volume 974, pages 315–322, 1988.
179. Demin Wang, L. Zhang, and A. Vincent. Improvement of JPEG 2000 using curved wavelet transform. In *Acoustics, Speech, and Signal Processing, 2005. Proceedings. (ICASSP '05). IEEE International Conference on*, volume 2, pages 365–368, 2005.
180. Demin Wang, L. Zhang, A. Vincent, and F. Speranza. Curved wavelet transform for image coding. *Image Processing, IEEE Transactions on*, 15(8):2413–2421, 2006.
181. Yu Liu and King Ngi Ngan. Weighted adaptive lifting-based wavelet transform for image coding. *Image Processing, IEEE Transactions on*, 17(4):500–511, 2008.
182. Weisheng Dong, Guangming Shi, and Jizheng Xu. Adaptive nonseparable interpolation for image compression with directional wavelet transform. *Signal Processing Letters, IEEE*, 15:233–236, 2008.
183. Chuo-Ling Chang and B. Girod. Direction-adaptive discrete wavelet transform for image compression. *Image Processing, IEEE Transactions on*, 16(5):1289–1302, 2007.
184. Bing Zeng and Jingjing Fu. Directional discrete cosine transforms — A new framework for image coding. *Circuits and Systems for Video Technology, IEEE Transactions on*, 18(3):305–313, 2008.
185. Hao Xu, Jizheng Xu, and Feng Wu. Lifting-based directional DCT-like transform for image coding. *Circuits and Systems for Video Technology, IEEE Transactions on*, 17(10):1325–1335, 2007.
186. Chuo-Ling Chang and B. Girod. Direction-adaptive partitioned block transform for image coding. In *Image Processing, 2008. ICIP 2008. 15th IEEE International Conference on*, pages 145–148, 2008.
187. O. N. Gerek and A. E. Cetin. Adaptive polyphase subband decomposition structures for image compression. *Image Processing, IEEE Transactions on*, 9(10):1649–1660, 2000.
188. Yan Ye and M. Karczewicz. Improved H.264 intra coding based on bi-directional intra prediction, directional transform, and adaptive coefficient scanning. In *Image Processing, 2008. ICIP 2008. 15th IEEE International Conference on*, pages 2116–2119, 2008.
189. T. K. Tan, C. S. Boon, and Y. Suzuki. Intra prediction by template matching. In *Image Processing, 2006 IEEE International Conference on*, pages 1693–1696, 2006.
190. Xiulian Peng, Feng Wu, and Jizheng Xu. Directional filtering transform. In *Multimedia and Expo, 2009. ICME 2009. IEEE International Conference on*, pages 1–4, 2009.
191. ITU-T. Key technical area software of the ITU-T, version jm11.0kta2.0, 2008. http://iphome.hhi.de/suehring/tml/download/KTA.
192. N. Jayant, J. Johnston, and R. Safranek. Signal compression based on models of human perception. *Proceedings of the IEEE*, 81(10):1385–1422, 1993.
193. I. Hontsch and L. J. Karam. Locally adaptive perceptual image coding. *Image Processing, IEEE Transactions on*, 9(9):1472–1483, 2000.
194. M. M. Reid, R. J. Millar, and N. D. Black. Second-generation image coding: An overview. *ACM Computing Surveys*, 29(1):2–29, 1997.
195. M. Tuceryan and A. K. Jain. Texture analysis. In *The Handbook of Pattern Recognition and Computer Vision* (2nd Edition), C. H. Chen, L. F. Pau, and P. S. P. Wang, Eds. World Scientific Publishing Co., pages 207–248, 1998.
196. Cheng en Guo, Song-Chun Zhu, and Ying Nian Wu. Efficient gathering of correlated data in sensor network. In *Computer Vision, 2003. Proceedings. Ninth IEEE International Conference on*, pages 1228–1235, 2003.
197. A. A. Efros and T. K. Leung. Texture synthesis by non-parametric sampling. In *Computer Vision, 1999. The Proceedings of the Seventh IEEE International Conference on*, volume 2, pages 1033–1038, 1999.

198. L.-Y. Wei and M. Levoy. Fast texture synthesis using tree-structured vector quantization. In *ACM SIGGRAPH*, pages 479–488, 2000.
199. M. Ashikhmin. Synthesizing natural textures. In *ACM Symposium on Interactive 3D Graphics*, pages 217–226, 2001.
200. A. Hertzmann, C. E. Jacobs, N. Oliver, B. Curless, and D. H. Salesin. Image analogies. In *ACM SIGGRAPH*, pages 327–340, 2001.
201. A. A. Efros and W. T. Freeman. Image quilting for texture synthesis and transfer. In *ACM SIGGRAPH*, pages 341–346, 2001.
202. L. Liang, C. Liu, Y.-Q. Xu, B. Guo, and H.-Y. Shum. Real-time texture synthesis by patch-based sampling. *Graphics, ACM Transactions on*, 20(3):127–150, 2001.
203. V. Kwatra, A. Schdl, I. Essa, G. Turk, and A. Bobick. Graphcut textures: Image and video synthesis using graph cuts. In *ACM SIGGRAPH*, pages 277–286, 2003.
204. S. Lefebvre and H. Hoppe. Parallel controllable texture synthesis. In *ACM SIGGRAPH*, pages 777–786, 2005.
205. V. Kwatra, I. Essa, A. Bobick, and N. Kwatra. Texture optimization for example-based synthesis. In *ACM SIGGRAPH*, pages 795–802, 2005.
206. M. Bertalmio, G. Sapiro, V. Caselles, and C. Ballester. Image inpainting. In *ACM SIGGRAPH*, pages 417–424, 2000.
207. T. F. Chan and J. Shen. Mathematical models for local nontexture inpaintings. 62(3):1019–1043, 2002.
208. C. Ballester, M. Bertalmio, V. Caselles, G. Sapiro, and J. Verdera. Filling-in by joint interpolation of vector fields and gray levels. *Image Processing, IEEE Transactions on*, 10(8):1200–1211, 2001.
209. T. F. Chan and J. Shen. Non-texture inpainting by curvature-driven diffusions (CDD). 12(4):436–449, 2001.
210. Jiaya Jia and Chi-Keung Tang. Image repairing: Robust image synthesis by adaptive ND tensor voting. In *Computer Vision and Pattern Recognition, 2003. Proceedings. 2003 IEEE Computer Society Conference on*, volume 1, pages 643–650, 2003.
211. I. Drori, D. Cohen-Or, and H. Yeshurun. Fragment-based image completion. In *ACM SIGGRAPH*, pages 303–312, 2003.
212. Antonio Criminisi, P. Perez, and K. Toyama. Region filling and object removal by exemplar-based image inpainting. *Image Processing, IEEE Transactions on*, 13(9):1200–1212, 2004.
213. M. Bertalmio, L. Vese, G. Sapiro, and S. Osher. Simultaneous structure and texture image inpainting. *Image Processing, IEEE Transactions on*, 12(8):882–889, 2003.
214. H. Grossauer. A combined PDE and texture synthesis approach to inpainting. In *Computer Vision, Euro. Conference on*, pages 214–224, 2004.
215. S. D. Rane, G. Sapiro, and M. Bertalmio. Structure and texture filling-in of missing image blocks in wireless transmission and compression applications. *Image Processing, IEEE Transactions on*, 12(3):296–303, 2003.
216. P. Prez, M. Gangnet, and A. Blake. Patchworks: Example-based region tiling for image editing, 2004. Microsoft Research, Tech. Rep. MSR-TR-2004-04.
217. J. Sun, L. Yuan, J. Jia, and H.-Y. Shum. Image completion with structure propagation. In *ACM SIGGRAPH*, pages 861–868, 2005.
218. L. Atzori and F. G. B. De Natale. Error concealment in video transmission over packet networks by a sketch-based approach. *Signal Processing: Image Communication*, 15(1-2):57–76, 1999.
219. X. Sun, F. Wu, and S. Li. Compression with vision technologies. In *Picture Coding Symposium*, 2006.
220. N. Jojic, B. J. Frey, and A. Kannan. Epitomic analysis of appearance and shape. In *Computer Vision, 2003. Proceedings. Ninth IEEE International Conference on*, pages 34–41, 2003.
221. V. Cheung, B. J. Frey, and Nebojsa Jojic. Video epitomes. In *Computer Vision and Pattern Recognition, 2005. CVPR 2005. IEEE Computer Society Conference on*, volume 1, pages 42–49, 2005.

222. Chen Wang, Xiaoyan Sun, Feng Wu, and Hongkai Xiong. Image compression with structure-aware inpainting. In *Circuits and Systems, 2006. ISCAS 2006. Proceedings. 2006 IEEE International Symposium on*, pages 1816–1819, 2006.
223. Wenjun Zeng and B. Liu. Geometric-structure-based error concealment with novel applications in block-based low-bit-rate coding. *Circuits and Systems for Video Technology, IEEE Transactions on*, 9(4):648–665, 1999.
224. C. A. Rothwell, J. L. Mundy, W. Hoffman, and V.-D. Nguyen. Driving vision by topology. In *Computer Vision. International Symposium on*, pages 395–400, 1995.
225. D. Tschumperle and R. Deriche. Vector-valued image regularization with PDES: A common framework for different applications. *Pattern Analysis and Machine Intelligence, IEEE Transactions on*, 27(4):506–517, 2005.
226. M. Armbrust, A. Fox, R. Griffith, A. D. Joseph, R. Katz, A. Konwinski, G. Lee, D. Patterson, A. Rabkin, I. Stoica, and M. Zaharia. A view of Cloud computing. *Comm. of the ACM*, 53:50–58, 2010.
227. Google. Streetview, 2010. https://developers.google.com/maps/documentation/streetview.
228. G. K. Wallace. The JPEG still picture compression standard. *Comm. of the ACM*, 34:30–44, 1991.
229. G. J. Sullivan and J. R. Ohm. Recent developments in standardization of high efficiency video coding (HEVC). In *SPIE Applications of Digital Image Processing XXXIII*, volume 7798, pages 1823–1826, 2010.
230. Y. Rui, T. S. Huang, and S. F. Chang. Image retrieval: Current techniques, promising directions, open issues. *Journal of Visual Communication and Image Representation*, 10:39–62, 1999.
231. M. S. Lew, N. Sebe, C. Djeraba, and R. Jain. Content-based multimedia information retrieval: State of the art and challenges. *Multimedia Computing, Comm. and Appl., ACM Transactions on*, 2:1–19, 2006.
232. J. R. Smith and S. F. Chang. Visualseek: A fully automated content-based image query system. In *ACM Multimedia*, page 8798, 1996.
233. C. H. Wang, Z. W. Li, and L. Zhang. Mindfinder: Image search by interactive sketching and tagging. In *Proc. of WWW*, pages 1309–1312, 2010.
234. Y. Ke, R. Sukthankar, and L. Huston. An efficient parts-based near-duplicate and sub-image retrieval system. In *ACM Multimedia*, pages 869–876, 2004.
235. Q. F. Zheng, W. Q. Wang, and W. Gao. Effective and efficient object-based image retrieval using visual phrases. In *ACM Multimedia*, pages 77–80, 2006.
236. Zhong Wu, Qifa Ke, M. Isard, and Jian Sun. Bundling features for large-scale partial-duplicate Web image search. In *Computer Vision and Pattern Recognition, 2009. CVPR 2009. IEEE Conference on*, pages 25–32, 2009.
237. W. G. Zhou, Y. J. Lu, H. Q. Li, Y. B. Song, and Q. Tian. Spatial coding for large-scale partial-duplicate Web image search. In *ACM Multimedia*, page 511–520, 2010.
238. J. Hays and A. A. Efros. Scene completion using millions of photographs. *Graphics, ACM Transactions on*, 26(3), 2007.
239. O. Whyte, J. Sivic, and A. Zisserman. Get out of my picture! Internet-based inpainting. In *British Machine Vision Conference*, 2009.
240. M. Eitz, K. Hildebrand, T. Boubekeur, and M. Alexa. Photosketch: A sketch based image query and composition. In *ACM SIGGRAPH*, 2009.
241. M. Eitz, R. Richter, K. Hildebrand, T. Boubekeur, and M. Alexa. Photosketcher: Interactive sketch-based image synthesis. *Computer Graphics and Applications, IEEE*, 31(6):56–66, 2011.
242. T. Chen, M. M. Cheng, P. Tan, A. Shamir, and S. M. Hu. Photosketch: Internet image montage. In *ACM SIGGRAPH Asia*, 2009.
243. M. K. Johnson, K. Dale, S. Avidan, H. Pfister, W. T. Freeman, and W. Matusik. Cg2Real: Improving the realism of computer generated images using a large collection of photographs. *Visualization and Computer Graphics, IEEE Transactions on*, 17(9):1273–1285, 2011.

244. P. Weinzaepfel, H. Jegou, and P. Perez. Reconstructing an image from its local descriptors. In *Computer Vision and Pattern Recognition (CVPR), 2011 IEEE Conference on*, pages 337–344, 2011.
245. M. Daneshi and J. Q. Guo. Image reconstruction based on local feature descriptors, 2011. http://www.stanford.edu/class/ee368/project 11/reports.
246. M. A. Fischler and R. C. Bolles. Random sample consensus: A paradigm for model fitting with applications to image analysis and automated cartography. *Comm. of the ACM*, 24:381–395, 1981.
247. D. G. Lowe. Distinctive image features from scale-invariant keypoints. *International Journal of Computer Vision*, 60:91–110, 2004.
248. S. Edelman, N. Intrator, and T. Poggio. Complex cells and object recognition, 1997. http://kybele.psych.cornell.edu.
249. Yan Ke and R. Sukthankar. PCA-SIFT: A more distinctive representation for local image descriptors. In *Computer Vision and Pattern Recognition, 2004. CVPR 2004. Proceedings of the 2004 IEEE Computer Society Conference on*, volume 2, pages 506–513, 2004.
250. Gang Hua, M. Brown, and S. Winder. Discriminant embedding for local image descriptors. In *Computer Vision, 2007. ICCV 2007. IEEE 11th International Conference on*, pages 1–8, 2007.
251. V. Chandrasekhar, G. Takacs, D. Chen, S. S. Tsai, J. Singh, and B. Girod. Transform coding of image feature descriptors. In *Visual Communication and Image Processing, SPIE Conference on*, volume 7257, 2009.
252. Chuohao Yeo, P. Ahammad, and K. Ramchandran. Rate-efficient visual correspondences using random projections. In *Image Processing, 2008. ICIP 2008. 15th IEEE International Conference on*, pages 217–220, 2008.
253. H. Jegou, M. Douze, and C. Schmid. Product quantization for nearest neighbor search. *Pattern Analysis and Machine Intelligence, IEEE Transactions on*, 33(1):117–128, 2011.
254. V. Chandrasekhar, G. Takacs, D. Chen, S. Tsai, R. Grzeszczuk, and B. Girod. CHoG: Compressed histogram of gradients a low bit-rate feature descriptor. In *Computer Vision and Pattern Recognition, 2009. CVPR 2009. IEEE Conference on*, pages 2504–2511, 2009.
255. V. Chandrasekhar, G. Takacs, D. Chen, S. S. Tsai, R. Grzeszczuk, and B. Girod. Compressed histogram of gradients: A low-bitrate descriptor. 96:384–399, 2012.
256. G. Francini, S. Lepsoy, and M. Balestri. Description of test model under consideration for CDVS, 2011. ISO/IEC JTC1/SC29/WG11, N12367, Geneva.
257. M. Makar, Chuo-Ling Chang, D. Chen, S.S. Tsai, and B. Girod. Compression of image patches for local feature extraction. In *Acoustics, Speech and Signal Processing, 2009. ICASSP 2009. IEEE International Conference on*, pages 821–824, 2009.
258. Jianshu Chao and E. Steinbach. Preserving Sift features in JPEG-encoded images. In *Image Processing (ICIP), 2011 18th IEEE International Conference on*, pages 301–304, 2011.
259. J. Sivic and A. Zisserman. Video Google: A text retrieval approach to object matching in videos. In *Computer Vision, 2003. Proceedings. 9th IEEE International Conference on*, pages 1470–1477, 2003.
260. J. Philbin, O. Chum, M. Isard, J. Sivic, and A. Zisserman. Object retrieval with large vocabularies and fast spatial matching. In *Computer Vision and Pattern Recognition, 2007. CVPR '07. IEEE Conference on*, pages 1–8, 2007.
261. O. Chum, J. Philbin, J. Sivic, M. Isard, and A. Zisserman. Total recall: Automatic query expansion with a generative feature model for object retrieval. In *Computer Vision, 2007. ICCV 2007. IEEE 11th International Conference on*, pages 1–8, 2007.
262. J. Philbin, O. Chum, M. Isard, J. Sivic, and A. Zisserman. Lost in quantization: Improving particular object retrieval in large-scale image databases. In *Computer Vision and Pattern Recognition, 2008. CVPR 2008. IEEE Conference on*, pages 1–8, 2008.
263. H. Jegou, M. Douze, and C. Schmid. Hamming embedding and weak geometric consistency for large-scale image search. In *Computer Vision, European Conference on*, 2008.
264. W. G. Zhou, Y. J. Lu, H. Q. Li, and Q. Tian. Scalar quantization for large-scale image search. In *ACM Multimedia*, 2012.

265. R. Talluri. Image alignment and stitching: A tutorial. *Foundations and Trends in Computer Graphics and Vision*, 2:1–104, 2012.
266. P. Torr and A. Zisserman. MLESAC: A new robust estimator with application to estimating image geometry. *Journal of Computer Vision and Image Understanding*, 78:138–156, 2000.
267. P. H. S. Torr and C. Davidson. IMPSAC: Synthesis of importance sampling and random sample consensus. *Pattern Analysis and Machine Intelligence, IEEE Transactions on*, 25(3):354–364, 2003.
268. O. Chum and J. Matas. Matching with PROSAC–Progressive sample consensus. In *Computer Vision and Pattern Recognition, 2005. CVPR 2005. IEEE Computer Society Conference on*, volume 1, pages 220–226, vol. 1, 2005.
269. H. Jegou and M. Douze. Inria holiday dataset, 2008. http://lear.inrialpes.fr/people/jegou/data.php.
270. P. Perez, M. Gangnet, and A. Blake. Poisson image editing. *Graphics, ACM Transactions on*, 22:313–318, 2003.
271. ITU-T. Methodology for the subjective assessment of the quality of television pictures, 2002. Recommendation ITU-R BT. 5000-11.
272. Z. Farbman, R. Fattal, and D. Lischinski. Convolution pyramids. In *ACM SIGGRAPH Asia*, 2011.
273. L. Dai, H. Yue, X. Sun, and F. Wu. IMShare: Instantly sharing your mobile landmark images by search-based reconstruction. In *ACM Multimedia*, pages 579–588, 2012.
274. H. Shao, T. Svoboda, and L. V. Gool. Zubud-Zurich buildings database for image based recognition, 2003. Technical Report 206, ETH Zurich.
275. Josh Ong. Picture this: Chinese Internet giant Tencent's Qzone social network now hosts over 150 billion photos, 2012. http://thenextweb.com/asia.
276. Microsoft. Windows 8 launch, 2012. http://www.microsoft.com.
277. Yurij S. Musatenko and Vitalij N. Kurashov. Correlated image set compression system based on new fast efficient algorithm of Karhunen-Loève transform. volume 3527, pages 518–529, 1998.
278. Kosmas Karadimitriou and John M. Tyler. The centroid method for compressing sets of similar images. *Pattern Recognition Letters*, 19(7):585–593, 1998.
279. Samy Ait-Aoudia and Abdelhalim Gabis. A comparison of set redundancy compression techniques. *EURASIP J. Appl. Signal Process.*, 2006:1–13, January 2006.
280. Chi-Ho Yeung, O. C. Au, Ketan Tang, Zhiding Yu, Enming Luo, Yannan Wu, and Shing-Fat Tu. Compressing similar image sets using low frequency template. In *Multimedia and Expo (ICME), 2011 IEEE International Conference on*, pages 1–6, 2011.
281. Cheng H., Li X., and Schmieder A. A study of clustering algorithm and validity for lossy image set compression. In *Image Processing, Computer Vision and Pattern Recognition, 2009 International Conference on*, pages 501–506, 2009.
282. Yang Lu, Tien-Tsin Wong, and Pheng-Ann Heng. Digital photo similarity analysis in frequency domain and photo album compression. In *Proceedings of the 3rd International Conference on Mobile and Ubiquitous Multimedia*, MUM '04, pages 237–244, 2004.
283. Chia-Ping Chen, Chu-Song Chen, Kuo-Liang Chung, H.-I. Lu, and G.Y. Tang. Image set compression through minimal-cost prediction structure. In *Image Processing, 2004. ICIP '04. 2004 International Conference on*, volume 2, pages 1289–1292, Vol. 2, 2004.
284. O. Au, Sijin Li, Ruobing Zou, Wei Dai, and Lin Sun. Digital photo album compression based on global motion compensation and intra/inter prediction. In *Audio, Language and Image Processing (ICALIP), 2012 International Conference on*, pages 84–90, 2012.
285. Ruobing Zou, Oscar C. Au, Guyue Zhou, Wei Dai, Wei Hu, and Pengfei Wan. Personal photo album compression and management. In *Circuits and Systems (ISCAS), 2013 IEEE International Symposium on*, pages 1428–1431, 2013.
286. A. W. M. Smeulders, M. Worring, S. Santini, A. Gupta, and R. Jain. Content-based image retrieval at the end of the early years. *Pattern Analysis and Machine Intelligence, IEEE Transactions on*, 22(12):1349–1380, 2000.

287. Yong Rui, T. S. Huang, M. Ortega, and S. Mehrotra. Relevance feedback: A power tool for interactive content-based image retrieval. *Circuits and Systems for Video Technology, IEEE Transactions on*, 8(5):644–655, 1998.
288. H. Bay, T. Tuytelaars, and L. Van Gool. Surf: Speeded up robust features. In *Computer Vision, 2006. European Conference on*, pages 404–417, 2006.
289. Huanjing Yue, Xiaoyan Sun, Jingyu Yang, and Feng Wu. Cloud-based image coding for mobile devices — Toward thousands to one compression. *Multimedia, IEEE Transactions on*, 15(4):845–857, 2013.
290. J. B. Kruskal. On the shortest spanning subtree of a graph and the traveling salesman problem. *Proceedings of American Mathematical Society*, 7:48–50, 1956.
291. Y. J. Chu and T. H. Liu. On the shortest arborescence of a directed graph. *Science Sinica*, 14:1396–1400, 1965.
292. H. Isack and Y. Boykov. Energy-based geometric multi-model fitting. *Computer Vision, International Journal of*, 97(2):123–147, 2012.
293. Y. Boykov, O. Veksler, and R. Zabih. Fast approximate energy minimization via graph cuts. *Pattern Analysis and Machine Intelligence, IEEE Transactions on*, 23(11):1222–1239, 2001.
294. Alhwarin, Faraj, Danijela Ristic-Durrant, and Axel Graser. VF-SIFT: Very fast SIFT feature matching. *Pattern Recognition, International Journal on*, pages 222–231, 2010.
295. D. L. Donoho. Compressed sensing. *Information Theory, IEEE Transactions on*, 52(4):1289–1306, 2006.
296. R. G. Baraniuk. Compressive sensing [lecture notes]. *Signal Processing Magazine, IEEE*, 24(4):118–121, 2007.
297. E. J. Candes and M. B. Wakin. An introduction to compressive sampling. *Signal Processing Magazine, IEEE*, 25(2):21–30, 2008.
298. R. Cristescu, B. Beferull-Lozano, M. Vetterli, and R. Wattenhofer. Network correlated data gathering with explicit communication: NP-completeness and algorithms. *Networking, IEEE/ACM Transactions on*, 14(1):41–54, 2006.
299. A. Ciancio, S. Pattem, A. Ortega, and B. Krishnamachari. Energy-efficient data representation and routing for wireless sensor networks based on a distributed wavelet compression algorithm. In *Information Processing in Sensor Networks, 2006. IPSN 2006. The Fifth International Conference on*, pages 309–316, 2006.
300. J. Acimovic, B. Beferull-Lozano, and R. Cristescu. Adaptive distributed algorithms for power-efficient data gathering in sensor networks. In *Wireless Networks, Communications and Mobile Computing, 2005 International Conference on*, volume 2, pages 946–951, vol. 2, 2005.
301. S. Yoon and C. Shahabi. The clustered aggregation (CAG) technique leveraging spatial and temporal correlations in wireless sensor networks. *Sensor Networks, ACM Transactions on*, 3(8), 2007.
302. H. Gupta, V. Navda, S. Das, and V. Chowdhary. Efficient gathering of correlated data in sensor network. *Sensor Networks, ACM Transactions on*, 4, 2008.
303. J. Chou, D. Petrovic, and Kannan Ramachandran. A distributed and adaptive signal processing approach to reducing energy consumption in sensor networks. In *INFOCOM 2003. 22nd Annual Joint Conference of the IEEE Computer and Communications. IEEE Societies*, volume 2, pages 1054–1062, 2003.
304. R. Cristescu, B. Beferull-Lozano, and M. Vetterli. On network correlated data gathering. In *INFOCOM 2004. 23rd Annual Joint Conference of the IEEE Computer and Communications Societies*, volume 4, pages 2571–2582, 2004.
305. G. Hua and W. Chen. Correlated data gathering in wireless sensor networks based on distributed source coding. *International Journal of Sensor Networks*, 4:13–22, 2008.
306. D. Slepian and J. K. Wolf. Noiseless coding of correlated information sources. *Information Theory, IEEE Transactions on*, 19(4):471–480, 1973.
307. K. Yuen, Ben Liang, and Baochun Li. A distributed framework for correlated data gathering in sensor networks. *Vehicular Technology, IEEE Transactions on*, 57(1):578–593, 2008.
308. W. Bajwa, J. Haupt, A. Sayeed, and R. Nowak. Compressive wireless sensing. In *Information Processing in Sensor Networks, 2006. IPSN 2006. The Fifth International Conference on*, pages 134–142, 2006.

309. M. Rabbat, J. Haupt, A. Singh, and R. Nowak. Decentralized compression and predistribution via randomized gossiping. In *Information Processing in Sensor Networks, 2006. IPSN 2006. The Fifth International Conference on*, pages 51–59, 2006.
310. J. Haupt, W. U. Bajwa, M. Rabbat, and R. Nowak. Compressed sensing for networked data. *Signal Processing Magazine, IEEE*, 25(2):92–101, 2008.
311. E. J. Candes, J. Romberg, and T. Tao. Robust uncertainty principles: Exact signal reconstruction from highly incomplete frequency information. *Information Theory, IEEE Transactions on*, 52(2):489–509, 2006.
312. D. L. Donoho, M. Elad, and V. N. Temlyakov. Stable recovery of sparse overcomplete representations in the presence of noise. *Information Theory, IEEE Transactions on*, 52(1):6–18, 2006.
313. P. Gupta and P. R. Kumar. The capacity of wireless networks. *Information Theory, IEEE Transactions on*, 46(2):388–404, 2000.
314. D. Marco, E. Duarte-Melo, M. Liu, and D. Neuhoff. On the many-to-one transport capacity of a dense wireless sensor network and the compressibility of its data. In *Information Processing in Sensor Networks, 2003. IPSN 2003. The Fifth International Conference on*, pages 1–16, 2003.
315. S. McCanne and S. Floyd. Network simulator NS-2. http://www.isi.edu/nsnam/ns/.
316. C. Chen and Y. Wang. Chain-type wireless sensor network for monitoring long range infrastructures: Architecture and protocols. *International Journal on Distributed Sensor Networks*, 4, 2008.
317. S. Tilak, N. Abu-Gahazaleh, and W. Heinzelman. Infrastructure trade-offs for sensor networks. In *Proc. of ACM IWWSNA*, pages 49–58, 2002.
318. NBDC. CTD data. http://tao.noaa.gov/refreshed/ctd delivery.php.
319. A. Martinez, A. Guillen i Fabregas, G. Caire, and F. M. J. Willems. Bit-interleaved coded modulation revisited: A mismatched decoding perspective. *Information Theory, IEEE Transactions on*, 55(6):2756–2765, 2009.
320. G. Holland, N. Vaidya, and P. Bahl. A rate-adaptive MAC protocol for multi-hop wireless networks. In *ACM MobiCom*, pages 236–251, 2001.
321. G. Judd, X. Wang, and P. Steenkiste. Efficient channel-aware rate adaptation in dynamic environments. In *ACM MobiSys*, pages 118–131, 2008.
322. R. T. Morris, J. C. Bicket, and J. C. Bicket. Bit-rate selection in wireless networks, 2005. Technical report, Masters thesis, MIT.
323. S. H. Y. Wong, H. Yang, S. Lu, and V. Bharghavan. Robust rate adaptation for 802.11 wireless networks. In *ACM MobiCom*, pages 146–157, 2006.
324. M. Vutukuru, H. Balakrishnan, and K. Jamieson. Cross-layer wireless bit rate adaptation. In *ACM SIGCOMM*, pages 3–14, 2009.
325. J. D. Brown, S. Pasupathy, and K. N. Plataniotis. Adaptive demodulation using rateless erasure codes. *Communications, IEEE Transactions on*, 54(9):1574–1585, 2006.
326. A. Shokrollahi. Raptor codes. *Information Theory, IEEE Transactions on*, 52(6):2551–2567, 2006.
327. G. Caire, Giorgio Taricco, and Ezio Biglieri. Bit-interleaved coded modulation. *Information Theory, IEEE Transactions on*, 44(3):927–946, 1998.
328. M. Luby. Lt codes. In *In FOCS '02: Proceedings of the 43rd Symposium on Foundations of Computer Science*, page 271, 2002.
329. D. Baron, Shriram Sarvotham, and R. G. Baraniuk. Bayesian compressive sensing via belief propagation. *Signal Processing, IEEE Transactions on*, 58(1):269–280, 2010.
330. Macro Pretti. A message-passing algorithm with damping. *Journal of Statistic Mech.*, 2005.
331. IEEE. IEEE standard for information technology — telecommunications and information exchange between systems — local and metropolitan area networks — specific requirements. part 11: Wireless LAN medium access control (MAC) and physical layer (PHY) specifications, 2007. IEEE Std 802.11-2007.
332. K. Tan, J. Zhang, J. Fang, H. Liu, Y. Ye, S. Wang, Y. Zhang, H. Wu, W. Wang, and G. M. Voelker. SORA: High performance software radio using general purpose multi-core processors. In *NSDI, USENIX Association*, pages 75–90, 2009.

333. G. Ungerboeck. On improving data-link performance by increasing channel alphabet and introducing sequence coding. In *IEEE International Symposium on Information Theory*, 1976.
334. G. Ungerboeck. Channel coding with multilevel/phase signals. *Information Theory, IEEE Transactions on*, 28(1):55–66, 1982.
335. H. Imai and S. Hirakawa. A new multilevel coding method using error-correcting codes. *Information Theory, IEEE Transaction on*, 23(3):371–377, 1977.
336. A. R. Calderbank. Multilevel codes and multistage decoding. *Communications, IEEE Transactions on*, 37(3):222–229, 1989.
337. U. Wachsmann, R. F. H. Fischer, and J. B. Huber. Multilevel codes: Theoretical concepts and practical design rules. *Information Theory, IEEE Transactions on*, 45(5):1361–1391, 1999.
338. E. Zehavi. 8-PSK trellis codes for a Rayleigh channel. *Communications, IEEE Transactions on*, 40(5):873–884, 1992.
339. Shihao Ji, Ya Xue, and L. Carin. Bayesian compressive sensing. *Signal Processing, IEEE Transactions on*, 56(6):2346–2356, 2008.
340. M. W. Seeger and H. Nickisch. Compressed sensing and Bayesian experimental design. In *ACM ICML*, pages 912–919, 2008.
341. P. Schniter, L. C. Potter, and J. Ziniel. Fast Bayesian matching pursuit. In *Information Theory and Applications Workshop, 2008*, pages 326–333, 2008.
342. Shriram Sarvotham, D. Baron, and R. G. Baraniuk. Sudocodes: Fast measurement and reconstruction of sparse signals. In *Information Theory, 2006 IEEE International Symposium on*, pages 2804–2808, 2006.
343. F. Wu, J. Fu, Z. Lin, and B. Zeng. Analysis on rate-distortion performance of compressive sensing for binary sparse source. In *Data Compression Conference*, pages 113–122, 2009.
344. Panagiotis Destounis, John D. Garofalakis, Panagiotis Kappos, and J. Tzimas. Measuring the mean Web page size and its compression to limit latency and improve download time. *Internet Research*, 11(1):10–17, 2001.
345. Hao Cui, Chong Luo, Kun Tan, Feng Wu, and Chang Wen Chen. Seamless rate adaptation for wireless networking. In *Proceedings of the 14th ACM MSWiM '11*, pages 437–446, 2011.
346. Aditya Gudipati and Sachin Katti. Strider: Automatic rate adaptation and collision handling. In *Proceedings of the ACM SIGCOMM 2011 Conference*, pages 158–169, 2011.
347. Jonathan Perry, Hari Balakrishnan, and Devavrat Shah. Rateless spinal codes. In *Proceedings of the 10th ACM Workshop on Hot Topics in Networks, HotNets '11*, pages 6:1–6:6, 2011.
348. R. Hamzaoui, V. Stankovic, and Zixiang Xiong. Optimized error protection of scalable image bit streams [advances in joint source-channel coding for images]. *Signal Processing Magazine, IEEE*, 22(6):91–107, November 2005.
349. L. P. Kondi, F. Ishtiaq, and A. K. Katsaggelos. Joint source-channel coding for motion-compensated DCT-based SNR scalable video. *Image Processing, IEEE Transactions on*, 11(9):1043–1052, September 2002.
350. Zhihai He, Jianfei Cai, and Chang Wen Chen. Joint source channel rate-distortion analysis for adaptive mode selection and rate control in wireless video coding. *Circuits and Systems for Video Technology, IEEE Transactions on*, 12(6):511–523, June 2002.
351. Qian Zhang, Wenwu Zhu, Zu Ji, and Ya-Qin Zhang. A power-optimized joint source channel coding for scalable video streaming over wireless channel. In *Proceedings of IEEE ISCAS 2001*, volume 5, pages 137–140 vol. 5, 2001.
352. S. S. Arslan, P. C. Cosman, and L. B. Milstein. Generalized unequal error protection lt codes for progressive data transmission. *Image Processing, IEEE Transactions on*, PP(99):1, 2012.
353. Xiang Pan, A. Cuhadar, and A. H. Banihashemi. Combined source and channel coding with JPEG 2000 and rate-compatible low-density parity-check codes. *Signal Processing, IEEE Transactions on*, 54(3):1160–1164, March 2006.
354. S. Aditya and S. Katti. Flexcast: Graceful wireless video streaming. In *Proceedings of the 17th Annual International Conference on Mobile Computing and Networking*, 2011.
355. O. Y. Bursalioglu, G. Caire, and D. Divsalar. Joint source-channel coding for deep space image transmission using rateless codes. In *Information Theory and Applications Workshop (ITA), 2011*, pages 1–10, February 2011.

356. A. D. Liveris, Zixiang Xiong, and C. N. Georghiades. Compression of binary sources with side information at the decoder using LDPC codes. *Communications Letters, IEEE*, 6(10):440–442, 2002.
357. M. Sartipi and F. Fekri. Source and channel coding in wireless sensor networks using LDPC codes. In *Proceedings of IEEE SECON 2004*, pages 309–316, October 2004.
358. Wei Zhong and Javier Garcia-Frias. LDGM codes for channel coding and joint source-channel coding of correlated sources. *EURASIP J. Appl. Signal Process.*, 2005:942–953, January 2005.
359. Qian Xu, V. Stankovic, and Zixiang Xiong. Distributed joint source-channel coding of video using raptor codes. *Selected Areas in Communications, IEEE Journal on*, 25(4):851–861, 2007.
360. G. Caire, S. Shamai, and S. Verdu. A new data compression algorithm for sources with memory based on error correcting codes. In *IEEE ITW*, pages 291–295, March 4-April 2003.
361. G. Caire, S. Shamai, A. Shokrollahi, and S. Verdu. Universal variable-length data compression of binary sources using fountain codes. In *IEEE ITW*, pages 123–128, October 2004.
362. J. DelSer, P. M. Crespo, I. Esnaola, and J. Garcia-Frias. Joint source-channel coding of sources with memory using Turbo codes and the Burrows-Wheeler Transform. *Communications, IEEE Transactions on*, 58(7):1984–1992, July 2010.
363. M. Fresia, F. Perez-Cruz, and H. V. Poor. Optimized concatenated LDPC codes for joint source-channel coding. In *IEEE ISIT*, pages 2131–2135, June 28-July 3, 2009.
364. Guang-Chong Zhu and F. Alajaji. Turbo codes for nonuniform memoryless sources over noisy channels. *Communications Letters, IEEE*, 6(2):64–66, February 2002.
365. F. Cabarcas, R. D. Souza, and J. Garcia-Frias. Turbo coding of strongly nonuniform memoryless sources with unequal energy allocation and pam signaling. *Signal Processing, IEEE Transactions on*, 54(5):1942–1946, May 2006.
366. S. Nanda, K. Balachandran, and S. Kumar. Adaptation techniques in wireless packet data services. *Communications Magazine, IEEE*, 38(1):54–64, January 2000.
367. Qiuyan Xia and M. Hamdi. Smart sender: A practical rate adaptation algorithm for multirate IEEE 802.11 WLANs. *Wireless Communications, IEEE Transactions on*, 7(5):1764–1775, May 2008.
368. Yang Song, Xiaoyan Zhu, Yuguang Fang, and Hailin Zhang. Threshold optimization for rate adaptation algorithms in IEEE 802.11 WLANs. *Wireless Communications, IEEE Transactions on*, 9(1):318–327, January 2010.
369. Su Min Kim, Wan Choi, Tae Won Ban, and Dan Keun Sung. Optimal rate adaptation for hybrid ARQ in time-correlated Rayleigh fading channels. *Wireless Communications, IEEE Transactions on*, 10(3):968–979, March 2011.
370. D. N. Rowitch and L. B. Milstein. On the performance of hybrid FEC/ARQ systems using rate compatible punctured turbo (RCPT) codes. *Communications, IEEE Transactions on*, 48(6):948–959, June 2000.
371. E. Soijanin, N. Varnica, and P. Whiting. Punctured vs. rateless codes for HARQ. In *IEEE ITW*, pages 155–159, March 2006.
372. E. Candes. Compressive sampling. In *Proc. of the Intl. Congress of Mathematicians*, pages 1433–1452, March 2006.
373. D. L. Donoho and Y. Tsaig. Fast solution of l1-norm minimization problems when the solution may be sparse. *IEEE TIT*, 54(11):4789–4812, November 2008.
374. J. A. Tropp and A. C. Gilbert. Signal recovery from random measurements via orthogonal matching pursuit. *IEEE TIT*, 53(12):4655–4666, December 2007.
375. Thomas Blumensath and Mike E. Davies. Iterative hard thresholding for compressed sensing. *Applied and Computational Harmonic Analysis*, 27(3):265–274, 2009.
376. Wei Dai and O. Milenkovic. Subspace pursuit for compressive sensing signal reconstruction. *IEEE TIT*, 55(5):2230–2249, May 2009.
377. Marco F. Duarte, Michael B. Wakin, Dror Baron, and Richard G. Baraniuk. Universal distributed sensing via random projections. In *IEEE IPSN*, pages 177–185, 2006.
378. W. Bajwa, J. Haupt, A. Sayeed, and R. Nowak. Joint source-channel communication for distributed estimation in sensor networks. *IEEE TIT*, 53(10):3629–3653, October 2007.

379. S. Feizi, M. Medard, and M. Effros. Compressive sensing over networks. In *48th Annual Allerton Conference on Communication, Control, and Computing*, pages 112–1136, October 2010.
380. S. Feizi and M. Medard. A power efficient sensing/communication scheme: Joint source-channel-network coding by using compressive sensing. In *49th Annual Allerton Conference on Comm., Control, and Comp.*, pages 1048–1054, September 2011.
381. F. Chen, F. Lim, O. Abari, A. Chandrakasan, and V. Stojanović. Energy-aware design of compressed sensing systems for wireless sensors under performance and reliability constraints. *Circuits and Systems I, IEEE Transactions on*, 2013.
382. Xiao Lin Liu, Chong Luo, and Feng Wu. Formulating binary compressive sensing decoding with asymmetrical property. In *Data Compression Conference (DCC), 2011*, pages 213–222, March 2011.
383. Giuseppe Caire, Shlomo Shitz Shamai, A Shokrollahi, and Sergio Verdu. Fountain codes for lossless data compression. In *DIMACS Series in Discrete Mathematics and Theoretical Computer Science*, number 68, December 2005.
384. S. Rangan. Generalized approximate message passing for estimation with random linear mixing. In *IEEE ISIT*, pages 2168–2172, 2011.
385. S. ten Brink. Convergence behavior of iteratively decoded parallel concatenated codes. *Communications, IEEE Transactions on*, 49(10):1727–1737, October 2001.
386. S. ten Brink, G. Kramer, and A. Ashikhmin. Design of low-density parity-check codes for modulation and detection. *Communications, IEEE Transactions on*, 52(4):670–678, April 2004.
387. IEEE Standard for Local and Metropolitan Area Networks—Part 16: Air Interface for Broadband Wireless Access Systems. *IEEE Std 802.16-2009*, May 2009.
388. Xi Chen, P. Gangwal, and D. Qiao. Practical rate adaptation in mobile environments. In *Pervasive Computing and Communications, 2009. PerCom 2009. IEEE International Conference on*, pages 1–10, 2009.
389. A. Aaron, Rui Zhang, and B. Girod. Wyner-Ziv coding of motion video. In *Signals, Systems and Computers, 2002. Conference Record of the 36th Asilomar Conference on*, volume 1, pages 240–244, 2002.
390. B. Girod, A. M. Aaron, S. Rane, and D. Rebollo-Monedero. Distributed video coding. *Proceedings of the IEEE*, 93(1):71–83, 2005.
391. R. Puri and K. Ramchandran. PRISM: A new robust video coding architecture based on distributed compression principles. In *the Annual Allerton Conference on Communication Control and Computing*, 2002.
392. Rohit Puri, A. Majumdar, and K. Ramchandran. PRISM: A video coding paradigm with motion estimation at the decoder. *Image Processing, IEEE Transactions on*, 16(10):2436–2448, 2007.
393. A. D. Wyner and J. Ziv. The rate-distortion function for source coding with side information at the decoder. *Information Theory, IEEE Transactions on*, 22(1):1–10, 1976.
394. Qian Xu and Zixiang Xiong. Layered Wyner-Ziv video coding. *Image Processing, IEEE Transactions on*, 15(12):3791–3803, 2006.
395. S. Jakubczak and D. Katabi. SoftCast: One-size-fits-all wireless video. In *ACM SIGCOMM Computer Communication Review*, 2010.
396. Szymon Jakubczak and Dina Katabi. A cross-layer design for scalable mobile video. In *Proc. of the 17th Annual International Conference on Mobile Computing and Networking*, pages 289–300. ACM, 2011.
397. Ruiqin Xiong, Jizheng Xu, Feng Wu, and Shipeng Li. Barbell-lifting based 3-D wavelet coding scheme. *Circuits and Systems for Video Technology, IEEE Transactions on*, 17(9):1256–1269, 2007.
398. Yixuan Zhang, Ce Zhu, and Kim-Hui Yap. A joint source-channel video coding scheme based on distributed source coding. *Multimedia, IEEE Transactions on*, 10(8):1648–1656, 2008.
399. Mei Guo, Zixiang Xiong, Feng Wu, Debin Zhao, Xiangyang Ji, and Wen Gao. Witsenhausen-Wyner video coding. *Circuits and Systems for Video Technology, IEEE Transactions on*, 21(8):1049–1060, 2011.

400. X. Fan, F. Wu, and D. Zhao. D-Cast: DSC based soft mobile video broadcast. In *Proceedings of the 10th International Conference on Mobile and Ubiquitous Multimedia*, 2011.
401. Xiaopeng Fan, Feng Wu, Debin Zhao, O. C. Au, and Wen Gao. Distributed soft video broadcast (DCAST) with explicit motion. In *Data Compression Conference (DCC), 2012*, pages 199–208, 2012.
402. J. Garcia-Frias. Compression of correlated binary sources using Turbo codes. *Communications Letters, IEEE*, 5(10):417–419, 2001.
403. A. Aaron, S. Rane, E. Setton, and B. Girod. Transform-domain Wyner-Ziv codec for video. In *SPIE Visual Communications and Image Processing*, 2004.
404. Xun Guo, Yan Lu, Feng Wu, Debin Zhao, and Wen Gao. Wyner-Ziv-Based multiview video coding. *Circuits and Systems for Video Technology, IEEE Transactions on*, 18(6):713–724, 2008.
405. M. Tagliasacchi, A. Trapanese, S. Tubaro, J. Ascenso, C. Brites, and F. Pereira. Intra mode decision based on spatio-temporal cues in pixel domain Wyner-Ziv video coding. In *Acoustics, Speech and Signal Processing, 2006. ICASSP 2006 Proceedings. 2006 IEEE International Conference on*, volume 2, pages II–II, 2006.
406. J. Slowack, S. Mys, J. Skorupa, N. Deligiannis, P. Lambert, A. Munteanu, and R. Van de Walle. Rate-distortion driven decoder-side bitplane mode decision for distributed video coding. *Signal Processing: Image Communication*, 25(9):660–673, 2010.
407. S. Benierbah and M. Khamadja. Generalized hybrid intra and Wyner-Ziv video coding. *Circuits and Systems for Video Technology, IEEE Transactions on*, 21(12):1929–1934, 2011.
408. A. Aaron, S. Rane, and B. Girod. Wyner-Ziv video coding with hash-based motion compensation at the receiver. In *Image Processing, 2004. ICIP '04. 2004 International Conference on*, volume 5, pages 3097–3100, 2004.
409. R. Martins, C. Brites, J. Ascenso, and F. Pereira. Refining side information for improved transform domain Wyner-Ziv video coding. *Circuits and Systems for Video Technology, IEEE Transactions on*, 19(9):1327–1341, 2009.
410. B. Macchiavello, D. Mukherjee, and R. L. De Queiroz. Iterative side-information generation in a mixed resolution Wyner-Ziv framework. *Circuits and Systems for Video Technology, IEEE Transactions on*, 19(10):1409–1423, 2009.
411. X. Fan, O. C. Au, N. M. Cheung, Y. Chen, , and J. Zhou. Successive refinement based Wyner-Ziv video compression. *Signal Processing: Image Communication*, 25(1):47–63, 2010.
412. Wei Liu, Lina Dong, and Wenjun Zeng. Motion refinement based progressive side-information estimation for Wyner-Ziv video coding. *Circuits and Systems for Video Technology, IEEE Transactions on*, 20(12):1863–1875, 2010.
413. C. Brites and F. Pereira. Correlation noise modeling for efficient pixel and transform domain Wyner-Ziv video coding. *Circuits and Systems for Video Technology, IEEE Transactions on*, 18(9):1177–1190, 2008.
414. Xiaopeng Fan, O. C. Au, and Ngai-Man Cheung. Transform-domain adaptive correlation estimation (trace) for Wyner-Ziv video coding. *Circuits and Systems for Video Technology, IEEE Transactions on*, 20(11):1423–1436, 2010.
415. N. Deligiannis, J. Barbarien, M. Jacobs, A. Munteanu, A. Skodras, and P. Schelkens. Side-information-dependent correlation channel estimation in hash-based distributed video coding. *Image Processing, IEEE Transactions on*, 21(4):1934–1949, 2012.
416. G. R. Esmaili and P. C. Cosman. Wyner-Ziv video coding with classified correlation noise estimation and key frame coding mode selection. *Image Processing, IEEE Transactions on*, 20(9):2463–2474, 2011.
417. Y. Kochman and R. Zamir. Joint Wyner-Ziv/Dirty-Paper coding by modulo-lattice modulation. *Information Theory, IEEE Transactions on*, 55(11):4878–4889, 2009.
418. T. Kratochvil. Hierarchical modulation in DVB-T/H mobile tv transmission. In *Multi-Carrier Systems & Solutions*, pages 333–341, 2009.
419. W. K. Pratt. Median filtering, 1975. Technical report, Image Processing Institute, University of Southern California.

420. K. Dabov, A. Foi, V. Katkovnik, and K. Egiazarian. Image denoising by sparse 3-D transform-domain collaborative filtering. *Image Processing, IEEE Transactions on*, 16(8):2080–2095, August 2007.
421. A. Buades, B. Coll, and J. Morel. A review of image denoising algorithms, with a new one. *Multiscale Modeling and Simulation*, 4(2):490–530, 2005.
422. A. Buades, B. Coll, and J. Morel. Image denoising methods. A new nonlocal principle. *SIAM Review*, 52(1):113–147, 2010.
423. S. Smith and J. Brady. SUSANA–A new approach to low level image processing. *International Journal of Computer Vision*, 23:45–78, 1997.
424. C. Tomasi and R. Manduchi. Bilateral filtering for gray and color images. In *Computer Vision, 1998. Sixth International Conference on*, pages 839–846, January 1998.
425. S. G. Chang, Bin Yu, and M. Vetterli. Adaptive wavelet thresholding for image denoising and compression. *Image Processing, IEEE Transactions on*, 9(9):1532–1546, September 2000.
426. Jean-Luc Starck, E. J. Candes, and D. L. Donoho. The curvelet transform for image denoising. *Image Processing, IEEE Transactions on*, 11(6):670–684, June 2002.
427. Kyong-Hwa Lee and D. Petersen. Optimal linear coding for vector channels. *Communications, IEEE Transactions on*, 24(12):1283–1290, December 1976.
428. Yiqiu Dong and Shufang Xu. A new directional weighted median filter for removal of random-valued impulse noise. *Signal Processing Letters, IEEE*, 14(3):193–196, March 2007.
429. X264, http://www.videolan.org/developers/x264.html.
430. BM3D algorithm and its extensions. Technical report, http://www.cs.tut.fi/ foi/GCF-BM3D/.
431. G. Cheung and A. Zakhor. Bit allocation for joint source/channel coding of scalable video. *Image Processing, IEEE Transactions on*, 9(3):340–356, March 2000.
432. Jr. Goblick, T. Theoretical limitations on the transmission of data from analog sources. *Information Theory, IEEE Transactions on*, 11(4):558–567, October 1965.
433. M. Gastpar, B. Rimoldi, and M. Vetterli. To code, or not to code: Lossy source-channel communication revisited. *Information Theory, IEEE Transactions on*, 49(5):1147–1158, May 2003.
434. M. Skoglund, N. Phamdo, and F. Alajaji. Design and performance of VQ-based hybrid digital-analog joint source-channel codes. *Information Theory, IEEE Transactions on*, 48(3):708–720, March 2002.
435. K. Narayanan, M. P. Wilson, and G. Caire. Joint source channel coding with side information using hybrid digital analog codes. *Information Theory, IEEE Transactions on*, 56:4922–4940, July 2010.
436. Y. Kochman and R. Zamir. Analog matching of colored sources to colored channels. *Information Theory, IEEE Transactions on*, 57(6):3180–3195, June 2011.
437. Z. Yang and Y. Zhao. Scalable video multicast over multi-antenna OFDM systems. *Wireless Personal Communications*, pages 1–18, 2012.
438. W. Ji, Z. Li, and Y. Chen. Joint source-channel coding and optimization for layered video broadcasting to heterogeneous devices. *IEEE Trans. on Multimedia*, 14(2):443–455, 2012.
439. IEEE Std. 802.11n-2009: Enhancements for Higher Throughput, 2009.
440. J. G. Andrews, A. Ghosh, and R. Muhamed. *Fundamentals of WiMAX: Understanding broadband wireless networking*. Prentice Hall PTR, 2007.
441. J. Lee, J. K. Han, and J. Zhang. MIMO technologies in 3GPP LTE and LTE-advanced. *EURASIP Journal on Wireless Communications and Networking*, 2009:3, 2009.
442. C. Nokes and J. Mitchell. Potential benefits of hierarchical modes of the DVB-T specification. In *IEE Digest*, volume 10, 1999.
443. Szymon Jakubczak and Dina Katabi. SoftCast: One-size-fits-all wireless video. In *HotNets*, 2009.
444. C.-H. Kuo and C.-C. J. Kuo. An embedded space-time coding (STC) scheme for broadcasting. *Broadcasting, IEEE Transactions on*, 53(1):48–58, March 2007.
445. C. H. Kuo, C. M. Wang, and J. L. Lin. Cooperative wireless broadcast for scalable video coding. *IEEE Trans. on Circuits and Systems for Video Technology*, 21(6):816–824, 2011.
446. C. Bilen, E. Erkip, and Y. Wang. Layered video multicast using diversity embedded space time codes. In *Sarnoff Symposium*, pages 1–5. IEEE, 2009.

447. S. Chang, M. Rim, P. Cosman, and L. Milstein. Superposition MIMO coding for the broadcast of layered sources. *IEEE Trans. on Communications*, (99):1–9, 2011.
448. E. J. Candes, M. B. Wakin, and S. P. Boyd. Enhancing sparsity by reweighted L1 minimization. *Journal of Fourier Analysis and Applications*, 14(5):877–905, 2008.
449. Xiao Lin Liu, Chong Luo, Wenjun Hu, and Feng Wu. Compressive broadcast in MIMO systems with receive antenna heterogeneity. In *IEEE INFOCOM'12*, pages 3011–3015, March 2012.
450. G. J. Foschini and M. J. Gans. On limits of wireless communications in a fading environment when using multiple antennas. *Wireless Personal Communications*, 6(3):311–335, 1998.
451. P. W. Wolniansky, G. J. Foschini, G. D. Golden, and R. A. Valenzuela. V-BLAST: An architecture for realizing very high data rates over the rich-scattering wireless channel. In *Proc. of URSI International Symposium on Signals, Systems, and Electronics*, pages 295–300. IEEE, 1998.
452. V. Tarokh, N. Seshadri, and A. R. Calderbank. Space-time codes for high data rate wireless communication: Performance criterion and code construction. *IEEE Trans. on Information Theory*, 44(2):744–765, 1998.
453. S. M. Alamouti. A simple transmit diversity technique for wireless communications. *IEEE JSAC*, 16(8):1451–1458, October 1998.
454. L. Zheng and D. N. C. Tse. Diversity and multiplexing: A fundamental trade-off in multiple-antenna channels. *IEEE Trans. on Information Theory*, 49(5):1073–1096, 2003.
455. H. Schwarz, D. Marpe, and T. Wiegand. Overview of the scalable video coding extension of the H.264/AVC standard. *IEEE Transactions on CSVT*, 17(9):1103–1120, September 2007.
456. M. W. Marcellin, M. J. Gormish, A. Bilgin, and M. P. Boliek. An overview of JPEG-2000. In *Proc. of Data Compression Conference*, pages 523–541. IEEE, 2000.
457. A. Skodras, C. Christopoulos, and T. Ebrahimi. The JPEG 2000 still image compression standard. *IEEE Signal Processing Magazine*, 18(5):36–58, 2001.
458. P. Luo. Hierarchical modulation for the downlink of MIMO multi-user channels. In *Proc. of 1st International Conference on CCSP with Special Track on Biomedical Engineering*, pages 77–80. IEEE, 2005.
459. A. M. C. Correia, J. C. M. Silva, N. M. B. Souto, L. A. C. Silva, A. B. Boal, and A. B. Soares. Multi-resolution broadcast/multicast systems for MBMS. *Broadcasting, IEEE Transactions on*, 53(1):224–234, March 2007.
460. J. Prades-Nebot, Y. Ma, and T. Huang. Distributed video coding using compressive sampling. In *Proc. of Picture Coding Symposium, 2009*, pages 1–4. IEEE, 2009.
461. C. Li, H. Jiang, P. Wilford, and Y. Zhang. Video coding using compressive sensing for wireless communications. In *Proc. of Wireless Communications and Networking Conference*, pages 2077–2082. IEEE, 2011.
462. S. Pudlewski, A. Prasanna, and T. Melodia. Compressed-sensing-enabled video streaming for wireless multimedia sensor networks. *IEEE Trans. on Mobile Computing*, (99), 2011.
463. H. Jiang, C. Li, R. Haimi-Cohen, P. A. Wilford, and Y. Zhang. Scalable video coding using compressive sensing. *Bell Labs Technical Journal*, 16(4):149–169, 2012.
464. Z. Xiong, K. Ramchandran, M. T. Orchard, and Y. Q. Zhang. A comparative study of DCT- and wavelet-based image coding. *IEEE Trans. on Circuits and Systems for Video Technology*, 9(5):692–695, 1999.
465. L. W. Kang and C. S. Lu. Distributed compressive video sensing. In *Proc. of IEEE International Conference on Acoustics, Speech and Signal Processing*, pages 1169–1172. IEEE, 2009.
466. Y. Yang, O. C. Au, L. Fang, X. Wen, and W. Tang. Perceptual compressive sensing for image signals. In *Proc. of IEEE International Conference on Multimedia and Expo*, pages 89–92. IEEE, 2009.
467. V. Cevher. Learning with compressible priors. *NIPS, Vancouver, BC, Canada*, pages 7–12, 2008.
468. J. N. Laska, P. T. Boufounos, M. A. Davenport, and R. G. Baraniuk. Democracy in action: Quantization, saturation, and compressive sensing. *Applied and Computational Harmonic Analysis*, 2011.

469. A. K. Fletcher, S. Rangan, and V. K. Goyal. On the rate-distortion performance of compressed sensing. In *Proc. of IEEE International Conference on Acoustics, Speech and Signal Processing*, volume 3, pages III–885. IEEE, 2007.
470. E. J. Candes, J. K. Romberg, and T. Tao. Stable signal recovery from incomplete and inaccurate measurements. *Communications on Pure and Applied Mathematics*, 59(8):1207–1223, 2006.
471. I. Daubechies. Orthonormal bases of compactly supported wavelets. *Communications on Pure and Applied Mathematics*, 41(7):909–996, 1988.
472. Xiao Lin Liu, Chong Luo, Wenjun Hu, and Feng Wu. Compressive broadcast in MIMO systems with receive antenna heterogeneity. In *INFOCOM*, pages 3011–3015. IEEE, March 2012.
473. E. Candès and T. Tao. Decoding by linear programming. *IEEE Trans. on Information Theory*, 51(12):4203–4215, December 2005.
474. L. Gan, T. Do, and T. D. Tran. Fast compressive imaging using scrambled block Hadamard ensemble. *Preprint*, 2008.
475. S. Haghighatshoar, E. Abbe, and E. Telatar. Adaptive sensing using deterministic partial Hadamard matrices. In *IEEE International Symposium on Information Theory Proceedings (ISIT)*, pages 1842–1846, 2012.
476. Jian Zhang, Debin Zhao, Chen Zhao, Ruiqin Xiong, Siwei Ma, and Wen Gao. Image compressive sensing recovery via collaborative sparsity. *IEEE Journal on Emerging and Selected Topics in Circuits and Systems*, 2(3):380–391, 2012.
477. Abbas El Gamal and Young-Han Kim. *Network Information Theory*. Cambridge University Press, New York, NY, 2012.
478. M. Brown and D. G. Lowe. Unsupervised 3d object recognition and reconstruction in unordered datasets. In *3-D Digital Imaging and Modeling, 2005. 3DIM 2005. Fifth International Conference on*, pages 56–63, 2005.
479. Noah Snavely, Steven M. Seitz, and Richard Szeliski. Photo tourism: Exploring photo collections in 3D. *ACM Trans. Graph.*, 25(3):835–846, July 2006.
480. Tianjia Shao, Weiwei Xu, Kun Zhou, Jingdong Wang, Dongping Li, and Baining Guo. An interactive approach to semantic modeling of indoor scenes with an RGBD camera. *ACM Trans. Graph.*, 31(6):136:1–136:11, November 2012.
481. Kwan-Jung Oh, A. Vetro, and Yo-Sung Ho. Depth coding using a boundary reconstruction filter for 3-D video systems. *Circuits and Systems for Video Technology, IEEE Transactions on*, 21(3):350–359, 2011.
482. Shujie Liu, PoLin Lai, Dong Tian, and Chang Wen Chen. New depth coding techniques with utilization of corresponding video. *Broadcasting, IEEE Transactions on*, 57(2):551–561, 2011.
483. Stefan Gumhold, Zachi Kami, Martin Isenburg, and Hans-Peter Seidel. Predictive point-cloud compression. In *ACM SIGGRAPH 2005 Sketches*, SIGGRAPH '05, 2005.
484. Ruwen Schnabel and Reinhard Klein. Octree-based point-cloud compression. In *Proceedings of the 3rd Eurographics / IEEE VGTC Conference on Point-Based Graphics, SPBG'06*, pages 111–121, 2006.
485. H. Yue, X. Sun, J. Yang, and F. Wu. Landmark image super-resolution by retrieving web images. *Image Processing, IEEE Transactions on*, PP(99):1–1, 2013.
486. Jian Sun, Nan-Ning Zheng, Hai Tao, and Heung-Yeung Shum. Image hallucination with primal sketch priors. In *Computer Vision and Pattern Recognition, 2003. Proceedings. 2003 IEEE Computer Society Conference on*, volume 2, pages II–729–736, 2003.
487. Jianchao Yang, J. Wright, T. S. Huang, and Yi Ma. Image super-resolution via sparse representation. *Image Processing, IEEE Transactions on*, 19(11):2861–2873, 2010.
488. M. C. Davey and D. J. C. MacKay. Low density parity check codes over GF(Q). In *Information Theory Workshop, 1998*, pages 70–71, 1998.

Index

N

O

P

Q